Water Science and Technology Library

Volume 111

The aim of the *Water Science and Technology Library* is to provide a forum for dissemination of the state-of-the-art of topics of current interest in the area of water science and technology. This is accomplished through publication of reference books and monographs, authored or edited. Occasionally also proceedings volumes are accepted for publication in the series. *Water Science and Technology Library* encompasses a wide range of topics dealing with science as well as socio-economic aspects of water, environment, and ecology. Both the water quantity and quality issues are relevant and are embraced by *Water Science and Technology Library*. The emphasis may be on either the scientific content, or techniques of solution, or both. There is increasing emphasis these days on processes and *Water Science and Technology Library* is committed to promoting this emphasis by publishing books emphasizing scientific discussions of physical, chemical, and/or biological aspects of water resources. Likewise, current or emerging solution techniques receive high priority. Interdisciplinary coverage is encouraged. Case studies contributing to our knowledge of water science and technology are also embraced by the series. Innovative ideas and novel techniques are of particular interest.

Comments or suggestions for future volumes are welcomed.

Vijay P. Singh, Department of Biological and Agricultural Engineering & Zachry Department of Civil and Environment Engineering, Texas A &M University, USA
Email: vsingh@tamu.edu

All contributions to an edited volume should undergo standard peer review to ensure high scientific quality, while monographs should also be reviewed by at least two experts in the field.

Manuscripts that have undergone successful review should then be prepared according to the Publisher's guidelines manuscripts: https://www.springer.com/gp/authors-editors/book-authors-editors/book-manuscript-guidelines

More information about this series at https://link.springer.com/bookseries/6689

Jai Prakash Narain Rai · Shweta Saraswat
Editors

Nano-biotechnology for Waste Water Treatment

Theory and Practices

Springer

Editors
Jai Prakash Narain Rai
Department of Environmental Sciences
Govind Ballabh Pant University
of Agriculture & Technology
Pantnagar, Uttarakhand, India

Shweta Saraswat
Department of Environmental Sciences
Govind Ballabh Pant University
of Agriculture & Technology
Pantnagar, Uttarakhand, India

ISSN 0921-092X ISSN 1872-4663 (electronic)
Water Science and Technology Library
ISBN 978-3-031-00814-6 ISBN 978-3-031-00812-2 (eBook)
https://doi.org/10.1007/978-3-031-00812-2

This Springer imprint is published by the registered company Springer Nature Switzerland AG
The registered company address is: Gewerbestrasse 11, 6330 Cham, Switzerland

Contents

About the Editors

Professor Jai Prakash Narain Rai is presently working as the Full-Time Professor, Department of Environmental Sciences, Gobind Ballabh Pant University of Agriculture & Technology, Pantnagar, Uttarakhand, India, besides chairing the department for eight years as Head of the department. He has more than two hundred publications to his credit, including research papers, review articles and book chapters in scientific journals of national and international repute, apart from being a versatile reviewer of several journals. He has enough teaching and research experience of thirty seven years in the field of environmental science in general and environmental biotechnology in particular. He has been successfully executed several projects for more than a decade in various fields of ecotechnology, bioenergy production, bioremediation of waste water treatment, terrestrial sites, nanobiomaterial synthesis and applications.

Dr. Shweta Saraswat is Assistant Professor in the Department of Environmental Sciences, G. B. Pant University of Agriculture & Technology, Pantnagar, Uttarakhand, India, and former Postdoctoral Fellow in various projects funded by Department of Science & Technology, Indian Council of Medical Research and Ministry of Environment, Forest and Climate Change, Government of India. 'She has twelve years of research and teaching experience to her credit.' Besides, she has published fifteen research papers including review articles and book chapters in reputed scientific journals in the field of bioremediation of terrestrial and aquatic systems, green technology development, environmental waste utilization and bioenergy production. Recently, she is working on integrating nanotechnology with bio-based systems for sensing, analyzing and treating the contaminated wastewater in an efficient and sustainable manner.

Chapter 1
Principles and Potentials of Nanobiotechnology

Jai Prakash Narain Rai and Shweta Saraswat

Abstract Due to rapid industrialization and uncontrolled urbanization worldwide, the wastewater pollution is overwhelming. The development of advanced eco-friendly technologies is therefore required for the treatment of wastewater and its reuse. In this context, application of nano-based technologies offers an excellent and multifunctional process for wastewater treatment and its purification. Nanobiotechnology (NBT) has great potential in waste treatment by eliminating the contaminants and pathogens either through being as stand-alone treating agents or by embedding into biological membranes and combined with the traditional treatment techniques. Owing to have unique optical, electrical, and magnetic characteristics, the nanoparticles (NPs) in conjunction with biomolecules of living cells can be used for the polluted water treatment. Its principles and characteristics could be utilized in biological processes to develop innovative products and techniques such as nanomembranes, nanofilters, nanobiosensors, nanofluids, and nanoproteomics. NBT has been touted as having the capacity to address issues such as water purification, food quality, bioremediation of contaminated sites, nano-fertilizer application, degradation, and scavenging of inorganic/organic/biological contaminants, antimicrobial agents, and other biodiversity-related problems. NBT offers additional advantages in the field of biomedical sciences, personal and healthcare products, food quality sensors and packaging, textiles products such as medical, anti-satin, and UV blocking textiles, besides energy production through photocatalysis and fuel additive catalysts. Various types of nanoparticles are also being utilized in biomedicines, solar cells, dyes, membranes, and automobile parts. Owing to have a vast sphere of applications, the current chapter presents a summary of the principles and vast potentials of nanobiotechnology in detail.

J. P. N. Rai (✉) · S. Saraswat
Department of Environmental Sciences, G. B. Pant University of Agriculture and Technology, Pantnagar, Uttarakhand 263145, India
e-mail: jamesraionline@gmail.com

J. P. N. Rai and S. Saraswat (eds.), *Nano-biotechnology for Waste Water Treatment*, Water Science and Technology Library 111,
https://doi.org/10.1007/978-3-031-00812-2_1

1.1 Nanotechnology: An Invisible Science

Nanotechnology (NT) has now been evolved as a cutting-edge technology having interdisciplinary relations with engineering, physics, chemistry, material science, and biology. The term "nano" comes from the Latin word "nanus", which means "dwarf". Nanotechnology refers to objects that are one billionth the size of a human hair, or about the breadth of six carbon atoms or ten water molecules. Atoms are smaller than 1 nm, but many compounds, including proteins, are 1 nm or larger. The basic concept of nanotechnology was formulated in the mid-twentieth century. Richard Feynman, an American physicist, regarded as the father of nanotechnology explained that how things behave at small scale (Feynman 1960). He also explained different techniques that make it possible to transform single atoms from one form to another minute form. Professor Norio Taniguchi of Tokyo Science University described "nanotechnology" as a branch of science concerned with altering matter at the atomic or molecular level. He opined that the nanotechnology is concerned with the separation, consolidation, and form alteration of materials by a single atom or molecule. Further, National Nanotechnology Initiative (NNI) regarded it as research and development in the sub-100 nm range at atom, molecule, and macromolecule levels to build structures, devices, and systems with novel functional features. The International Organization for Standardization (ISO) established the inaugural Technical Committee 229 Nanotechnologies (TC 229) in 2005 to focus on the study and management of matter and processes at the nanoscale, often but not solely below 100 nm in one or more dimensions. ISO/TC 229 was also able to use nanoscale materials' properties, which happen to be other than individual atoms, molecules, and bulk matters.

In general, nanotechnology offers innovative materials with highly enhanced quantum, catalytic, optical, antimicrobial, electrical, and absorptive properties by controlling the morphology, size, and chemical structure of materials (Alvarez et al. 2018). Nanoparticles have various physical and chemical properties than larger-scale matter. At nanometric size, the quantum effect becomes dominant by reaching 100 nm or less. Subsequently, various physical (mechanical, electrical, optical, etc.) properties alter while moving from macro- to microsystems (Thomas et al. 2015). For instance, gold is normally solid, but it becomes liquid at the nanoscale at room temperature. Furthermore, on the nanoscale, matter properties change relative to particle size. The relative surface area of a particle increases as it gets smaller, as do the electrical surface charges. Properties such as electrical conductivity, magnetic permeability, chemical reactivity, fluorescence, and melting point change due to varied particle size as it is reduced to the nanoscale. The unique features of nanoparticles allow multi-dimensional applications in colourful fields due to quantum effects and huge surface area. Investigations at the nanoscale allow researchers to harness the benefit of the extraordinary physical, mechanical, chemical, and optical features of nanoparticles (NPs) which occur naturally. Furthermore, because most biological activities take place at the nanoscale, scientists have configurations and templates to work with.

1.2 Classification of Nanoparticles

Nanomaterials (NMs) are categorized according to origin (natural or man-made), chemical nature (organic and inorganic), production (atmospheric, geogenic, biogenic, and anthropogenic), dimensions and features, and research and industrial applications.

1.2.1 Based on Origin

1.2.1.1 Natural Nanoparticles

Natural nanoparticles (NNPs) are ubiquitous and occur naturally in all components of environment (i.e. atmosphere, lithosphere, hydrosphere, and biosphere), due to volcanic activity, production of cosmic dust, sea sprays, radioactive decay of radon gas, (bio)chemical weathering of minerals, biological mineralization, and fragmentation (Fig. 1.1). NNPs are produced spontaneously in various environmental compartments by a variety of physical, mechanical, thermal, photochemical, chemical, and biological processes undergoing alone or in combination (Hochella et al. 2019). NNPs, e.g. humic acids, fulvic acids, minerals (e.g. Au, Ag), oxides and oxyhydroxides of iron, manganese, and aluminium and aluminosilicates (e.g. FeS_2, Fe_2O_3, SiO_2, Al_2O_3, MnO_2, etc.), are generated from weathering processes. The organic materials, e.g. peptidoglycans, carbohydrates, polysaccharides, enzymes, and similar

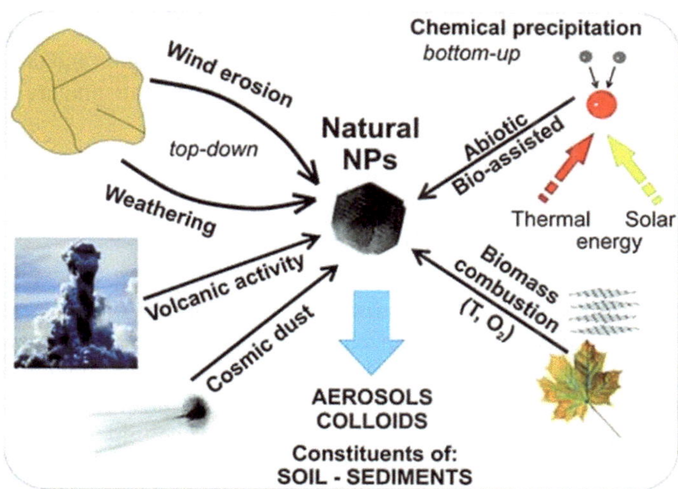

Fig. 1.1 Naturally occurring nanoparticles. *Source* Hochella et al. (2012)

biomolecules present in natural waters, are also natural nanoparticles. Moreover, viruses, nucleic acids, proteins (bovine serum albumin, insulin), lipid, carbohydrates, blood, lymph, wax crystals over lotus or nasturtium leaf, and butterfly wing scales are all natural organic nanomaterials.

1.2.1.2 Incidental Nanoparticles

Incidental nanoparticles (INPs) are accidentally formed as a by-product of any human action or process, whether direct or indirect. Vehicle exhausts, combustion operations, smelting, welding fumes, home solid fuel heating, wear and corrosion by-products, and so on are all sources of accidental nanoparticles. Besides, soot nanoparticles, consist mainly of carbon, are commonly found in the atmosphere as incidental nanoparticles. During mining activities, the Fe oxides produced as an indirect product are also considered as INPs.

1.2.1.3 Engineered/Anthropogenic Nanoparticles (ENPs/ANPs)

Combustion, mining, tillage, demolition/construction, atmospheric release, and nucleation are all examples of anthropogenic activities and/or processes that produce these particles (Lee et al. 2019). At the cellular, atomic, and molecular levels, such designed nanoparticles have distinct properties that give unique functionality such as catalytic performance, enhanced strength, greater electrical and thermal conductivity, and regulated delivery of host molecules. Carbon in designed nanomaterials like fullerenes, carbon nanotubes, and graphene has a more regulated, geometric shape and structure than natural soot NPs. ENPs have potential applications in health, industry, the environment, agriculture, food packaging, and a variety of other consumer products due to their excellent physicochemical features and intrinsic size. The TiO_2 ENPs, for example, have a regulated surface structure and are reduced in size.

1.2.2 Based on Morphology/Shape

Each nanoparticle is a three-layered structure which made up of surface, shell, and core. The surface layer is functionalized with metal ions, polymers, surfactants, and small molecules; the shell layer is distinct chemically from the core material; and the core layer is located at the centre which is referred as the nanoparticle itself. Nanoparticles are categorized as zero-dimensional (0D), one-dimensional (1D), two-dimensional (2D), and three-dimensional (3D) nanoparticles (Tiwari et al. 2012) based on their size and length scale of dimensions (Fig. 1.2). All three dimensions (x, y, and z) of 0D NMs are measured on the nanoscale. As a result, 0D NMs are limited to three dimensions with sizes ranging from a few nanometres to 100 nm. NPs,

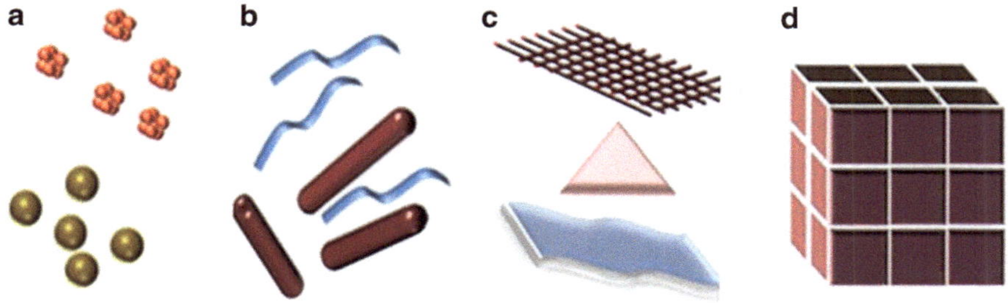

Fig. 1.2 Nanomaterials' classification (a) 0D spheres and clusters; (b) 1D nanofibers, wires, and rods; (c) 2D films, plates, and networks; and (d) 3D nanomaterials. *Source* Alagarasi (2009)

nanoshells, fullerenes, and nanorings are all examples of nanomaterials. Regardless, 2D NMs exhibit one nanoscale and another macroscale dimension. These structures are recognized as the thinnest material with large surface-to-volume area. They bear unique mechanical, chemical, and optical characteristics, enabling them fit for diverse applications. Graphene, nanofilms, nanolayers, and nanocoatings are examples of 2D nanomaterials with plate-like structures. There are no nanoscale dimensions in three-dimensional (3D) structures; instead, all dimensions are macroscale (>100 nm). 3-D nanomaterials can contain bulk powders, dispersions of nanoparticles, bundles of nanowires, and nanotubes, multi-nanolayers, bulk materials composed of individual blocks such as fullerites; skeletons of fibers and nanotubes; honeycombs, and foams; layer-fiber skeletons; composites of layers, fibers, and particles in a matrix; and membranes and powder fiber or nanoparticles (Pokropivny and Skorokhod 2007).

1.2.3 *Based on Physicochemical Characteristics*

The NPs are classified as follows based on distinctive physical and chemical features such as large surface area-to-volume ratio, greater electrical conductivity, and magnetic characteristics:

1.2.3.1 Metallic NPs and Their Clusters

Metallic NPs are formed up of alkali metal precursors and noble metals such as Ag, Au, and Cu. Metallic NPs (5–100 nm) have unique physicochemical, optical, and electronic properties at nanoscale. Because of their functional ease, large surface area-to-volume ratio, distinctive redox properties, electrical conductivity, and spectral and optical properties, along with luminescence, noble metal NPs outperform other nanomaterials. These NPs have rare optical and electrical characteristics due to well-known localized feature of surface plasmon resonance. They've been successfully exploited as transducing objects in the manufacture of sensors analysing food pollutants, poisons, and pathogens (Saha et al. 2012). Metal nanoclusters of Au, Ag, Pt, and Cu also have high photostability and quantum yield, as well as low toxicity and biocompatibility for conjugation, making them ideal for use in cell-line imaging and nanosensing, where they can replace traditional fluorescent dyes and semiconductor quantum dots (Shang et al. 2011).

1.2.3.2 Carbon-Based NPs

C-based nanomaterials are composed of carbon nanotubes, fullerenes, graphene and its derivatives, and nanospheres, among others. The geometrical structure of C-based nanomaterials is used to classify them as tube, horn, sphere, or ellipsoid in shape. Carbon nanotubes (CNTs) and fullerenes are two major groups of C-based NPs. The hollow globular cage is used to make fullerenes, which are allotropic carbon. Because of electrical conductivity, electron affinity, high strength, and adaptability, fullerenes offer a wide range of commercial applications such as nanocomposites (Saeed and Khan 2016), effective gas adsorbents for environmental clean-up (Ngoy et al. 2014), and besides as a support medium for various inorganic and organic catalysts (Mabena et al. 2011).

1.2.3.3 Dendrimers

Dendrimers are symmetrical polymeric molecules that are nanoscale in size and contain a tiny atom or its group encircled by symmetric branches called dendrons. Dendrimers have multiple chains on their surfaces that can be changed to perform various activities. Application of dendrimers is in the field of medicals, including drug delivery, anticancer drugs formulation, and photodynamic therapy, as well as for environmental usage, such as the decontamination of toxic metal ions and organic solutes from cosmetics.

1.2.3.4 Semiconductor NPs

Semiconductor NPs used in biosensing are typically metal-nonmetallic alloys. QDs are an example of widely used semiconductor nanoparticles. They are fluorescent and have higher quantum yields and stability than fluorescent dyes, as well as size-controlled emission (Gill et al. 2008). They have certain unique and exciting thermal, electrical, and optical properties due to their quantum confinement, such as crisp and symmetrical emission spectra, distinctive chemical characteristics, and excellent photostability. Photocatalysis and electrical devices all use these materials (Sun 2000). Owing to their distinctive properties, semiconductor NPs hold potential applications in bioimaging, electronic optics, genosensors, and enzyme-based sensors, gas sensors, and sensors for various organic and inorganic substances (Khan et al. 2017).

1.2.3.5 Nanocomposites

Nanocomposites are unique materials and consist of two phases of different materials which retain their properties independently during its production. They are termed as multiphase materials containing at least one of the phases of nanoscale dimension. The attributes of nanocomposites are determined by the constituent phases' properties, their relative amounts, the dispersed phase's geometry, their shape and size, dispersion, and orientation, e.g. colloids, gels, carbon–metal oxides, and copolymers. Quality of nanocomposite is determined by the attributes of each component, their relative amounts, and the overall shape of the composite. Nanocomposites have different phases in nanoscale size and are incorporated in a metal, ceramic, or polymer matrix. High mechanical and thermal stability, multifunctionality, and a large interphase zone are additional features of these materials. As a result, they can be employed in a variety of disciplines of science and technology, such as catalysis, the fabrication of nanosensors and nanoprobes, the sorption process, fuel cells, nonlinear optics, bioceramics, higher-power batteries, environmental protection, and anticorrosion agents.

1.2.3.6 Polymeric NPs

These are mostly organic NPs in the form of nanocapsules or nanospheres. The latter are matrix particles with a solid mass, whereas the other molecules are adsorbed at the outer edge of the spherical surface. In the former case, the solid mass is completely enclosed within the particle (Rao and Geckeler 2011). Polymeric NPs have a wide range of applications in the literature due to their ease of functionalization (Abd Ellah and Abouelmagd 2016).

1.2.3.7 Lipid-Based NPs

These NPs contain lipid moieties and can be used in a wide range of biomedical applications. A lipid NP has a spherical shape and a diameter ranging from 10 to 1000 nm. Lipid nanoparticles, like polymeric nanoparticles, have a solid lipid core and a matrix containing soluble lipophilic compounds. Lipid nanotechnology deals with design and fabrication of lipid nanoparticles for a variety of applications such as drug carriers, transport, and RNA release in cancer treatment (Gujrati et al. 2014).

1.3 Physicochemical Properties of Nanoparticles

Because of the size decrease, the properties of macroscopic materials quickly shift to some remarkable properties. For example, Japanese scientists discovered in the

1970s that metals with electrical and thermal conductivities, such as copper and silver, lose their natural character when lowered to the nanoscale and become nonconductive and nonthermal (Shi et al. 2015). The extraordinary characteristics of NPs are due to this high surface area-to-volume ratio phenomenon. Physical features such as melting temperatures and optical adsorption change dramatically as the nanoscale gets smaller. Because of the increased ratio of corner and edge sites to terrace atoms in nanomaterials, they have enhanced catalytic activity in chemical reactions (Navalon and Garcia 2016). When compared to typical microstructures, nanocrystalline materials with very tiny grain sizes exhibit dramatically different mechanical characteristics such as superplasticity, extraordinary hardness, and strength. Nanoparticles are better equipped to dramatically influence their interactions with biomolecules/ cells due to their tiny size, solubility, surface morphology, and electrical charge. Engineered nanoparticles have also exhibited great potential as high-efficiency medication, protein, siRNA, enzyme, and biomolecule delivery transporters into cells (Prabha et al. 2016).

1.3.1 Optical Properties

Absorption, transmission, reflection, and light emission are all dynamic optical properties of nanomaterials that are directly related to their electrical and electronic properties. The optical properties can confine their electrical properties to produce a quantum effect, with variations in form, size, or type potentially influencing the colour they produce. Spherical gold nanomaterials with a diameter of 25 nm seem green, whereas those with a diameter of 100 nm appear orange. A spherical gold NM of 100 nm diameter appears orange, while a spherical silver NM with 100 nm diameter appears yellow. The produced surface plasmon resonance effect due to the resonance of outer electron band of nanomaterial with light wavelengths is responsible for the colour variation. When photons excite the outer electrons of the particles, the metal particles' outer electrons jitter at specific wavelengths and absorb light matching to that resonance. Applications based on optical properties of nanomaterials include optical detectors, phosphors, sensors, displays, lasers, solar cells, photocatalysis, photoelectrochemistry, and biomedicine.

1.3.2 Magnetic Properties

Surface effects (Koksharov 2009) and finite-size effects dominate the two main magnetic properties of NPs. Magnetic anisotropy, magnetic moment per atom, Curie temperature, and the coercivity field of NPs, as well as other main magnetic properties, can all differ from those of bulk material (Koksharov 2009; Issa et al. 2013). Magnetic nanoparticles that are superparamagnetic are favoured for their stability. A NP with a large magnetic moment acts like a huge paramagnetic atom with very quick response to magnetic fields and low remanence and coercivity when it is superparamagnetic. Geometrical properties, intra-particle magnetic interactions, inter-particle magnetic interactions, magnetic interactions between NPs and the matrix material and particle-applied magnetic field interactions are the five main determinants of magnetic properties of nanoparticles (Obaidat et al. 2015). The geometrical features of nanoparticles include: (a) particle sizes and shapes; (b) particle size distribution; and (c) anisotropy axes distribution. Magnetic nanoparticles have been claimed to have technological uses in biomedicine, medication delivery, magnetic resonance imaging, magnetic hyperthermia, and data storage systems.

1.3.3 Electrical Properties

Electrical qualities such as conductivity and resistivity, which tend to alter at the nanoscale level, are similar to optical properties. The conductivity of a bulk or huge substance is unaffected by characteristics such as diameter or cross-sectional area, as well as fold in the conducting wire. However, it has been discovered that the conductivity of CNTs changes as the cross-section area changes. When a shear strain (in plain terms, a twist) is applied to a nanotube, the conductivity changes as well. Carbon nanotubes, like graphite, can behave as good conductors or semiconductors in behaviour. A multi-walled carbon nanotube's conductivity differs from that of a single nanotube of the same size.

1.3.4 Electronic Properties

The electronic band composition of NPs, i.e. the state and behaviour of electrons in the NMs in terms of their energy and momentum, is described by their electronic characteristics. The electronic properties, in turn, are inextricably linked to electric qualities like conductivity and dielectric response, as well as optical characteristics like damping constant, refractive index, and absorbance, which define the material's response to electromagnetic radiation.

1.3.5 Mechanical Properties

Nanoparticles have different magnetic properties than microparticles and bulk materials, such as hardness, toughness, yield strength, tensile strength, elastic modulus, adhesion, and friction, allowing for more effective surface modification of many devices in terms of mechanical strength, or to improve nanomanufacturing/nanofabrication quality, for example. NMs have a mechanical strength that is one or two orders of magnitude more than single crystals in aggregate form. The nanoscale conversion of materials improves crystal completeness or reduces flaws, resulting in a mechanical strength boost. The increases in metal hardness linearly enhance grain size, while in NMs, the hardness increases linearly as the particle size decreases.

1.3.6 Thermal Properties

Heat capacity, thermal conductivity, thermal effusivity and diffusivity, and material density are all thermal properties of nanoparticles. Heat transmission in NMs is mostly dependent on energy conduction caused by electrons and phonons (lattice vibrations), as well as scattering effects. Thermal and electrical conductivity, thermoelectric power or Seebeck coefficient, and heat capacity are all useful indicators of nanomaterials' significant physical properties.

1.4 Synthesis of Nanoparticles

NMs offer remarkable potential in different fields due to their enormous surface-to-volume ratio; however, several nanoparticles have also exhibited toxicity at the nanoscale dimension. Many biosynthetic routes for nature-friendly nanobioparticle fabrication using organisms such as plants, fungi, and bacteria have been developed to overcome the problem of toxicity, revealing significant benefits to natural environment through environmentally benign methods (Duan et al. 2015). The choice of a green or environment friendly solvent, a good reducing agent, and a safe material for stabilization are three most important parameters for the production of NPs. NPs production methods are separated into two broad categories, i.e. top-down and bottom-up approaches (Fig. 1.3).

1.4.1 Top-Down Approach

Nanoparticles are generated via size reduction using appropriate lithographic processes such as crushing, spitting, and milling to mould the material into desired model and micro-fabrication in the top-down approach (destructive approach). Temperature, pH, plant extract concentration, metal salt solution, and incubation period can all be used to influence the stability, shape, and size of nanoparticles. The most common approach in the top-down method starts with the selection of appropriate material and then shaping the functionality from inner material. Another method is laser pyrolysis, an easy and effective method of synthesizing nanopowders. The energy current between the carbon dioxide laser and the reagent increases the temperature in the reactive zone and causes the vibration of molecules. The above

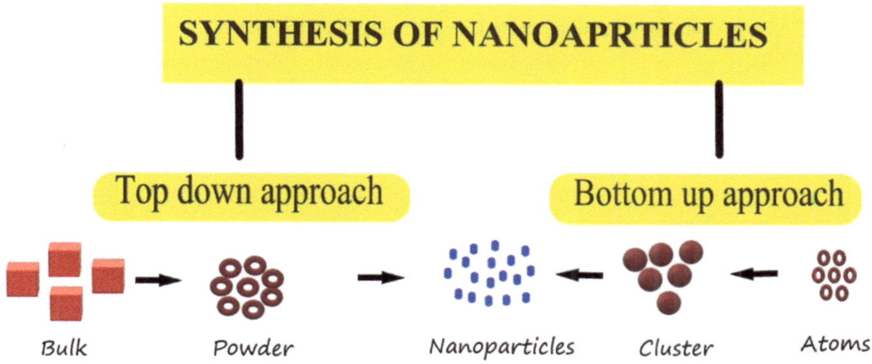

Fig. 1.3 Top-down and Bottom-up approaches of nanoparticles synthesis

method is a simple way of particle synthesis in the range of 15–20 nm. Other top-down methods for creating nanoparticles include ionic/electronic irradiation, etching, mechanical milling, and sputtering. However, due to crystallographic injury during nanostructure development, this technique harms the surface structure.

1.4.2 Bottom-Up Approach

In the bottom-up approach (constructive method), the NPs are synthesized employing minor entities, such as atoms and molecules, for which the oxidation/reduction reactions are the main proponents. The bottom-up approach utilizes chemical characteristics of individual substances to form nanoparticles employing different methods such as supercritical fluid, sol–gel, aerosol-based processes, plasma spraying, laser pyrolysis, atomic condensation, electrochemical precipitation, vapour deposition, green synthesis, and so on where the formation of nanoparticles begins at the atomic or molecular level. Bacteria, fungi, plants, actinomycetes, and other organisms that contain biochemicals, e.g. acids, polyphenols, polysaccharides, terpenoids, flavonoids, alkaloids, etc., have been discovered to be effective reducing, capping, and stabilizing agents in green synthesis of NPs. Uni- and multicellular organisms are allowed to react in these syntheses.

1.4.2.1 Biological Approaches for the Synthesis of ZnO-Nanobioparticles

Bacteria-Mediated Synthesis of Nanobioparticles

NPs biosynthesis utilizing bacteria has arisen as a fast-growing study topic in green nanotechnology, with a variety of biological objects being used in the synthesis of

Table 1.1 Green synthesis of nanobioparticles employing bacterial species

S. No.	Bacteria	Nanoparticle	Size (nm)	References
1	*Proteus vulgaris ATCC-29905*	Fe	19.23	Majeed et al. (2020)
2	*Bacillus subtilis*	Se	50–400	Wang et al. (2010)
3	*Bacillus megaterium*	Pb, Ag, Cd	10–20	Prakash et al. (2010)
4	*Bacillus licheniformis*	Ag	40–50	Kalishwaralal et al. (2009)
5	*Pseudomonas fluorescens*	Au	50–70	Rajasree and Suman (2012)
6	*Pseudomonas aeruginosa JP2*	Ag	17.2	Zainab et al. (2016)
7	*Streptomyces sp. HBUM171191*	Zn, Mn	10–20	Waghmare et al. (2011)
8	*Bacillus cereus*	Ag	10–30	Prakash et al. (2011)
9	*Pseudomonas fluorescens*	Ag	85.46	Silambarasan and Jayanthi (2013)
10	*Planomicrobium sp*	TiO_2	8.89	Malarkodi et al. (2013)
11	*Lactobacillus sp. strain LCM5*	Ag	13.84	Matei et al. (2020)
12	*Pseudomonas aeruginosa JP1*	Ag	13.44	Ali et al. (2017)
13	*Serratia nematodiphila*	ZnO	24.79	Jain et al. (2020)

NPs on a regular basis, providing an appealing alternative to traditional chemical and physical processes. Metal pollution and wastes can be remedied using microbial metal reduction as a technique. Bacteria are considered as biofactory for the production of NPs of Au, Ag, Pt, Pd, Ti, TiO_2, Fe_3O_4, and CdS. Magnetotactic bacteria and S-layer bacteria are two well-known examples of bacteria that synthesize inorganic compounds. Metal nanobioparticles can be made using bacteria, such as E. coli. Bacillus subtilis (Sundaram et al. 2012) has a long and illustrious history. Silver nanoparticles (AgNPs) have also been synthesized using Escherichia coli (Gurunathan et al. 2009). The generated AgNPs had a homogeneous distribution and a size of 50 nm on average. Juibari et al. (2011) investigated the role of the extremophilic *Ureibacillus thermosphaericus* in the production of AgNPs at high temperatures. The bacteria *Staphylococcus aureus*, a Gram-positive pathogen, was employed to make AgNPs (Prakash et al. 2010). Table 1.1 lists the many bacterial species involved in nanoparticle biosynthesis.

Fungi-Mediated NPs

Fungi are superior biogenic agents than other biological systems utilized for synthesis because of their diversity, cultivability, and better growth control. It can produce NPs both extra- and intracellularly, while the specific mechanisms are unknown. Binding of heavy metals to fungal cell wall (proteins or enzymes) via electrostatic interactions is one of the possible mechanisms during intracellular production. Further, metal ions are reduced by fungal cell wall enzymes, resulting in metal ion aggregation and nanoparticle production. Extracellular synthesis implied metal ion contact and enzyme release, mostly reductase, followed by the production of nanoparticles in solution. Kumar et al. (2007) demonstrated the particular role of alpha NADPH-dependent nitrate reductase in AgNPs production. The reduction of Ag+ ions by nitrate reductase results in the creation of AgNPs with diameters of 10–20 mm. Srivastava et al. (2013) also generated AgNPs with a diameter of 50 nm from Halococcus salifodinae, confirming the participation of intracellular nitrate reductase. In addition, silver nanoparticle formation mediated by *Trichoderma virens* was discovered to require silver ion reduction via nitrate reductase (Devi et al. 2013). Das et al. (2009) also discovered that Au NPs made from *Rhizopus oryzae* have a high ability for adsorption of the Rhodamine B dye. Rajan et al. (2016) have demonstrated ZnO NPs formation using the fungus *Aspergillus fumigatus*. Several other metallic nanoparticles were synthesized using fungi, including nanosized Fe_3O_4 by *Mucor javanicus*, *F. oxysporum*, and *Verticellum* sp., CdSe QDs by *F. oxysporum*, Se NPs by *A. alternata*, $SrCO_3$ nanocrystals by *F. oxysporum*, Si NPs by *F. oxysporum*, and Ti NPs by *F. oxysporum. Penicillium verrucosum* has lately emerged as a promising contender for the synthesis of Ag NPs (Yassin et al. 2021).

Algae Mediated NPs

Algae are first-rate sources for the production of metallic NPs due to their predominance and ease of access. Chlorophyceae, Phaeophyceae, Cyanophyceae, Rhodophyceae, and other algae are involved in the biosynthesis of NPs (diatoms and euglenoids). Algae's unique ability to capture metals and decrease metal ion concentration enables them an excellent option for NP production. Algae are also very appropriate and easy to handle, with various other advantages such as synthesis at low temperature with greater energy efficiency, less toxicity, and minimum environmental risks. The synthesis of NPs by algae takes a fraction of the time that other biological approaches do. Many algae like *Chlorella vulgaris, Lyngbya majuscule*, and *Spirulina platensis* were used as a cheap method for synthesis of Ag NPs (El-Sheekh and El-Kassas 2016). *S. platensis* has been used for the extracellular synthesis of Au, Ag, and Au/Ag composite NPs (Chakraborty et al. 2009). Microalgae, e.g. diatoms (*Navicula atomus*, *Diadesmis gallica*), can also synthesize Au NPs and gold–silica nanobiocomposites (Schrofel et al. 2011). So far, several seaweeds, e.g. *Sargassum wightii* and *Fucus vesiculosus,* have been traced for synthesizing AgNPs of varied sizes and shapes (Singaravelu et al. 2007; Kannan et al. 2013). Several algae were employed to make silver NPs, including *Lyngbya majuscule*, *Spirulina platensis*, and *Chlorella vulgaris* (El-Sheekh and El-Kassas 2016).

Antifungal, antibacterial, and anticancer characteristics of algal-derived NPs have been investigated. For example, NPs of Ag, Au, and the bimetallic alloy Ag–Au were generated by *Gracilaria* sp., a marine red alga that showed antibacterial action against *S. aureus* (Gram-positive) and Gram-negative (*K. pneumonia*) bacteria (Ramakritinan et al. 2013). Cotton Fusarium wilts (*Fusarium oxysporum* f. sp. *vasinfectum*) and bacterial leaf blight (*Xanthomonas campestris pv. malvacearum*) were inhibited by extracellular production of AgNPs from the broth of the marine algae thallus of *Padina pavonica* (Linn) (Sahayaraj et al. 2012). Antifungal properties of AgNPs biosynthesized with the red sea weed *Gelidiella acerosa* as the reducing agent have been observed against *Fusarium dimerum, Humicola insolens, Mucor indicus,* and *Trichoderma reesei* (Suganya et al. 2015).

Plant-Mediated NMs

Plants' biomolecules, e.g. carbohydrates and proteins derived from plant parts have, an exceptional ability to synthesize metal NPs. Owing to nature-friendly, low-cost, and relatively safe for human therapeutic use, plant-based nanoparticle production is best suited and convenient. NPs of silver from $AgNO_3$, gold from $AuCl_2$, zinc oxide from $Zn(NO_3)_2$ and Zn $(CH_3COO)_2$, cadmium sulphide and zinc sulphide from $CdSO_4$ and $ZnSO_4$, and other nanoparticles were manufactured using various plants and their parts. In contrast to microbial systems, plant-mediated gold nanoparticle synthesis requires less incubation time and could be easily adopted for bulk demands. Sukri et al. (2019) used *Punica granatum* fruit peel extract to biosynthesize zinc oxide nanoparticles (ZnO NPs) at various temperatures. At a concentration of 31.25 g/ml, smaller ZnO NPs were observed as more effective in the cytotoxic experiment. Nava et al. (2017) investigated the photocatalytic activity of ZnO NPs produced by green synthesis utilizing fruit extracts of *Citrus sinensis*, *C. paradisi*, *C. aurantifolia,* and *Lycopersicon esculentum,* as reducing agents. Because of the diverse chemical compositions of fruit extract, the surface shape of the nanoparticles differed. In prior investigations, different plant parts were employed to produce diverse nanobioparticles, as shown in Table 1.2.

1.5 Natural Compounds for Nano Biomaterials' Synthesis

In nanotechnological investigations, natural polysaccharides, proteins, aromatic peptides, enzymes, DNA, amyloid fibrils, actin filaments, minerals, bacteriophages, and viruses are commonly employed. Polysaccharides including gelatin, chitosan, and albumin are among the best-known natural polymers which are utilized to make nanoparticles, mostly to reduce the risk of negative effects during drug

Table 1.2 Synthesis of metal nanobioparticles using various plants and their parts

Plant source	Synthesized nanoparticles	Remarks	References
Azadirachta indica leaves	Zn	18 nm	Elumalai and Velmurugan (2015)
Trifolium pretense flower	Zn	60–70 nm	Dobrucka and Długaszewska (2016)
Aloe barbadensis Miller	In_2O_3	5–50 nm	Maensiri et al. (2008)
Pulicaria glutinosa	Pd	Applied for catalytic action	Borah et al. (2015)
Piper betle leaves	Pd	Studies for antifungal activity	Koduru et al. (2013)
Jatropha curcas L	Pb	10–12.5 nm	Joglekar et al. (2011)
Euphorbiaceae latex	Cu	antimicrobial activity	Valodkar et al. (2011)
Catharanthus roseus leaf	TiO_2	*Hippobosca maculata* and *Bovicola ovis* (flies) repellant	Velayutham et al. (2010)
Mimosa pudica leaf	Au	12.5 nm	Suganya et al. (2015)
Macrotyloma uniflorum root and stem	Ag	–	Mutahir et al. (2015)
Sorghum bran	Fe, Ag	–	Njagi et al. (2011)
Capsicum annuum	Se	Se with protein nanocomposites	Li et al. (2007)
Diopyros kaki	Pt	15–19 nm	Song et al. (2010)
Anacardium occidentale leaf	Pt	Thermal and catalytic utilities	Sheny et al. (2013)

delivery and tissue engineering discovery. Following are the major naturally occurring polysaccharides and proteins that are used in synthesis and functionalization of NBMs/NBPs:

1.5.1 Amyloid Fibrils

Amyloid fibrils (7–10 nm) are organized nanostructures that self-assemble from a variety of polypeptides and proteins and have been linked to more than two dozen human disorders, including Alzheimer's, Prions, and Parkinson's diseases. Amyloid fibrils are extremely durable and resistant to proteases as well as other severe environmental factors. Amyloid fibrils provide a platform for the manufacturing of NMs using prion proteins and have now emerged as a novel, extremely effective vital mineral delivery mechanism.

1.5.2 Actin Filaments

Actin filaments are 7 nm auto-assembled fibrils with a fairly homogeneous structure. Many of the mechanical properties of cytoplasm, a complex, viscoelastic substance, are accounted for by actin filaments. When cytoplasm is viscoelastic, it can resist flow like a viscous liquid (such as molasses) while also storing mechanical energy when stretched or squeezed. The capacity of actin filaments to govern polymerization and reverse in the presence of ATP nucleotides is one of their advantages. Because the actin monomers' crystal structure is known, additional functions could be logically generated and inserted into the fibrilar system, which could then be used as a platform for nanoscale engineering.

1.5.3 Peptides

Aromatic peptides are hollow with a few tens of nanometres of inner diameter. With peptide nanotubes as templates, silver nanowires can be formed. Within the tube lumen, ionic silver is converted to metallic silver. Enzymatic degradation is then used to eliminate the peptide envelope.

1.5.4 Bacteriophages

Bacteriophages are bacteria-infecting viruses. The capacity of phage particles to manifest numerous copies of specific proteins on their surface is a significant benefit. The phage coat proteins revealed small peptides having affinity for semiconductors and metals, which were exploited to fabricate more complicated assemblies with nanoscale order.

1.5.5 Minerals

The structure of inorganic components underpins the skeletal tissues such as bones and teeth to provide strength. Calcium, in its various forms, is by far the most well-known mineral. Bones and teeth created artificially using biological templates in NT could revolutionize modern medicine. The fragment of peptide, e.g. silaffin1, a replicated protein which mediates silica organization, is one example of such a method. Likewise, the peptide fragment containing repeat sequence of the amino acid has the ability to control and influence silica organization morphology.

1.5.6 *Viruses*

Viruses have played an essential role in virus nanotechnology as a source of natural nanoparticles for biomedical applications. Viruses consist of a genome and a capsid, with some being enclosed. The majority of virus capsids have a diameter of 20–500 nm. For example, mammalian viruses are being researched as vectors for gene transfer, while bacteriophages and plant viruses have been used in drug delivery and imaging applications, as well as vaccinations and therapeutic intervention.

1.5.7 *Enzymes and Nucleic Acids*

In biotechnology, enzymes and nucleic acids are utilized in diagnostics. Antibodies containing home pregnancy test kits, which quantify precisely the human chorionic gonadotropin hormone, and the determination of HIV diagnostic kits meant for assessing infections of hepatitis virus are examples of diagnostic immunoassays. More advanced equipment incorporates a nano-device that uses an electrochemical reaction to assess glucose levels and nanoelectrodes are attached to a controlled delivery system that distributes insulin as per a pre-programmed profile. For individuals suffering from Type I and Type II diabetes, such a nano-device will be capable of improving the parts of the pancreas to restore its normal activities. Enzymes also act as catalysts in a variety of culinary, biotechnology, and related processes and applications. Nanotechnology has promise for enzyme immobilization and, as a result, improved enzyme functionality. Nanoparticles not only provide thermostability, but also act as activity boosters for enzymes that have low activity in their original condition. Amylase, galactosidase, cellulase, lipase, and other enzymes have been nanoimmobilized and used in food systems. Further, the unique features of nucleic acid (e.g. DNA) and their capacity to interact with a variety of proteins justify its usage as a nanoscale building material with direct nanotechnology applications. DNA is employed as a nanowire that has special recognition information. DNA-based wires can be used as a bearer of address data, allowing the DNA strand to be directed to specified spots, which is an important stage in the production of auto-assembled electrical circuits.

1.5.8 *Other Natural Polymeric Compounds as Nanobioparticles (NBPs)*

In various nanotechnological researches, natural polysaccharides and proteins are commonly employed.

1.5.8.1 Polysaccharides

Polysaccharides (e.g. dextrin, chitosan, and alginate) are complex carbohydrates made up of monosaccharides and are present in all microbes, animals, and plants. Being biodegradable, biocompatible, and non-toxic, polysaccharides are commonly employed in drug delivery research. Because of their reactive functional groups, they connect to target tissues and form noncovalent interactions with epithelial cells. They produce a neutral coating with low surface energy and allow nonspecific protein adsorption due to their capacity to recognize specific receptors. They have a lot of hydroxyl groups, which allow specific ligands to attach to the structure.

Chitosan

Chitosan is a non-toxic, biocompatible, cost-effective natural polymer obtained from deacetylation of chitin, sourced from outer skeleton of lobster, shellfish, crab, and shrimp. Chitosan-based nanoparticles are widely used as adhesives in medical and surgical treatments including wound healing and drug delivery which are relatively safe for human consumption as well (Mohammed et al. 2017). The structure of chitosan can be chemically and physically altered to produce unique properties useful in medicinal delivery and tissue engineering. The biomedical field has benefited greatly from the tiny size, tailored surface, increased solubility, multifunctionality, and tissue adhesion capabilities of chitosan-based nanomaterials.

Dextran

Dextran is a linear 1,6-linked d-glucopyranose polymer produced by bacteria extracellularly. *Leuconostoc mesenteroides* fermented sucrose in sugar cane or beet syrups to produce this bacterial polysaccharide can also be produced by the organs of *Lactobacillus, Leuconostoc*, *Streptococcus* sp., and other bacteria. Because of their biocompatibility and biodegradability, dextran-based nanoparticles are ideal for application in nanomedicine. Dextran has also been utilized to conjugate bioactive molecules (e.g. medicines, enzymes, hormones, and antibodies) to prolong circulation lives, boost in vivo stability, or suppress antigenicity due to its hydrophilic nature. Dextran is also widely utilized for surface modification in biosensor construction since it is a hydrophilic and functionalized polymer. Dextran coatings on sensor chips can substantially reduce nonspecific analyte adsorption and allow surface immobilization of ligand, boosting biosensor sensitivity. This water-soluble polysaccharide is frequently utilized as a thickening, emulsion forming, stabilizer, and packaging in the food sector. Dextran-coated nanoparticles are also used in the pharmaceutical, chemical, therapeutic, environmental, and food industries for protein separation and purification, as well as the preparation of customized drug delivery systems. Dextran-coated nanoparticles have good adsorptive properties and could be utilized for heavy metal removal from wastewater.

Alginate

Alginate NPs (anionic polysaccharides abundantly distributed in brown algal cell walls that in aqueous state produce a viscous gum) have emerged as best-known biomaterials for drug delivery and pin point various administration routes. Their benefits encompass not only diverse physicochemical features, enabling chemical transformations for targeting-specific sites, but also their biocompatibility, mucoadhesiveness, and biodegradation profiles. Further, mixing of alginate NPs with other polymers could tailor the surface with specific moieties and also adjust cross-linking, mechanical strength, cell affinity, and gelation (Severino et al. 2019). Like chitosan, alginate has less toxicity and inflammation, biodegradability, adjustable porosity, and capable of forming linkages with bioactive molecules and thus utilized widely in wound healing and pharmaceutical and food applications (Reddy et al. 2015; Yang et al. 2015). The ability of sodium alginate to form a simple aqueous gel with cations having bivalent electrons, such as calcium (Ca^{2+}), makes it ideal for drug delivery applications. The engulfment of cells within alginate particles boosts the cells' growth and proliferation, and thus facilitates its application in tissue engineering investigations.

1.5.8.2 Proteins

Proteins are flexible proteins made up of amino acid chains that are essential for biological processes like molecular transportation and DNA replication. These biodegradable polymers have a low efficiency and high cost of manufacture. Soybean, wheat, and corn proteins sourced from plants and gelatin, casein, collagen, and keratin from animals can be used in the development of materials, the construction of nanoscale devices, and the biosensing of nanowires. Signal transduction is used in biosensing to detect biological and chemical substances at the nanoscale. Cross-linking chemicals immobilize a protein enzyme onto a protein nanotube or nanofiber. The electrode is then connected to this structure. The covalent attachment of glucose oxidase to the protein nanotube complex allows the biosensing device to be used for applications such as glucose sensing. Protein nanoparticles have the potential to be used as a drug delivery mechanism for parenteral, peroral, and ocular administration, as well as a vaccine adjuvant. Proteins are more favourable than other colloidal carriers such as liposomes due to their superior stability and ease of manufacture.

Collagen

The extracellular matrix of the human body is mostly made up of collagen, a natural protein. It gives skin, bone, tendon, cornea, and other tissues mechanical stability as well as appropriate elasticity and strength. The exogenous collagen is used widely in bioscaffolds, skin, bone, vascular grafting, nerve tissue, and articular cartilage regeneration, especially in the field of medicines and pharmaceuticals, while nanocollagen is a novel material in NT with a lot of untapped potential (Lo and Fauzi 2021). It has benefits over traditional 3D-collagen designs, mainly due to its nano-size that offers higher surface area-to-volume ratio, allowing it to withstand large loads with little tension. It can be produced in many ways, including using the electrospinning technique to create nanocollagen strands that look like natural extracellular matrix.

Gelatin

Gelatin, a protein made up of hydrolysed collagen, is a biodegradable substance with its much application in NBT and nanopharmaceutics. Due to their biocompatibility and easy biodegradation ability, gelatin NPs have been exploited as medication and gene carriers to targeted sick tissues such as cancer, tuberculosis, HIV infection, and the therapy of vasospasm and restenosis. Coating quantum dots with gelatin, for example, reduces their cytotoxicity. Further, gelatin NPs have been observed to pass the blood–brain barrier, making them a prospective choice for treating brain ailments. Application of gelatin NPs to target macrophages for treating several diseases has been documented in previous years. Gelatin is widely used in tissue engineering to create biological and long-lasting 3D scaffolds for the manufacture of bio-artificial tissues and organs. Nanofibres of gelatin are also employed in gene therapy (Yasmin et al. 2017).

Albumin

Albumin is additional biopolymer used extensively to prepare nanocapsules and nanospheres. Due to availability in pure state, an easy biodegradation and nontoxicity has preferred use of albumin for NP production. Besides having functional moieties such as thiol, amino, and carboxylic groups on its surface, this could be more conveniently modified and bonded with ligands. It is an important plasma protein (e.g. bovine serum albumin and human serum albumin) which can easily give rise to albumin NPs of desired size.

Lactide and Glycolide Copolymers

Lactide and glycolic acids are used to make these aliphatic polyester copolymers. Because of its synthetic malleability and biocompatibility, poly lactide-co-glycolide (PLGA) has become common in the biomedical area and has found widespread uses in nanomedicine and drug delivery formulations. Due to their managed biodegradability, outstanding biocompatibility, adaptable degradation rates, mechanical characteristics, and thermal processability, PLGA, copolymer of polylactic and polyglycolic acid, has been opted in the development of many drug carrier systems and therapeutic devices such as scaffolds, grafted tissues, surgical sutures, and bone tissue engineering (Ulery et al. 2011). It is very versatile and may be modified into a variety of preparations, such as NPs, microparticles, hydrogels, membranes, scaffolds, and sponges (Sun et al. 2017). PLGA has also been approved for human therapy by US Food and Drug Administration (FDA).

1.6 Characterization of Bionanomaterials

Surface analysis and chemical characterization techniques are used to characterize nanomaterials and nanostructures. Small-angle X-ray scattering, X-ray diffraction, transmission electron microscopy, scanning electron microscopy, and scanning probe microscopy are some of the most widely used structural characterization methods for studying the composite morphology, phases of crystal, and their average size. The surface characterization of atoms, compounds, and their spatial distributions is determined chemically. Electron spectroscopy, ionic spectrometry, optical spectroscopy, absorption and transmission spectroscopy, fluorescence and mass spectrometry, photoluminescence, and colorimetry are few examples of these techniques. Fourier-transform infrared spectroscopy (FTIR), Raman spectroscopy, X-ray photoelectron spectroscopy, and energy-dispersive X-ray spectroscopy can all be used to look for functional groups on the surface of nanotubes. The thermogravimeter can be used to investigate thermal stability. Physisorption and chemisorption analysers can determine the surface area, porosity, and size and distribution of pores. By measuring the amount of gas desorbed at various temperatures, temperature-programmed desorption (TPD) analyses the number, type, and strength of active sites available on the surface of a NMs. If there are multiple active components present on NM, the gas desorbs at varied temperatures. The amount of reducible metal species found in the NM and the temperature at which reduction occurs are determined via temperature-programmed reduction (TPR). This can be estimated by determining the amount of temperature linked analysis gas (such as hydrogen) with the catalyst. The amount to which NMls can reoxidize is determined by temperature-programmed oxidation (TPO).

1.7 Current Applications and Future Prospects of Nanobiotechnology

Considering the unique electronic, magnetic, optical, catalytic, large surface area, and energy characteristics, scientists have been attracted to biosynthesize green NPs due to their eco-friendly nature and less toxicity. NBT encompassing NPs, nanowires, nanofibers, and nanotubes showed a multitude of potentials in the field of medical science, anticancer therapy, biosensing, molecular imaging, information technology, food, energy generation, environmental remediation, biotechnology, agriculture, robotics, aerospace, transportation, and other industrial applications. This section highlights the prospective applications of nanobioparticles and/or nanobiotechnology in various fields, although more explorations and research are still required to inflate the possibilities and methodologies in agriculture, environment, energy, and biomedical sciences as well. Several applications of NBT are discussed hereunder:

1.7.1 Medical Applications

In medical field, tissue engineering, controlled drug delivery, imaging of target sites, DNA structures' probing, gene transport, biomolecular sensing, cells and tissues' photothermal ablation, etc., have all been accomplished with NPs. NPs are also used in the treatment of diabetes, cancer, inflammation, allergies, and infections in a variety of therapies. NBT has aided in finding innovative methods for early diseases diagnosis, using less expensive materials and more advanced equipment. Pharmaceuticals, medical imaging, illness detection and treatment, implantable materials, tissue regeneration, and multifunctional platform integration have all been made possible by nanotechnology. AgNPs' properties have also been studied in tissue engineering and regenerative medicine, mostly for wound dressing purposes. Furthermore, the rising occurrence of infections caused by multidrug-resistant microorganisms is a serious public health issue that necessitates the inclusion of antimicrobial components in the wound healing scaffold design. These particles have been used in scaffolds constructed of a variety of materials, including polyvinyl alcohol, gelatin, chitosan-alginate, and cellulose acetate. All of the scaffolds that resulted have a lot of antibacterial activity.

Magnetite and maghemite (Fe_3O_4 and Fe_2O_3, respectively) are most commonly used forms of iron oxide NPs in biological applications including enhancement of MRI contrast, tissue healing, immunoassay, detoxification of biological fluid, drug administration, and separation of cells (Laurent et al. 2009). In addition, Au NPs have been successfully employed in cancer detection and treatment, protein and immunoassay and capillary electrophoresis. Besides, RNA nanoparticles are currently being researched for use in tumour diagnosis. Homogenous RNA and quantum dots as functionalized nanoparticles were created to diagnose diseases such as cancer and visualize biological events respectively in vivo, without accumulating in tissues and organs in real time, noninvasively.

NBT can be used to provide more accurate diagnoses of early-stage cancers, which is crucial for improving survival rates. Technologies like nanoparticle-based bio-bar codes have also been developed for ultrasensitive detection, enabling detection of very small concentrations of components like proteins, thereby increasing the chances of early-stage detection and diagnosis of diseases. Nanotechnology can also help pave the way for personalized medicine, where a person's unique characteristics are taken into account to formulate more effective treatments. NT also has applications in neuroscience, and it may be possible to noninvasively map the detailed workings of the brain to an extent that hasn't yet been achieved.

The development of biodegradable NPs as effective medication delivery systems has piqued interest (Zhang and Saltzman 2013). Drug delivery research has used a variety of lipids and polymers because they can effectively transfer medications to the target region, enhancing therapeutic benefit while limiting side effects. Nanoparticles containing drugs are coated with targeting agents such as conjugated antibodies which circulate through the blood vessels and reach the target cells.

Nanotechnological interventions have also been found in tissue engineering (TE) and organ transplantation. To repair or reconstruct missing or injured tissue, TE uses growth hormones, cell therapy, injectable biopolymers, and biomaterials. Electrospinning is a process in which a solution is injected into the system via a syringe and then exposed to a voltage difference, resulting in solid fibre after the operation. Surface conjugation and conducting characteristics of Au NPs, antimicrobial properties of Ag and other metallic NPs and metal oxides, fluorescence properties of QDs, and the unusual electromechanical characteristics of CNTs have all made them very useful in tissue engineering. Similarly, researches on cell mechanotransduction, cell patterning, gene delivery, and formulation of 3D tissues have exploited magnetic NPs (Hasan et al. 2018).

Nanomaterials, which could be used for gene transfer, have also gotten a lot of attention. Gene therapy is a relatively recent approach to treating or preventing genetic disorders by replacing or correcting disease-causing genes. The most common strategy for fixing broken genes is to insert a normal gene into an unspecific section of the genome to replace a nonfunctional gene. Selective reverse mutation can also be used to replace a defective gene with a normal gene, restoring the gene's function. Mammalian cells are typically a few microns in diameter, with organelles that are nanometres in size. Nanoscale objects could be utilized to successfully allocate repaired genes or to substitute flawed genes as being able to enter cells more easily than larger devices, allowing them to interact with cells effectively or at least in a different way. Protection of nucleic acids from nucleases and its surrounding is facilitated by specific structure of NPs and thus minimizing its side effects by instructing/accommodating nucleic acid at appropriate site of action. Various types of nanoparticles, including lipid-based nanoparticles (Pensado et al. 2014) and polymer-based nanoparticles, have been studied as gene carriers (Gao et al. 2015). Because of their qualities related to safety, non-immunogenicity, controllability, and low cost, nanoparticles have recently been studied for use as gene delivery devices to overcome delivery hurdles (Chen et al. 2016). In phase 1 clinical trials against malignancies, Davis et al. (2010) announced the first nanoparticle-based gene delivery method, CALAA-01. A recent experiment showed the use of nanoparticles to successfully perform gene editing in the liver to reduce blood cholesterol. Another team showed how nanoparticles can successfully turn off genes in the bone marrow, which could be beneficial in the treatment of heart disease or stem cell deficiencies.

Nanotechnology breakthroughs have led to the creation of nanoparticle-based simple assays for the precise detection of clinically relevant analytes. For diagnostic reasons, gold nanoparticles with unique optical characteristics and a large surface area are often used. Au nanoparticles in colloidal state exhibit colour variability, which depends on shape, size, and magnitude of aggregation of NPs. Bacteria, poisons, proteins, and nucleic acids can now be coupled to these nanoparticles due to advancements in nanomaterial manipulation. High reactivity and excellent chemical characteristics of NPs pave way for its usage as biosensors in nucleic acid and protein detection. Early identification of tuberculosis has also been achieved using nanoparticle-based platforms, allowing for better infection control.

Apart from other attributes, large surface area of NPs permits for the immobilization of a greater number of biomolecules by way of increasing available reaction sites for interaction of target moieties. NMs are perfect for detection and fabrication of biosensors with improved sensitivity and fast response. They're ideal for soluble chemical applications which escape several limitations associated with cell-based tests. Biosensors are especially valuable in the field of receptor research as they do not necessitate removal of receptor from the lipid membrane of the cell. Nonspecific binding on nanotubes, which often occurs with several nonspecific proteins, is removed through immobilization of polyethylene oxide chains. Functionalized QDs have been applied as labels in FISH assays and for DNA probing of genomic DNA as a replacement for traditional fluorescent dyes.

1.7.2 Environmental Applications

At present, nanobiotechnology has explored many ways of detecting and clean-up of environmental contaminants present in high concentrations through biologically synthesized nanoparticles. The use of nanostructured materials as adsorbents or catalysts to remove hazardous and dangerous compounds from air, industrial effluent, groundwater, and soil, as well as sensing and detecting toxicants and pathogens in the environment, has been the focus of recent studies. To fulfil the objectives, NBMs should satisfy certain criteria i.e. they should be non-toxic, eco-friendly, selective for low concentration contaminants, high adsorption capacity, facile elimination of contaminants adsorbed on the surface of nanomaterial, and recyclable. Various environmental applications of NBPs are highlighted hereunder.

For water purification, multifunctional distributed treatment plants can be empowered by engineered nanomaterials. These treatment systems are easily deployable, small, and reduce water loss, contamination problems, and energy requirements as well. Multiple treatment functions (for instance, adsorption, disinfection, and catalytic degradation) can also be facilitated with engineered nanomaterials that can target different contaminants simultaneously with high removal efficiency. Metallic nanoadsorbents, such as TiO_2, Al_2O_3 ZnO, and iron oxides, have been used widely to eliminate metals from wastewater, as they are cheap and efficient. The efficient nanoadsorbents are magnetic nanoparticles (MNPs), which comprise hematite (–Fe_2O_3), magnetite (Fe_3O_4), and maghemite (–Fe_2O_3). Metal impurities can be removed and separated using superparamagnetic iron oxide NPs. A wide range of heavy metals in their ionic forms including chromium, lead, cobalt, copper, nickel, etc., has also been removed by using magnetic nanoparticles (Lei et al. 2014). Carbon nanotubes synthesized from poultry litter waste in the presence of Ni/Mo/MgO as catalyst have also been deployed for Cr removal from tannery wastewaters (Haleem et al. 2021). These carbons are employed as filters in filtering, as adsorbents in water purification, and to remove hydrogen, nitrogen, smells, vinyl chloride, and other contaminants.

Despite this, iron nanoparticles have been observed to be effective in groundwater treatment, especially for pesticides and polychlorinated biphenyls, dyes, nitrates, and phosphates. The reduction of a dye developer 4-nitrophenol to 4-aminophenol was similarly catalysed by gold nanoparticles recovered by *Gymnocladus assamicus* (Tamuly et al. 2013). However, Au nanoparticles produced from *Parkia roxburghii, Sesbania grandiflora*, and *Pogestemon benghalensis* performed exceptionally well in reducing the methylene blue dye (Paul et al. 2015, 2016). Using Au nanoparticles produced from *Eucommia ulmoides*, Congo red and reactive yellow 179 dyes were degraded by photocatalytic activity (Guo et al. 2015). Arifin et al. (2017) produced iron oxide nanoparticles of size 53.76 nm from mill scale and recorded 99% adsorption of dyes from wastewater. The large surface area and recyclability of TiO_2 nanoparticles make them a potential resource for the degradation of nitroarene compounds and toxic dyes. Green TiO_2 nanoparticles were also reported to remove Cr by 76.48% and COD by 82.86% from tannery wastewater (Goutam et al. 2018). Palladium particles made biosynthetically are also used as nanocatalysts in the degradation of dyes such as methyl orange, Coomassie brilliant blue G-250, methylene blue, and 4-nitrophenol (Kora and Rastogi 2018). Peres et al. (2018) also used corn husk waste to make silica nanoparticles, which they used as a catalyst to remove methyl blue dye by 80%. Zhang et al. (2020) used saccharin-containing wastewater to generate NiFe2O4 nanoparticles, which were then combined with a mesoporous magnet NiFe2O4/ ZnCuCr-LDH combination to remove Congo red up to 97%.

The combination of magnetic and sorption properties in nanomaterials led to the development of advanced nanosorbents allowing selective and fast recovery of organic pollutants by employing magnetic separation techniques. Nanotechnology-enabled biosensors can now detect and characterize pollutants of chemical and biological origin in wastewater, air, and soil with utmost sensitivity. Chemically and physiologically, neutral iron oxide nanoparticles (Fe_3O_4) coated with enzymes, catalysts, and even antibodies are used as biosensors. Chauhan et al. (2016) employed poly(indole-5-carboxylic acid) to modify Fe_3O_4 nanoparticles and created a nanobiocomposite that may be used as a biosensor to detect pesticides like malathion and chlorpyrifos. Researchers are looking into particles like auto-assembled monolayers on mesoporous support medium (SAMMSTM), carbon nanotubes, and dendrimers to find the ways for how they may be used for toxic site clean-up. Previously, scientists developed low-cost optical sensors for mercury (Hg^{2+}) detection utilizing recycled carbon nanoparticles from pomelo peel waste. There have also been a few investigations on the capture of air pollutants using nanobioparticles. With adsorption capacity of 159 mg/g at 75 °C under 15% CO_2, Zeng and Bai (2016) developed porous silica nanoparticles as good sorbents from rice husks to capture post-combustion CO_2.

To assess the air quality around flames, NASA has also created unique sensor as a smartphone extension for fire-fighters.

There have also been a number of researches on the identification and eradication of pathogens in wastewater using NBMs. Using the photocatalytic capabilities of zinc sulphide (ZnS) nanoparticles, sulphate-reducing bacteria were selectively discovered (Qi et al. 2013). These nanoparticles exhibited potential antibacterial activity against infectious species such as *Staphylococcus aureus*, *Vibrio cholerae*, *Bacillus subtilis*, *Syphillis typhus*, *Pseudomonas aeruginosa*, and *Escherichia coli*. TiO_2 nanoparticles obtained from *Hibiscus rosa sinensis* shown excellent antibacterial efficacy against both Gram-negative and Gram-positive strains of bacteria (Kumar et al. 2014).

Additionally, the potential of green synthesized Al_2O_3 NPs in development of inorganic membranes and as adsorbent in removal of organic and inorganic contaminants including chromium, fluoride, and arsenic (Goutam et al. 2018). Similarly, silica has been a good desiccant and relative humidity maintainer in gas drying, during fabrication of inorganic membranes, along with adsorbent for remediation of other contaminants. Nanozeolites are employed in wastewater treatment, liquid and gas sweetening, CO_2 recovery, and SOx and NOx pollution control by desulfurization and denitrogenation. Nanoclay is used to purify water and oils, as well as to refine mineral oils. Membrane development and water treatment both involve polymers. For efficient desalination, engineers developed a thin-film membrane incorporating nanopores. Such membrane filtered two to five times more water than contemporary conventional filters. Nanofabric "paper towel" made of ultra-thin potassium manganese oxide wires that can absorb twenty times its weight in oil for clean-up purposes has been developed. Further, the applications of magnets to mechanically remove oil from spills with magnetic water-repellent nanoparticles have been devised. Many aeroplane cabins and air filters use NT-enabled "mechanical filtration technique", wherein nanoscale pores in the fibres catch particles larger than their pores' size. Charcoal layers may also be present in the filters, which help to eliminate odours.

1.7.3 Energy Production

Nanotechnology is often being used to improve alternative energy solutions in order to fulfil the world's growing energy requirements. Scientists are finding techniques to produce clean, renewable, and economical energy sources to reduce energy usage and environmental injuries. Significant advances have been made in designing and synthesis of nanomaterials (catalysts, adsorbents, membranes, etc.) and processes for emission control of nitrogen oxides, hydrocarbons, volatile organic compounds, hydrogen sulfide, carbon monoxide, soot particles etc. from mobile and stationary sources, along with CO_2 capture and transformation to methane and renewable fuels to generate clean energy. Through improved catalysis, efficient fuel synthesis from crude petroleum is being facilitated by NT. Due to enhanced combustion efficiency and reduced friction, it also reduces fuel consumption in automobiles and power plants. NT used in oil and gas extraction has been clearly exemplified by application of NT-enabled gas lift valves in offshore operations and detection of fractures in down-well oil pipeline. Besides, applications of CNTs, membranes, and scrubbers in extraction of CO_2 from power plant exhaust are another arena of its significance. NT can also be used into solar panels to upgrade the efficiency of sunlight being converted to energy, resulting in more affordable solar power for days to come. Nanofabricated solar cells have been manufactured using print-like methods and flexible rolls in place of discrete panels, which could be less expensive for production and installation. Novel batteries are already being developed using NT that are lighter in weight, charge faster, and have better power density for longer duration. Apart from that, epoxy-containing CNTs are in use to create longer, stronger, and lighter windmill blades, and thus allowing windmills to generate more electricity.

In photocatalytic applications, NPs are frequently used to generate energy from photoelectrochemical and electrochemical splitting of water in the field of energy harvesting. Apart from water splitting, advanced energy generation alternatives include electrochemical CO_2 reduction to fuel precursors and piezoelectric generators. NPs are also employed in energy storage applications at the nanoscale to store energy in various forms. Nanogenerators that transform mechanical energy into electricity using piezoelectric technology have recently been developed; this is a novel way of generating energy. Similarly, numerous NT-based possibilities for transforming waste heat in autos, homes, power plants, computers, and other devices to useful electrical power are being explored. NT is facilitating more efficient lighting systems, lighter and stronger car chassis materials, less energy usage in electronics and light-responsive glass coatings, etc. NBT has aided in the development of batteries with more energy storage for electric automobiles, as well as effective solar panels that convert more sunlight into electricity, in order to address different issues posed by climate change and the energy crisis. Nanotexturing or nanomaterials are employed in transformation of flat surface into a three-dimensional structure with larger surface area, allowing more space for reactions and thus enhancing efficiency. Innovative nanomaterials and concepts open vistas to produce energy with high conversion efficiency from light, movement, temperature fluctuations, glucose, and so on.

1.7.4 Industrial Applications

As per industry analysts, NT is expected to make "game-changing" breakthroughs in renewable energy, pollution clean-up, communications, electronics, medicine, textiles, sporting equipment, cosmetics, sunscreens, batteries, agriculture, food packaging, dietary supplements, and many more. Lighter and stronger materials, like used in the car industry, will be extremely useful to aircraft manufacturers, resulting in improved performance. Spacecraft will benefit as well, as weight is a crucial consideration. As a result, nanotechnology may be able to help minimize the size of the equipment and, as a result, the amount of fuel necessary to make it airborne.

Nanotechnology, which is used in nanolubrication, is also a rapidly emerging scientific field. Nanoparticles may improve tribological qualities such as anti-friction, anti-wear, severe pressure, flash point, and high-temperature resistance when added to a base oil or water. Metals such as Au, Ag, and Cu, as well as metal oxides such as Si, Ti, and ZrO_2, are the most often utilized nanoparticles. The direct effect of the protective nanoparticle film, the ball-bearing effect, and surface augmentation through nanoparticle deposition, which artificially smooths up the surface, are all credited with their lubricating action.

NT is being used to develop self-healing and corrosion-resistant coatings for insulation purposes. Coatings are widely used in construction to paint walls, doors, and windows. These coatings being hydrophobic repel water molecules from metal pipes and protect them from corrosion. Furthermore, systems based on nanoparticles can give improved adherence and transparency. By using a photocatalytic process, the TiO_2 coating absorbs all sorts of air pollutants, making roads more environment friendly. NT-enabled construction could be made quicker, cheaper, safer, and more varied. In a nutshell, NT could be used to detect faults in architectural foundations and dispatch nanobots to repair them. NP-mediated cement will open up a plethora of new possibilities in ceramics, high-strength composites, and electrical applications. The application of nanotechnology to glass, another essential construction material, is also being studied. Because TiO_2 NPs have sterilizing and anti-fouling capabilities, they are employed to cover glazing. Organic contaminants, volatile organic chemicals, and bacterial membranes are all broken down by the particles, which catalyse strong reactions. Being hydrophilic, TiO_2 can attract rainfall, which subsequently washes the filth away. In addition, NT has evolved light- and fire-resistant glasses likely to be used in windows and other infrastructural development.

NPs are also employed to improve the efficacy of cosmetic and personal care products. In cosmetics, nanosomes, fullerenes, liposomes, solid lipid nanoparticles, etc., are used to improve colour and quality, and provide higher UV protection, deeper skin penetration, long-lasting effects, etc. These nanoparticles are relatively felt light on skin and block UV radiation effectively. TiO_2 and ZnO are two forms of NPs that are often used in sunscreen. Pupil correction and laser eye surgery could be more precise because of nano-optics. Nanotechnological research has also provided sunglasses with ultra-thin polymer coatings that serve as protective and anti-reflective.

The engineered nanofibers also find their applications in the textile industry to fabricate advanced clothes having antibacterial, antimicrobial, stain-repellent, wrinkle-free, water-resistance, and protective properties. Nanotechnologically finished textiles may be washed at extremely low temperatures. NT-enabled textiles with nanocarbon particles embedded into the membrane ensure full-surface protection of the person from electrostatic charges. Silica NPs can aid in the preparation of textile with stain-resistant, hydrophobic fabrics than can also repel other liquids. Liquid creates small beads on waterproof clothes, which roll off the fabric instead of being absorbed.

Sports equipment and products such as soccer, football, tennis balls, and baseball can all benefit from nanotechnology. Nanotechnology aids with the retention of bounce in these balls as well as the strengthening of tennis racquets. New sporting shoe materials may be developed to make the shoe lighter. Carbon nanotubes incorporated in resin are used to produce baseball bats, which make them lighter and more efficient. Antimicrobial nanotechnology has been utilized to create other goods for players, e.g. sports towels and mats for yoga and exercises, to avoid infections caused by germs such as methicillin-resistant Staphylococcus aureus (MRSA).

Apart from that, the term "nano-weapon" refers to military technology that is currently in development and aims to harness the strength of NT in present day (Altmann 2004). The nanoparticles are utilized to make coated polymer threads that may be weaved into troops' uniforms to keep them light and comfortable while stopping bullets, defending against poisons, monitoring vital signs, and giving first aid. Such polymer threads might be utilized as a means of communication among soldiers, with coded messages relayed by flashing lights. The uniforms' threads may be adjusted to distinct light wavelengths, making it impossible for anybody else to listen in. Silent communication between soldiers is possible because of these specially coated polymer threads weaved into the outfit. This would reduce the chances of anything being intercepted by unintended listeners. Furthermore, unique molecules known as dendrimers are used to guard against chemical and biological threats. The hazardous compounds cling to dendrimers and are made inert.

Auto-cleaning surfaces on glassware or ceramics are the poplar application of NT at home. Nanoceramic particles improve the heat resistance and smoothness of daily used items like flat irons. Similarly, NT is helpful in the production of flameless furniture. Coating of upholstered furniture foam with CNFs may lower flammability up to 35%. Adhesives can also benefit from nanotechnology. Surprisingly, the application of nano-glue in place of ordinary glue can better withstand and enhance stickiness at higher surrounding temperature. Likewise, nanocomposite coatings provide scratch-resistant surfaces.

1.7.5 Applications in Electronics

1D semiconductors and metals have extraordinary structural, electrical, and optical characteristics, enabling them a significant structural component in next-generation electronics, sensors, and photonic materials. The development of printed electronics is gaining popularity as a low-cost, large-area electronics solution for flexible monitors and sensors. Printed electronics using various functional inks incorporating NPs, e.g. organic electronic molecules, metallic NPs, CNTs, and ceramic NPs, were projected to become a mass production procedure for innovative electronic equipment (Kosmala et al. 2011). The use of CNTs in semiconductor chips and photovoltaics; use of light-emitting diodes (LEDs) and organic light-emitting diodes (OLEDs); use of QDs in lasers; use of lithium-ion batteries; and the development of solder-free assembly technology are some of the conventional and modern NMs' uses in electronic industry. Aside from electronics, several nanoparticles are employed as surface coatings in a variety of electrical items, primarily for antimicrobial qualities.

Researchers have also looked into using DNA-based computer circuits, or biological circuits, made of synthetic DNA as a computing medium. It sounds promising since DNA has such a high data density: it contains all of the information and instructions needed to develop and run a human body. Long texts have been encoded in DNA, and the molecule has also been used to form simple logic gates and circuits, the fundamental building blocks of computing. DNA computing has a broad range of medical applications, including cancer diagnosis and the detection of internal injuries like traumatic brain damage, haemorrhagic shock, and more. Synthetic biological circuits can also be utilized to deliver medications into cells precisely and at precise concentrations.

1.7.6 Agricultural Applications

Nanotechnology has paved way to improve and encourage sustainable forms of agricultural production. Preliminary research shows that nanotechnologies can improve seed germination and growth, as well as crop productivity by using nano-pesticides and fertilizers. Similarly, using nanomaterials such as SiO_2, TiO_2, and carbon nanotubes, nano zeolites and hydrogels have enhanced soil quality while accelerating plant development, plant protection, disease detection, and pesticide/herbicide residue detection. Aside from that, smart crop monitoring using nanosensors has been a benefit to it (Khot et al. 2012). Nanotechnology could help soils utilize nutrients more efficiently, resulting in increased productivity and improved environmental security. Nano fertilizers are gaining traction as a viable alternative to conventional fertilizers, allowing for regular nutrient accumulation in soils and therefore lessening the problem of eutrophication in aquatic water bodies, as well as contamination of drinking water. It could also facilitate the nutrient transport processes, i.e. ion exchange, adsorption–desorption, and solubility-precipitation in soil–plant systems, and would ensure nutrients availability to plants best suited to it.

To boost agricultural production, new nanomaterials based on inorganic, polymeric, and lipid nanoparticles have been produced using a variety of processes (emulsification, ionic gelation, polymerization, oxidation–reduction, and so on). In order to have hazard-free usage, easy storage, and need-based delivery, nanoformulated materials with plant nutrients can be employed in hydrogel forms and aqueous suspension. Similarly, the use of zerovalent Fe NPs could be used for cleansing of soils polluted with heavy metals, pesticides, and radionuclides. Iron nanoparticles, like calcium carbonate nanoparticles, have excellent soil binding characteristics that aid in the development of soil micro- and macroaggregates. Plant genetic improvement, allocation of genes and drug molecules at particular site in cells, plant gene expression to overcome stress, as well as creation of sensors and protocols involved in precision farming, food contaminants and pathogen detection, smart processing, packaging, and monitoring in food industries are added advantages of NTs.

Nano-formulations-based pesticides and fertilizers have been found to control the populations of insects, pests, and host pathogens which in return improve yields with little or no toxicity. Encapsulation of pesticide at nanoscale has improved the solubility, selectivity, permeability, and stability of the pesticide's release and thus controlled-release towards target organisms (Prasad et al. 2017), along with increased resistance against hydrolysis and photodecomposition (Mishra and Singh 2015; Grillo et al. 2016). Despite this, Oh et al. (2006) found that Ag-SiO_2 NPs have a strong antifungal effect against *Botrytis cinerea*, and Medda et al. (2015) found that Ag NPs synthesized from *Aloe vera* leaf had antifungal activity against *Rhizopus stolonifer* and *Aspergillus* sp., inhibiting hyphal growth, budding, and conidial germination. Antifungal activity of Ag NPs against *Fusarium oxysporum*, *Colletotrichum gloeosporioides*, *Bipolaris sorokiniana*, and *Magnaporthe grisea* was also discovered by various researchers. Besides, nanoparticles are useful for the detection of toxins in the soil as well as for crop disease and their diagnosis. The sensitivity of nanodevices can also be used for fast, economical, and efficient detection of phytopathogens, thereby reducing the chance of large-scale destruction of crops. Nanotechnology is used for better breeding results, monitoring and improving the health of animals, as well as for better production. Table 1.3 lists the agricultural applications of several nanoparticles.

1.7.7 Food Processing and Safety

NT can tackle a complicated set of scientific and engineering issues in the field of food bioprocessing for effective production of quality and safe food. Modern researches are focused on adding antioxidants, antimicrobials, biosensors, and other nanomaterials to food packaging. Identification of bacteria, biosensing of food quality, smart systems of food packaging, and nanoscale encapsulation of bioactive food ingredients are some of the novel NT uses in the food processing, safety, and packaging fields. Nanocomposite coating with antimicrobial agents could improve food packaging and mechanical and thermal resistance, and reduce oxygen transfer rates and fluctuations in gas permeability of various fillers as required for various goods. The three new/nanofoods, viz. Canola Active Oil, a tea called Nanotea, and Nanoceuticals Slim Smoothie Chocolate, a chocolate diet shake, are the outcomes of the Project on Emerging Nanotechnologies (PEN). As per PEN's website, Shemen Industries' canola oil contains a "nanodrops" ingredient that helps vitamins, minerals, and phytochemicals. The shake industry, viz. RBC Life Sciences Inc., employs cocoa-infused "NanoClusters" to increase the taste and health advantages of cocoa with no additional sweetener.

Silver and similar metal nanoparticles have been used in a variety of nano-based commercial goods due to their antimicrobial properties. According to previous studies, an increased surface area/smaller particle size improves antibacterial efficacy. According to some scientists, other nanoparticles, such as zinc and sulphur, also have antibacterial properties. Silver nanoparticles will be combined with water-soluble biopolymers to develop new antimicrobials. Gum acacia, starch, gelatin, sodium alginate, and carboxymethylcellulose were among the natural polymers

Table 1.3 Types, synthesis methods, and application of nanoagroparticles in agriculture

Type of NPs	Production	Uses	References
Crop protection			
Cu	Chemical	Fungicide	Kanhed et al. (2014)
Cu and Ag	Chemical	Fungicide	Ouda (2014)
Ag	Biological	Fungicide	Khadri et al. (2013)
Ag	Biological	Bactericide	El-Rahman and Mohammad (2013)
Ag	Biological	Pesticide/insecticide	Ali et al. (2015)
ZnO	Chemical	Bactericide	Hafez et al. (2014)
Seed treatment			
Ag	Biological	dressing of seeds	Anand and Kulothungan (2014)
Ag	Biological	Surface sterilize of seed crops	Morsy et al. (2014)
Herbicide formulation			
Carboxymethyl chitosan	*Chemical*	Herbicide	Yu et al. (2015)
Chitosan	Chemical	Herbicide	Grillo et al. (2014)
Polyepsilon-caprolactone	Chemical	Herbicide	Pereira et al. (2014)
Biosensing			
Carbon	Chemical	Herbicide detection	Luo et al. (2014)
Graphene	Chemical	Herbicide detection	Zhao et al. (2014)
Au	Chemical	Herbicide detection	Boro et al. (2011)
Au	Chemical	Organophosphates detection	Kang et al. (2010)
Ag	Chemical	Herbicide detection	Dubas and Pimpan (2008)
Plant growth promotion			
Fe, Co, Cu	Chemical	Increase germination rate	Ngo et al. (2014)
Ag	Chemical	Plant growth	Syu et al. (2014)
ZnO	Chemical	Flowering and seed productivity	Laware and Raskar (2014)

Source Baker et al. (2017)

employed to create biocompatible polymeric silver nanocomposites. Chitosan, a natural polymer composed of poly(-(1- 4)-2-amino-2-deoxy-D-glucose is another most common structural polysaccharide after cellulose in nature. Chitosan readily interacts with bacteria, binding to DNA, glycosaminoglycans, and the majority of proteins, boosting the antibacterial activity of silver nanoparticles (Vimala et al. 2011). They also looked at incorporating curcumin into a chitosan-PVA nanoparticles film to see whether it could boost the film's therapeutic antibacterial activity.

Furthermore, some study indicates a viable strategy for creating novel antibacterial drugs by combining silver nanocomposites with a natural polyphenol (curcumin). These compounds could be used in antimicrobial packaging materials. Curcumin, having antioxidative, anti-inflammatory, and anticancer effects, is isolated from the rhizomes of the Curcuma species and used to make a nanogel. Further, the usage of natural polymeric nanocomposite films in rapid detection of food-borne pathogens, nanoimmunosensors array for microbial pathogen detection, food packaging for safe storage, CNTs, and nanoscale TiO_2 particles as a UV light protecting agent in plastic packaging are some of the applications of NT.

Furthermore, nanosensors, undoubtedly, provide smart sensing of food components and contaminants, as well as make alert to consumers about the condition of the food quality inside. Sensors can warn before spoilage of food product and provide information on the nutritional quality of the foodstuff. Nanosensors can also monitor physical parameters (e.g. humidity, pH, temperature, light exposure), gas mixtures (e.g. O_2 and CO_2), pathogens and toxins, freshness (e.g. ethanol, lactic acid, acetic acid), and decomposition (e.g. putrescine, cadaverine) in intelligent food packaging.

1.8 Challenges Ahead

Although nanotechnology may hold the key to healing severe diseases and unlocking and manipulating the mysteries of the genome, there is still a long way to go before its full promise is fulfilled. Engineered nanobioparticles are increasingly being included into consumer items, with the potential to be changed, dispersed, and carried throughout their life cycle in water, soil, sediment, and air environments. Such modes of transportation may have a greater impact on NPs' environmental destiny, bioavailability, and toxicity, as well as human health. The antibacterial effect of nanomaterials causes toxicity in living organisms. Various microorganisms, including fungus and bacteria, produce toxins. Microbial toxins' health consequences are determined by the type of exposure (ingestion, inhalation, and skin contact) as well as the toxin's chemical structure and physicochemical features. Unfortunately, despite their widespread use and distribution, engineered nanoparticles' toxicity and potential for harm have yet to be properly explored and/or removed. As a result, determining the fate of nanoparticles and investigating any associated possible dangers become critical. Studies at the nano-bio interaction are needed to better understand the risks of materials and design safer nanomaterials. The impact of physicochemical qualities on cell bioavailability, uptake, and bioprocessing is being investigated in these studies. Furthermore, the complicated structure of manufactured nanoparticles made of various materials would very certainly necessitate novel detection systems to determine their degrees of presence in the environment. Nanotechnologies raise ethical concerns, such as a lack of knowledge about the potential toxicity and environmental fate of nanobiomaterials. Furthermore, the lack of cost-effective methods for mass production of nanoparticles and materials greatly limits the reach of nanotechnology. With the growth of nanotechnology, there is also the possibility of large-scale employment losses in the agriculture industry. Nanotechnology's potential for genetic engineering raises ethical concerns. As a result, a scientific evidence-based strategy is required to comprehend the health, environmental, and safety risks of nanoparticles.

1.9 Conclusion

Nanobiotechnology offers great opportunities for developing novel biomaterials and methods that could augment the capacity to solve various environmental problems. Researchers are increasingly focused on green nanoparticle synthesis, using eco-friendly techniques and green synthetic routes for nanomaterial synthesis using plants, microbes, and other natural resources. Due to their cost-effectiveness, harmless method, easy availability, and environmentally friendly character, several studies on the synthesis of biogenic nanoparticles and their prospective uses in diverse industries are now underway. However, because these nanoparticles might affect the living biota, including human health, it is critical to analyse the ecological risk and repercussions of their use. Furthermore, research on innovative nano-bioproducts, as well as their toxicokinetics, must be reviewed in order to prevent these nanoparticles from having a negative impact on the environment's flora and fauna. Before these nanoparticles are used on a large scale, it is necessary to gain a better knowledge of their environmental fate. Furthermore, the successful and long-term deployment of nanotechnology will be contingent on the selection of low-cost, environmentally safe nanotechnology for applications in biomedicine, agriculture, and bioremediation, among other fields.

References

Abd Ellah NH, Abouelmagd SA (2016) Surface functionalization of polymeric nanoparticles for tumor drug delivery: approaches and challenges. Expert Opin Drug Deliv 14(2):201–214

Alagarasi A (2009) Introduction to nanomaterials. In: Viswanathan B. (ed) Nanomaterials 2–25. Narosa Publishing House, Mumbai

Ali J, Ali N, Jamil SUU, Waseem H, Khan K, Pan G (2017) Insight into eco-friendly fabrication of silver nanoparticles by Pseudomonas aeruginosa and its potential impacts. J Environ Chem Eng 5(4):3266–3272

Ali SM, Yousef NMH, Nafady NA (2015) Application of biosynthesized silver nanoparticles for the control of land snail Eobania vermiculata and some plant pathogenic fungi. J Nanomater 2015:Article ID 218904. https://doi.org/10.1155/2015/218904

Altmann J (2004) Military uses of nanotechnology: perspectives and concerns. Secur Dialogue 35:61–79

Alvarez PJJ, Chan CK, Elimelech M, Halas NJ, Villagrán D (2018) Emerging opportunities for nanotechnology to enhance water security. Nature Nanotech 13:634–641

Anand R, Kulothungan S (2014) Silver mediated bacterial nanoparticles as seed dressing against crown rot pathogen of groundnut. Arch Appl Sci Res 6:109–113
Arifin SARA, Ismail I, Halim AA, Nabilah SF, Nazlan R, Riati II (2017) Iron oxide nanoparticles derived from mill scale waste as potential scavenging agent in dye wastewater treatment for batik industry. Solid State Phenom 268:393–398
Baker S, Volova T, Prudnikova SV, Satish S, Prasad MNN (2017) Nanoagroparticles emerging trends and future prospect in modern agriculture system. Environ Toxicol Pharmacol 53:10–17
Borah R, Saikia H, Mahanta A, Das V, Bora U, Thakur AJ (2015) Biosynthesis of poly(ethylene glycol)-supported palladium nanoparticles using Colocasia esculenta leaf extract and their catalytic activity for Suzuki-Miyaura cross-coupling reactions. RSC Adv 89(5):72453–72457
Boro RC, Kaushal J, Nangia Y, Wangoo N, Bhasinc A, Suri CR (2011) Gold nanoparticles catalyzed chemiluminescence immunoassay for detection of herbicide 2,4-dichlorophenoxyacetic acid. Analyst 136:2125–2130
Chakraborty N, Banerjee A, Lahiri S, Panda A, Ghosh AN, Pal R (2009) Biorecovery of gold using cyanobacteria and an eukaryotic alga with special reference to nanogold formation—a novel phenomenon. J Appl Phycol 21:145–152
Chauhan N, Narang J, Jain U (2016) Amperometric acetylcholinesterase biosensor for pesticides monitoring utilising iron oxide nanoparticles and poly(indole-5-carboxylic acid). J Exp Nanosci 11(2):111–122
Chen J, Guo Z, Tian H, Chen X (2016) Production and clinical development of nanoparticles for gene delivery. Mol Ther: Methods Clin Dev 3:16023. https://doi.org/10.1038/mtm.2016.23
Das SK, Das AR, Guha AK (2009) Gold nanoparticles: microbial synthesis and application in water hygiene management. Langmuir 25(14):8192–8199
Davis ME, Zuckerman JE, Choi CH, Seligson D, Tolcher A, Alabi CA et al (2010) Evidence of RNAi in humans from systemically administered siRNA via targeted nanoparticles. Nature 464:1067–1070
Devi TP, Kulanthaivel S, Kamil D, Borah JL, Prabhakaran N, Srinivasa N (2013) Biosynthesis of silver nanoparticles from Trichoderma species. Indian J Exp Biol 51(7):543–547
Dobrucka R, Długaszewska J (2016) Biosynthesis and antibacterial activity of ZnO nanoparticles using Trifolium pratense flower extract. Saudi J Biol Sci 23(4):517–523
Duan H, Wang D, Li Y (2015) Green chemistry for nanoparticle synthesis. Chem Soc Rev 44. https://doi.org/10.1039/c4cs00363b
Dubas ST, Pimpan V (2008) Humic acid assisted synthesis of silver nanoparticles and its application to herbicide detection. Mater Lett 62:2661–2663
El-Rahman AFA, Mohammad TGM (2013) Green synthesis of silver nanoparticle using Eucalyptus globulus leaf extract and its antibacterial activity. J Appl Sci Res 9(10):6437–6440
El-Sheekh MM, El-Kassas HY (2016) Algal production of nano-silver and gold: their antimicrobial and cytotoxic activities: a review. J Genetic Eng Biotechnol 14(2):299–310
Elumalai K, Velmurugan S (2015) Green synthesis, characterization and antimicrobial activities of zinc oxide nanoparticles from the leaf extract of *Azadirachta indica* (L.). Appl Surf Sci 345:329–336
Feynman RP (1960) There's plenty of room at the bottom. Eng Sci 23:22–36
Gao S, Tian H, Guo Y, Li Y, Guo Z, Zhu X et al (2015) miRNA oligonucleotide and sponge for miRNA-21 inhibition mediated by PEI-PLL in breast cancer therapy. Acta Biomater 25:184–193
Gill R, Zayats M, Willner I (2008) Semiconductor quantum dots for bioanalysis. Angew Chem Int Ed Engl 47:7602–7625
Goutam SP, Saxena G, Singh V, Yadav AK, Bharagava RN, Thapa KB (2018) Green synthesis of TiO_2 nanoparticles using leaf extract of Jatropha curcas L. for photocatalytic degradation of tannery wastewater. Chem Eng J 336:386–396
Grillo R, Abhilash PC, Fraceto LF (2016) Nanotechnology applied to bio-encapsulation of pesticides. J Nanosci Nanotechnol 16:1231–1234

Grillo R, Pereira AES, Nishisaka CS, Lima RD, Oehlke K, Greiner R et al (2014) Chitosan/tripolyphosphate nanoparticles loaded with paraquat herbicide: an environmentally safer alternative for weed control. J Hazard Mater 278:163–171

Gujrati V, Kim S, Kim SH, Min JJ, Choy HE, Kim SC et al (2014) Bioengineered bacterial outer membrane vesicles as cell-specific drug-delivery vehicles for cancer therapy. ACS Nano 8(2):1525–1537

Guo M, Li W, Yang F, Liu H (2015) Controllable biosynthesis of gold nanoparticles from a *Eucommia ulmoides* bark aqueous extract. Spectrochim Acta A Mol Biomol Spectrosc 142:73–79

Gurunathan S, Kalishwaralal K, Vaidyanathan R, Venkataraman D, Pandian SR, Muniyandi J (2009) Biosynthesis, purification and characterization of silver nanoparticles using Escherichia coli. Colloids Surf B Biointerfaces 74(1):328–335

Hafez EE, Hassan HS, Elkady M, Salama E (2014) Assessment of antibacterial activity for synthesized zinc oxide nanorods against plant pathogenic strains. Int J Sci Technol Res 3:318–324

Haleem N, Jamal Y, Khan SN, Baig MA, Wahab M, Yang X (2021) Synthesis of carbon nanotubes (CNTs) from poultry litter for removal of chromium (Cr (VI)) from wastewater. Materials 14:5195

Hasan A, Morshed M, Memic A, Hassan S, Webster TJ, Marei HE (2018) Nanoparticles in tissue engineering: applications, challenges and prospects. Int J Nanomedicine 13:5637–5655

Hochella M, Aruguete D, Kim B, Elwood Madden A (2012) Naturally occurring inorganic nanoparticles: general assessment and a global budget for one of earth's last unexplored major geochemical components. In Nature's Nanostructures. Pan Stanford Publishing Pte. Ltd., pp 1–42

Hochella MF, Mogk DW, Ranville J, Allen IC, Luther GW, Marr LC et al (2019) Natural, incidental, and engineered nanomaterials and their impacts on the Earth system. Science 363:6434

Issa B, Obaidat IM, Albiss BA, Haik Y (2013) Magnetic nanoparticles: surface effects and properties related to biomedicine applications. Int J Mol Sci 14:21266–21305

Jain D, Shivani BAA, Singh H, Daima HK, Singh M et al (2020) Microbial fabrication of zinc oxide nanoparticles and evaluation of their antimicrobial and photocatalytic properties. Front Chem 8:778

Joglekar S, Kodam K, Dhaygude M, Hudlikar M (2011) Novel route for rapid biosynthesis of lead nanoparticles using aqueous extract of Jatropha curcas L. latex. Mater Lett 65:3170–3172

Juibari MM, Abbasalizadeh S, Jouzani GhS, Noruzi M (2011) Intensified biosynthesis of silver nanoparticles using a native extremophilic Ureibacillus thermosphaericus strain. Mater Lett 65(6):1014–1017

Kalishwaralal K, Banumathi E, Pandian SRK, Deepak V, Muniyandi J, Eom SH (2009) Silver nanoparticles inhibit VEGF induced cell proliferation and migration in bovine retinal endothelial cells. Colloids Surf B Biointerfaces 73:51–57

Kang T, Wang F, Lu L, Zhang Y, Liu T (2010) Methyl parathion sensors based on gold nanoparticles and Nafion film modified glassy carbon electrodes. Sens Actuators B 145:104–109

Kanhed P, Birla S, Gaikwad S, Gade A, Seabra AB, Rubilar O et al (2014) In vitro antifungal efficacy of copper nanoparticles against selected crop pathogenic fungi. Mater Lett 115:13–17

Kannan RRR, Stirk WA, Staden JV (2013) Synthesis of silver nanoparticles using the seaweed Codium capitatum P.C. Silva (Chlorophyceae). S Afr J Sci Bot 86:1–4

Khadri H, Alzohairy M, Janardhan A, Kumar AP, Narasimha G (2013) Green synthesis of silver nanoparticles with high fungicidal activity from olive seed extract. Adv Nanopart 2:241–246

Khot LR, Sankaran S, Maja JM, Ehsani R, Schuster EW (2012) Applications of nanomaterials in agricultural production and crop protection: a review. Crop Prot 35:64–70

Khan NT, Khan MJ, Jameel J, Jameel N, Rheman SUA (2017) An overview: biological organisms that serves as nanofactories for metallic nanoparticles synthesis and fungi being the most appropriate. Bioceram Dev Appl 7:1–4

Koduru M, Nannepaga J, Reddy BVS, Golla N (2013) Palladium nanoparticles: Single-step plant-mediated green chemical procedure using Piper betle leaves broth and their anti-fungal studies. Int J Chem Anal Sci 4(1). https://doi.org/10.1016/j.ijcas.2013.03.006

Koksharov YA (2009) Magnetism of nanoparticles: effects of size, shape, and interactions. In: Gubin SP (ed) Magnetic nanoparticles, 1st edn. Wiley-VCH: Berlin, Germany, pp 228–229

Kora AJ, Rastogi L (2018) Green synthesis of palladium nanoparticles using gum ghatti (Anogeissus latifolia) and its application as an antioxidant and catalyst. Arab J Chem 11(7):1097–1106
Kosmala A, Wright R, Zhang Q, Kirby P (2011) Synthesis of silver nano particles and fabrication of aqueous Ag inks for inkjet printing. Mater Chem Phys 129(3):1075–1080
Kumar PSM, Francis AP, Devasena T (2014) Biosynthesized and chemically synthesized Titania nanoparticles: Comparative Analysis of Antibacterial Activity. J Environ Nanotechnol 3(3):73–81
Kumar S, Abyaneh M, Gosavi S, Kulkarni S, Pasricha R, Ahmad A, Khan M (2007) Nitrate reductase-mediated synthesis of silver nanoparticles from $AgNO_3$. Biotech Lett 29:439–445
Laurent S, Boutry S, Mahieu I, Vander Elst L, Muller RN (2009) Iron oxide based MR contrast agents: from chemistry to cell labeling. Curr Med Chem 16(35):4712–4727
Laware S, Raskar S (2014) Influence of zinc oxide nanoparticles on growth, flowering and seed productivity in onion. Int J Curr Microbiol Appl Sci 3:874–881
Lee SH, Gordon H, Yu H, Lehtipalo K, Haley R, Li Y et al (2019) New particle formation in the atmosphere: from molecular clusters to global climate. J Geophys Res Atmos 124:7098–7146
Lei Y, Chen F, Luo Y, Zhang L (2014) Three-dimensional magnetic graphene oxide foam/Fe_3O_4 nanocomposite as an efficient absorbent for Cr(VI) removal. J Mater Sci 49(12). https://doi.org/10.1007/s10853-014-8118-2
Li S, Shen Y, Xie A, Yu X, Zhang X, Yang L et al (2007) Rapid, room-temperature synthesis of amorphous selenium/protein composites using *Capsicum annuum* L extract. Nanotechnology 18:405101. https://doi.org/10.1088/0957-4484/18/40/405101
Lo S, Fauzi MB (2021) Current update of collagen nanomaterials-fabrication, characterisation and its applications: a review. Pharmaceutics 13(3):316
Luo M, Liu D, Zhao L, Han J, Liang Y, Wang P et al (2014) A novel magnetic ionic liquid modified carbon nanotube for the simultaneous determination of aryloxyphenoxy-propionate herbicides and their metabolites in water. Anal Chim Acta 852:88–96
Mabena LF, Ray SS, Mhlanga SD, Coville NJ (2011) Nitrogen-doped carbon nanotubes as a metal catalyst support. Appl Nanosci 1:67–77
Maensiri S, Laokul P, Klinkaewnarong J, Phokha S, Promarak V, Seraphin S (2008) Indium oxide (In_2O_3) nanoparticles using Aloe vera plant extract: synthesis and optical properties. Optoelectron Adv Mater Rapid Commun 2(3):161–165
Majeed S, Danish M, Mohamad I, Mohamad N, Sekeri SH, Ansari MT, Nanda A, Ahmad G (2020) Bacteria mediated synthesis of iron oxide nanoparticles and their antibacterial, antioxidant, cytocompatibility properties. J Cluster Science 32:1083–1094
Malarkodi C, Chitra K, Rajeshkumar S, Gnanajobitha G, Paulkumar K, Vanaja M et al (2013) Novel eco-friendly synthesis of titanium oxide nanoparticles by using *Planomicrobium* sp. and its antimicrobial evaluation. Pelagia Res Lib Der Pharma Sinica 4(3):59–66
Matei A, Matei S, Matei GM, Cogălniceanu G, Cornea CP (2020) Biosynthesis of silver nanoparticles mediated by culture filtrate of lactic acid bacteria, characterization and antifungal activity. The EuroBiotech J 4(2):97–103
Medda S, Hajra A, Dey U, Bose P, Mondal NK (2015) Biosynthesis of silver nanoparticles from aloe vera leaf extract and antifungal activity against *Rhizopus* sp. and *Aspergillus* sp. Appl Nanosci 5:875–880
Mishra S, Singh HB (2015) Biosynthesized silver nanoparticles as a nanoweapon against phytopathogens: exploring their scope and potential in agriculture. Appl Microbiol Biotechnol 99:1097–1107
Mohammed MA, Syeda JTM, Wasan KM, Wasan EK (2017) An overview of chitosan nanoparticles and its application in non-parenteral drug delivery. Pharmaceutics 9(4):53
Morsy MK, Khalaf HH, Sharoba AM, El-Tanahi HH, Cutter CN (2014) Incorporation of essential oils and nanoparticles in pullulan films to control foodborne pathogens on meat and poultry products. J Food Sci 79:675–682
Mutahir A, Chandra H, Baunthiyal M (2015) Green synthesis of silver nanoparticles from different parts of the plant Macrotyloma uniflorum. Octa J Bioscience 3:58–61

Nava OJ, Soto-Robles CA, Gómez C, Vilchis-Nestor A, Castro-Beltrán A, Olivas A et al (2017) Fruit peel extract mediated green synthesis of zinc oxide nanoparticles. J Mol Struct 1147:1–6
Navalon S, Garcia H (2016) Nanoparticles for catalysis. Nanomaterials 6(7):123
Ngo QB, Dao TH, Nguyen HC, Tran XT, Nguten TV, Khuu TD, Huynh TH (2014) Effects of nanocrystalline powders (Fe, Co and Cu) on the germination, growth, crop yield and product quality of soybean (Vietnamese species DT-51). Adv Nat Sci Nanosci Nanotechnol 5:1–7
Ngoy JM, Wagner N, Riboldi L, Bolland O (2014) A CO_2 capture technology using multi-walled carbon nanotubes with polyaspartamide surfactant. Energy Procedia 63:2230–2248
Njagi E, Huang H, Stafford L, Genuino H, Galindo H, Collins J, Hoag G, Suib S (2011) Biosynthesis of iron and silver nanoparticles at room temperature using aqueous sorghum bran extracts. Langmuir: the ACS J Surf Coll 27:264–271
Obaidat IM, Issa B, Haik Y (2015) Magnetic properties of magnetic nanoparticles for efficient hyperthermia. Nanomaterials 5:63–89
Oh SD, Lee S, Choi SH, Lee IS, Lee YM, Chun JH et al (2006) Synthesis of Ag and Ag-SiO_2 nanoparticles by γ-irradiation and their antibacterial and antifungal efficiency against *Salmonella enterica* serovar Typhimurium and Botrytis cinerea. Colloids Surf A: Physicochemical Eng Aspects 275:228–233
Ouda SM (2014) Antifungal activity of silver and copper nanoparticles on two plant pathogens, Alternaria alternata and Botrytis cinerea. Res J Microbiol 9:34–42
Paul B, Bhuyan B, Purkayastha DD, Dey M, Dhar SS (2015) Green synthesis of gold nanoparticles using *Pogestemon benghalensis* (B) O. Ktz. leaf extract and studies of their photocatalytic activity in degradation of methylene blue. Mater Lett 148:37–40
Paul B, Bl B, Purkayastha DD, Dhar SS (2016) Photocatalytic and antibacterial activities of gold and silver nanoparticles synthesized using biomass of *Parkia roxburghii* leaf. J Photochem Photobiol B Biol 154:1–7
Pensado A, Seijo B, Sanchez A (2014) Current strategies for DNA therapy based on lipid nanocarriers. Expert Opin Drug Deliv 11:1721–1731
Pereira AES, Grillo R, Mello NFS, Rosa AH, Fraceto LF (2014) Application of poly(epsilon-caprolactone) nanoparticles containing atrazine herbicide as an alternative technique to control weeds and reduce damage to the environment. J Haz Mater 268:207–215
Peres FF, Lima AC, Hallak JEC, Crippa JA, Silva RH, Abílio VC (2018) Cannabidiol as a promising strategy to treat and prevent movement disorders? Front Pharmacol 9:482. https://doi.org/10.3389/fphar.2018.00482
Pokropivny VV, Skorokhod VV (2007) Classification of nanostructures by dimensionality and concept of surface forms engineering in nanomaterial science. Mater Sci Eng C 27:990–993
Prabha S, Arya G, Chandra R, Ahmed B, Nimesh S (2016) Effect of size on biological properties of nanoparticles employed in gene delivery. Artif Cells Nanomed Biotechnol 44:83–91
Prakash A, Sharma S, Ahmad N, Ghosh A, Sinha P (2010) Bacteria mediated extracellular synthesis of metallic nanoparticles. Int Res J Biotechnol 1(5):71–79
Prakash A, Sharma S, Ahmad N, Ghosh A, Sinha P (2011) Synthesis of AgNPs by *Bacillus cereus* bacteria and their antimicrobial potential. J Biomater Nanobiotechnol 2:156–162
Prasad R, Bhattacharyya A, Nguyen QD (2017) Nanotechnology in sustainable agriculture: recent developments, challenges, and perspectives. Front Microbiol 8:1014. https://doi.org/10.3389/fmicb.2017.01014
Qi P, Zhang D, Wan Y (2013) Sulfate-reducing bacteria detection based on the photocatalytic property of microbial synthesized ZnS nanoparticles. Anal Chim Acta 800:65–70
Rajan A, Cherian E, Baskar G (2016) Biosynthesis of zinc oxide nanoparticles using *Aspergillus fumigatus* JCF and its antibacterial activity. Int J Modern Sci Technol 1(2):52–57
Rajasree SRR, Suman TY (2012) Extracellular biosynthesis of gold nanoparticles using a Gram negative bacterium *Pseudomonas fluorescens*. Asi Pacific J Tropic Dis 2(2):5796–5799
Ramakritinan CM, Kaarunya E, Shankar S, Kumaraguru AK (2013) Antibacterial effects of Ag, Au and bimetallic (Ag–Au) nanoparticles synthesized from red algae. Solid State Phenom 201:211–230

Rao JP, Geckeler KE (2011) Polymer nanoparticles: preparation techniques and size-control parameters. Prog Polym Sci 36:887–913

Reddy YD, Dhachinamoorthi D, Chandra Sekhar KB (2015) A brief review on polymeric nanoparticles for drug delivery and targeting. J Med Pharma Innov 2:19–32

Saeed K, Khan I (2016) Preparation and characterization of single walled carbon nanotube/nylon 6,6 nanocomposites. Instrum Sci Technol 44:435–444

Saha K, Agasti SS, Kim C, Li X, Rotello VM (2012) Gold nanoparticles in chemical and biological sensing. Chem Rev 112:2739–2779

Sahayaraj K, Rajesh S, Rathi JM (2012) Silver nanoparticles biosynthesis using marinealga Padina pavonica (Linn.) and its microbicidal activity. Dig J Nanomater Biostruct 7:1557–1567

Schrofel A, Kratošová G, Krautová M, Dobročka E, Vávra I (2011) Biosynthesis of gold nanoparticles using diatoms—silica-gold and EPS-gold bionanocomposite formation. Nanopart Res 13:3207–3216

Severino P, da Silva CF, Andrade LN, de Lima OD, Campos J, Souto EB (2019) Alginate nanoparticles for drug delivery and targeting. Curr Pharm Des 25(11):1312–1334

Shang L, Dong S, Nienhaus GU (2011) Ultra-small fluorescent metal nanoclusters: synthesis and biological applications. Nano Today 6:401–418

Sheny DS, Philip D, Mathew J (2013) Synthesis of platinum nanoparticles using dried Anacardium occidentale leaf and its catalytic and thermal applications. Spectrochim Acta A: Mol Biomol Spectrosc 114:267–271

Shi D, Guo Z, Bedford N (2015) Nanomaterials and devices. Oxford, UK; William Andrew, Waltham, MA, 347 pages. ISBN 9781455777549. https://doi.org/10.1016/C2012-0-03204-8

Silambarasan S, Jayanthi A (2013) Biosynthesis of silver nanoparticles using *Pseudomonas fluorescens*. Res J Biotechnol 8:71–75

Singaravelu G, Arockiyamari J, Ganesh Kumar V, Govindaraju K (2007) A novel extracellular synthesis of monodisperse gold nanoparticles using marine alga, Sargassum wightii Greville. Colloids Surf B Biointerfaces 57:97–101

Song JY, Kwon EY, Kim BS (2010) Biological synthesis of platinum nanoparticles using Diopyros kaki leaf extract. Bioprocess Biosyst Eng 33(1):159–164

Srivastava P, Bragança J, Ramanan SR, Kowshik M (2013) Synthesis of silver nanoparticles using haloarchaeal isolate Halococcus salifodinae BK3. Extremophiles 17(5):821–831

Suganya KU, Govindaraju K, Kumar VG, Dhas TS, Karthick V, Singaravelu G, Elanchezhiyan M (2015) Blue green alga mediated synthesis of gold nanoparticles and its antibacterial efficacy against Gram positive organisms. Mater Sci Eng C 47:351–356

Sukri SNAM, Kamyar S, Wong M, Teow SY, Chew J, Ismail N (2019) Cytotoxicity and antibacterial activities of plant-mediated synthesized zinc oxide (ZnO) nanoparticles using *Punica granatum* (pomegranate) fruit peels extract. J Molecular Struc 1189:57–65

Sun X, Xu C, Wu G, Ye Q, Wang C (2017) Poly(lactic-co-glycolic acid): applications and future prospects for periodontal tissue regeneration. Polymers 9(6):189

Sun S (2000) Monodisperse FePt nanoparticles and ferromagnetic FePt nanocrystal superlattices. Science 80(287):1989–1992

Sundaram P, Augustine R, Marikani K (2012) Extracellular biosynthesis of iron oxide nanoparticles by *Bacillus subtilis* strains isolated from rhizosphere soil. Biotechnol Bioprocess Eng 17(4). https://doi.org/10.1007/s12257-011-0582-9

Syu Y, Hung J, Chen JC, Chuang H (2014) Impacts of size and shape of silver nanoparticles on Arabidopsis plant growth and gene expression. Plant Physiol Biochem 83:57–64

Tamuly C, Hazarika M, Bordoloi M (2013) Biosynthesis of Au nanoparticles by Gymnocladus assamicus and its catalytic activity. Mater Lett 108:276–279

Thomas S, Harshita BSP, Mishra P, Talegaonkar S (2015) Ceramic nanoparticles: fabrication methods and applications in drug delivery. Curr Pharm Des 21:6165–6188

Tiwari JN, Tiwari RN, Kim KS (2012) Zero-dimensional, one-dimensional, two-dimensional and three-dimensional nanostructured materials for advanced electrochemical energy devices. Prog Mater Sci 57:724–803

Ulery BD, Nair LS, Laurencin CT (2011) Biomedical applications of biodegradable polymers. J Polym Sci Part B: Polym Phys 49:832–864

Valodkar M, Modi S, Pal A, Thakore S (2011) Synthesis and anti-bacterial activity of Cu, Ag and Cu–Ag alloy nanoparticles: a green approach. Mater Res Bull 46:384–389

Velayutham K, Rahuman AA, Rajakumar G, Santhoshkumar T, Marimuthu S, Jayaseelan C (2010) Evaluation of *Catharanthus roseus* leaf extract-mediated biosynthesis of titanium dioxide nanoparticles against *Hippobosca maculata* and *Bovicola ovis*. Parasitol Res 111(6):2329–2337

Vimala K, Mohan YM, Varaprasad K, Redd NN, Ravindra S, Naidu NS, Raju KM (2011) Fabrication of curcumin encapsulated chitosan-PVA silver nanocomposite films for improved antimicrobial activity. J Biomaterials Nanobiotechnol 2:55–64

Waghmare SS, Deshmukh AM, Kulkarni SW, Oswaldo LA (2011) Biosynthesis and characterization of manganese and zinc nanoparticles. Univer J Enviro Res Technol 1:64–69

Wang T, Yang L, Zhang B, Liu J (2010) Extracellular biosynthesis and transformation of selenium nanoparticles and application in H_2O_2 biosensor. Colloids Surf B Biointerfaces 80:94–102

Yang J, Han S, Zheng H, Dong H, Liu J (2015) Preparation and application of micro/nanoparticles based on natural polysaccharides. Carbohyd Polym 123:53–66

Yasmin R, Shah M, Khan SA, Ali R (2017) Gelatin nanoparticles: a potential candidate for medical applications. Nanotechnol Rev 6(2):191–207

Yassin MA, Abdallah ME, El-Samawaty AE-RMA, Almunqedhi BMA (2021) Biosynthesis of silver nanoparticles using Penicillium verrucosum and analysis of their antifungal activity. Saudi J Biol Sci 28(4):2123–2127

Yu Z, Sun X, Song H, Wang W, Ye Z, Shi L, Ding K (2015) Glutathione-responsive carboxymethyl chitosan nanoparticles for controlled release of herbicides. Mater Sci Appl 6:591–604

Zainab S, Ali N, Ahmed S, Ali J, Ali M, Hameed A (2016) Role of catalytic protein and stabilizing agents in transformation of Ag ions to nanoparticles by Pseudomonas aeruginosa. IET Nanobiotechnol 10(5):295–300

Zeng W, Bai H (2016) High-performance CO_2 capture on amine-functionalized hierarchically porous silica nanoparticles prepared by a simple template-free method. Adsorption 22:117–127

Zhang H, Xia B, Wang P, Wang Y, Li Z, Wang Y et al (2020) From waste to waste treatment: mesoporous magnetic $NiFe_2O_4$/ZnCuCr-layered double hydroxide composite for wastewater treatment. J Alloys Compounds 819:153053

Zhang J, Saltzman M (2013) Engineering biodegradable nanoparticles for drug and gene delivery. Chem Eng Prog 109:25–30

Zhao L, Seth A, Wibowo N, Zhao CX, Mitter N, Yu C et al (2014) Nanoparticle vaccine. Vaccine 32:327–337

Chapter 2
Application of Nano-biotechnology in Wastewater Treatment: An Overview

Shweta Saraswat

Abstract Rapid industrial growth, urbanization and increased population are exaggerating the problem of polluted wastewater generation. Due to constant rise in wastewater pollution, the associated health risks to human beings, flora and fauna are also growing simultaneously. Applications of nanobiotechnology-based methods have great potential in remediating polluted water systems and improving their treatment efficiency as well. The selection of cost-effective and appropriate nanobiomaterials having unique characteristics increases the efficiency of their applications in vivid fields. In the present chapter, novel, fast and efficient plant- and microbes-mediated nanobio-based technologies have been discussed for effective and nature-friendly treatment of wastewater containing heavy metals, dyes, pesticides, pathogens and so on. The prospects and challenges ahead for large-scale industrial processes have also been emphasized in this chapter.

2.1 Introduction

Providing a clean, hygienic and affordable water supply to the world's ever-growing population is a major challenge in today's world. Despite this, an extensive discharge of contaminated wastewater into streams and rivers due to rapid industrialization and modern agricultural practices has also escalated the problem of clean water. Modern water treatment technologies, e.g. ion exchange, adsorption, membrane filtration, coagulation, photocatalysis, microwave and ultrasonic irradiation, reverse osmosis, etc., are continuously being used to serve the purpose of contaminated wastewater treatment, but their non-specificity, low efficiency, high energy requirement and operational cost limit the large-scale applications. Therefore, novel technologies capable of sensing, monitoring, preventing and treating large amounts of contaminants simultaneously are required. In this regard, nanotechnology holds enormous potential to

S. Saraswat (✉)
Department of Environmental Sciences, Govind Ballabh Pant University of Agriculture & Technology, Pantnagar, Udham Singh Nagar, Uttarakhand 263145, India
e-mail: shwetasara@gmail.com

J. P. N. Rai and S. Saraswat (eds.), *Nano-biotechnology for Waste Water Treatment*, Water Science and Technology Library 111,
https://doi.org/10.1007/978-3-031-00812-2_2

purify wastewaters and improve the efficiency of treatment plants. Nanoparticles' unique characteristics, such as shape and size, vast surface area, mechanical, electrical and magnetic capabilities (Das et al. 2015), make them effective environmental pollutants sensors and adsorbents. Researchers have synthesized various types of nanoparticles, nanotubes and nanocomposites through different physico-chemical processes for fulfilling the objectives of wastewater treatment, but their conventional ways of synthesis utilize hazardous and volatile chemicals which further creates secondary pollution. Therefore, attempts have been made to develop biogenic pathways and procedures employing plants and microbes, which are comparatively environmentally safe and inexpensive. In general, biomolecules such as carbohydrates, enzymes, proteins, polymers, flavonoids, alkaloids, antioxidants and functional groups such as carboxyl groups, aldehydes, polyphenols and others obtained from plants, bacteria, algae and fungi are used in the oxidation and reduction of metallic ions in the biosynthesis of nanoparticles. Various biomolecules have proven their efficiency and usefulness as capping and reducing agents in the fabrication of nanobiomaterials. The green/bio-synthesized nanoparticles offer several benefits such as eco-friendly, cheaper, nontoxic, high stability and reusability, feasible large-scale synthesis. Various plant parts and extracts have been used as reducing and capping agents for nanobioparticle production, e.g. *Catharanthus roseus* leaf (Mukunthan et al. 2011), *Nephelium lappaceum* peel (Kumar et al. 2015), *Azadirachta indica* leaf (Ahmed et al. 2016; Roy et al. 2017) and *Cissus quadrangularis* (Dhand et al. 2016; Devipriya and Roopan 2017). Several biopolymers, e.g. alginate, cellulose, starch, dextran, chitosan and so on have received specific attention in the development of nanoadsorbents and nanocatalysts for the clean-up of a wide range of pollutants (Mittal and Ray 2016). Potential applications of biosynthesized nanoparticles in wastewater purification, as well as associated restrictions, have been thoroughly examined in this chapter. Table 2.1 shows the use of various biogenic nanomaterials (NMs) for the removal of contaminants from wastewater.

Table 2.1 Role of nanotechnology in wastewater treatment (Chaturvedi et al. 2012)

S. No	Pollutants	Nano-based techniques used
1	Dye, organic contaminants	Carbon nanotubes
2	Heavy metals, radionuclides, dyes	Nanoscale metal oxide
3	Pesticides, polychlorinated biphenyls, azo dyes, etc.	Nanocatalyst
4	Organic pollutants, microorganisms inactivation	Nanostructured catalyst
5	Pathogens	Bioactive nanoparticle
6	Salts	Biomimetic membranes

2.2 Nano-based Methodologies for Wastewater Treatment

Because of their persistent and stubborn nature, the various organic and inorganic pollutants prevalent in waterbodies have now become a serious concern worldwide. Organic pollutants are mainly composed of carbon and hydrogen, with minor amounts of halogens, nitrogen, phosphorus and sulphur. Medicines, aromatic hydrocarbons, halogenated compounds, synthetic dyes, pesticides, insecticides and phenols are examples of xenobiotics, which are highly poisonous, non-biodegradable, and detrimental to our environment (Lapworth et al. 2012). Industrial and commercial discharge, domestic sewage, and agricultural runoff, transport persistent pollutants to bodies of water, where they bioaccumulate in animal and plant tissues, posing major health risks, such as cancer and immune system disruption. Nanotechnology has developed a number of successful approaches to clean filthy wastewater in a more precise and appropriate manner using a variety of nanomaterials to handle this. Nano-based technologies offer a wide range of novel nano-adsorbents, catalysts, and coatings for removing organic contaminants effectively. Synthetic azo dyes are of particular concern among organic pollutants because they are carcinogenic, mutagenic and allergic to living organisms, as well as blocking light penetration into aquatic bodies, resulting in reduced photosynthesis and the creation of anaerobic conditions in the aquatic environment. Bioactive nanoparticles, biomimetic membranes, nanosorbents, nanocatalysts, nanostructured catalytic membranes and molecularly imprinted polymers are some of the nanomaterials currently being used to remove inorganic and organic solutes, toxic metal ions, disease-causing microbes and other contaminants from polluted wastewater (Chaturvedi et al. 2012). The following are some examples of nanomaterials that are utilized in water purification and treatment.

2.2.1 Nanosorbents (Nano-adsorbents)

Nanoadsorbents are nanoparticles having unique physicochemical characteristics and high specific surface area with linked sorption sites, which facilitate high chemical reactivity and adsorption capacity. Varied nature of nanoadsorbents include carbon-based (carbon nanotubes, carbon nanosheets), silica nanotubes (SiNTs), metal and metal oxide-based, magnetic polymeric, nanofiltration membranes, nanocatalysts, fullerene, nano-zerovalent Iron (nZVI) nanosorbent (Panahi et al. 2018), etc. Earlier studies demonstrated that Ag nanocrystals can remove Ag^+ ions from wastewater (Li et al. 2010), carbon-based nanosorbents can remove nickel (Lee et al. 2012), toxic dyes and phosphorus content can be eliminated using nanoclays, and organic pollutants can be treated using the magnetic nature of nanosorbents (Campos et al. 2012). Dendrimers, mesoporous silica and chitosan are also good nanoadsorbents for removing toxic metal ions from contaminated water because of their high mechanical and chemical resistance and specific surface area and (Vunain et al. 2016). The following sections highlight various types of nanoadsorbents used in wastewater treatment.

2.2.1.1 C-Based Nano-adsorbents

C-based NMs with increased adsorption potential are used to generate nanoadsorbents such as C-nanoparticles (CNPs), C-nanosheets and C-nanotubes (CNTs). Because of their superior chemical resistance, mechanical strength, electrical characteristics, large adsorption capacity and surface area, CNTs are favoured over other materials (Lee et al. 2012). CNTs are of two types: single-walled CNTs (SWCNTs) and multi-walled CNTs (MWCNTs). SWCNTs are cylindrical and made up of a monolayer of graphene, whereas MWCNTs are made up of multiple graphene layers and are relatively better than SWCNTs in terms of cost, mass production and chemical and thermal stability. During functionalization of SWCNTs, the structural faults occurred by C=C bond breakage, as a result the mechanical and electrical characteristics may change. However, chemical modification of MWCNTs at surface being facilitated by excessive exposure of MWCNTs walls to modifiers, which can preserve the intrinsic features of carbon nanotubes.

Surface modification of carbon nanotubes is done to provide them new qualities for extremely particular applications that often require organic solvent or water-solubilization, functionality, dispersion and compatibility, or lessening the toxicity of CNTs. Common functionalized CNTs, such as MWNT-COOH (Product No. 755125), are made by oxidizing them with acids, ozone, or plasma, which results in the formation of additional functional groups of oxygen (e.g. –OH, –C=O). Exfoliation of CNT bundles is aided by the presence of oxygen-containing groups, which also improves solubility in polar fluids and chemical affinity with ester-containing substances like polyesters. COOH groups on nanotube surfaces can be used to modify them further. Through the formation of amide and ester linkages, many compounds, such as synthetic and natural polymers, can be grafted.

Aggregates of CNT with interstitial grooves and spaces provide high amount of energy for adsorption to bind organic moieties in aqueous solutions because to the hydrophobicity of graphitic surfaces. In this process, especially large organic molecules are trapped due to high adsorption capacity and large pore sizes of CNTs. Besides, electrostatic forces, covalent bonding, hydrogen bonding, and pi–pi stacking are other motives for the separation of polar aromatic compounds and polycyclic aromatic hydrocarbons in wastewater (Paszkiewicz et al. 2018). Magnetic forces can easily remove modified magnetic CNTs from wastewater because they have a high dispersion and regeneration ability (Madrakian et al. 2011).

After modification of functional groups on their surfaces, CNTs are highly useful in heavy metal removal from wastewater, such as Ni, Cd, Cr, Pb, Hg, Cu, Zn and Co, as reported by Ren et al. (2011), Bassyouni et al. (2020) and Wang et al (2021).

Earlier Tang et al. (2012) recorded Cu(II) removal using multi-walled CNTs. Tarigh and Shemirani (2013) also supported the application of multi-walled CNTs for Pb(II) and Mn(II) removal.

2.2.1.2 Metal and Its Oxide-Based Nanoadsorbents

Metal oxide-dependent nanoadsorbents are popularly known for their affinity with metal ions in contaminated wastewater, making them ideal for wastewater treatment. MgO, TiO_2, MnO_2, Fe_3O_4/Fe_2O_3, MnO_2, Al_2O_3 and CeO_2 are examples of metal oxide nanoparticles that have been studied for their applications in water treatment by a number of researchers. Strong adsorption, a substantial retardation factor and resistant desorption are features of iron oxides having large surface area (Shipley et al. 2010) make them effective for pollution remediation. To optimize their activity and avoid aggregation, which might reduce adsorption efficiency, it's critical to change their surface. Attachment of appropriate active groups such as carboxylic and phosphoric acid, silanol, amine, thio, polymer, biomolecules and other tiny organic moieties and metal NPs can be used to modify the material (Dave and Chopda 2014). Another strategy to boost their performance is to have them coexist with other supporting materials. Another green approach was used to make a novel Fe_3O_4 nanosorbent utilizing *Cucumis sativus*, *Citrus limon* and *Vitis vinifera* plant extracts (Venkateswarlu et al. 2013). These Fe_3O_4 (cucumber), Fe_3O_4 (lemon) and Fe_3O_4 (grapes) nanosorbents were used to remove seven antibiotics from water bodies, including sulfamethoxazole, piperacillin, tazobactam, tetracycline, erythromycin, trimethoprim and ampicillin. Zn, having weaker redox potential, has been widely employed in treatment of wastewater (Yan et al. 2013; Fu et al. 2014). Under ambient settings, Zn NPs have also been shown to efficiently degrade octachlorodibenzo-p-dioxin into its corresponding congeners of chlorinated nature (Bokare et al. 2013). In addition, three prescription medications, Diclofenac, Ibuprofen and Naproxen, were removed from wastewater samples utilizing biosynthesized Cu NPs made from aqueous Tilia extract residues, where 10.0 mg of Cu NPs exhibited elimination efficiency 91.4, 74.4 and 86.9%, respectively at pH 4.5 and temperature 298 k for 60 min (Husein et al. 2019). Various biosynthesized metal oxide nanoparticles for organic pollutants removal are exemplified in Table 2.2.

2.2.1.3 Polymeric Nanoadsorbents

Large surface area, improved durability, processability, economic feasibility and selectivity to remove varied pollutants from wastewater are all advantages of polymeric nanoadsorbents (Pandey et al. 2017). The most prominent advantages of polymeric nanoadsorbents are their high adsorption capacity and excellent heat stability over a wide pH range. Polysaccharides, such as nano-magnetic polymers, cyclodextrin, extracellular polymeric compounds, covalent organic polymers and other polymeric adsorbents are all inexpensive and widely employed (Reddy and

Table 2.2 Biosynthesized metal oxide nanoparticles for organic pollutants removal

Biosynthesized metal oxide nanoparticles	Size range (nm)	Organic pollutant	Exposure time (min)	Efficiency (%)	References
ZnO from *Moringa oleifera* leaf	52	Titan yellow dye	60	96	Lee et al. (2016)
ZnO from *Catharanthus roseus*	38	Phenol red dye	480	–	Hassan et al. (2013)
ZnO from *Ulva lactusa*	10–50	Methylene blue dye	120	90.4	Glory et al. (2019)
ZnO from *Allium sativum*	14–70	Methylene blue dye	180	100	Raja et al. (2018)
TiO_2 from *Tamarindus indica*	20–40	Titan yellow dye	120	95	Suresh et al. (2015)
TiO_2 from *Euphorbia hirta* leaves	20–50	Methylene blue, Methyl orange and crystal violet dye	360	95.8, 77.5 and 86.7, respectively	Hiremath et al. (2018)
FeO from *Amaranthus spinosus* leaves	91	Methylene blue and methyl orange	–	–	Muthukumar and Matheswaran (2015)
CuO from *Citrus grandis* peel extracts	22–27	Methyl red	135	96%	Sinha and Ahmaruzzaman (2015a)

Lee 2013; Alaba et al. 2018). Nanocellulosics have the characteristics of being ubiquitous, nontoxic, easily adsorbable and modifiable on the surface, making them ideal for wastewater treatment (Abdi and Abedini 2020). The creation of lignin-derived nanoparticles has recently been introduced, and it has shown great promise for wastewater treatment. They've been discovered as efficient catalytic degraders of nitroarenes and dyes and removal of heavy metal (Chen et al. 2018).

Nanocellulose (NC)

Because of its remarkable adsorptive qualities, NC has sparked a lot of attention. NC is a large potential component in water filtration membranes due to its intrinsic fibrous character and amazing mechanical capabilities, including low cost and biocompatibility (Carpenter et al. 2015). NC is made serially with the steps including mechanical fibrillation of cellulose, chemical purification or removal of hemicellulose, lignin and other impurities. Strong acids are used to hydrolyze alkali-treated, delignified cellulose fibres to produce cellulose nanocrystals (CNCs). Cellulose, on the other hand, is chemically pre-treated before being sheared to produce nanoscale cellulose fibres, also known as cellulose nanofibrils (CNFs). Based on origin and preparation technique, cellulose nanocrystals (CNC) and nanofibrils (CNF) are prepared by acid hydrolysis employing HCL and H_2SO_4 d into rod-like NPs having length range from 100 to 2000 nm and widths from 2 to 20 nm. H_2SO_4 combines with the –OH groups on the cellulose surface to generate sulphate halfesters, which result in negatively charged CNC nanoparticles that are electrostatically stabilized. When hydrochloric acid is used, the surface hydroxyl groups remain unaltered, resulting in CNC particles with an extremely low surface charge density. Unlike CNFs and CNCs, bacterial nanocellulose (BNC) is produced through bacterial synthesis and hence devoid of lignin or hemicellulose. Owing to their low cost, high surface-area-to-volume ratio, abundance and intrinsic inertness, NCs are efficient polymeric adsorbents for heavy metals, dyes, viruses and other contaminants (Gopakumar et al. 2016). Further, the surface NC possesses highly functionalized –OH groups, which makes it easier to incorporate chemical moieties that could improve pollutant binding effectiveness to nanocellulosic materials (Carpenter et al. 2015). Asper et al. (2015) demonstrated that green nanocellulose-made filter paper having high retention value (i.e. LRV > 5.25) could eliminate xenotropic leukaemia virus by size exclusion against the virus. This suggested that biosynthesized nanocellulosic filter paper could be efficient in removal of endogenous rodent retroviruses and like particles during recombinant protein production. Biosynthesized CNFs prepared from *Cladophora glomerata* algae and bleached *Eucalyptus* pulp were also used to make the nanofibres (Xiang et al. 2016). Nanocellulose isolation from renewable sources is becoming a hot topic among scientists all over the world. CNFs have also been extracted mechanically from various plants' cell using techniques such as grinding, cryocrushing, high-pressure homogenization and high-intensity ultrasonic treatments for cleaving cellulose fibres transversely. Zhang et al. (2014a) employed (poly)acrylic acid grafting to functionalize the surface of bamboo CNFs and used to remove Cu(II) from aqueous solutions. Modified bamboo CNFs have a threefold greater adsorption capacity than pure bamboo CNFs. For Cd and Pb adsorption, Jalali and Aboulghazi (2013) employed cellulose produced from sunflower stalks. Liu et al. (2015) found that enzymatically phosphorylated CNCs and CNFs had a high adsorption capacity for Ag(I), Cu(II) and Fe(III) in aqueous medium. In other study, Ma et al. (2011, 2012) reported that CNCs loaded electrospun nanofibrous membranes could be employed to remove bacteria (*Brevundimonas diminuta* and *Escherichia coli*) and virus model (bacteriophage MS2).

Nanochitin (NCH)

Chitin, the second most natural predominant polymer, is present in the structural components of exoskeleton of arthropods, and cell walls of fungi and yeast (Barikani et al. 2014). Chitin is extracted from crab and shrimp shells and degraded by chitinase to generate NCH with a significant surface area (Goodrich and Winter 2007). Chitin is an antibacterial, nonallergic, nontoxic, biodegradable and insoluble in water. Chitin and cellulose are polysaccharide molecules that are almost identical, with the distinction that cellulose has –OH groups while chitin has $-C_2H_5NO$ group. Chitin is a high-molecular-weight polymer with high crystallinity and mechanical strength. As a result, the surface area of NCH and the abundance of functional groups (–NHCO– and OH) make it an efficient chelating agent for heavy metal removal from polluted water (Liu et al. 2013; Lofrano et al. 2016). Nanochitin is an excellent candidate for reinforcing polymer nanocomposites. Earlier studies reported that NCH has been isolated from a number of sources, including shellfish, fungi and insects. These biopolymers have been used to remove dyes and pigments from wastewater, as well as metal ions, such as Hg, Cu, Fe, Ni, Cr, Pb, Zn, Cd, Ag and Co. Wu et al. (2019) developed nano-hybridized hydrogels with high ability of high metal ion removal and potential for remote magnetic separation by encapsulating nanochitin (i.e. metal ion adsorbent) and magnetic microparticles (i.e. magnetic controller) into poly(vinyl alcohol)-enhanced chitosan hydrogels (PVA/CTS). They also demonstrated chitosan hydrogel beads containing biodegradable nanochitin for quick and effective Cu(II) removal from aqueous solution. As a consequence, nano-surface modified nanopolysaccharides prove as excellent biosorbents for trapping heavy metal ions from wastewater, which has wide application in industries to further improve next-generation water purification technologies.

2.2.1.4 Carbon Nanotubes (CNTs)

CNTs have a wide range of purity, length, functionalization and surface interactions with plasma proteins, which can influence their cellular toxicity. CNTs exhibit antibacterial activity primarily through two mechanisms: chemically interacting with pathogens and physically preventing pathogens from passing through filters. CNTs have been proven to kill microbial cells either by physical interaction or oxidative stress, which affects integrity of cell membrane and causes disintegration. By physically preventing the movement of pathogens such as bacteria and even viruses across filters and membranes, CNTs can be very successful at filtering them out. SWCNTs (single walled carbon nanotubes) with very small dimensions ranging from 2 to 5 nm are capable of filtering out practically all known diseases. CNTs have good antimicrobial activity against various microorganisms, including both gram-negative and gram-positive bacteria, i.e. *E. coli*, *Staphylococcus aureus* and *Salmonella enterica*, as well as other fungal strains. Through physical, physiological and genetic alterations, CNTs can kill or limit the growth of microbes (Amiri et al. 2016). Despite this, CNTs are frequently used as an electrode in biosensor applications.

CNTs can damage a cell's plasma membrane, compromising cell integrity and impairing the selective transport of nutrients across the cell membrane (Amiri et al. 2016; Morsi et al. 2017). CNTs can kill bacteria by destroying important biological molecules like coenzymes and cell wall proteins (Ong et al. 2010). CNTs also act as strong oxidants in microbial cells, causing oxidative stress among critical proteins both outside and inside the cell membrane (Geng et al. 2014). Thus, the functional group of CNTs interact with molecules present on the cell membrane and suppress microbial growth (Yang et al. 2013). At intracellular level, CNTs interact with the bacterial cells' nucleic acids (RNA and DNA) and destroy genetic components. CNTs have also gained a lot of attention as effective adsorbents for a several organic molecules, e.g. dyes, dioxin, polyaromatic hydrocarbons, polybrominated diphenyl ethers, chloro-phenols and benzenes, trihalomethane, phthalate esters, insecticides (DDT and its metabolites, thiamethoxam, imidacloprid and acetamiprid), and herbicides (e.g. sulfuron derivatives, atrazine, dicamba, nonylphenol and bisphenol A) (Yang et al. 2006; Gotovac et al. 2006; Yang and Xing 2007).

2.2.1.5 Magnetic Nanobioparticles

Magnetic NMs have also been proposed as a means of removing water pollutants such as dyes, pesticides, aromatics, chlorinated hydrocarbons, heavy metals and microbiological contaminants. Magnetic NPs with good magnetic properties showed significant promise for water resource rehabilitation, which might allow dispersion in massive water systems followed by retrieval and potential reuse using a magnetic field. While investigating appropriate magnetic NPs and nanocomposites for prospective use in wastewater treatment, the characteristics such as strong magnetic response, high contaminant removal rate, quick recovery and reuse, stability should be addressed. Strong magnetic response allows the quick recovery of NM from suspension and prevents its further loss to the environment. High contaminant removal rate could be owing to the contaminant's efficient adsorption or high enzymatic activity against the contaminant species. NMs can be quickly recovered and reused with minimal processing. Adsorbent nanoparticles may require adsorbed pollutants to be removed or desorbed by elution or other procedures, whereas catalytic nanomaterials may be reusable right away. It is also required that the NMs should remain stable and functional even after multiple recycling stages.

Biosynthesized magnetic NPs produced from several plant species have been used in wastewater treatment processes to remove hazardous textile dyes, heavy metals, pigments, pharmaceuticals, cosmetics, pesticides and organic contaminants from household wastewater (Venkateswarlu et al. 2016; Lingamdinne et al. 2017; Srivastava et al. 2017). Owing to have functional groups (e.g. amino acids, polyphenols, sugars, proteins, terpenoids, alkaloids, carboxyl, carbonyl and polysaccharides), unique morphology and superparamagnetic behaviour, the plant-derived magnetic nanoparticles (PDMNPs) undergoes electrostatic attractions and thus helpful in contaminant removal and heavy metal ions recovery from wastewater. PDMNPs have been used to adsorb heavy metals such as Ni(II), Pb(II), Cr(III), Cr(VI), Co(II), Cu(II), As(III), As(V), Hg(II) and Cd^{2+} in recent investigations (Fato et al. 2019). Different researches looked into the adsorption capability and removal of various organic pollutants, total phosphorous (TP), phosphates ($PO4^{3-}$), ammoniacal nitrogen (NH_3–N), nitrates, chemical oxygen demand (COD) and malachite green (MG) (Ramasahayam et al. 2012; Wang et al. 2014; Lingamdinne et al. 2016). Using *Tridax* sp. leaf extract, biogenically produced maghemite nanoparticles (Fe_2O_3) removed 85 and 96% of Pb in 2 and 24 h respectively. In addition, considerable Cd elimination was detected, with levels falling below the detection level (0.01 ppm in 2 h) (Yadav and Fulekar 2018). According to Ramasahayam et al. (2014), phosphorus (P) was mostly adsorbed through an integration of adsorption and surface precipitation mechanisms, resulting in randomly distributed multilayer adsorption. On the other hand, MG adsorption and degradation were caused by a chemical diffusion mechanism and breaking of benzene ring bonds, respectively (Weng et al. 2013). Despite this, Wang et al. (2014) observed the nitrate adsorptive removal from residential wastewater due to chemisorption mechanism. PDMNPs were also employed to remove TP, NH_{3}_N, COD and $PO4^{3-}$ from household wastewater.

Magnetic nanoadsorbents of metals are also appealing as they can be easily retained and recovered from treated water. In the literature, Fe and its oxide NPs have been extensively reported for decontamination of various heavy metals, including Co^{2+} (Hooshyar et al. 2013), Ni^{2-}, Cu^{2+} (Poguberovic et al. 2016) and Cd^{2+} (Ebrahim et al. 2016), and the remediation of organic solvents of chlorinated nature (Han and Yan 2016; Guo et al. 2017). Donation of electron from the zerovalent Fe core can decrease both chlorinated chemicals and heavy metals. In addition, the shell part of the NPs can help with the decontamination of pollutants viz. heavy metals with a larger standard reduction potential (E^0) than the Fe^{2+}/Fe pair.

2.2.2 *Nano-photocatalysis*

Nano-Photocatalysis is a novel green technology for water decontamination which depends upon advanced oxidation processes (AOPs) and capable of generating various reactive species such as reactive nitrogen species (RNS), reactive oxygen species (ROS), hydroxyl radicals (OH•) and oxygen radicals (•O–$2O^{2-}$). It shows potential application for degradation of organic wastes present in wastewater (Cavassin et al. 2015; Waiskopf et al. 2016; Nosaka and Nosaka 2017). Photocatalysis is defined as "Alteration in chemical reaction rate or its initiation in the presence of a substance i.e. the photocatalyst which absorbs the light and chemically transform the reaction partners under the action of UV, visible, or IR radiation." A semiconductor material serves as the catalyst medium in a conventional photocatalysis system, forming an electron–hole (e–h) pair when exposed to light with a wavelength larger than its bandgap energy. The e–h pair generated in presence of light in water gives rise to most reactive oxidizing and/or reducing radicals like superoxides (•O2), hydroxyl ions (•OH) and others. The produced hydroxyl radical is a powerful oxidizing agent that totally destroys the dye into non-hazardous compounds (CO_2, H_2O, etc.). When the electrons or holes are generated in response to light are transmitted directly from the catalyst to contaminant moieties, water pollutants can be broken down. As most of the photo-genic electrons and holes are present at the surface of the nano-photocatalyst, nanostructured semiconductor materials are more suitable for photocatalysis than their bulk counterparts. To enhance the efficiency of photocatalyst, the semiconductor should have a broad bandgap so as to produce an e–h pair with sufficient energy to undertake secondary processes, including recombination of an e–h pair up to the minimum extent possible. High photoactivity, photostability, biological and chemical inertness leading to nontoxicity, and with low cost should be vital characteristics of an ideal photocatalyst. Some examples of commonly used nanostructured semiconductor photocatalysts include TiO_2, ZnO, Fe_2O_3 and sulphides of Zn and Cd.

Small ($\leq$10 nm), medium ($\leq$50 nm) and large ($\leq$100 nm) biogenic NMs have been shown to interact with light in some situations and adsorb UV–V is, which can be defined by plasmon bands and surface effects. The photocatalytic activity of composite NPs like TiO_2 and ZnO is dependent on their ability to absorb light and generate reactive oxygen species on their surfaces. For generating photocatalytic effects, solar radiation is more efficient than other radiation sources. The electrons on the surface of NPs are excited and move from valence to the conducting band on exposure to the desired wavelength of radiation. Charge carriers are electrons that migrate and create a deficiency as a result of their migration. The excited electrons are accepted by the H_2O molecules adsorbed on the NP surface, which are transformed into –OH radicals, which breakdown organic pollutants into simpler inorganic molecules. In the presence of UV radiation, TiO_2 exhibits excellent photocatalytic activity. The antibacterial activity of TiO_2 is mostly due to the production of ROS, particularly hydroxyl free radicals and peroxide, which are produced by sequential oxidation and reduction processes within a cell on UV exposure. One of the most essential characteristics of TiO_2 nanoparticle disinfection is their capacity to demonstrate photocatalytic activity even in the presence of visible light. The photocatalytic degradation of dyes viz. Eosin Y, methyl orange, methylene blue, Bromophenol blue and Orange G was found to be time-dependent employing Ag NPs made using *Hibiscus sabdariffa* extract fixed on cellulose. Nanoscale materials are advantageous because of having high surface area to volume ratio which facilitates trapping of organic and microbial contaminants even at extremely low NM concentrations. Silver anobioparticles undergo photocatalytic activity in colloidal water, where photons interact with AgNPs and activate electrons at their surface. The excited electrons are accepted by the H_2O–O_2 molecules and transformed into oxygen anion radicals, which have a wide range of interactions with organic moieties and thus break down of organic dyes takes place into simpler inorganic molecules, causing fast degradation and decolourization of dyes. As photocatalysis is primarily surface phenomena, charge carriers are generated on the surface of the NM after interaction with the proper wavelength of radiation. Charge carriers react with O_2 or H_2O molecules adsorbed on the photocatalyst's surface to form reactive oxygen species, which aid in the redox process of organic pollutant destruction. Presence of surfactants, capping agents or other chemical moieties on the surface of nanobiomaterials might compete with O_2 or H_2O molecules for charge carrier interaction, thereby preventing the photocatalytic process. Table 2.3 lists some applications of nanobiomaterials for photocatalytic decontamination of organic pollutants. Due to

Table 2.3 Biogenic nanomaterials used for photocatalytic degradation of organic pollutants

Contaminant	Biogenic nanoparticles	Outcomes	References
Methylene blue	*Ocimum tenuiflorum* Ag/ZnO nanocomposites	Methylene blue degradation ranged 49–74% based on Ag/ZnO dose	Panchal et al. (2020)
Azo dyes	*Escherichia sp.* Cu nanoparticles (22–39 nm)	Congo red degradation up to 97%, malachite green up to 90.5%, direct blue-1 up to 88.4% and reactive black-5 up to 83.6%	Noman et al. (2020)
Evans blue dye	*Limonia acidissima* NiO NPs (20 nm)	Evans blue degradation up to 91%	Kannan et al. (2020)
Rhodamine B dye	*Ferulago angulata* ZnO and CuO NPs (44 nm)	Rhodamine B degradation up to 83% and 93% by using CuO and ZnO NPs, respectively	Mehr et al. (2018)
Methylene blue dye	*Abelmoschus esculentus* CdO NPs (18–89 nm)	dye discoloration ranged 70–85%	Somasundaram and Rajan (2018)
Methylene blue and 4-Chlorophenol dye	*Capsicum annuum L.* TiO_2–C–O nanohybrids (8–15 nm)	Methylene blue degradation up to 75% and 4-chlorophenol degradation up to 70%	Rajeswari et al. (2018)
Rhodamine B dye and chlorpyrifos pesticide	*Fennel vulgare* Ag–ZnO nanocomposites	>90% photocatalytic degradation of Rhodamine B and chlorpyrifos within 35 min	Choudhary et al. (2019)

Source Bandala et al. (2020)

the photocatalytic impact of visible light, Ag nanocomposite doped with ZnO demonstrated improved methyl orange degradation, with approximately 65% dye elimination after an hour (Zhang and Zeng 2010). Methyl orange was degraded photocatalytically by 90% utilizing Ag NBPs prepared from unriped *Solanum nigrum* fruit extract (Malaikozhundan et al. 2017). In another study, green production of silver nanoparticles using Sapindus mukorossi extract soaked into chitosan-based hydrogels resulted in photocatalytic methyl orange decomposition. Further, dyes like Methylene Blue, Methyl violet 6B and Rose Bengal were found to be degraded >95% in the presence of sunlight using Ag NBPs and Au–Ag bimetallic nanostructures synthesized from *Anas platyrhynchos* finely powdered eggshell extract (Sinha and Ahmaruzzaman 2015b). Methyl orange and 4-nitrophenol degradation by Ag NBPs adsorbed on biosynthesized nanocomposites Rostami-Vartooni and Moradi-Saadatmand (2019), while that of Methylene blue by ZnO–Ag nanocustard apple mono-dispersion was recorded up to 95% efficiency (Kaviya and Prasad 2015).

The destructive effects of photocatalysis on a wide range of microorganisms, including bacteria, viruses, fungus and protozoa, demonstrate its utility for water disinfection (Ray et al. 2017; Laxma Reddy et al. 2017). The efficiency of photocatalytic disinfection is affected by factors viz. pH, the chemical composition of the bacterial suspension medium, type of photocatalyst, size, surface morphology, specific surface area, high surface energy, zeta potential and atomic ligand deficiency, light intensity, concentration and exposure duration. When microorganisms undergo photocatalysis, the cell wall material is exposed with –OH radical created during catalysis. These radicals, as the most powerful agent, can disrupt various organic covalent bonds in biomolecules like carbohydrates, amino acids, proteins, nucleic acid and even DNA, such as C–C, C–H, C–N, C–O and H–O (Wu et al. 2016; Laxma Reddy et al. 2017; Xu et al. 2018).

Gram-negative and gram-positive bacteria are varied in their peptidoglycan composition. The reactive species formed during photocatalysis assaults the phospholipid bilayer of the cell membrane, causing ions leakage, lipid peroxidation, structural and functional degradation of enzyme and nuclear material damage as per Saravanan et al. (2021). Photocatalytic disinfection involves lipid peroxidation of phospholipid bilayers especially made up of fatty acids that are susceptible to peroxidation. In intact cell membrane, the possibility of radicals penetrating intracellular components is minimized. Once generated, such radicals would attack the target moieties directly. Peroxide radical, in addition to hydroxide radical, has been proven to be more prominent owing to its long-lasting properties. Table 2.4 highlights the applicability of several nanobioparticles for microbial disinfection.

2.2.3 Nanocatalysts

Nanocatalysts utilized in wastewater treatment have a large surface area and thus exhibit high catalytic activity in terms of degradation and pollutants reactivity rate. Nanocatalysts such as zero-valence metals, semiconductor materials and bimetallic NPs can degrade environmental toxicants, e.g. azo dyes, polychlorinated biphenyls, halogenated aliphatics, nitro-aromatics, halogenated herbicides and organo-chlorine pesticides (Xin et al. 2011). Microbial pollutants are efficiently degraded by Ag nanocatalysts, ZrO_2 NPs and N-doped TiO_2 catalysts. They also have the advantage of being reusable nanocatalysts (Chaturvedi et al. 2012). Cr(IV) removal from wastewater is done by using TiO_2-AGs as their absorption band shifted from UV to visible. Other recalcitrant contaminants such as halogenated organic compounds are treated with Pd nanocatalysts prior to their biodegradation in the treatment plant. As the nanocatalyst utilized has ferromagnetism, it can be easily removed from the recycled (Hildebrand et al. 2008). WO3 nanocatalysts and palladium-incorporated ZnO nanoparticles were used to eliminate *E. coli* cells (Gondal et al. 2009; Khalil et al. 2011). In other study, Omole et al. (2009) also showed that Pd NPs can reduce

Table 2.4 Photocatalytic disinfection and/or antibacterial action using biogenic nanoparticles

S. No	NBM type/biogenic source	Photocatalytic disinfection of microbial species	Study outcomes	References
1	Ag NPs from *Helicteres isora* root extract	*E. coli, V. cholera, S. typhi, P. aeruginosa, B. subtilis, M. luteus*	Non-photocatalyzed antibacterial activity of Ag NPs against Gram-negative and gram-positive bacteria. *S. typhi* and *P. aeruginosa*	Bhakya et al. (2016)
2	*Ocimum tenuiflorum* Ag/ZnO nanocomposites	*E. coli*	*E. coli* was completely photocatalytically inactivated	Panchal et al. (2020)
3	*Camellia sinensis* Ag NPs	*S. aureus, P. aeruginosa, Klebsiella pneumoniae, E. coli, S. enterica*	Non-photocatalyzed antibacterial activity of Ag NPs against *S. enterica* was the highest, with no damage to HaCaT mammalian cells	Rolim et al. (2019)
4	*Cinnamomum verun* Mn NPs	*S. aureus, E. coli*	Nonphotocatalytic antibacterial activity of Mn NPs was assessed in petri dishes as an inhibitory zone	Kamran et al. (2019)
5	*Abelmoschus esculentus* CdO NPs	*S. aureus, B. subtilis, E. coli, S. paratiphi, A. niger, C. albicans, A. fumigates*	Nonphotocatalytic antifungal activity of CdO NPs was higher than antibacterial activity assessed as an inhibitory zone in petri dishes	Somasundaram and Rajan (2018)
6	*Abutilon indicum* CuO NPs	*E. coli, Klebsiella, B. subtilis, S. aureus*	Inhibitory zone created in petri dishes due to bactericidal action	Ijaz et al. (2017)
7	ZnO NPs from *Monsonia burkeana*	*S. aureus, E. faecalis, E. coli, P. aeruginosa*	Non-photocatalyzed antibacterial activity was found in ZnO NPs against gram-negative and gram-positive bacteria, as well as an inhibitory effect on A549 lung cancer cell lines	Ngoepe et al. (2018)
8	*Theobroma cacao* Ag NPs	*B. subtilis, E. coli*	Nonphotocatalytic antifungal and antibacterial activity of Ag NPs was assessed in petri dishes as an inhibitory zone	Thatikayala et al. (2019)

(continued)

Table 2.4 (continued)

S. No	NBM type/biogenic source	Photocatalytic disinfection of microbial species	Study outcomes	References
9	Se NPs from *Monascus purpureus*	*P. aeruginosa, K. pneumoniae, E. coli, S. aureus, C. albicans, A. solani, A. niger, F. oxysporum*	When compared to amoxicillin/clavulanic acid, Se NPs had nonphotocatalytic antibacterial action and reduced microbial growth including antifungal activity against examined fungus species	El-Sayed et al. (2020)

Source Bandala et al. (2020)

Cr(IV) to Cr(III). Nanocatalysts could be coupled with nanosorbents for pollutants sorption and degradation. Activation of Ag and amidoxime fibre nanocatalysts with tetrahydrofuran has also been adopted for remediation of organic dyes (Wu et al. 2010).

2.2.3.1 Nanoscale Catalytic Membranes (NCMs)

NCMs are preferred because of their ability to optimize, their restricted catalyst contact time, consistent catalytic sites, ease of industrial scale-up and the capacity to perform sequential reactions. Membranes exposed to UV radiation and nanostructured TiO_2 coatings help in inactivating the microbes, reducing biofouling activity, separation of contaminants and breakdown of organic pollutant in wastewater (Choi et al. 2009). Metallic NPs could be immobilized on a variety of membranes including chitosan, cellulose acetate, polyvinylidene fluoride, polysulfone and so on. Excellent reactivity, lack of agglomeration, organic portioning and reduced surface passivation are some of the advantages of immobilized metallic NPs. Nanocomposite films prepared from palladium acetate and polyetherimide were investigated for interactions between H_2 and Pd-dependent NPs, demonstrating their capability in polluted water treatment (Sandra et al. 2010). Pollutants detection in water samples have also been done utilizing nanosensors based on TiO_2 nanowires or Pd NPs.

2.2.3.2 Magnetically Separable Nanocatalysts

Magnetically separable nanocatalysts provide a simple and cost-effective approach for enhanced water treatment by allowing the catalyst to be recovered and reused after being exposed to a magnetic field. Despite this, the application and recovery of these catalysts in water treatment systems are now a design issue. Magnetic separation plays a key role in improvement of such methods. Magnetic separation enables the safe deployment of active NPs ans NMs by coating these compounds on a magnetic particle core process, which pave the way for an efficient, cost-effective and ecofriendly water purification. Magnetic separability, which allows for the application of catalytic or adsorptive materials, is a significant technology adopted in advanced water treatment systems. In the realm of water treatment, magnetic separation has observed as a vital attribute for the easy deployment of sub-micro and microparticles into solution. The problems arise after post-treatment can be overcome by attaching such materials to magnetic compounds, as extraction of active materials from contaminated water merely requires magnetic field application. Magnetic separation technique will allow the use of contaminant removing chemicals and materials to save enough time and operating cost. Magnetic separation promises to be a significant feature in these improved approaches, permitting the safe deployment of active materials.

2.3 Nanocomposite Membranes and Nanofilters

Advanced water purification, desalination and sewage treatment technologies rely heavily on filter and membrane-based processes. Membrane fouling, particularly organic and biofouling, is a major drawbacks of such processes. Biofouling is caused by the addition and proliferation of microbial strains on the membrane, which eventually form a biopolymeric environment on the membrane shell referred as a biofilm. Nanocomposites are divided into groups based on their extraordinary structure and type of matrix. The nanocomposites are of two types: polymer-based and non-polymer-based (Paul and Robeson 2008). Metal-polymer nanocomposites have been widely used among various forms of polymer-based nanocomposites. Polymer-based nanocomposites are typically made up of organic polymers with an embedded inorganic nanofiller. Some metallic nanocomposites are being researched extensively owing to be multifunctional, ecofriendly, biodegradable and biocompatible (Wang et al. 2015; Al-Jumaili et al. 2019). Implanting hydrophilic NPs in membrane models to alter the membrane shell characteristics and reducing fouling is one of the acknowledged methods. Ag and TiO_2 are widely used NPs for reducing biofouling. AgNPs were employed to construct minimal biofouling RO membranes using a phase inversion technique, according to a few studies (Basri et al. 2011; Mollahosseini et al. 2012; Koseoglu-Imer et al. 2013). Membranes impregnated with photoactive NPs become reactive rather than a simple physical barrier, allowing several treatment goals to be accomplished in one reactor while minimizing fouling. Shawky et al. (2020) also tested thin-film composite membranes with AgNPs in the polysulfone support layer (PSF) for non-adherence and repellence of bacteria and found that thin-film nanocomposite Ag/PSf membranes have huge potential for usage as a promising anti-biofouling membrane for groundwater desalination.

2.4 Potential Applications of Nanobioparticles in Water Treatment

2.4.1 Pesticide Remediation Employing Nanobioparticles

Herbicides, insecticides, fungicides, weedicides and rodenticides repellents are most extensively used agrochemicals in developing nations, while in developed countries herbicides are used widely (Peng et al. 2020). Rani and Shanker (2018) developed green metal hexacyanoferrate NPs incorporating aqueous extract of Sapindus mukorossi (family Sapindeae) and showed chlorpyrifos remediation of 85–98% under photocatalytic reduction. Budarz et al. (2019) observed the photo-reactivity of TiO_2 NPs during the photodegradation of chlorpyrifos (CPF) employing photoreactive TiO_2 NPs as a photocatalyst. Because of its self-regenerating and reusable qualities, TiO_2 was exploited as a photocatalyst. After 24 h, 80% of the chlorpyrifos was photodegraded to chlorpyrifos oxon or TCP (3,5,6-trichloropyridinol). Das et al. (2017) assessed CPF degradation by 99% utilizing laccase (a microbial enzyme) mounted on magnetic Fe NPs at pH 7 and 60 °C temperature with continuous stirring upto 12 h. Laccase immobilized on magnetic Fe NPs also showed good reusability, with 95% activity after five regeneration cycles. Apart from that, lindane, an organic contaminant found in drinking water, is degraded and scavenged using FeS-NP of size 200 nm (Paknikar et al. 2005). FeS-NPs were biosynthesized using a wet chemical method and stabilized using basidiomycetes derived polymer (Ramasamy et al. 2010). After few hours, the stabilized NPs decomposed lindane substantially at a concentration of 5 mg/L. In subsequent stages, the remaining lindane, its partially degraded components and stabilizing polymers were entirely destroyed by microbial treatment.

Further, Samy et al. (2020) developed a photocatalytic reactor that uses a coating of zirconium vanadate (ZrV_2O_7) and graphene nanoplatelets (GNPs) mixture to destroy CPF. ZrV_2O_7 was used as photocatalyst due to its narrow band gap and good absorber of visible light. The use of graphene nanoparticles together with ZrV_2O_7 has solved the problem of semiconductor recombination. CPF can be transformed into less hazardous end products using the reactive oxygen species created during the photocatalytic process. The GNP/ZrV2O7 nanocomposite had a high chlorpyrifos breakdown efficiency of 96.8%, even after five repeated cycles, demonstrating its high durability and pesticide remediation potential. Hydro-dehalogenation on Pd-based NPs can be used to biodegrade halogenated organic contaminants (Nutt et al. 2005). Pd NPs and bimetallic NPs such as Pd/Au have also been discovered to function admirably in the hydrodechlorination of trichloroethylene from aqueous solutions. Immobilized NPs, such as metalloporphyrinogens, have also been utilized to dehalogenate chlorinated organic molecules in a reductive manner. Another potential method for removing persistent pollutants has been the use of nanoparticles to degrade them. Herbicides and pesticides viz. chlorpyrifos, atrazine and molinate have been reported to be destroyed utilizing iron nZVI (Tripathi et al. 2018). Thus, nanobiotechnology proves its potential to detoxify pesticide-contaminated water in the future.

2.4.2 Dye Removal

Many countries are grappling with high dye concentrations in natural water sources and industrial effluent streams. The dyes are known for their long-term stability, high toxicity and carcinogenic impurities, as well as vomiting, mutagenic and reproductive problems in humans. Anionic, cationic and nonionic dyes are the three types of dyes. Anionic dyes are divided into direct, acid and reactive dyes that dissociate in aqueous solution to produce a -ve charged ion, whereas cationic dyes are basic in nature which dissociate into positively charged ions. Cationic dyes interact with the negatively charged groups on fibre molecules to form salts, which can be firmly attached to the fibres for staining. For treatment of dye-laden wastewater, chemical degradation, membrane separation and adsorption technologies are being used at present day, but the high maintenance and operating costs hamper their suitability for macroscale applications. As a result, for dye environmental cleanup, unique and low-cost technologies with adequate removal performance are required.

Lignin is a natural aromatic polymer that comprises ether linkages, aromatic hydroxyl, aldehyde and methoxyl groups, which make it feasible for a variety of alterations and potentially interacting with dye molecules for future dye removal. Many researchers have recently demonstrated that lignin, in either its natural or its modified form, can be used to remove dyes (Zhang et al. 2016; Fang et al. 2021). Lignin NPs are considered as good adsorbents owing to have high surface area, but their high stability in solution makes separation difficult, limiting their use as an efficient adsorbent. Enrichment of LNPs with important functional groups could destabilize the LNP and encourage their separation and adsorption. Fan et al. (2009) observed 90% dye removal within an hour exposing nZVI with 200 ml of dye at pH 9. Chen et al. (2011) also recorded 99.75% dye degradation in 6 h on exposure of 25 ml dye at pH 8.3 in the presence of nZVI.

2.4.3 Heavy Metal Removal

Heavy metal pollution in water reservoirs is one of the world's most susceptible environmental issues since it cannot be easily reduced in situ to nontoxic end products. Toxic metal ions viz. Cd(II), Pb(II), Ni(II), Hg(II), Cr(III), Cr(VI), Co(II), Cu(II), Ag(I), As(III), As(V), etc. contaminate water, posing enormous risks on public health and environment. The oxidized CNTs have a high adsorption capacity for heavy metal ions. The functional groups (e.g. carboxyl, hydroxyl and phenol) present on CNTs' surface are the primary locations for metal ion adsorption, which takes place by electrostatic attraction and chemical bonding. Owing to the high accessibility of adsorption sites and low diffusion distance within the particle, CNTs have a quick adsorption kinetics. Concerning this, Jain et al. (2015) biosynthesized Se NPs utilizing anaerobic granular sludge for Zn(II) reduction in wastewater. Within a short time span, approximately 70% of Zn(II) get adsorbed with maximum removal at an equilibrium period of 4 h. Suganthi and Kandasamy (2017) used biosynthesized Fe NPs generated from *Streptomyces thermolineatus* culture supernatant to remove Cu ions from pigment industry effluents by 85%. Srivastava et al. (2015) used biocompatible *Acacia* gum to make flower-shaped magnesium oxide nanoparticles (MgO NPs) and proved their capacity to remove numerous divalent heavy metals from synthetic wastewater, including Cd(II), Pb(II), Co(II), Cu(II), Zn(II), Mn(II) and Ni(II).

Earlier, Kostal et al. (2005) demonstrated the usage of NBT to generate functional materials and design recyclable protein-based NMs with metal-binding domains. The production of biopolymers containing polyhistidine tail as the metal-binding agent for Cd(II) and a metal regulatory protein to bind Hg has been emphasized by workers. By simply heating and chilling water solutions containing metal ion-laden biopolymers, the biopolymer chelating agents can be recovered and recycled. Ehrampoush et al. (2015) used tangerine peel extract to biosynthesize Fe_2O_3 NPs of size 50 nm and observed Cd(II) removal by 90% from wastewater at pH 4 and 0.4 g/100 ml adsorbent dose.

Fazlzadeh et al. (2017) biosynthesized ZnO NPs employing *Peganum harmala* seed extract to remove Cr(VI) from wastewater upto 97.59% within 30 min at pH 2, 1000 mg/l initial concentration of Cd and 2 g/l load of ZnO NPs. Shikha and Sharma (2017) also biosynthesized CuO NPs using *Calotropis procera* latex and recorded substantial removal of Cr(VI) from aqueous solution. CuONPs were shown to have a 99.99% removal effectiveness at an adsorbent dosage of 18.84 g/L. By utilizing zero valent Fe NPs ranged 50–80 nm and *Eucalyptus globules* leaf extract, Madhavi et al. (2013) removed Cr(VI) by 98.1%. Lingamdinne et al. (2017) used *Cnidiummonnieri* (L.) *cuss* seed extract for biosynthesis of magnetic Fe NPs and applied for significant Cr(III) and Pb(II) removal from aqueous systems. Lunge et al. (2014) also removed As(III) and As(V) through magnetic FeO NPs synthesized from tea waste with maximum adsorption capacities, i.e. 188.69 mg g1 and 153.8 mg g^{-1}, respectively. Furthermore, bio-reducing agents obtained from eucalyptus extract were also used to prepare Fe_2O_3 NPs, which were embedded in chitosan to construct a reusable organic-nanoiron hybrid of magnetic nature for the remediation of arsenic from contaminated water (Martínez-Cabanas et al. 2016). Mukherjee et al. (2016) used Aloe vera leaf extract to biologically manufacture rod-shaped Fe_2O_3 NPs that were used to sequester arsenic in the aqueous phase. The active function of anthraquinones and polysaccharides derived from *Aloe vera* extract for Fe_2O_3 NPs production was validated by FTIR analysis. Andjelkovic et al. (2017) used a chemoautotrophic bacteria *Mariprofundus ferrooxydanson* to naturally generate iron oxide nanowires in a bacterial biofilm and used as adsorbents for As(III) and As(V) removal with maximum adsorption capabilities as 104.53 mg g^{-1} and 48.06 mg g^{-1}, respectively. Silveira et al. (2017) also synthesized Fe_2O_3 NPs for adsorption of fluoride ion from wastewater utilizing leaf extract of *Moringa oleifera* as a template. During investigation, the adsorption of fluoride ion was found to be favourable with maximum adsorption at pH 7. It's now become obvious that a single sort of bioengineered NP may eliminate multi-metals.

2.4.4 Disinfection and Microbial Control Through Nanobioparticles

The disinfection procedures employed in drinking water purification can efficiently reduce microbiological infections, but produce hazardous by-products (DBPs). Strong oxidants employed in drinking water purification, such as ozone, free chlorine, chloramines react with diverse elements present in wastewater to generate DBPs (e.g. trihalomethanes, haloacetic acids and aldehydes), which are carcinogenic. As some infections are resistant to standard chemical disinfectants, exceptionally high disinfection dosages are required, resulting in increased DBP production. Therefore, novel replacement methods that improve disinfection reliability and resilience while preventing DBP generation are required. Nanoparticles are also known to be effective antibacterial agents that are low in oxidants and largely inert in water. As a result, they don't produce any toxic DBPs. The activity of antibacterial nanoparticles is dependent on their ability to destroy germs or halt their growth without causing harm to the cells around them. Bacterial responses to antibiotic agents range from one bacterial strain to the next. Their antibacterial efficacy is dependent on the bacterial species, initial bacterial concentration, physicochemical features of nanoparticles and nanoparticle concentration. NPs can act upon microbial cells, disrupt or penetrate the cell envelope, interrupt trans-membrane transfer of electrons, oxidize cell constituents and even produce cell damaging secondary products (e.g. ROS and/or dissolved heavy metal ions). Biosynthesized NPs made up of natural chitosan and peptides could be useful in low-cost disinfection of aquatic systems. Natural peptides act upon bacteria by causing osmotic collapse in bacterial cell membranes through creation of nanoscale channels. Various methods have been proposed to explain the antibacterial activities of chitosan nanoparticles. The main antibacterial mechanism of NPs involves the contact of positively charged chitosan NPs with negatively charged cell membranes, which cause an increase in permeability of cell membrane leading to the disruption and leakage of intracellular components. Another concept suggests that chitosan made NP penetrates cell membranes, binds to DNA, inhibits RNA synthesis in cells and consequently check microbial growth. Nanoscale chitosan and peptides could be used for inside coating of water storage tanks and as antibacterial agents in sponges and membranes.

Further, many modern water filtration and disinfection systems use nanoscale Ag-impregnated membranes. Due to strong antibacterial activity, less toxicity and ease of use, nano-Ag is a popular antimicrobial NM for wastewater disinfection. AgNPs, with a diameter of <10 nm, have been reported to be highly poisonous to *E.*

coli and *Pseudomonas aeruginosa* (Salomoni et al. 2017). By interacting with thiol groups in enzymes, Ag^+ ions produce ROS within the cells, rendering respiratory enzymes ineffective and resulting in cell death. Under UV exposure, Ag^+ ions exhibit photocatalytic activity, which is important for disinfection of bacteria. Apart from this, Ag^+ can also bind to the glycoproteins of viruses which block the viruses from binding to host cells. Comparatively, smaller and triangular AgNPs have shown stronger antibacterial effects than Ag nanorods and nanospheres, suggesting the importance of NPs size as well as their chemical form in eliciting desirable effects (Morones et al. 2005).

The antibacterial effectiveness of *Chenopodium murale* leaf extract-based silver nanoparticles against *S. aureus* (gram +ve) was also demonstrated (Abelaziz et al. 2014). As reviewed by Jain et al. (2021), various antibacterial and antifungal activities of biosynthesized NPs have been tested for bacterial species viz. *Escherichia coli, Enterococcus faecalis, S aureus, Streptococcus pneumoniae, Pseudomonas aeruginosa, Bacillus subtilis, K. pneumoniae, Staphylococcus enterica, Salmonella paratyphi* and some fungal strains *Aspergillus niger, Alternaria alternata, Rhizopus oryzae, Rhizopus stolonifer Phomopsis asparagi*, etc. (Zhong et al. 2007; Guerra-Sanchez et al. 2009). Leela and Vivekanandan (2008) have previously investigated Ag bio-reduction utilizing several plant extracts such as *Oryza sativa, Helianthus annus, Saccharum officinarum, Basella alba, Zea mays* and *Sorghum bicolor*. Some weeds viz. *Enhydra fluctuans, Ipomoea aquatica, Parthenium hysterophorus, Chenopodium murale* and *Ludwigia adscendens* were also employed as precursors.

Ag nanoparticles attached to the bacterial cell surface of *Lactobacillus fermentum* have been shown to have antibacterial and anti-biofouling capabilities (Zhang et al. 2014b). Murugesan et al. (2017) used the disc diffusion method to assess the antibacterial activity of AgNPs developed from *Salpa fusiformis* against bacterial pathogens, i.e. *S. aureus, E. coli, P. aeruginosa* and *K. pneumoniae*, with inhibition zone corresponding to the AgNP's antibacterial activity. In other experiments, the extracellularly generated AgNPs containing *Pseudomonas deceptionensis* DC5 showed excellent antibacterial potential against *Vibrio parahaemolyticus*, followed by *Candida albicans, S. aureus, S. enterica* and *Bacillus anthracis* (Murugesan et al. 2017).

Similar to AgNPs, TiO_2 NPs may produce highly reactive oxidants such hydroxyl radicals, which can be used to sterilize microorganisms including fungi, bacteria, viruses and algae (Foster et al. 2011). Owing to have relatively low photocatalytic activity, TiO_2 is doped with other transition metal ion to improve its activity. In addition, nanoscale ZnO has a high photocatalytic activity and UV absorption efficiency. The formation of hydrogen peroxide within the cells, which oxidizes the cell components, is one of the key processes of photocatalytic degradation by ZnO. Another theory proposes that interaction with ZnO NPs inhibits bacterial growth through cell envelope penetration and disruption of bacterial membrane. Bhuyan et al. (2015) used leaf extract of *Azadirachta indica* to synthesize ZnO NPs and assessed their antibacterial efficacy. The study revealed that ZnO NPs were extremely efficient against bacterial sp., e.g. *S. aureus* and *Streptococcus pyogenes*. The highly soluble nature of ZnO, however, limits its use in drinking water treatment because it is toxic to aquatic species.

Further, Sankar et al. (2014) observed that CuO NPs produced from *Origanum vulgare* leaf extract inhibited the colonial cyanobacterium, e.g. *Microcystis aeruginosa* up to 89.7% with 50 mg l^{-1} NP dose. The Biosynthetic Fe NPs have a lot of potential for removing As(V) from the environment (Wu et al. 2019). Furthermore, MgO NPs and cellulose acetate fibres with embedded Ag NPs exhibited highly efficient biocidal activity against bacterial spores, gram-positive and gram-negative bacterial species (Nora and Mamadou 2005).

2.4.5 Ammonia and Phosphate Removal

Excessive addition of phosphates and nitrates in natural water bodies leads to eutrophication, causing significant degradation of freshwater ecosystems. This situation necessitates the development of innovative and cost-effective restoration technologies for affected aquatic ecosystems. In an study, biologically produced Fe_2O_3 NPs by *Eucalyptus* leaf extracts doped onto zeolite were used to simultaneously eliminate 99.8% $PO4^{-3}$phosphate and 43.3% NH^{4+} from aqueous solutions at an initial concentration of 10 mg l^{-1} for each of the two co-existing ions. The maximum adsorption capacity of the Fe_2O_3 NPs for NH^{4+} and $PO_4{}^{3-}$ was 3.47 and 38.91 mg g^{-1}, respectively (Xu et al. 2020).

2.5 Sensing and Monitoring of Pollutants by Nano-bioparticles

Presence of extremely low concentrations of contaminants, absence of rapid detection of pathogens and the great complexity of the wastewater matrices make water quality monitoring a difficult task. Novel sensors with high sensitivity and selectivity, as well as being cost-effective, small and environmentally friendly, are needed to address these concerns. Using innovative nanofabrication and green synthesis processes, nanotechnology produces leading biosensors. Nanobiosensors (NBSs) detect contaminants with extremely high sensitivity and response quickly during real-time analysis. The contaminant detection limits at nano to picomolar range have been well described in the literature. Nanosensors are devices or materials that are very sensitive to environmental changes, such as concentration, heat, volume, chemical and mechanical stress, gravitational and magnetic forces, electrical forces and are employed to transmit information regarding behaviour and properties of NPs (Vikesland and Wigginton 2010; Kurbanoglu and Ozkan 2018). The construction of

NBSs for environmental monitoring takes place by utilization of biological entities in combination with functionalized NMs. NBSs involve a bioreceptor (enzyme, protein, antibody, bacterium and DNA impregnated nanometals, nanotubes, nanowires, NPs and so on) is coupled to a transducer (amperometric, voltammetric, conductometric, spectrophotometric, etc.) and a display for real-time monitoring. Biosensors are categorized into three groups on the basis of transduction methods: optical, electrochemical, calorimetric and mass-based biosensors (Riquelme et al. 2017; Talari et al. 2021). The nanomaterial-related signal transduction achieves sensitivity and rapid responsiveness when a recognition event occurs. Antibodies, aptamers, polysaccharides and antimicrobial peptides have all been created as recognition agents (Vikesland and Wigginton 2010). The selectivity is provided by recognition agents that engage specifically with antigens or other epitopes on the pathogen surface. Nanomaterials' unique physicochemical features, particularly electrochemical, optical and magnetic capabilities boost detection sensitivity and speed, allowing for multiplex target detection. These sensors can identify entire cells as well as macromolecules (Vikesland and Wigginton 2010; Theron et al. 2010). Magnetic NPs, Quantum dots, dye-doped NPs and CNTs are the extensively employed NMs in pathogen detection, while for sample concentration and purification purposes, magnetic NPs and CNTs are used. Dynabeads, a commercial magnetic nanocomposite, have been developed recently that could be used for various disease detection assays.

Further, antibiotic-resistant microorganisms and the antibiotic-resistant genes (ARGs) are continuously increasing in wastewater, posing a serious threat to public health. Methicillin-resistant *S. aureus* (MRSA) is a tainted antibiotic-resistant organism, in which mecA, an ARG, provide antibiotic resistance. For environmental monitoring of mecA ARG, Au NPs were functionalized with complementary oligonucleotides (Riquelme et al. 2017). The detection of ARG spikes using synthetic nanosensors revealed a high selectivity for ARG, with a detection limit of 70 ppm (Riquelme et al. 2017).

Pesticides (organophosphorus chemicals) have also become more concentrated in agricultural runoff, industry effluent and waterways. Their presence in water, even in tiny amounts, is dangerous since they are neurotoxic. Talari et al. (2021) used reduced grapheme oxide quantum dots (rGQDs) and MWCNTs to develop aptamer-based optical nanosensors to detect organophosphorus chemicals (e.g. diazinon) in potable water, agricultural discharge and rivers. In another investigation, the nanosensors showed diazinon selectivity and were able to detect it quickly and accurately. Triclosan (TCS), a common antifungal and antibacterial compound, is found in cosmetics and cleaning products. Its widespread usage and subsequent discharge into sediments, surface water and other aquatic sources causes toxicity to aquatic flora and fauna. Concerning this, Atar et al. (2015) developed a chemical nanosensor based on molecular-imprinted surface plasmon resonance (SPR) to detect TCS in wastewater. Further, caffeine is a medication and personal care product that is widely distributed in the environment through pharmaceutical wastes, tea, coffee beans, colas and drugs, for which Hu et al. (2018) developed a chemical nanosensor by doping AgNPs in molecularly imprinted polymers (MIP) with the detection limit of AgNPs@MIP as 100 ng l^{-1}.

The strong discrimination of various metal ions (Pt^{2+}, Pd^{2+}, Co^{2+} and Pb^{2+}) except ionic mercury (Hg^{2+}) was reported in a study for metal ion detection utilizing peptide-AuNPs probes, assisting the use of Flg-A3 fusion peptide (-Asp-Tyr-Lys-Asp-Asp-Asp-Asp-Lys-Pro-Ala-Tyr-Ser-Ser-Gly-Pro-Ala-Pro-Pro-Met-Pro-Pro-Phe-) as a multitasking agent. For production and stabilization of Au NPs from tetrachloroauric acid (HAuCl4), the Flg-A3 (AYSSGPAPPMPPF) method was used. At the same time, the peptide's N-terminal portion (-Asp-Tyr-Lys-Asp-Asp-Asp-Lys-) allowed non-covalent interactions to form complexes with metal ions (Slocik et al. 2005). A colorimetric technique was also described for detection of Hg^{2+} ions by Alam et al. (2015), which consists the combination of a NBM prepared from Ag NP and a *Baccaurea ramiflora* extract as a biological agent. The method involves slow addition of NPs to an ionic mercury-containing solution in a wide pH range (3.73–11.18) at room temperature, and the presence of ionic mercury was depicted by change in colour. The use of Au NPs stabilized by *Xanthoceras sorbifola* extract is another method for recognition of ionic chromium (Cr^{3+}). Ha et al. (2014) presented a method that has an ionic chromium detection limit of 3 M concentrations. In addition, the nanobiosensor proved to be very successful in detecting Cr^{3+} in biological and environmental samples. An extract of *Anacardium occidentale* leaves was used to develop a novel Cr^{3+} detection method (Balavigneswaran et al. 2014). The results showed that a NBS may be highly specific and sensitive to the presence of Cr^{3+} and Cr^{+6} ions, with a detection limit of 1 M. Cr^{3+}detection was carried out by secondary metabolites present in the extract which reduced Cr^{6+} ions in Cr^{3+}. The presence of Cr^{3+} is indicated by a change in the sample's hue. Recently, Akhondi and Jamalizadeh (2020) also demonstrated the synthesis of silver NPs, sodium dodecyl sulphonate-AgNPs and sodium dodecyl sulphonate-β-cyclodextrin modified AgNPs as simple, fast, sensitive and selective colorimetric sensors with excellent potential for quick on-site detection of Mn^{2+} and Cr^{2+} ions in aquatic systems.

Further, a wide range of single enzyme nanoparticles (SEMs) are also available for specific contaminants detection along with no by-products generation as no microorganisms are involved. Chemical conversions, biosensing and bioremediation are just a few of the applications for SENs. They are far more effective than synthetic catalysts because of their targeted effectiveness and specificity. However, enzymes are short-lived and unstable in harsh environments, their direct application in contamination detection is restricted. Nanotechnology is critical in enhancing the stability of SENs. SENs are enzymes that are protected by a nanometre-thick protective cage. The synthesis of SEN is a three-step procedure which involves covalent modification of the surface of enzyme through vinyl groups creation and dissolution in a non-polar solvent followed by addition of silane monomers with free trimethoxysilane groups to form vinyl group polymers attached to the enzyme surface and finally hydrolysis of trimethoxysilane groups to form a thin silicate shell. Such shell extends the enzyme's life by sheltering it from harsh environments while enabling some enzyme active sites to bind with target pollutants. In terms of pollutant recognition, nanobiosensors have demonstrated great performance and dependable data.

2.6 Challenges and Future Prospects

NBT is progressing in the field of remediation of harmful contaminants including dyes, heavy metal ions and even pathogenic microorganisms from water and wastewaters. Application of NBT for wastewater treatment is promising for low-income nations, as it can achieve zero effluent discharge while consuming little energy and expense. Novel NPs having broad spectrum of organic functional groups must be synthesized by altering plant metabolites for selective and sensitive multi-contaminant removal simultaneously from contaminated waterways without compromising the quality. Various biological entities viz. plant, bacteria, fungi and algae have yet to be investigated for their ability to produce NPs. To improve specificity, yield, cost-effectiveness and removal efficiency for a variety of pollutants, extensive research on optimizing biosynthetic production of NMs and their application in wastewater decontamination is required. Besides, the mechanistic features, cost-effectiveness, high adsorption capacity, high selectivity of fabrication processes, and recyclability and reusability of green-fabricated nanocatalysts and nanomaterials are still need to explored critically. Before using desorbed pollutants, the toxicity of long-term exposure to biosynthesized NPs to humans should be evaluated. Despite the straightforward synthesis of NBPs, the volume of extract and solution, temperature, pH, solvent type, functional groups and precursor strength from plant metabolites need to be well optimized to evade any alterations in the magnetic behaviour and saturation magnetization value of NBPs. Further, some crucial and critical concerns relating to toxicity and biosafety, such as the effects of NPs in different mediums, pH levels and reaction products with various pollutants, must be taken into account. CNTs, Ag and TiO_2 NPs, in particular, are the most suitable for water treatment applications, but are most poisonous to the cells. The safe use of such NMs in water purification must be approved and governed by regulatory organizations along with general public acceptance for their usage. In order to commercialize the use of biosynthesized NPs and methods in the detoxification and purification of contaminated water and wastewater systems, more research in the area of cost-benefit analysis is required. As such, NPs have enormous potential for wastewater treatment and purification by developing innovative nanostructures, improving existing treatment methods, increasing process efficiency and reusability of NMs, all of which can reduce the operating cost of the treatment plant at great extent.

2.7 Conclusions

The use of biosynthesized NBPs and NBSs to eliminate organic, inorganic and microbial pollutants from wastewaters has shown to be very promising. The biologically produced nanoparticles are long-lasting, low-cost, energy-efficient and environmentally friendly, and they can be used to decontaminate both drinking water and industrial effluent. Nanotechnology applications, in general, emphasize on the development of novel functionalized materials with innovative capabilities and specialized affinity for specific pollutants in a cost-effective and environmentally benign manner. Even in the presence of co-existing cations or anions, the biosynthesized NBPs could work effectively under wide temperature and pH range, and could be regenerated without losing much of their catalytic or sorptive activity. The manufacturing procedures of NBPs also reduce environmental impacts and serve as a valuable tool in the field of water treatment. However, improving and developing innovative nanobiomaterials in terms of selectivity, stability, affinity, size control, aggregation and toxicity remains a major problem before they can be used in large-scale wastewater treatment under any operating conditions. Further, the application of NBSs to detect target pollutants and toxic materials, such as heavy metals, pesticides and pathogens in wastewater, has demonstrated their good performance in terms of high sensitivity, cost-effectiveness, size and portability.

References

Abdi J, Abedini H (2020) MOF-based polymeric nanocomposite beads as an efficient adsorbent for wastewater treatment in batch and continuous systems: modelling and experiment. Chem Eng J 400:125862

Abelaziz M, Shaheen M, El-nekeety A, Abdel-Wahhab M (2014) Antioxidant and antibacterial activity of silver nanoparticles biosynthesized using *Chenopodium murale* leaf extract. J of Saudi Chem Soc 18(4):356–363

Ahmed S, Saifullah AM, Swami BL, Ikram S (2016) Green synthesis of silver nanoparticles using *Azadirachta indica* aqueous leaf extract. J Radiat Res Appl Sci 9:1–7

Akhondi M, Jamalizadeh E (2020) Selective colorimetric detection of Mn^{2+} and Cr^{2+} Ions using Silver nanoparticles modified with sodium dodecyl sulfonate and β-cyclodextrin. Acta Chim Slov 67(2):537–550

Alaba PA, Oladoja NA, Sani YM, Ayodele OB, Mohammed IY, Olupinla SF, Daud WMW (2018) Insight into wastewater decontamination using polymeric adsorbents. J Environ Chem Eng 6:1651–1672

Alam MN, Chatterjee A, Das S, Batuta S, Mandal D, Begum NA (2015) Burmese grape fruit juice can trigger the "logic gate"-like colorimetric sensing behavior of Ag nanoparticles towards toxic metal ions. RSC Adv 5:23419–23430

Al-Jumaili A, Mulvey P, Kumar A, Prasad K, Bazaka K, Warner J, Jacob MV (2019) Eco-friendly nanocomposites derived from geranium oil and zinc oxide in one step approach. Sci Rep 9:5973

Amiri A, Zare-Zardini H, Shanbedi M, Kazi SN, Taheri-Kafrani A, Chew BT, Zarrabi A (2016) Microbial toxicity of different functional groups-treated carbon nanotubes. In: Alexandru MG (ed) Surface chemistry of nanobiomaterials. Elsevier, Amsterdam, The Netherlands, pp 33–70.

Andjelkovic I, Azari S, Erkelens M, Forward P, Lambert MF, Losic D (2017) Bacterial iron-oxide nanowires from biofilm waste as a new adsorbent for the removal of arsenic from water. RSC Adv 7(7):3941–3948

Asper M, Hanrieder T, Quellmalz A, Mihranyan A (2015) Removal of xenotropic murine leukemia virus by nanocellulose based filter paper. Biologicals 43(6):452–456

Atar N, Eren T, Yola ML, Wang S (2015) A sensitive molecular imprinted surface plasmon resonance nanosensor for selective determination of trace triclosan in wastewater. Sens Actuators B Chem 216:638–644

Balavigneswaran CK, Kumar TSJ, Packiaraj RM, Prakash S (2014) Rapid detection of Cr(VI) by AgNPs probe produced by *Anacardium occidentale* fresh leaf extracts. Appl Nanosci 4(3):367–378

Bandala E, Stanisic D, Tasic L (2020) Biogenic nanomaterials for photocatalytic degradation and water disinfection: a review. Environ Sci: Water Res Technol 6:3195–3213

Barikani M, Oliaei E, Seddiqi H, Honarkar H (2014) Preparation and application of chitin and its derivatives: a review. Iran Polym J Eng Ed 23(4):307–326

Basri H, Ismail AF, Aziz M (2011) Polyethersulfone (PES)–silver composite UF membrane: effect of silver loading and PVP molecular weight on membrane morphology and antibacterial activity. Desalination 273:72–80

Bassyouni M, Mansi AE, Elgabry A, Ibrahim BA, Kassem OA, Alhebeshy R (2020) Utilization of carbon nanotubes in removal of heavy metals from wastewater: a review of the CNTs' potential and current challenges. Appl Phys A 126:38

Bhakya S, Muthukrishnan S, Sukumaran M, Muthukumar M (2016) Biogenic synthesis of silver nanoparticles and their antioxidant and antibacterial activity. Appl Nanosci 6(5):755–766

Bhuyan T, Mishra K, Khanuja M, Prasad R, Varma A (2015) Biosynthesis of zinc oxide nanoparticles from *Azadirachta indica* for antibacterial and photocatalytic applications. Mater Sci Semicond Process 32:55–61

Bokare V, Jung JL, Chang YY, Chang YS (2013) Reductive dechlorination of octachlorodibenzo-p-dioxin by nanosized zero-valent zinc: modeling of rate kinetics and congener profile. J Hazard Mater 250:397–402

Budarz JF, Cooper EM, Gardner C, Hodzic E, Ferguson PL, Gunsch CK et al (2019) Chlorpyrifos degradation via photoreactive TiO_2 nanoparticles: assessing the impact of a multi-component degradation scenario. J Hazard Mater 372:61–68

Campos AFC, Aquino R, Cotta TAPG, Tourinho FA, Depeyrot J (2012) Using speciation diagrams to improve synthesis of magnetic nanosorbents for environmental applications. Bull Mater Sci 34(7):1357–1361

Carpenter AW, Lannoy CFD, Wiesner MR (2015) Cellulose nanomaterials in water treatment technologies. Environ Sci Technol 49(9):5277–5287

Cavassin ED, de Figueiredo LF, Otoch JP, Seckler MM, de Oliveira RA, Franco FF et al (2015) Comparison of methods to detect the in vitro activity of silver nanoparticles (AgNP) against multidrug resistant bacteria. J Nanobiotechnol 13:64

Chaturvedi S, Dave PN, Shah NK (2012) Applications of nano-catalyst in new era. J Saudi Chem Soc 16(3):307–325

Chen B, Chen S, Zhao H, Liu Y, Long F, Pan XA (2018) A versatile β-cyclodextrin and polyethyleneimine bi-functionalized magnetic nanoadsorbent for simultaneous capture of methyl orange and Pb(II) from complex wastewater. Chemosphere 216:605–616

Chen ZX, Jin XY, Chen Z, Megharaj M, Naidu R (2011) Removal of methyl orange from aqueous solution using bentonite-supported nanoscale zero-valent iron. J Colloid Interface Sci 363(2):601–607

Choi H, Al-Abed SR, Dionysiou DD (2009) Nanostructured titanium oxide film- and membrane-based photocatalysis for water treatment. In: Savage N, Diallo M, Duncan J, Street A, Sustich R (ed) Nanotechnology applications for clean water, Chapter 3. William Andrew Publishing, Norwich, NY, pp 39–46.

Choudhary MK, Kataria J, Bhardwaj VK, Sharma S (2019) Green biomimetic preparation of efficient Ag–ZnO heterojunctions with excellent photocatalytic performance under solar light irradiation: a novel biogenic-deposition-precipitation approach. Nanoscale Adv 1:1035–1044

Das A, Singh J, Yogalakshmi KN (2017) Laccase immobilized magnetic iron nanoparticles: fabrication and its performance evaluation in chlorpyrifos degradation. Int Biodeterior Biodegrad 117:183–189

Das S, Sen B, Debnath N (2015) Recent trends in nanomaterials applications in environmental monitoring and remediation. Environ Sci Pollut Res 22:18333–18344

Dave PN, Chopda LV (2014) Application of iron oxide nanomaterials for the removal of heavy metals. J Nanotechnol, 1–14, Article ID 398569

Devipriya D, Roopan SM (2017) Cissus quadrangularis mediated ecofriendly synthesis of copper oxide nanoparticles and its antifungal studies against *Aspergillus niger*, *Aspergillus flavus*. Mater Sci Eng C Mater Biol Appl 80:38–44

Dhand V, Soumya L, Bharadwaj S, Chakra S, Bhatt D, Sreedhar B (2016) Green synthesis of silver nanoparticles using *Coffea arabica* seed extract and its antibacterial activity. Mater Sci Eng C Mater Biol Appl 58:36–43

Ebrahim SE, Sulaymon AH, Saad Alhares H (2016) Competitive removal of Cu^{2+}, Cd^{2+}, Zn^{2+}, and Ni^{2+} ions onto iron oxide nanoparticles from wastewater. Desalin Water Treat 57:20915–20929

Ehrampoush MH, Miria M, Salmani MH, Mahvi AH (2015) Cadmium removal from aqueous solution by green synthesis iron oxide nanoparticles with *tangerine* peel extract. J Environ Heal Sci Eng 13:1–7

El-Sayed ESR, Abdelhakim HK, Ahmed AS (2020) Solid-state fermentation for enhanced production of selenium nanoparticles by gamma-irradiated *Monascus purpureus* and their biological evaluation and photocatalytic activities. Bioprocess Biosyst Eng 43:797–809

Fan J, Guo Y, Wang J, Fan M (2009) Rapid decolorization of azo dye methyl orange in aqueous solution by nanoscale zerovalent iron particles. J Hazard Mater 166(2–3):904–910

Fang L, Wu H, Shi Y, Tao Y, Yong Q (2021) Preparation of lignin-based magnetic adsorbent from kraft lignin for adsorbing the Congo red. Front Bioeng Biotechnol 9:691528. https://doi.org/10.3389/fbioe.2021.691528

Fato TP, Li DW, Zhao LJ, Qiu K, Long YT (2019) Simultaneous removal of multiple heavy metal ions from river water using ultrafine mesoporous magnetite nanoparticles. ACS Omega 4(4):7543–7549

Fazlzadeh M, Khosravi R, Zarei A (2017) Green synthesis of zinc oxide nanoparticles using *Peganum harmala* seed extract, and loaded on *Peganum harmala* seed powdered activated carbon as new adsorbent for removal of Cr(VI) from aqueous solution. Ecol Eng 103:180–190

Foster H, Ditta I, Varghese S, Steele A (2011) Photocatalytic disinfection using titanium dioxide: spectrum and mechanism of antimicrobial activity. Appl Microbiol Biotechnol 90:1847–1868

Fu F, Dionysiou DD, Liu H (2014) The use of zero-valent iron for groundwater remediation and wastewater treatment: a review. J Hazard Mater 267:194–205

Geng J, Kim K, Zhang J, Escalada A, Tunuguntla R, Comolli LR, Allen FI, Shnyrova AV, Cho KR, Munoz D (2014) Stochastic transport through carbon nanotubes in lipid bilayers and live cell membranes. Nature 514:612–615

Glory LM, Mbita Z, Ntsendwana B, Mathipa MM, Mketo N, Hintsho-Mbita NC (2019) ZnO nanoparticles via Sutherlandia frutescens plant extract: physical and biological properties. Mater Res Express 6(8):14. https://doi.org/10.1088/2053-1591/ab1afa

Gondal MA, Dastageer MA, Khalil A (2009) Synthesis of nano-WO3 and its catalytic activity for enhanced antimicrobial process for water purification using laser induced photo-catalysis. Catal Commun 11(3):214–219

Goodrich J, Winter W (2007) R-chitin nanocrystals prepared from shrimp shells and their specific surface area measurement. Biomacromolecules 8:252–257

Gopakumar DA, Thomas S, Grohens Y (2016) Nanocelluloses as innovative polymers for membrane applications. In: Puglia D, Fortunati E, Kenny JM (eds) Multifunctional polymeric nanocomposites based on cellulosic reinforcements. William Andrew Publishing, pp 253–275

Gotovac S, Hattori Y, Noguchi D, Miyamoto JI, Kanamaru M, Utsumi S, Kanoh H, Kaneko K (2006) Phenanthrene adsorption from solution on single wall carbon nanotubes. J Phys Chem 110:16219–16224

Guo M, Weng X, Wang T, Chen Z (2017) Biosynthesized iron-based nanoparticles used as a heterogeneous catalyst for the removal of 2,4-dichlorophenol. Sep Purif Technol 175:222–228

Guerra-Sanchez MG, Vega-Pérez J, Velazquez-Del Valle MG, Hernandez-Lauzardo AN (2009) Antifungal activity and release of compounds on *Rhizopus stolonifer* (Ehrenb.: Fr.) Vuill. by effect of chitosan with different molecular weights. Pestic Biochem Physiol 93(1):18–22.

Ha W, Yu J, Wang R, Chen J, Shi YP (2014) Green colorimetric assay for the selective detection of trivalent chromium based on *Xanthoceras sorbifolia* tannin attached to gold nanoparticles. Anal Methods 6:5720–5726
Han Y, Yan W (2016) Reductive dechlorination of trichloroethene by zero-valent iron nanoparticles: reactivity enhancement through sulfidation treatment. Environ Sci Technol 50:12992–13001
Hassan NK, Hashim MR, Bououdina M (2013) One-dimensional ZnO nanostructure growth prepared by thermal evaporation on different substrates: ultraviolet emission as a function of size and dimensionality. Ceram Int 39:7439–7444
Hildebrand H, Mackenzie K, Kopinke FD (2008) Novel nano-catalysts for wastewater treatment. Global Nest J 10(1):47–53
Hiremath S, Antony Raj MAL, Chandra Prabha MN, Vidya C (2018) Tamarindus indica mediated biosynthesis of nano TiO_2 and its application in photocatalytic degradation of Titan yellow. J Environ Chem Eng 6(6):7338–7346
Hooshyar Z, Bardajee GR Ghayeb Y (2013) Sonication enhanced removal of nickel and cobalt ions from polluted water using an iron based sorbent. J Chem 2013:786954. https://doi.org/10.1155/2013/786954
Hu R, Tang R, Xu J, Lu F (2018) Chemical nanosensors based on molecularly-imprinted polymers doped with silver nanoparticles for the rapid detection of caffeine in wastewater. Anal Chim Acta 1034:176–183
Husein DZ, Hassanien R, Al-Hakkani MF (2019) Green-synthesized copper nano-adsorbent for the removal of pharmaceutical pollutants from real wastewater samples. Heliyon 5:e02339
Ijaz F, Shahid S, Khan SA, Ahmad W (2017). Green synthesis of copper oxide nanoparticles using Abutilon indicum leaf extract: antimicrobial, antioxidant and photocatalytic dye degradation activities. Trop J Pharm Res 16(4):743–753.
Jain AS, Pawar PS, Sarkar A, Junnuthula V, Dyawanapelly S (2021) Bionanofactories for green synthesis of silver nanoparticles: toward antimicrobial applications. Int J Mol Sci 22(21):11993. https://doi.org/10.3390/ijms222111993
Jain R, Jordan N, Schild D, van Hullebusch ED, Weiss S, Franzen C et al (2015) Adsorption of zinc by biogenic elemental selenium nanoparticles. Chem Eng J 260:855–863
Jalali M, Aboulghazi F (2013) Sunflower stalk, an agricultural waste, as an adsorbent for the removal of lead and cadmium from aqueous solutions. J Mat Cycles Waste Manage 15(4):548–555
Kamran U, Nawaz H, Iqbal M, Jamil S, Zahid M (2019) Biogenic synthesis, characterization and investigation of photocatalytic and antimicrobial activity of manganese nanoparticles synthesized from Cinnamomum verum bark extract. J Mol Struct 1179:532–539
Kannan K, Radhika D, Nikolova MP, Sadasivuni KK, Mahdizadeh H, Verma U (2020) Structural studies of bio-mediated NiO nanoparticles for photocatalytic and antibacterial activities. Inorg Chem Commun 2020:113:107755
Kaviya S, Prasad E (2015) Biogenic synthesis of ZnO–Ag nano custard apples for efficient photocatalytic degradation of methylene blue by sunlight irradiation. RSC Adv 5(22):17179–17185
Khalil A, Gondal MA, Dastageer MA (2011) Augmented photocatalytic activity of palladium incorporated ZnO nanoparticles in the disinfection of *Escherichia coli* microorganism from water. Appl Catal A 402(1):162–167
Koseoglu-Imer DY, Kose B, Altinbas M, Koyuncu I (2013) The production of polysulfone (PS) membrane with silver nanoparticles (AgNP): physical properties, filtration performances, and biofouling resistances of membranes. J Membr Sci 428:620–628
Kostal J, Prabhukumar G, Lao UL, Chen A, Matsumoto M, Mulchandani A, Chen W (2005) Customizable biopolymers for heavy metal remediation. J Nanopart Res 7:517–523
Kumar B, Smita K, Cumbal L, Angulo Y (2015) Fabrication of silver nanoplates using *Nephelium lappaceum* (Rambutan) peel: a sustainable approach. J Mol Liq 211:476–480
Kurbanoglu S, Ozkan SA (2018) Electrochemical carbon based nanosensors: a promising tool in pharmaceutical and biomedical analysis. J Pharm Biomed Anal 147:439–457
Lapworth DJ, Baran N, Stuart ME, Ward RS (2012) Emerging organic contaminants in groundwater: a review of sources, fate and occurrence. Environ Pollut 163:287–303

Laxma Reddy PV, Kavitha B, Kumar Reddy PA, Kim KH (2017) TiO_2-based photocatalytic disinfection of microbes in aqueous media: a review. Environ Res 154:296–303

Lee XJ, Lee LY, Foo LPY, Tan KW, Hassell DG (2012) Evaluation of carbon-based nanosorbents synthesised by ethylene decomposition on stainless steel substrates as potential sequestrating materials for nickel ions in aqueous solution. J Environ Sci 24:1559–1568

Lee KM, Lai CW, Ngai KS, Juan JC (2016) Recent developments of zinc oxide based photocatalyst in water treatment technology: a review. Water Res 88:428–448

Leela A, Vivekanandan M (2008) Tapping the unexploited plant resources for the synthesis of silver nanoparticles. Afr J Biotechnol 7:3162–3165

Li XG, Feng H, Huang MR (2010) Redox sorption and recovery of silver ions as silver nanocrystals on poly(aniline-co-5-sulfo-2-anisidine) nanosorbents. Chem-A Eur J 16(33):10113–10123

Lingamdinne LP, Yang JK, Chang YY, Koduru JR (2016) Low-cost magnetized Lonicera japonica flower biomass for the sorption removal of heavy metals. Hydrometallurgy 165:81–89

Lingamdinne LP, Chang YY, Yang JK, Singh J, Choi EH, Shiratani M, Koduru JR, Attri P (2017) Biogenic reductive preparation of magnetic inverse spinel iron oxide nanoparticles for the adsorption removal of heavy metals. Chem Eng J 307:74–84

Liu P, Sehaqui H, Tingaut P, Wichser A, Oksman K, Mathew AP (2013) Cellulose and chitin nanomaterials for capturing silver ions (Ag^+) from water via surface adsorption. Cellulose 21:449–461

Liu P, Borrell PF, Božič M, Kokol V, Oksman K, Mathew AP (2015) Nanocelluloses and their phosphorylated derivatives for selective adsorption of Ag^+, Cu^{2+} and Fe^{3+} from industrial effluents. J Hazard Mater 294:177–185

Lofrano G, Carotenuto M, Libralato G, Domingos RF, Markus A, Dini L, Meric S (2016) Polymer functionalized nanocomposites for metals removal from water and wastewater: an overview. Water Res 92(1):22–37

Lunge S, Singh S, Sinha A (2014) Magnetic iron oxide (Fe_3O_4) nanoparticles from tea waste for arsenic removal. J Magn Magn Mater 356:21–31

Ma H, Burger C, Hsiao BS, Chu B (2011) Ultra-fine cellulose nanofibers: new nano-scale materials for water purification. J Mater Chem 21:7507–7510

Ma H, Burger C, Hsiao BS, Chu B (2012) Nanofibrous microfiltration membrane based on cellulose nanowhiskers. Biomacromolecules 13:180–186

Madhavi V, Prasad TN, Reddy AV, Ravindra Reddy B, Madhavi G (2013) Application of phytogenic zerovalent iron nanoparticles in the adsorption of hexavalent chromium. Spectrochim Acta A Mol Biomol Spectrosc 116:17–25

Madrakian T, Afkhami A, Ahmadi M, Bagheri H (2011) Removal of some cationic dyes from aqueous solutions using magnetic-modified multi-walled carbon nanotubes. J Hazard Mater 196:109–114

Malaikozhundan B, Vijayakumar S, Vaseeharan B, Jenifer AA, Chitra P, Prabhu NM, Kannapiran E (2017) Two potential uses for silver nanoparticles coated with *Solanum nigrum* unripe fruit extract: biofilm inhibition and photodegradation of dye effluent. Microb Pathog 111:316–324

Martínez-Cabanas M, López-García M, Barriada JL, Herrero R, Sastre de Vicente ME (2016) Green synthesis of iron oxide nanoparticles: development of magnetic hybrid materials for efficient As(V) removal. Chem Eng J 301:83–91

Mehr ES, Sorbium M, Ramazani A, Fardood T (2018) Plant-mediated synthesis of zinc oxide and copper oxide nanoparticles by using ferulago angulata (schlecht) boiss extract and comparison of their photocatalytic degradation of Rhodamine B (RhB) under visible light irradiation. J Mater Sci Mater Electron 29:1333–1340

Mittal H, Ray SS (2016) A study on the adsorption of methylene blue onto gum ghatti/TiO_2 nanoparticles-based hydrogel nanocomposite. Int J Biol Macromol 88:66–80

Mollahosseini A, Rahimpour A, Jahamshahi M, Peyravi M, Khavarpour M (2012) The effect of silver nanoparticle size on performance and antibacteriality of polysulfone ultrafiltration membrane. Desalination 306:41–50

Morones JR, Elechiguerra JL, Camacho A, Holt K, Kouri JB, Ramírez JT, Yacaman MJ (2005) The bactericidal effect of silver nanoparticles. Nanotechnology 16:2346–2353

Morsi RE, Alsabagh AM, Nasr SA, Zaki MM (2017) Multifunctional nanocomposites of chitosan, silver nanoparticles, copper nanoparticles and carbon nanotubes for water treatment: antimicrobial characteristics. Int J Biol Macromol 97:264–269

Mukherjee D, Ghosh S, Majumdar S, Annapurna K (2016) Green synthesis of α-Fe_2O_3 nanoparticles for arsenic (V) remediation with a novel aspect for sludge management. J Environ Chem Eng 4:639–650

Mukunthan KS, Elumalai EK, Patel TN, Murty VR (2011) *Catharanthus roseus*: a natural source for the synthesis of silver nanoparticles. Asian Pacific J Trop Biomed 1:270–274

Murugesan S, Bhuvaneswari S, Sivamurugan V (2017) Green synthesis, characterization of silver nanoparticles of a marine red alga *Spyridia fusiformis* and their antibacterial activity. Int J Pharm Pharm Sci 9:192–197

Muthukumar H, Matheswaran M (2015) *Amaranthus spinosus* leaf extract mediated FeO nanoparticles: physicochemical traits, photocatalytic and antioxidant activity. ACS Sustain Chem Eng 3(12):3149–3156

Ngoepe NM, Mbita Z, Mathipa M, Mketo N, Ntsendwana B, Hintsho-Mbita NC (2018) Biogenic synthesis of ZnO nanoparticles using *Monsonia burkeana* for use in photocatalytic, antibacterial and anticancer applications. Ceram Int 44(14):16999–17006

Noman M, Shahid M, Ahmed T, Niazi MBK, Hussain S, Song F, Manzoor I (2020) Use of biogenic copper nanoparticles synthesized from a native *Escherichia sp.* as photocatalysts for azo dye degradation and treatment of textile effluents. Environ Pollut 257:113514. https://doi.org/10.1016/j.envpol.2019.113514

Nora S, Mamadou SD (2005) Nanomaterials and water purification: opportunities and challenges. J Nanoparticle Res 7:331–342

Nosaka Y, Nosaka AY (2017) Generation and detection of reactive oxygen species in photocatalysis. Chem Rev 117:11302–11336

Nutt MO, Joseph BH, Michael SW (2005) Designing Pd-on-Au bimetallic nanoparticle catalysts for trichloroethene hydrodechlorination. Environ Sci Technol 39:1346–1353

Omole MA, K'Owino I, Sadik OA (2009) Nanostructured materials for improving water quality: potentials and risks. In: Savage N, Mamadou D, Duncan J, Street A, Sustich R (eds) Nanotechnology applications for clean water: micro and nano technologies. William Andrew Publishing, pp 233–247. ISBN 9780815515784, https://doi.org/10.1016/B978-0-8155-1578-4.50026-3

Ong YT, Ahmad AL, Zein SHS, Tan SH (2010) A review on carbon nanotubes in an environmental protection and green engineering perspective. Braz J Chem Eng 27:227–242

Paknikar KM, Nagpal V, Pethkar AV, Rajwade JM (2005) Degradation of lindane from aqueous solutions using iron sulfide nanoparticles stabilized by biopolymers. Sci Technol Adv Mat 6:370–374

Panahi Y, Mellatyar H, Farshbaf M, Sabet Z, Fattahi T, Akbarzadehe A (2018) Biotechnological applications of nanomaterials for air pollution and water/wastewater treatment. Mater Today: Proc Part 3 5(7):15550–15558

Panchal P, Paul DR, Sharma A, Choudhary P, Meena P, Nehra SP (2020) Biogenic mediated Ag/ZnO nanocomposites for photocatalytic and antibacterial activities towards disinfection of water. J Colloid Interface Sci 563:370–380

Pandey N, Shukla SK, Singh NB (2017) Water purification by polymer nanocomposites: an overview. Nanocomposites 3:47–66

Paul DR, Robeson LM (2008) Polymer nanotechnology: nanocomposites. Polymer 49:3187–3204

Peng W, Lam SS, Christian S (2020) Support Austria's glyphosate ban. Science 367(6475):257–258

Poguberovic SS, Krcmar DM, Maletic SP, Konya Z, Pilipovic DDT, Kerkez DV, Roncevic SD (2016) Removal of As(III) and Cr(VI) from aqueous solutions using "green" zero-valent iron nanoparticles produced by oak, mulberry and cherry leaf extracts. Ecol Eng 90:42–49

Paszkiewicz M, Sikorska C, Leszczyńska D, Stepnowski P (2018) Helical multi-walled carbon nanotubes as an efficient material for the dispersive solid-phase extraction of low and high molecular weight polycyclic aromatic hydrocarbons from water samples: theoretical study. Water Air Soil Pollut 229(8):253

Raja A, Ashokkumar S, Marthandam RP, Jayachandiran J, Khatiwada CP, Kaviyarasu K et al (2018) Eco-friendly preparation of zinc oxide nanoparticles using Tabernaemontana divaricata and its photocatalytic and antimicrobial activity. J Photochem Photobiol b: Biology 181:53–58

Rajeswari PV, Tiwari B, Ram S, Pradhan D (2018) A biogenic TiO_2-C-O nanohybrid grown from a Ti^{4+}-polymer complex in green tissues of chilis, interface bonding, and tailored photocatalytic properties. J Mater Sci 53(5):3131–3148

Ramasahayam SK, Guzman L, Gunawan G, Viswanathan T (2014) A comprehensive review of phosphorus removal technologies and processes. J Macromol Sci Part a: Pure Appl Chem 51(6):538–545

Ramasahayam SK, Gunawan G, Finlay C, Tito V (2012) Renewable resource-based magnetic nanocomposites for removal and recovery of phosphorous from contaminated waters. Water Air Soil Pollut 223(8):4853–4863

Ramasamy K, Malik MA, Helliwell M, Tuna F, O'Brien P (2010) Iron thiobiurets: single-source precursors for iron sulfide thin films. Inorg Chem 49:8495–8503

Rani M, Shanker U (2018) Removal of chlorpyrifos, thiamethoxam, and tebuconazole from water using green synthesized metal hexacyanoferrate nanoparticles. Environ Sci Pollut Res 25(11):10878–10893

Ray SK, Dhakal D, Pandey RP, Lee SW (2017) Ag-$BaMoO_4$: $Er3^+$/$Yb3^+$ photocatalyst for antibacterial application. Mater Sci Eng C Mater Biol Appl 78:1164–1171

Reddy DHK, Lee SM (2013) Application of magnetic chitosan composites for the removal of toxic metal and dyes from aqueous solutions. Adv Coll Interface Sci 201–202:68–93

Ren X, Chena C, Nagatsu M, Wang X (2011) Carbon nanotubes as adsorbents in environmental pollution management: a review. J Chem Eng 170:395–410

Riquelme MV, Leng W, Carzolio M, Pruden A, Vikesland P (2017) Stable oligonucleotide-functionalized gold nanosensors for environmental biocontaminant monitoring. J Environ Sci 62:49–59

Rolim WR, Pelegrino MT, De AB, Ferraz LS, Costa FN, Bernardes JS et al (2019) Green tea extract mediated biogenic synthesis of silver nanoparticles: Characterization, cytotoxicity evaluation and antibacterial activity. Appl Surf Sci 463:66–74

Rostami-Vartooni A, Moradi-Saadatmand A (2019) Green synthesis of magnetically recoverable Fe_3O_4/HZSM-5 and its Ag nanocomposite using *Juglans regia* L. leaf extract and their evaluation as catalysts for reduction of organic pollutants. IET Nanobiotechnol 13(4):407–415.

Roy P, Das B, Mohanty A, Mohapatra S (2017) Green synthesis of silver nanoparticles using *Azadirachta indica* leaf extract and its antimicrobial study. Appl Nanosci 7:843–850

Salomoni R, Léo P, Montemor AF, Rinaldi BG, Rodrigues M (2017) Antibacterial effect of silver nanoparticles in *Pseudomonas aeruginosa.* Nanotechnol Sci Appl 10:115–121

Samy M, Gamal EIM, Gar Alalm M, Fujii M, Diab K, Elkady M (2020) Innovative photocatalytic reactor for the degradation of chlorpyrifos using a coated composite of ZrV_2O_7 and graphene nano-platelets. Chem Eng J 395:124974

Sandra S, Eliane E, Laurent D, Didier L (2010) Nanocomposite membranes of polyetherimide nanostructured with palladium particles: processing route, morphology and functional properties. J Membrane Sci 361:167–175

Sankar R, Prasath BB, Nandakumar R, Santhanam P, Shivashangari KS, Ravikumar V (2014) Growth inhibition of bloom forming *cyanobacterium Microcystis aeruginosa* by green route fabricated copper oxide nanoparticles. Env Sci Pollut Res 21:1–9

Saravanan A, Senthil Kumar P, Jeevanantham S, Karishma S, Kiruthika AR (2021) Photocatalytic disinfection of micro-organisms: mechanisms and applications. Environ Technol Innov 24:101909

Shawky H, Yaseen R, Kotp Y, Eissa D (2020) Biosynthesis of silver nanoparticles and its effect on TFC RO membrane for groundwater desalination. Desalin Water Treat 193:34–47

Shikha D, Sharma CY (2017) *Calotropis procera* mediated one pot green synthesis of Cupric oxide nanoparticles (CuONPs) for adsorptive removal of Cr(VI) from aqueous solutions. Appl Organometal Chem 31:1–15

Shipley H, Engates K, Guettner A (2010) Study of iron oxide nanoparticles in soil for remediation of arsenic. J Nanoparticle Res 13:2387–2397

Silveira C, Shimabuku Q, Silva M, Bergamasco R (2017) Iron-oxide nanoparticles by green synthesis method using *Moringa oleifera* leaf extract for fluoride removal. Environ Technol 39:1–40

Sinha T, Ahmaruzzaman M (2015a) High-value utilization of egg shell to synthesize silver and gold–silver core shell nanoparticles and their application for the degradation of hazardous dyes from aqueous phase-A green approach. J Colloid Interface Sci 453:115–131

Sinha T, Ahmaruzzaman M (2015b) Green synthesis of copper nanoparticles for the efficient removal (degradation) of dye from aqueous phase. Environ Sci Pollut Res 22(24):20092–20100

Slocik JM, Stone MO, Naik RR (2005) Synthesis of gold nanoparticles using multifunctional peptides. Small 1(1):1048–1052

Somasundaram G, Rajan J (2018) Effectual role of *Abelmoschus esculentus* (Okra) extract on morphology, microbial and photocatalytic activities of CdO Tetrahedral Clogs. J Organomet Polym 28:152–167

Srivastava S, Agrawal SB, Synthesis MMK (2017) characterization and application of Lagerstroemia speciosa embedded magnetic nanoparticle for Cr(VI) adsorption from aqueous solution. J Environ Sci (china) 55:283–293

Srivastava V, Sharma YC, Sillanpaa M (2015) Green synthesis of magnesium oxide nanoflower and its application for the removal of divalent metallic species from synthetic wastewater. Ceramic Int 41(5). https://doi.org/10.1016/j.ceramint.2015.01.112

Suganthi H, Kandasamy R (2017) A novel single step synthesis and surface functionalization of iron oxide magnetic nanoparticles and thereof for the copper removal from pigment industry effluent. Seperation Purific Technol 188:458–467

Suresh D, Nethravathi PC, Rajanaika H, Nagabhushana H, Sharma SC (2015) Green synthesis of multifunctional zinc oxide (ZnO) nanoparticles using Cassia fistula plant extract and their photodegradative, antioxidant and antibacterial activities. Mater Sci Semicond Process 31:446–454

Talari FF, Bozorg A, Faridbod F, Vossoughi M (2021) A novel sensitive aptamer-based nanosensor using rGQDs and MWCNTs for rapid detection of diazinon pesticide. J Environ Chem Eng 9(1):104878. https://doi.org/10.1016/j.jece.2020.104878

Tang WW, Zeng GM, Gong JL, Liu Y, Wang XY, Liu YY et al (2012) Simultaneous adsorption of atrazine and Cu(II) from wastewater by magnetic multi-walled carbon nanotube. Chem Eng J 211:470–478

Tarigh GD, Shemirani F (2013) Magnetic multi-wall carbon nanotube nanocomposite as an adsorbent for preconcentration and determination of lead (II) and manganese (II) in various matrices. Talanta 115:744–750

Thatikayala D, Jayarambabu N, Banothu V, Ballipalli CB, Park J, Rao KV (2019) Biogenic synthesis of silver nanoparticles mediated by Theobroma cacao extract: enhanced antibacterial and photocatalytic activities. J Mater Sci Mater Electron 30(18):17303–17313

Theron J, Cloete TE, de Kwaadsteniet M (2010) Current molecular and emerging nanobiotechnology approaches for the detection of microbial pathogens. Crit Rev Microbiol 36(4):318–339

Tripathi S, Sanjeevi R, Anuradha J, Chauhan DS, Rathoure AK (2018) Nano-bioremediation: nanotechnology and bioremediation. In Rathoure A. (ed.) Biostimulation remediation technologies for groundwater contaminants. IGI Global, pp 202–219. https://doi.org/10.4018/978-1-5225-4162-2.ch012.

Venkateswarlu S, Lee D, Yoon M (2016) Bioinspired 2D-carbon flakes and Fe_3O_4 nanoparticles composite for arsenite removal. ACS Appl Mater Interfaces 8(36):23876–23885

Venkateswarlu S, Yakkate S, Balaji T, Prathima B, Jyothi NV (2013) Biogenic synthesis of Fe_3O_4 magnetic nanoparticles using plantain peel extract. Mater Lett 100:241–244

Vikesland PJ, Wigginton KR (2010) Nanomaterial enabled biosensors for pathogen monitoring—a review. Environ Sci Technol 44:3656–3669

Vunain E, Mishra A, Mamba B (2016) Dendrimers, mesoporous silicas and chitosan-based nanosorbents for the removal of heavy-metal ions: a review. Int J Biol Macromol 86:570–586

Waiskopf N, BenShahar Y, Galchenko M, Carmel I, Moshitzky G, Soreq H et al (2016) Photocatalytic reactive oxygen species formation by semiconductor–metal hybrid nanoparticles toward light-induced modulation of biological processes. Nano Lett 16:4266–4273

Wang L, Wang C, Yang C, Hsieh C, Chen S, Shen C, Wang J, Huang KS (2015) Synthesis and antifungal effect of silver nanoparticles–chitosan composite particles. Int J Nanomed 10(1):2685–2696

Wang T, Lin J, Chen Z, Megharaj M, Naidu R (2014) Green synthesized iron nanoparticles by green tea and eucalyptus leaves extracts used for removal of nitrate in aqueous solution. J Clean Prod 83:413–419

Wang Z, Xu W, Jie F, Zhao Z, Zhou K, Liu H (2021) The selective adsorption performance and mechanism of multiwall magnetic carbon nanotubes for heavy metals in wastewater. Sci Rep 11:16878

Weng X, Huang L, Chen Z, Megharaj M, Naidu R (2013) Synthesis of iron-based nanoparticles by green tea extract and their degradation of malachite. Ind Crops Prod 51:342–347

Wu J, Cheng X, Li Y, Yang G (2019) Constructing biodegradable nanochitin-contained chitosan hydrogel beads for fast and efficient removal of Cu(II) from aqueous solution. Carbohyd Polym 211:152–160

Wu D, Wang W, Ng TW, Huang G, Xia D, Yip HY et al (2016) Visible-light-driven photocatalytic bacterial inactivation and mechanism of zinc oxysulfide under LED light irradiation. J Mater Chem A 4:1052–1059

Wu ZC, Zhang Y, Tao T, Zhang L, Fong H (2010) Silver nanoparticles on amidoxime fibers for photo-catalytic degradation of organic dyes in waste water. Appl Surf Sci 257:1092–1097

Xiang Z, Gao W, Chen L, Lan W, Zhu JY, Runge T (2016) A comparison of cellulose nanofibrils produced from *Cladophora glomerata* algae and bleached eucalyptus pulp. Cellulose 23:493–503

Xin Z, Lu L, Bingcai P, Weiming Z, Shujuan Z, Quanxing Z (2011) Polymer-supported nanocomposites for environmental application: a review. Chem Eng J 170(2–3):381–394

Xu Q, Li W, Ma L, Cao D, Owens G, Chen Z (2020) Simultaneous removal of ammonia and phosphate using green synthesized iron oxide nanoparticles dispersed onto zeolite. Sci Tot Environ 703:135002. https://doi.org/10.1016/j.scitotenv.2019.135002

Xu Y, Liu Q, Liu C, Zhai Y, Xie M, Huang L et al (2018) Visible-light-driven $Ag/AgBr/ZnFe_2O_4$ composites with excellent photocatalytic activity for *E. coli* disinfection and organic pollutant degradation. J Colloid Interface Sci 512:555–566

Yadav KV, Fulekar MH (2018) Biogenic synthesis of maghemite nanoparticles (γ-Fe_2O_3) using *Tridax* leaf extract and its application for removal of fly ash heavy metals (Pb, Cd). Mater Today 5:20704–20710

Yan W, Lien HL, Koel BE, Zhang WX (2013) Iron nanoparticles for environmental clean-up: recent developments and future outlook. Environ Sci Process Impacts 15:63–77

Yang H, Tong M, Kim H (2013) Effect of carbon nanotubes on the transport and retention of bacteria in saturated porous media. Environ Sci Technol 47:11537–11544

Yang K, Wang X, Zhu L, Xing B (2006) Competitive sorption of pyrene, phenanthrene, and naphthalene on multiwalled carbon nanotubes. Environ Sci Technol 40:5804–5810

Yang K, Xing B (2007) Desorption of polycyclic aromatic hydrocarbons from carbon nanomaterials in water. Environ Pollut 145:529–537

Zhang D, Zeng F (2010) Synthesis of an Ag–ZnO nanocomposite catalyst for visible light-assisted degradation of a textile dye in aqueous solution. Res Chem Intermed 36(9):1055–1063

Zhang M, Zhang K, Gusseme BD, Verstraete W, Field R (2014a) The antibacterial and anti-biofouling performance of biogenic silver nanoparticles by *Lactobacillus fermentum*. Biofoul 30(3):347–357

Zhang S, Wang Z, Zhang Y, Pan H, Tao L (2016) Adsorption of methylene blue on organosolv lignin from rice straw. Procedia Environ Sci 31:3–11

Zhang X, Zhao J, Cheng L, Lu C, Wang Y, He X, Zhang W (2014b) Acrylic acid grafted and acrylic acid/sodium humate grafted bamboo cellulose nanofibers for Cu^{2+} adsorption. RSC Adv 4:55195–55201

Zhong Z, Chen R, Xing R, Chen X, Liu S, Guo Z et al (2007) Synthesis and antifungal properties of sulfanilamide derivatives of chitosan. Carbohyd Res 342(16):2390–2395

Chapter 3
Emerging Nano-Bio Material for Pollutant Removal from Wastewater

Dheeraj Rathore, Anoop Singh, Shiv Prasad, Piyush Malaviya, and Surajbhan Sevda

Abstract Industrial and domestic processes are generating huge quantity of organic and inorganic pollutants into the environment through releasing wastewater into the environment. The unimpeded dumping of these pollutants not only risks the human and animal health but also deteriorates land and water quality. This generates demand to remove/utilize these pollutants in sustainable manner before releasing wastewater into the environment. Several physical–chemical and biological technologies are available to remove the organic and inorganic pollutants from wastewater. However, some of these technologies are not so efficient in removing the pollutants from wastewater, whereas other technologies either not economically viable or not too suitable to the environment. Development of nanotechnology has provided new paradigm to the several fields including wastewater treatment technologies due to its unique and exceptional characteristics such a superhydrophilicity, high surface area, surface functionalities. Nanomaterials showed a great potential to address all the flaws in removal of pollutants from wastewater. The present chapter enlightens the applicability of bionanomaterials for removal of organic pollutants from wastewater and its sustainability.

D. Rathore
School of Environment and Sustainable Development, Central University of Gujrat, Gandhinagar 382030, Gujrat, India

A. Singh (✉)
Department of Scientific and Industrial Research, Ministry of Science and Technology, Government of India, Technology Bhawan, New Mehrauli Road, New Delhi 110016, India
e-mail: apsinghenv@gmail.com

S. Prasad
Centre for Environment Science and Climate Resilient Agriculture (CESCRA), Indian Agricultural Research Institute, New Delhi 110012, India

P. Malaviya
Department of Environmental Sciences, University of Jammu, Jammu 180006, Jammu and Kashmir, India

S. Sevda
Department of Biotechnology, National Institute of Technology Warangal, Warangal, Telangana 506004, India

J. P. N. Rai and S. Saraswat (eds.), *Nano-biotechnology for Waste Water Treatment*, Water Science and Technology Library 111,
https://doi.org/10.1007/978-3-031-00812-2_3

Keywords Nano-biomaterials · Pollutant · Wastewater · Treatment · Sustainability

3.1 Introduction

Freshwater is a fundamental necessity for existence of the life on the earth and sustainable development of society (Sajid et al. 2018). The accessibility to clean water is a global environmental problem since it becomes serious confutation to supply clean and inexpensive drinking water to all (Sharabati et al. 2021). The proliferation of industries and urbanization has created inevitable effects on the environment (Singh et al. 2002) as a consequence of amplified contamination with hazardous materials such as acidic and basic pollutants including persistent inorganic and organic compounds and heavy metals (Singh and Rathore 2019, 2021; Chung et al. 2022), which directly and indirectly harms the water quality. The potential hazards caused due to unchecked disposal of industrial effluent have prompted to limit the untreated effluent discharge in the environment (Singh and Rathore 2018, 2020). The availability of clean/potable water is a challenge for all water supply practices and the regularization in standards to maintain the water quality has developed new paradigm for current water treatment and distribution system (Mehta et al. 2021).

Available traditional wastewater treatment regime is confined to three-step process viz. primary, secondary, and tertiary treatment. Prior to primary treatment screening of the wastewater is required to prepare the wastewater for primary anaerobic treatment by maintaining pH or temperature and to filter out the larger sized particles which can go for further treatment at other facility or land filling. Primary treatment process involves microbial degradation of organic waste material under anaerobic condition, while secondary treatment process depends on the microbial decomposition process that removes suspended solids from the wastewater under aerobic condition. Tertiary treatment was carried out to further improve the water quality by microbial activity before discharging into the environment or recycled for utilization to other purposes such as agriculture, industrial, or household purpose (Singh 2021). Advanced oxidation, biodegradation, flocculation, use of bioadsorbents or activated carbon adsorption, coagulation, ion exchange, membrane separation, etc. are some of the widely used processes (Abhinav et al. 2021).

Current available techniques used for treatment of wastewater have limitation to offer potable or palatable class of water that can be used for different domestic or industrial uses to reduce the burden on natural water resource and reduced commotion to environment. Advancement on the traditional wastewater treatment technologies is a continuous process to get an improved and efficient treatment technology (Mehta et al. 2021). Being environmental friendly, low cost, simple in design, operational fluidity and higher effectiveness for removal of pollutants from the wastewater streams (Kubra et al. 2021) made adsorption process highly promising and sustainable way for wastewater treatment (Zhao et al. 2019; Sharabati et al. 2021). The pollutants present on the wastewater streams adsorb on the surface of the adsorbing material and thus removed from the wastewater stream. Several absorbent materials such as activated carbon, zeolite, clay minerals, chitosan, etc. are actively in use for treatment of wastewater (Yu and Han 2015; Al Bsoul et al. 2021). Among these, the activated carbon is oldest known and the most popular and commonly used absorbent. Its large surface area and porous structure is highly efficient to remove the pollutants (Yu and Han 2015; Al Bsoul et al. 2021). Its surface area can reach 500–2000 m^2 g^{-1} and is efficient to adsorb various heavy metals present in wastewater (Carrott et al., 1997; Gabaldon et al. 2006). However, higher cost, low competency for several other metals and poor removal of hydrophilic organic pollutants, etc. restrict its industrial use for wastewater treatment (Alsbaiee et al. 2016; Sharabati et al. 2021). Similarly, low permeability and pretreatment requirement of zeolite and lower efficiency of clay particles restrict its industrial use (Stojakovic et al. 2011; Yu and Han 2015). Therefore, the researchers are focusing on the developing absorbent with improved performance with low cost and wider availability to improve the effectiveness of water treatment (Kurniawan et al. 2006; Sharabati et al. 2021). A superlative adsorbent ought to have multiple characteristics, e.g., large surface area that increases the adsorption capacity, economical, environmental friendly, reusable, well-suited, and high selectivity toward water pollutants (Nasar and Mashkoor 2019) which increase the efficiency of water treatment and thus improve its industrial application. Nanotechnology is expected to provide the highly efficient techno-economical elucidation for wastewater treatment problems (Mehta et al. 2021).

3.2 Extraordinary Properties

Nanomaterials have gained prominence in research and development due to their exceptional tunable properties (Sadegh and Ali 2018). The importance of this class of materials was realized when scientist found that the size can influence the physic-chemical properties of material (Khan et al., 2019). The unique characteristics of nanomaterials, i.e., size in nm, exceptionally large surface to volume ratio, higher capacity and efficiency than the material itself, simplicity of fictionalization, stability, reuse potential, and antimicrobial structure make them attractive substances to be used in different industrial and engineering applications including wastewater treatment (Ponnusamy and Saththasivam 2021; Sharabati et al. 2021). The efforts have been made in this chapter to summarize the recent advancements in nano-bio materials for wastewater treatment in sustainable way.

3.3 Nano-Bio Materials

Focus of the nanotechnology is on the design of new materials, structures, and devices that having minimum one dimension fitting the nanoscale, i.e., less than 100 nm (Mohmood et al. 2013; Rienzie et al. 2019), which could be an effective means to treat wastewater contaminated with pollutant ranging from metallic ions to organic and inorganic solutes, and microbial pathogens (Aguilar-Perez et al. 2020), because nanomaterials contains unique physical and chemical properties due to their nanoscale size. Unique physic-chemical properties of nanomaterial included large surface to aspect ratio, topography, aggregation state, strong solution mobility, surface chemistry such as optical property, reactivity, catalytic potential, porosity, strong adsorption, dispersibility, and enhanced redox ratio (Mohmood et al. 2013; Yaqoob et al. 2020), which enables them to interact with the pollutants through increased binding sites causing them inactive or degrade (Saharan et al. 2014; Aguilar-Perez et al. 2020). The nanomaterial-based water treatment applications of nanotechnology can be grouped in to three categories, viz. water filtration, water remediation, and purification or disinfection of water (Fig. 3.1). Water filtration is mainly based on the use of semi-permeable nanoporous membranes to remove the organic and micropollutants. Membrane used for this technology is highly stable, easy in operation, and low cost and low energy consuming (Santhosh et al. 2016; Ali et al. 2019a). In recent years fullerene, carbon nanotubes and graphene-based material is gaining attention for water filtration and desalination (Bodzek et al. 2019). Focus of the water remediation is on the adsorption phenomenon of the nanosorbents to remove a broad range of pollutant from the wastewater streams with high compatibility to

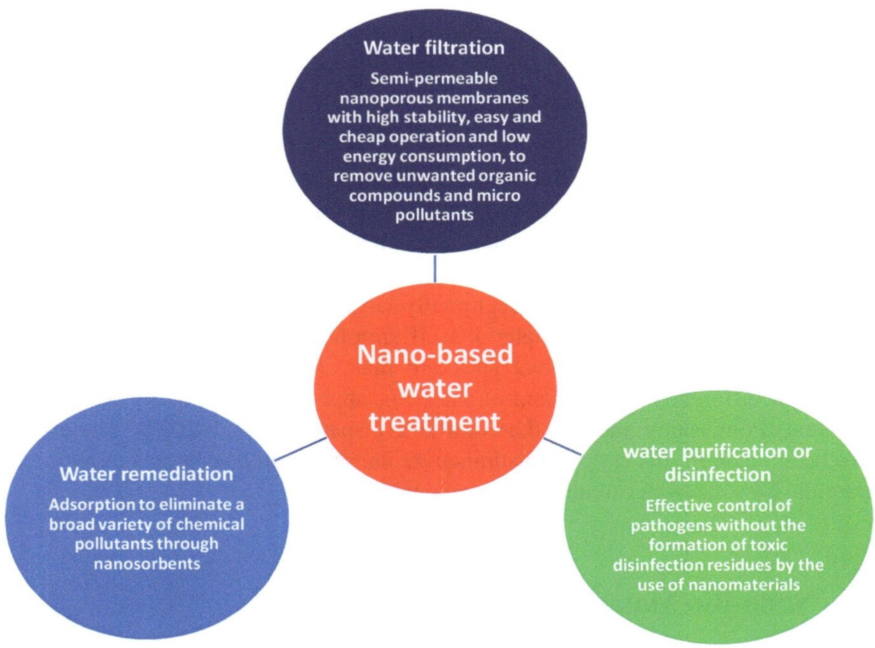

Fig. 3.1 Classification of nanotechnology-based water treatment application (adopted from Aguilar-Perez et al., 2020)

specific organic or inorganic pollutants (Cincinelli et al. 2015; Santhosh et al. 2016). Recently, numerous studies reported promising results for zeolites, carbon-based, bio-based, and metal oxide-based materials (El-Sayed 2020). Water purification or disinfection targeted on the effective control of microbial pathogens without generation of harmful residues by the use of nanomaterials with antimicrobial activity as graphene-based materials, metallic, and bioactive composites (Mohmood et al. 2013; Morsi et al. 2017; Aguilar-Perez et al. 2020).

Synthesis of metallic nanoparticles can be exercised using physical processes such as evaporation–condensation, laser ablation, electrolysis, pyrolysis, diffusion plasma arcing, sputter deposition, and high energy balling, chemical processes such as chemical reduction, microemulsion, electrochemical, thermal decomposition, sol–gel process and coprecipitation, and biological processes such as exploitation of biological mass, e.g., plant, fungi, microbes, etc. (Aguilar-Perez et al. 2020). Number of metallic nanoparticles-based nanosorbent materials is being produced through physical and chemical methods to remove dyes and antibiotics such as Rhodamine, amoxicillin, doxorubicin (Aguilar-Perez et al. 2020). However, biological methods are preferred over other methods because these methods are easy, cost-effective, ecofriendly, and convenient. Further, due to slow kinetic reaction, it is convenient to control the growth and stabilization (Singh et al. 2021).

The investigation of biomaterials up to the nanoscale turn out to be practicable due to the methodical improvement in sophisticated instrumentation technologies, especially on neutron, X-ray, light scattering, diffraction methods, and spectroscopic techniques, including atomic force microscopy/spectroscopy, or Raman spectroscopy (Gadomski 2008). Bionanomaterials are molecular material composed partially or completely from biological active molecules (Honek 2013), are multifaceted and generally nonlinear viscoelastic and stochastic structure (Gadomski 2008) obtained from plants, animals, fungi, peptide, nucleic acid, agricultural waste, etc. (Singh et al. 2021; Barhoum et al. 2022).

The formation of nanoparticles from or with the presence of bioactive molecules such as peptides, enzymes, vitamins, alkaloids, phenolics, along with others derived from variable biological sources (e.g., plants, bacteria, fungi, etc.) is depending on the character of the precursor material and the end use (Ali et al. 2019b). Debnath et al. (2020) have successfully synthesized zirconia nanoparticles using microbial green technology which can be used for remediation of tetracycline from wastewater. They reported that bacteria *Pseudomonas aeruginosa* enable the production of secondary metabolites during the bacterial growth in the synthesis activity, that bestowed the capping and stabilizing agents for the development of zinconia nanoparticle, which showed high adsorption capacity for tetracycline and enable them suitable for practical applications.

3.4 Utilizations of Nano-Bio Materials in Wastewater Treatment

Manna et al. (2019) synthesized a value-added nanomaterial from carbonaceous industrial waste. They coated the carbonaceous industrial waste with graphene oxide and activated it at 800 °C and characterized the graphene oxide and graphene oxide-coated biochar with various microscopy, spectroscopy, and x-ray diffraction methods. This graphene oxide-coated biochar was tested for phenol removal capacity in wastewater. They reported that the graphene oxide-coated biochar (nanocomposites) showed improved separation potential than that for the graphene oxide. They also claimed that the availability of salt showed positive effect on phenol severance by the nanocomposites process. They concluded that the nanocomposites have promising separation capacity for wide range of sorption parameters.

Goswami et al. (2020) assessed the potential of an up-flow packed bed bioreactor (UFPBBR) with immobilized bacteria *Rhodococcus opacus* onto biochar loaded polyurethane foam (PUF) as the packing material for treating biomass gasification wastewater. They reported that a maximum 81 $\pm$ 2.65% COD removal was obtained with 1820 mg L^{-1} of influent COD concentration in 24 h of HRT when the bioreactor was operated with PUF only as the support material, while more than 95 $\pm$ 1.27% COD removal was obtained when biochar loaded PUF was used as the support material. They also concluded 96.2% detoxification by the application of novel biochar-based bionanomaterial for biomass gasification wastewater remediation.

During his research Zheng et al (2021) recently explained functional microbiome linkages and method implicated in aromatic ring breakdown on coke enhanced nano-Fe_3O_4 activated bio-system (FEBS) under limited-oxygen state for proficient remediation of aromatic organic compounds in coal pyrolysis wastewater. They also identified that the efficiency of biodegradation with the FEBS supplementation presented excellent organic removal (average 92.29%) and biodegradability continuance (>40%) then control reactors. They concluded that the enhanced biodegradation in the FEBS is principally due to enriched functional species. Another study by Muthukumar et al. (2020) found photocatalytic dilapidation of caffeine in synthetically produced wastewater of coffee industry by silver ferrite nanoparticles produced with extract of *Amaranthus blitum* leaf. Furthermore, Abbasi et al. (2020) found effective cationic dye remediation of crystal violet from aqueous solution using Keratine nanoparticles obtained from human hair. This could also be a suitable use of wasted hairs.

In a recent study, Deng et al. (2022) designed a highly efficient and stable microsphere-immobilized bacterial strain of *Bacillus velezensis* to remediate organic substance and suppress the growth of unsafe microorganism the course of slaughter wastewater treatment. Immobilization of *Bacillus velezensis* was done at the surface of sodium alginate (SA)/Polyvinyl alcohol (PVA)/Nano-Zinc Oxide (Nano-ZnO) microsphere. Adhesion of the bacterial strain was managed by direct physical adsorption. The researchers reported inhibition of *Escherichia coli* on SA/PVA/ZnO and SA/ZnO microspheres with addition of 0.15 g/L nano-ZnO without affecting *Bacillus velezensis* strain. Process reported 16.99% removal rate of the chemical oxygen demand (COD) of SA/PVA/ZnO microsphere-immobilized cells, followed by 13.69% COD removal rate of SA/ZnO and free bacteria (7.61%) from 50% concentration slaughter wastewater within 24 h on 37 ∘C temperature and 7.0 pH at 120 rpm. From the results of the study, Deng et al. (2022) also reported that the strategy design could enhance the degradation potential to a greater extent, inhibit the growth of other bacteria without affecting the activity of protease in slaughter wastewater. Thus it was concluded that the wastewater treatment system of nano-ZnO hydrogel immobilization *Bacillus velezensis* is an imperative alternative technique to remediate pollutants from slaughter wastewater with a novel and ecofriendly and low-priced investment.

3.5 Sustainability and Future Perspectives

Nanotechnology is an economical, efficient, and trustworthy approach to improve the eminence and reliability of wastewater treatment. Nanomaterials have already been successfully utilized in several area, like medical, catalysis, agriculture, etc. owing to their exceptional attributes like nano in size, increased surface area than mass, higher reactivity, mobility, sturdy mechanical strength, highly porous surface, hydrophilic and dispersible nature (Khan et al. 2018; Yaqoob and Ibrahim 2019; Wu et al. 2019). Environmental experts have designated application of nanomaterials for wastewater remediation as a superior strategy than the other individual classical wastewater treatment techniques due to its potential to eradicate hazardous elements like Pb, Mo, Co, other organic and inorganic contaminants, and pathogenic microorganisms harmful for human health, present in water (Umar, 2018, Mohammad et al., 2019, Sekoai et al. 2019; Khan et al. 2021). Nanotechnology sector is expected to expand further with the increase in its application for domestic and industrial purposes due to lower energy consumption and reduced production cost, boosting quality, and better environmental control (Patil 2015).

Nanotechnology is established its success in a variety of manufacturing applications including wastewater treatment, and it has capability to enhance wastewater treatment as the materials may be controlled and altered at the nanoscale. Further the nanomaterial obtained through biological sources could be controlled and stabilized. It also has several unique qualities, e. g., low size-specific surface energy, affinity and reactivity with other material, stability, and transferrable atomicity. These features are peculiar for decontamination of drinking water and improved wastewater treatment technologies (Khan and Sengül 2016, Samanta et al. 2016, Khan et al. 2018, Khan et al. 2021). Further, the nanoscale bioadsorbents obtained from waste biomass sources are renewable, sustainable, cost-effective, and environmental friendly material for wastewater treatment (Solangi et al. 2021). Potential to increase efficiency of wastewater treatment plant contribute in development of nanotechnology sector and its application and research for domestic and industrial use.

Bionanomaterials may composed of proteins, DNA, or other biological origin material, and although these biomolecules are "natural" compounds, they may have harmful effects ranging from no toxicity to high toxicity (Honek 2013). Although nanoparticles in water have no instantaneous consequence on human health but there is a possibility of nanoparticle ingestion through bio-magnification, hence the effect of nanomaterials on aquatic organisms is essential to address (Khan et al. 2021). Pernicious effects of nanomaterials in biological world are by the fact that such exposed nanomaterials are contemplated as the potential body "invaders" alike viruses or bacteria, nonetheless, their baneful effects depends on the interactions to immune cells. Interactions of the nanomaterials with biomolecules may lead to inflammation. However, no mandatory guidelines were specified by the regulatory agencies specifically for nanomaterials, its abrupt use could pretense threat to the environment and society. Thus the sustainable nanotechnological development may be ensured by compelling different regulatory guidelines suggested by Environmental Protection Agency (EPA), Toxic Substances Control Act (TSCA), and Federal Insecticide, Fungicide, and Rodenticide Act (FIFRA). The commercialization of nanomaterials requires an inclusive considerate on Ecological and societal, effects of the exposed engineered materials predominantly at bio-nanointerface (Karak 2019; Goodman et al. 2004).

Aguilar-Perez et al. (2020) concluded that nanotechnology is attracting the researchers for wastewater treatment because of its imperative properties like small size and large surface area, which changed the physic-chemical property of the material including chemical reactivity, catalytic property, adsorption capacity, etc., that enables them to potentially decontaminate wastewater from several common and specific organic and inorganic pollutants. The biogenic nanomaterials are considered as ecofriendly and they can also be produced from waste. There is an inevitability to evaluate the performance of bionanomaterials under controlled environments. Further, more thorough and careful research ventures are also require to optimize its contrive. The challenges and issues with the bionanomaterials are to address the fate and its toxicity in the environment and the administrating policies and protocols for appropriate disposing or recycling.

References

Abbasi F, Yaraki MT, Farrokhnia A, Bamdad M (2020) Keratin nanoparticles obtained from human hair for removal of crystal violet from aqueous solution: optimized by Taguchi method. Int J Biol Macromol 143:492–500

Abhinav S, Pradeep KR, Mathur V, Pravin S (2021) Studies of wastewater treatment techniques using low-cost biosorbents. In: Microbial ecology of wastewater treatment plants, pp 395–410

Aguilar-Perez KM, Aviles-Castrillo JI, Ruiz-Pulido G (2020) Nano-sorbent materials for pharmaceutical-based wastewater effluents—an overview. Case Stud Chem Environ Eng 2:100028

Al Bsoul A et al (2021) Efficient removal of phenol compounds from water environment using Ziziphus leaves adsorbent. Sci Total Environ 761:143229

Ali I, Peng C, Khan ZM, Naz I, Sultan M, Ali M, Abbasi IA, Islam T, Ye T (2019a) Overview of microbes based fabricated biogenic nanoparticles for water and wastewater treatment. J Environ Manage 230:128–150

Ali S, Rehman SAU, Luan H-Y, Farid MU, Huang H (2019b) Challenges and opportunities in functional carbon nanotubes for membrane-based water treatment and desalination. Sci Total Environ 646:1126–1139

Alsbaiee A, Smith BJ, Xiao L, Ling Y, Helbling DE, Dichtel WR (2016) Rapid removal of organic micropollutants from water by a porous β-cyclodextrin polymer. Nature 529(7585):190–194

Barhoum A, Jeevaandam J, Danquah MK (2022) Fundamentals of bionanomaterials (Ist ed.). Elsevier

Bodzek M, Konieczny K, Kwiecińska-Mydlak A (2019) Nanotechnology in water and wastewater treatment. Graphene—the nanomaterial for next generation of semipermeable membranes. Critical Rev Environ Sci Technol, 1–65. https://doi.org/10.1080/10643389.2019.1664258

Carrott PJM, Carrott MMLR, Nabais JMV et al (1997) Influence of surface ionization on the adsorption of aqueous zinc species by activated carbons. Carbon 35(3):403–410

Chung WJ, Shim J, Ravindran B (2022) Application of wheat bran based biomaterials and nanocatalyst in textile wastewater. J King Saud Univ Sci 34:101775

Cincinelli A, Martellini T, Coppini E, Fibbi D, Katsoyiannis A (2015) Nanotechnologies for removal of pharmaceuticals and personal care products from water and wastewater. Rev J Nanosci Nanotechnol 15(5):3333–3347

Debnath B, Majumdar M, Bhowmik M, Bhowmik KL, Debnath A, Roy DN (2020) The effective adsorption of tetracycline onto zirconia nanoparticles synthesized by novel microbial green technology. J Environ Manage 261:110235

Deng J, Jia M, Zeng YQ, Li W, He JT, Ren J, Bai J, Zhang L, Li J, Yang S (2022). Enhanced treatment of organic matter in slaughter wastewater through live Bacillus velezensis strain using nano zinc oxide microsphere. Environ Pollut 292(Part A):118306

El-sayed MEA (2020) Nanoadsorbents for water and wastewater remediation. Sci Total Environ 739:139903. https://doi.org/10.1016/j.scitotenv.2020.139903

Gabaldon C, Marzal P, Alvarez-Hornos FJ (2006) Modelling Cd(II) removal from aqueous solutions by adsorption on a highly mineralized peat. Batch and fixed-bed column experiments. J Chem Technol Biotechnol 81(7):1107–1112

Gadomski A (2008) Editorial introduction to the special issue on bio(nano)materials with structure–property relationship. BioSystems 94:191–192

Goodman CM, McCusker CD, Yilmaz T, Rotello VM (2004) Toxicity of gold nanoparticles functionalized with cationic and anionic side chains. Bioconjug Chem 15:897–900

Goswami L, Pakshirajan K, Pugazhenthi G (2020) Biological treatment of biomass gasification wastewater using hydrocarbonoclastic bacterium *Rhodococcus opacus* in an up-flow packed bed bioreactor with a novel waste-derived nano-biochar based bio-support material. J Clean Prod 256:120253

Honek JF (2013) Bionanotechnology and bionanomaterials: John Honek explains the good things that can come in very small packages. BMC Biochem 14:29. https://doi.org/10.1186/1471-2091-14-29

Karak N (2019) Fundamentals of sustainable nanostructural materials at bio-nano interface. In: Dynamics of advanced sustainable nanomaterials and their related nanocomposites at the bio-nano interface. In: Karak N (ed) Elsevier, pp 1–24

Khan I, Saeed K, Khan I (2019) Nanoparticles: properties, applications and toxicities. Arab J Chem 12:908–931

Khan S, Naushad M, Al-Gheethi A, Iqbal J (2021) Engineered nanoparticles for removal of pollutants from wastewater: current status and future prospects of nanotechnology for remediation strategies. J Environ Chem Eng 9:106160

Khan S, Sengül H, Dan Z (2018) Transport of TiO_2 nanoparticles and their effects on the mobility of Cu in soil media. Desalin Water Treat 131:230–237

Khan S, Sengül H (2016) Experimental investigation of stability and transport of TiO_2 nanoparticles in real soil columns. Desalin Water Treat 57(2016):26196–26203

Kubra KT, Salman MS, Hasan MN (2021) Enhanced toxic dye removal from wastewater using biodegradable polymeric natural adsorbent. J Mol Liq 328:115468

Kurniawan TA, Chan GYS, Lo WH, Babel, S (2006) Comparisons of low-cost adsorbents for treating wastewaters laden with heavy metals. Sci Total Environ 366(2–3):409–426

Manna S, Prakash S, Das P (2019) Synthesis of graphene oxide nano-materials coated bio-char using carbonaceous industrial waste for phenol separation from water. Colloids Surf A Physicochem Eng Asp 581:123818

Mehta KP, Sharma R, Haldar S, Kumar A (2021) Advancement in treatment of wastewater with nano technology. Mater Today: Proc. https://doi.org/10.1016/j.matpr.2021.07.253.

Mohammad AT, Abdulhameed AS, Jawad AH (2019) Box-Behnken design to optimize the synthesis of new crosslinked chitosan-glyoxal/TiO_2 nanocomposite: methyl orange adsorption and mechanism studies. Int J Biol Macromol 129:98–109

Mohmood I, Lopes CB, Lopes I, Ahmad I, Duarte E, Pereira AC (2013) Nanoscale materials and their use in water contaminants removal—a review. Environ Sci Pollut Res 20:1239–1260

Morsi RE, Alsabagh AM, Nasr SA, Zaki MM (2017) Multifunctional nanocomposites of chitosan, silver nanoparticles, copper nanoparticles and carbon nanotubes for water treatment: antimicrobial characteristics. Int J Biol Macromol 97:264–269

Muthukumar H, Shanmugam MK, Gummadi SN (2020) Caffeine degradation in synthetic coffee wastewater using silver ferrite nanoparticles fabricated via green route using Amaranthus blitum leaf aqueous extract. J Water Process Eng 36:101382

Nasar A, Mashkoor F (2019) Application of polyaniline-based adsorbents for dye removal from water and wastewater—a review. Environ Sci Pollut Control Ser 26(6):5333–5356

Patil T (2015) Wastewater treatment using nanoparticles. J Adv Chem Eng 5:131. https://doi.org/10.4172/2090-4568.1000131

Ponnusamy G, Saththasivam J (2021) Applications of nanomaterials for water disinfection. Springer, Singapore, pp 311–329

Rienzie R, Ramanayaka S, Adassooriya NM (2019) Nanotechnology applications for the removal of environmental contaminants from pharmaceuticals and personal care products. In: Prasad MNV, Vithanage M, Kapley A (eds) Pharmaceuticals and personal care products: waste management and treatment technology emerging contaminants and micro pollutants. Elsevier, Butterworth-Heinemann, pp 279–296

Sadegh H, Ali GAM (2018) Potential applications of nanomaterials in wastewater treatment, pp 51–61

Saharan P, Chaudhary GR, Mehta SK, Umar A (2014) Removal of water contaminants by iron oxide nanomaterials. J Nanosci Nanotechnol 14(1):627–643

Sajid M, Nazal MK, Ihsanullah BN, Osman AM (2018) Removal of heavy metals and organic pollutants from water using dendritic polymers based adsorbents: a critical review. Sep Purif Technol 191:400–423

Samanta H, Das R, Bhattachajee C (2016) Influence of nanoparticles for wastewater treatment-a short review. Austin Chem Eng 3:1036

Santhosh C, Velmurugan V, Jacob G, Jeong SK, Grace AN, Bhatnagar A (2016) Role of nanomaterials in water treatment applications: a review. Chem Eng J 306:1116–1137

Sekoai PT, Ouma CNM, Du Preez SP, Modisha P, Engelbrecht N, Bessarabov DG, Ghimire A (2019) Application of nanoparticles in biofuels: an overview. Fuel 237:380–397

Sharabati MA, Abokwiek R, Al-Othman A, Tawalbeh M, Karaman C, Orooji Y, Karimi F (2021) Biodegradable polymers and their nano-composites for the removal of endocrine-disrupting chemicals (EDCs) from wastewater: a review. Environ Res 202:111694

Singh A (2021) A review of wastewater irrigation: environmental implications. Resour Conserv Recycl 168:105454

Singh A, Agrawal SB, Rai JPN, Singh P (2002) Assessment of the pulp and paper mill effluent on growth, yield and nutrient quality of wheat (*Triticum aestivum* L.). J Environ Biol 23(3):283–288

Singh KRB, Nayak B, Singh RP (2021) Introduction to bionanomaterials: an overview. In: Singh RP, Singh KRB (eds) Bionanomaterials: fundamentals and biomedical applications. IOP Science, pp 1–22

Singh R, Rathore D (2018) Oxidative stress defence responses of wheat (*Triticum aestivum* L.) and chilli (*Capsicum annum* L.) cultivars grown under textile effluent fertilization. Plant Physiol Biochem 123:342–358. https://doi.org/10.1016/j.plaphy.2017.12.027

Singh R, Rathore D (2019) Impact assessment of azulene and chromium on growth and metabolites of wheat and chilli cultivars under biosurfactant augmentation. Ecotoxicol Environ Saf 186:109789. https://doi.org/10.1016/j.ecoenv.2019.109789

Singh R, Rathore D (2020) Role of transitory starch on growth, development and metal accumulation of *Triticum aestivum* cultivars grown under textile effluent fertilization. Environ Sci Pollut Res 27:24201–24217. https://doi.org/10.1007/s11356-020-08735-y

Singh R, Rathore D (2021) Effects of fertilization with textile effluent on germination, growth and metabolites of chilli (*Capsicum annum* L) cultivars. Environ Process 8:1249–1266. https://doi.org/10.1007/s40710-021-00531-1

Solangi NH, Kumar J, Mazari SA, Ahmed S, Fatima N, Mubarak NM (2021) Development of fruit waste derived bio-adsorbents for wastewater treatment: a review. J Hazard Mater 416:125848

Stojakovic D, Milenkovic J, Daneu N et al (2011) A Study of the removal of copper ions from aqueous solution using clinoptilolite from Serbia. Clays Clay Miner 59(3):277–285

Umar K (2018) Water contamination by organic-pollutants: TiO_2 photocatalysis. In: Oves M, Khan MZ, Ismail IMI (eds) Modern age environmental problems and their remediation. Springer, Switzerland, pp 95–109

Wu Y, Pang H, Liu Y, Wang X, Yu S, Fu D, Chen J, Wang X (2019) Environmental remediation of heavy metal ions by novel-nanomaterials: a review. Environ Pollut 246:608–620

Yaqoob AA, Ibrahim MNM (2019) A review article of nanoparticles; synthetic approaches and wastewater treatment methods. Int Res J Eng Technol 6:1–7

Yaqoob AA, Parveen T, Umar K, Ibrahim MNM (2020) Role of nanomaterials in the treatment of wastewater: a review. Water 12(2):495. https://doi.org/10.3390/w12020495

Yu C, Han X (2015) Adsorbent material used in water treatment-a review. 2nd International Workshop on Materials Engineering and Computer Sciences (IWMECS 2015). pp 290–323

Zhao L et al (2019) Adsorptive decontamination of Cu^{2+}-contaminated water and soil by carboxylated graphene oxide/chitosan/cellulose composite beads. Environ Res 179:108779

Zheng M, Han H, Shi J, Zhang Z, Ma W, Xu C (2021) Metagenomic analysis of aromatic ring-cleavage mechanism in nano-Fe_3O_4@activated coke enhanced bio-system for coal pyrolysis wastewater treatment. J Hazard Mater 414:125387

Chapter 4
Application of Plant-Based Nanoparticles in Wastewater Decontamination

Sudhir Kumar and Jaspal Singh Chauhan

Abstract Wastewater treatment is an important challenge of this century as it safeguards the health of our environment and living being. Wastewater is always regarded as a significant source of environmental pollution due to its potential to harm both living and non-living beings. Many physical, biological, and chemical modes of treatment are implied to comply with the standards of wastewater discharge, given by competent national agencies for protecting the environment. Researchers from all over the world have recently become more interested in the synthesis of plant nanoparticles and their application for wastewater decontamination since it is an environmentally benign, cost-effective, and efficient technology. Parts and extracts of various plants are being explored for the synthesis of nanoparticles. Green synthesized nanoparticles are highly efficient for recycling and removal of toxic contaminants from wastewaters and make it reusable in different aspects. However, synthesis, regeneration, and reusability are the major obstacles that must be addressed before the technology transferred from laboratory to commercial applications. In this chapter, we focused on the different approaches of plant-based nanoparticle synthesis and their applications in wastewater treatment. Further, important challenges in the field of plant-based nanoparticles in the wastewater treatment are also discussed.

Keywords Plant-based nanoparticles · Metal ion reduction · Wastewater treatment · Protein based nanoparticles

---

S. Kumar
Department of Biotechnology, School of Life Sciences, H.N.B. Garhwal University (A Central University), Srinagar Garhwal, Uttarakhand, India

J. S. Chauhan (✉)
Department of Himalayan Aquatic Biodiversity, School of Life Sciences, H.N.B. Garhwal University (A Central University), Srinagar Garhwal, Uttarakhand, India
e-mail: jaspal.env@gmail.com

J. P. N. Rai and S. Saraswat (eds.), *Nano-biotechnology for Waste Water Treatment*, Water Science and Technology Library 111,
https://doi.org/10.1007/978-3-031-00812-2_4

4.1 Introduction

Water is the most important as well as valuable environmental resource required for the existence of various living organism on the earth. The amount of water present on the earth is approximately 70% and further divided into Freshwater, Marine water, and Saline water. The freshwater system is categoried into two groups, Lentic system (lakes, ponds) and Lotic system (streams, rivers, springs) which are the important source of water to satisfy the daily need for mankind (Mishra et al. 2016, 2015). In present time, the water quality is influenced by both natural and manmade effect such as local climatic conditions, irrigation practice, topography and geology, etc. (Harman et al. 2012). Physicochemical properties of water such as temperature, dissolved oxygen, pH, transparency, conductivity, water current and some chemical parameters as nitrate, phosphate, heavy metals, organic matter, etc., are important in distinguishing effects of pollution on the water quality (Awasthi and Tiwari 2004) and it provides a direct and indirect manifestation of stream condition which support communities of various biological entities (Harman et al. 2012).

At present, the world is in the phase of meeting the rising demands of clean water for various purposes like drinking, agriculture and industrial use. The major reasons for depletion of fresh water are extended droughts, population growth and pollution. Unavailability of sufficient clean water which is free from any contamination is leading to deteriorated health of living organisms especially human beings. Industrialization coupled with improper or inefficient treatment process has led to water pollution with a number of organic and inorganic pollutants, mostly persistent, toxic, and hazardous to the living organisms. Developing countries are the major one to face the problem due to water contamination by wastewater. The unavailability of suitable technology for wastewater treatment seems to be the prime reason for contamination of water. High operational and maintenance cost, high energy requirement for conventional methods of wastewater treatment which includes adsorption and biosorption, chemical precipitation, chemical oxidation or reduction, filtration, electrochemical treatment, reverse osmosis, ion exchange, membrane technologies, and evaporation recovery are the main constraints in the developing countries (Agency 2011; Dimapilis et al. 2018; Kunduru et al. 2017).

Ensuring safe and clean water by proper treatment of water and wastewater is a prime concern of the present era. Nanotechnology is considered as the new promising technology opted for removing the contaminants of organic and inorganic nature from water. Nanotechnological approaches have shown potential in improving water and wastewater treatment with a little financial change. Many types of nanoparticles are derived from various sources like plant, bacteria, and algae, which keep the synthesis in environmentally friendly mode. In the current scenario, nanoparticles like SiO_2, TiO_2 and ZnO are widely advocated for the wastewater treatment process. The properties like larger specific surface area, nontoxic, stability, efficient, environment-friendly, regeneration, and reusability make nanoparticles suitable for environmental applications (Venkateswarlu et al. 2016). Till date nanoparticles are successfully practiced in wastewater treatment for detection and removal of various categories of pollutants. They have been found efficient in removing both organic and inorganic pollutant from wastewater which otherwise could pose serious threats to living organisms and environment. The nanoparticles derived from various sources have shown better results when implemented in different process of water treatment like photocatalysis, adsorption, filtration, and electrochemical oxidation. These nanoparticles are used as nanowire membranes, carbon nanotubes, polymer membranes, ceramic membranes, submicron nanopowder, metal oxides, magnetic nanoparticles, boron doped diamond to improve water quality (Qu et al. 2013; Sushma and Richa 2015).

The chemically derived nanoparticles are harmful to environment and living organisms. Further, their synthesis requires highly complex methodology and hence results in expensive process (Saha et al. 2013). Contrastingly green nanoparticles synthesized from biogenic sources are eco-friendly, less toxic, biodegradable, and less expensive in nature. These are basically developed from the parts of plant or microbes (Saha et al. 2013; Singh et al. 2019). Among these biogenic nanoparticles plant extracted nanoparticles are considered to be a better option for solving problems of environmental pollution (Singh et al. 2019). A list of plant-based nanoparticles used for treatment of contaminants of environment is given in Table 4.1.

4.2 Green Nanotechnology

Green nanotechnology includes the use of environmentally affable methods for the manufacturing of nanomaterials. The chemical synthesis of nanomaterials is generally not environmental friendly and produces harmful byproducts (Nasrollahzadeh et al. 2019). The benefits of the nanoparticles are particularly dwarfed by the potentially dangerous byproducts that may form upon absorption by humans, animals, and plants, etc. It is therefore important that the environmentally sustainable and natural nanomaterials are produced which reduce the burden of pollution as well as risk of side effects (Corradini 2010). The course of manufacturing of nanoparticles is based on heavy use of unsafe chemicals which are extremely harmful to the environment. Therefore, the technology using plants and other biological entities for synthesis of nanoparticles is one way to reduce this problem (Shanker et al. 2016). The biological intervention to generate the nanoparticles leaves no harmful material. The technologies employing green chemistry also rely on biodegradable materials which do not cause any harmful side effect to the environment (Hutchison 2008). Chemical methods of nanoparticle synthesis use hazardous reducing and capping agents which are generally non-biodegradable and remain in environment for long time. Since most of these chemicals are hydrophobic they need organic solvents for their dissolution which are harmful to both flora and fauna including humans. Apart from the synthesis, some of the heavy metal NPs like mercury, lead, tin, etc., themselves are dangerous for the environment (Khan et al. 2019). Metal NPs also tend to cause health conditions in plants, animals, and humans (Hristozov and Malsch 2009).

Table 4.1 List of plant-based nanoparticles used in removal of pollutants

S. No	Plant/plant part	Nanoparticle	Pollutant removed	Efficiency	References
1.	*Camellia sinensis and Punica granatum*/plant leaf extract	Fe NPs	Textile wastewater	95% color removal and almost 80% dissolved organic carbon	Ozkan et al. (2018)
2.	*Jatropha curcas* L/ plant leaf extract	TiO_2 NPs	Photocatalytic treatment of tannery wastewater	COD removal, 82.26%; Cr removal, 76.48%	Goutam et al. (2018)
3.	*Ficus benjamina*/ plant leaf extract	Ag NPs	Cd removal from aqueous solution,	85%	Al-Qahtani (2017)
4.	*Citrus grandis*/ peel extract	Cu NPs	Degradation of methyl red	96%	Sinha and Ahmaruzzaman (2015)
5.	*Eucalyptus sp.*/ plant leaf extract	Fe NPs	Treatment of eutrophic wastewater	N-71.7%; P-30.4%; COD-84.5%	Wang et al. (2014)
6.	*Punica granatum*/ peel extract	Nickel–iron nanocomposite	Tetracycline removal	99%	Ravi kumar et al. (2020)
7.	Tangerine/ peel extract	Iron oxide NPs	Cd removal	90%	Ehrampoush et al. (2015)
8.	*Piliostigma thonningii*/ plant leaf extract	Ag NPs	Heavy metal removal activity	Iron-96.9%; copper-89%; lead-97.89%; magnesium-93.6%	Shittu and Ihebunna (2017)
9.	*Azadirachta indica, Magnolia champaca, Mangifera indica, and Murraya koenigii* /	Fe NPs	Treatment of domestic wastewater	Phosphate-98.1%; ammonium nitrogen-84.3% COD-82.4%	Devatha et al. (2016)
10.	*Padina pavonica/beads*	Fe_3O_4-NPs	Pb solution	91% removal of Pb	El-Kassas et al. (2016)

These engineered nanoparticles are now routinely been used in cosmetic products, ready to eat food products and their packaging, drug delivery systems, therapeutics, biosensors, etc. and hence a large number of population is exposed to them on daily basis (Contado 2015; Patra et al. 2018). The effects of this exposure are not fully understood and may need more detailed studies (Ray et al. 2009). Olga V. Morozova reviewed the toxicity of silver nanoparticles involving biological macromolecules like RNA and DNA by "Trojan horse," inductive, and quantum mechanical modes

(Morozova 2021). Kowalczyk et al. also reviewed the toxicity of silver nanoparticles in animal tissues, bacteria and viruses and calls for detailed safety studies in light of their extensive use (Kowalczyk et al. 2021).

4.3 Plant-Derived Nanoparticles

Generally, metal is used for the generation of nanoparticles. Many metals like gold (Shankar et al. 2014), silver (Krithiga et al. 2015), copper (Murthy et al. 2020), titanium oxide (Dobrucka 2017), zinc oxide (Naseer et al. 2020), iron oxide (Bhuiyan et al. 2020), etc., are used. Different parts of the plants and herbs which include stems, leaves, fruit, flowers, and roots have been found useful for this purpose (Fig. 4.1). Plants can reduce the metal ions in cells at various locations like leaves, flowers, fruits, etc., and have been used for phytomining, a plant-based bio-sequestering technique. It was noticed that during bioaccumulation in plants, majority of metals are stored

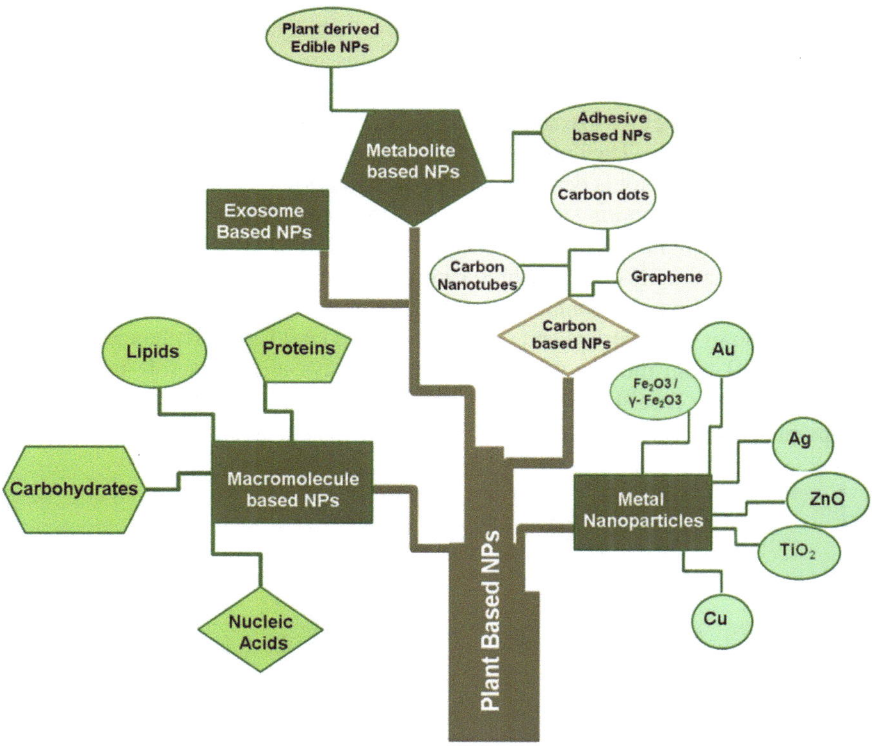

Fig. 4.1 Plant-based nanoparticles. Plant extracts have the potential to reduce many metal ions into the nanoparticles in environmental friendly way. Nanoparticles can also be derived from macromolecules and metabolites produced by plants

as nanoparticles. The plants can take up the metal ions and deposit them at various places which can then be harvested. The major obstacle with this technique is that the size, shape, and quantity of the nanoparticles cannot be controlled. Therefore, the invitro reduction technique using the plant extracts has been developed which is more flexible, fast, and high yielding (Makarov et al. 2014). Metabolites like ployphenols (flavonoids) and phenolic acids, alkaloids, terpeniods, etc., and proteins, enzymes can potentially catalyze reduction of metal ions into stable nanoparticles of varying shape and size (Begum et al. 2009; Mittal et al. 2013; Shamaila et al. 2016; Velusamy et al. 2016).

4.4 Mechanism

There are two ways of biological synthesis of nanoparticles, a) Bio-reduction and b) Bio-absorption or Biosorption; the former being used frequently due to its ease of operation (Al-khattaf 2021). The first step in nanoparticle formation is reduction of ions followed by the cluster formation and growth (Jain and Mehata 2017). As already mentioned, polyphenols, terpenoids, carbohydrates, alkaloids, phenolic acids, and other plant metabolites including proteins are potential bioreducing agents that can reduce metal ions into nanoparticles. Singh A. K. et al. suggest that Eugeniol, a terpenoid found in clove extract, can reduce AU and Ag ions and proposed a mechanism for this reaction. They proposed that the hydroxyl group (-OH) attached to the benzene ring of eugeniol is deprotonated to anionic form (O^-). This anionic form is stabilized by the resonating structures, which are then stabilized by the release of 2 electrons which go on to reduce Au^{1+}/Ag^{4+} ions (Singh et al. 2010). The flavonoids like quercetin and taxifolin have three ringed structure named A, B and C where B ring is a catechol moiety containing –OH groups at 3' and 4' positions (Fig. 4.2). Quercetin shows tautomerism and exists in keto and enol forms. It changes into keto form as the H from the 3'-OH is removed and plays important role in reduction of nanoparticles (Trouillas et al. 2006). Quercetin having armed with these –OH groups can chelate metal ions like Pb^{2+}, Fe^{2+}, Cu^{2+}, Zn^{2+}, etc. (Makarov et al. 2014).

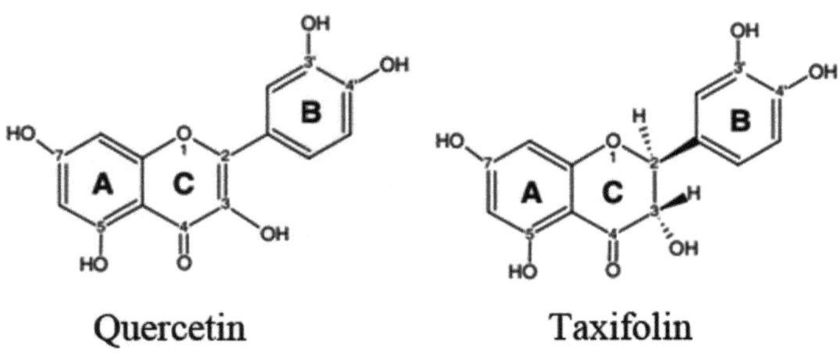

Fig. 4.2 Chemical structure of flavonoids quercetin and taxifolin

Monosaccharide like glucose and fructose also has the capacity to reduce metal ions to nanoparticles. The conversion of aldehyde group of sugar into a carboxyl group by addition of -OH leads to the reduction of metal ions into nanoparticles (Panigrahi et al. 2004). Huang J. et al. observed that there was a substantial decrease in the concentration of flavonoids and monosaccharides in solution after the reduction of $AgNO_3$ during silver nanoparticles formation (Huang et al. 2011) signifying their role in the reduction.

Amino acids with charged side chain like Lysine and arginine (+ve charge), Methionine and Cysteine (–SH group) can bind and potentially reduce metals to form nanoparticles (Clem Gruen 1975). Aspartate (Mandal et al. 2002) and tryptophan (Tan et al. 2010) were also found to be suitable for reduction of chloroauric acid ($HAuCl_4$) into Au nanoparticles. Individual amino acids showed different behavior than the peptides in reduction of the metals. Amino acids that can bind strongly to the metal ions can reduce them efficiently but excess binding inhibits the reduction efficiency (Makarov et al. 2014; Tan et al. 2010). Apart from the nature of macromolecules, physical conditions like pH, time of exposure, temperature, and concentration have substantial effect on the nanoparticles formation by plant extracts (Dada et al. 2018; Jeevanandam et al. 2020; Kredy 2018). The reduction of metals in plant extracts generally proceeds very fast and is completed within few minutes (Dada et al. 2018). As the pH approach alkaline range, nanostructures become more stable with lower tendency to aggregate. This might be attributed to the repulsive electrostatic/electrosteric interactions that come into play as the nanoparticles clusters are completely charged (Traiwatcharanon et al. 2015).

Silver: $AgNO_3$ solution (1-mm) is taken and is mixed with aqueous extract from leaves (Banerjee et al. 2014; Begum et al. 2009; Hemlata et al. 2020; Krishnaraj et al. 2010b), fruits (Lakshmanan et al. 2018; Masum et al. 2019; Niluxsshun et al. 2021), roots (Dangi et al. 2020; Ilahi et al. 2021; Singh et al. 2020), flowers (Devanesan and AlSalhi 2021; Mahmoodi Esfanddarani et al. 2018; Mathew et al. 2020; Padalia et al. 2015), and stems (Bharathi et al. 2018; Karthiga 2018; Subha et al. 2018). The mixture is kept in dark for incubation to reduce chances of photoactivation of $AgNo_3$. As the $AgNO_3$ is reduced to metallic silver nanoparticles, the color of the mixture changes to colloidal brown. In some cases, the nanoparticles appear as fast as 30 min of the incubation (Krishnaraj et al. 2010a). The nanoparticles are characterized by XRD, FTIR, SEM, etc.

Gold NPs: Similar to the silver nanoparticles, gold is also the favored choice for the nanoparticle formation because of its inert nature. Extracts from several plants have been used for the reductive formation of the AuNPs of varying shape and sizes (Aboyewa et al. 2021; De Canha et al. 2021; Fanoro et al. 2021; Hassanisaadi et al. 2021; Rahman et al. 2021). These include *Ipomoea carnea* (Abbasi et al. 2015), *Cassia fistula* (Daisy and Saipriya 2012), citrus plants (Rao et al. 2017; Sujitha and Kannan 2013; Yuan et al. 2017), *Azadirachta indica* (Dharmatti et al. 2014; Pandey et al. 2013; Rao et al. 2017), *Cocculus hirsutus* (Bar et al. 2012), etc. J. L. Gardea-Torresdey et al. utilized the bio-absorption for fabrication of AuNPs by growing *alfa alfa* plants in $AuCl_4$ rich medium (Gardea-Torresdey et al. 2002). Plant extracts contain metabolites like flavonoids, phenols, protein, etc., which can reduce Au_4+

ions to form gold nanoparticles. Shankar et al. used geranium leaves extract for production of gold NPs from $AuCl_4$ and hypothesized that possible reducing agent in extract could be citronellol and geraniol, terpenoids present in many aromatic plants (Shankar et al. 2003).

4.5 Plant-Based Nanoparticles

Macromolecules present in plants are gaining popularity to be used as nanomaterials as they are readily available and biodegradable. Some of the important plant-derived nanomaterials include (but not limited to) protein based NPs, polysaccharide based NPs, carbon based NPs, exosome-like NPs, adhesive based NPs and lipids based NPs (Mohammadinejad et al. 2016) (Table 4.2). Plant-derived nanostructures and nanomaterials have been proposed as therapeutic agents (Zhang et al. 2016) and many have found their way into potential anti-cancer treatments (Verma et al. 2021). Apart from these their use and efficacy as antimicrobial agents (Mostafa 2021), anti-larval and crop protection agents (Karny et al. 2018; Shang et al. 2019), targeted drug delivery and therapeutics using miRNAs (Iravani and Varma 2019), and biosensors (Cavalcante et al. 2021) are well documented (Verma et al. 2021).

4.6 Plant Proteins as Nanoparticles

Proteins have unique properties that make them a popular choice for nanoparticles synthesis. Protein nanoparticles are biodegradable and biocompatible (Lohcharoenkal et al. 2014). Proteins sourced from plants which include Zein (corn storage protein), gliadin (wheat gluten derived protein), legumin (legume storage protein), etc., have been used to make nanoparticles with varying uses. Joye et al. prepared gliadin NPs using an antisolvent precipitation method in which gliadin extract (in 70% ethanol (v/v)) was mixed into water in 1:5 ratio at room temperature. The NPs were later hardened with glutaraldehyde treatment before use (Joye et al. 2015). Chen et al. modified this protocol to get gliadin—phospholipid hybrid nanoparticles which were more stable and have better foaming capacity (Chen et al. 2019). Xu et al. fabricated hollow Zein nanoparticles which can be loaded with drug of choice for intracellular delivery (Xu et al. 2011). Mirshahi et al. prepared legumin protein nanoparticles and studied their loading with methylene blue as hydrophilic drug model. They observed that glutaraldehyde cross linking increased the stability of the legumin NPs but decreased the release of the dye (Mirshahi et al. 1996). Nanoparticles prepared from other proteins like fibroins, albumin, gelatine, and ferritin proteins have also been prepared for diverse uses (Hong et al. 2020). The applications of protein nanoparticles range from food industry to drug delivery and targeted therapies.

Table 4.2 Plant-derived nanomaterials (Adapted from Mohammadinejad et al. 2016)

Macromolecule	Source	Applications	Reference
Protein Based NPs			
Zein	Corn	Pharmaceutical film coatings	Patel et al. (2010)
		Drug Loading and Delivery	Lee et al. (2013b), Paliwal and Palakurthi (2014), and Xu et al. (2011)
		Gene Therapy and siRNA Delivery	Karthikeyan et al. (2015)
		Controlled delivery of Anti-cancer drug	Xu et al. (2015)
Gliadin	Wheat	α-tocopherol encapsulation	Duclairoir et al. (2002)
		Delivery vehicles for oral immunization	Kajal and Misra (2011)
		Anti-cancer drug carrier	Gulfam et al. (2012)
Legumin	Soyabean	Drug delivery system	Mirshahi et al. (1996)
Carbohydrate based NPs			
Cellulose	Plant cell wall	Contaminant removal from wastewater	Tshikovhi et al. (2020)
		Cell culture	Bhattacharya et al. (2012)
		Drug delivery systems	Lin and Dufresne (2014)
		Substitute/medical biomaterial	Lin and Dufresne (2014)
Lignin	Plant cell wall	Nanocomposites and nanohybrids	Lizundia et al. (2021)
		Nanoparticle reinforcer	Low et al. (2021)
Carbon Based NPs			
Graphene	Plants	Antimicrobial, antifungal	Li et al. (2017)

(continued)

Table 4.2 (continued)

Macromolecule	Source	Applications	Reference
		Biomedical applications	Magne et al. (2021)
Carbon Dots	Kitchen waste	Plant growth	Waghmare et al. (2021)
Lipid based NPs			
Plant-derived edible nanoparticles (PDNPs)	Edible plants	Nanomedicine and drug delivery	Yang et al. (2018)
		Nutrient delivery to plants	Karny et al. (2018)
		Plant growth induction	Fincheira et al. (2020)
Exosome based NPs			
Plant-derived exosome-like nanoparticles (PDENs)	Plants	Food and health related applications	Suharta et al. (2021)
		Therapeutics and drug delivery	Di Gioia et al. (2020) and Kim et al. (2021)

4.7 Plant Carbohydrates as Nanoparticles

Cellulose, a polysaccharide that is in abundance in plants has been extensively studied for its use as nanomaterial. Cellulose is composed of microfibrils which in turn are made of nanofibrils about 2–20 nm wide and few micrometers long (Rojas 2015). Lin et al. documented the studies in nanocellulose, which is extracted from cellulose (Lin and Dufresne 2014). Nanocelluose is extensively used in drug related studies (Hasan et al. 2020; Salimi et al. 2019), tissue substitutes (Pereira et al. 2018), and antimicrobial materials (Kupnik et al. 2020). Lignin, which is technically not a carbohydrate, is also used extensively in nanoparticles formations for its applications as nano-reinforcer for increasing the antibacterial, anti-oxidant, radiation protector, and catalytic properties of various nanomaterials (Low et al. 2021).

4.8 Other Plant-Based Nanoparticles

Plants and plant waste can also be used to synthesize specialized structures like graphene, carbon nanotubes, and carbon dots. Supriya et al. established a simple method of graphene formation using rice husk and KOH (Muramatsu et al. 2014). Magnabosco et al. reported a successful use of *A. thaliana* plant as bioreactor for the entrapment of single wall carboxylated carbon nanotubes (SWCNT-COOH) by roots and fabrication into bionic nanocomposite (Magnabosco et al. 2020). Waghmare et al. demonstrated the synthesis of water soluble carbon dots by pyrolysis of kitchen tea waste and found that it has positive effect on overall growth of plants (Waghmare et al. 2021).

Exosomes which are small nanosized microvesicles (50–100 nm) secreted by variety of cells and can potentially activate or modulate endocrine system have potential applications in many fields (Ju et al. 2013). Suharta et al. reviewed the preparation methods, characterization, and applications of the plant-derived exosome-like nanoparticles (PDENs) and observed their potential in health and food industry (Suharta et al. 2021). Gioia et al. mentioned the potential use of PDENs in variety of fields like tumor growth suppression, anti-inflammation, drug and gene delivery, etc., (Di Gioia et al. 2020). Differential ultracentrifugation is preferred method for the isolation and purification of the exosomes in which the plant extract undergoes series of centrifugation with gradual increase in speed and time to remove the larger and higher-density components leaving behind the exosome nanoparticles (Akuma et al. 2019). Exosomes secreted by tumor cells can be modified by injecting then with adjuvants which potentially show anti-tumor properties (Théry et al. 2009).

Additionally, lipids derived from plants are another favorite source for nanoparticles that can be used as potential drug carriers. Nano scale vesicles isolated from edible plants known as Plant-derived edible nanoparticles (PDNPs) are finding their way into the nanomedicine field. These PDNPs are preferred because they do not show any noticeable toxicity or immune reaction in body and have efficient uptake capacity (Yang et al. 2018). Tenchov et al. recorded the use of lipids as nanoparticles and their diverse applications ranging from liposomes to m-RNA vaccine delivery (Tenchov et al. 2021). Lipid based nanoparticles are also a medium of choice when the drug delivery has to be done across blood brain barrier (Fernandes et al. 2021). Glutathione PEGylated nanoliposomes based on egg yolk or hydrogenated soy phosphatidylcholine were developed by Hu et al. for delivery of methotrexate to the brain (Hu et al. 2019).

4.9 Application of Plant-Based Nanoparticles in Wastewater Treatment

Wastewater is a major culprit responsible for contamination of water resources. Treatment or removal of pollutants (organic and inorganic) released from the industries and domestic sources are major concern of scientists. For this scientist are exploring plant-based nanoparticles for their efficiency to treat the different pollutants of wastewater. Conventionally wastewater from different sources is treated in the treatment plants, which follows preliminary screening, primary, secondary, and tertiary treatment. The basic process of wastewater treatment plant is illustrated in Fig. 4.3.

The primary treatment is the physical method of removing most of the heavier solids that will be settled in the settling or sedimentation tank due to gravity. Secondary treatment of wastewater is basically the biological methods designed to degrade the biologically active content of the waste by means of aerobic processes. The final phase is the tertiary treatment in which the quality of the water is improved up to the standard of domestic and industrial use or to meet specific requirements so that the safe discharge of treated water can be done. The disinfection of treated water is also conducted in tertiary treatment to remove the pathogens, which makes that water safe for drinking purposes.

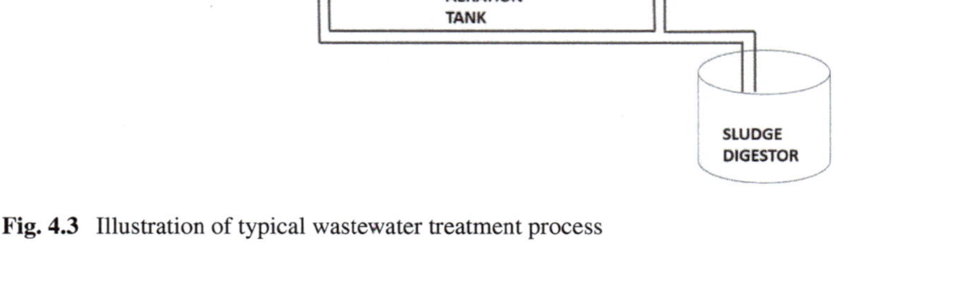

Fig. 4.3 Illustration of typical wastewater treatment process

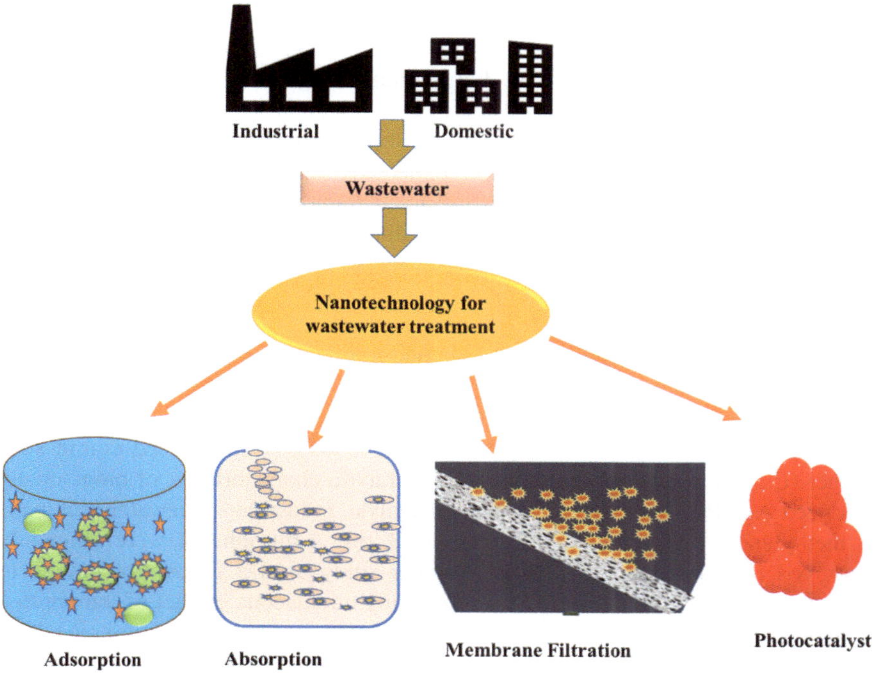

Fig. 4.4 Nanotechnological processes in the wastewater treatment

Nanotechnology has provided some of the efficient techniques that can improve the wastewater treatment process. The reasons for efficient treatment include high surface to volume ratio, small size, organized structure, stability, competence of filtration, and reusability (Gawande et al. 2011). Some of the nanotechnology methods used in treatment of wastewater are Nano adsorbents, Nanomembranes, and Nano-catalysts (Fig. 4.4).

4.10 Nano-Catalysts

Nano-catalysts are photocatalyst that involves photocatalytic activities, which include light energy interacting with metallic nanoparticles. By reacting with hydroxyl radicals, photocatalytic activities reduce micro-organisms (bacteria) and organic compounds. Inorganic materials such as metal oxides and semiconductors are commonly employed in nano-catalysts. In recent years, photocatalysts have been widely used in the degradation of pollutants in the environment or aquatic systems owing to its sustainable nature (Bagheri et al. 2017; Shak et al. 2018; Wang et al. 2018). Various photocatalysts are used for the degradation of pollutants from the different ecosystems of environment (Yao et al. 2021), like Zalerionmaritimum (Paço et al. 2017), Titania nanotubes (Ali et al. 2021), ZnO (Tofa et al. 2019a), N-doped TiO_2 (Ariza-Tarazona et al. 2020), Pt/ZnO (Tofa et al. 2019b), Graphitic carbon nitride (g-C_3N_4) (Liu et al. 2021), etc. Photocatalytic degradation of pollutants mainly occurs due to the absorbance of light energy by semiconductor to become excited and produce electrons (e^-) and holes (h^+) pairs. Both the electrons and holes react with water (H_2O) and oxygen (O_2) or hydroxyl ions (OH^-) to generate superoxide anions ($\bullet O^{2-}$) and hydroxyl radicals ($^\bullet OH$). These anions/radicals are called reactive oxygen species (ROS) and are very efficient in degrading organic pollutants. Recent studies on the pollutant degradation illuminated that photocatalysts can be ideal choices for treatment of wastewater (Pham et al. 2021; Sharma et al. 2021).

The surface functionalization of TiO_2 by Au NPs, which aids in improved charge separation and improves the interfacial charge-transfer process (Dawson and Kamat 2001), can enable photocatalytic oxidation of water contaminants such as thiocyanate ions (Dawson and Kamat 2001; Subramanian et al. 2004), azo dyes (Arabatzis et al. 2003), and 4-chlorophenol (Orlov et al. 2004). A study used cumin seed extract (*Cuminum cyminum*) to assess the catalytic characteristics of biogenically produced AgNPs. Using AgNPs as a nanocatalyst, industrial wastewater pollutants as methylene blue, methyl red, rhodamine-B, and 4-nitrophenol were efficiently destroyed (Choudhary et al. 2018). Another study used a greener technique to make Anatase TiO_2 NPs doped with iron, using an aqueous extract of lemongrass (*Cymbopogon citratus*), and these Fe-TiO_2 NPs showed promise for photocatalytic wastewater treatment and organic pollutant degradation (Solano et al. 2019).

4.11 Nano Adsorbents

This treatment employs highly adsorbent organic or inorganic nanoparticles. Nano adsorbents are extremely effective in removing a wide range of pollutants. The ideal absorbent is always small in size, has a large surface area, wide catalytic potential, and high reactivity. Metallic nanoparticles, magnetic nanoparticles, nanostructured mixed oxides, and metallic oxide nanoparticles are all types of nano adsorbents characterized by their adsorption method. For example, green nanoparticles were made from eucalyptus leaf extract and were used to treat swine water with high levels of nitrogen and phosphorus. Iron nanoparticles effectively eliminated total nitrogen (71.7%), total phosphorus (30.4%), and chemical oxygen demand (COD) (84.5%) from water, preventing eutrophication (Wang et al. 2014). Similarly, the ability of green-fabricated amorphous iron NPs to remove highly poisonous and carcinogenic arsenic (As) from damaged resources was also investigated. As a result, arsenate was shown to be equally adsorbed on the surfaces of iron NPs. This indicated that green synthesis of iron NPs has a great application potential for removing As(V) while also being simple to make (Wu et al. 2019).

Another work used polyphenol-rich pomegranate (*Punica granatum*) peel extract to make nickel–iron nanocomposite, which was then immobilized on to biocompatible and biodegradable alginate to make nanocomposite beads (GS-NiFe beads). In a batch reactor, 99 percent removal of tetracycline was achieved utilizing optimal conditions, with adsorption and degradation processes in the cleanup (Ravikumar et al. 2020). A green technique was used to make a novel Fe_3O_4 nanosorbent utilizing cucumber (*Cucumis sativus*), lemon (*Citrus limon*), and black grapes (*Vitis vinifera*) plant extracts. The as-prepared Fe_3O_4 (cum), Fe_3O_4 (lem), and Fe_3O_4 (grp) nanosorbents were used to remove seven antibiotics from water bodies, including piperacillin, sulfamethoxazole, tetracycline, trimethoprim, erythromycin, and ampicillin. The adsorption of selected antibiotics was observed, with an excellent clearance rate of >90% for the majority of these antibiotics (Stan et al. 2017).

4.12 Nanomembranes

Membranes are physical barriers that use their size to separate solutes from solvents. Furthermore, because wastewater comprises a variety of solid components of varying dimensions, membrane processes aid in the more efficient purification of water. Nanomembranes are used to remove contaminants from wastewater during the treatment. Heavy metals, dyes, and other pollutants are frequently removed with these devices. Nanotubes, nanoribbons, and nanofibers are some of the most frequent nanomembranes used. Membrane fouling, on the other hand, is a significant negative in the membrane filtration process, posing a substantial problem that calls into question the feasibility of membrane use (Nasrollahzadeh et al. 2021). Silver nanoparticles are antibacterial agents used in the treatment of wastewater that contains a lot of germs. Biofouling prevention (disable bacterial cells) and microbial attachment reduction (due to their hydrophilic nature) are two functions of silver nanoparticles and graphene oxide nanoparticles (Yu et al. 2021).

Depending on the pore size and molecule size, contaminants can be eliminated. Filtration capacity can be increased by employing the right chemicals, nanoparticles, or a mix of the two. For yeast filtration, Parham et al. (Parham et al. 2013) developed a CNT-ceramic composite filter. The constructed filter had a high filtration effectiveness for yeast (98%) and nearly 100% removal efficiency for heavy metal ions from water (Nair et al. 2012). Electrospun chitosan nanofibers can be used to remove arsenate traces from water. Electrospinning a polymer, poly (ethylene oxide), Fe_3+, and chitosan combination, followed by crosslinking with ammonia vapor, produces nanofibers. At neutral pH, this iron-doped chitosan nanofiber shows a high adsorption affinity for arsenic, i.e., As (V). The surface area of chitosan is increased by fabricating it over nanofibers. The adsorption rate of these nanofibers was boosted by the inclusion of functional groups such as –NH, –OH, and C–O. This is easily regenerated with NaOH solution and can be reused (Lee et al. 2013a, b). Another study found that electrospinning a single layer of polystyrene (PS) nanofibers onto a stainless-steel mesh produced superhydrophobic-superoleophilic membranes that help separate oils like mineral oil, diesel, and gasoline from water. Because of the entanglement of the nanofibers, all of the holes were interconnected, and porosity was increased. The non-wetting of water is enhanced by this three-dimensional network structure with low free energy PS, while the inherent wetting of oil is maintained. As a result, it is easier to separate oil from water (Min et al. 2016).

4.13 Conclusion

Nanoparticles can significantly influence the mode of wastewater treatment as these are efficient and low energy option in comparison to conventional chemical methods. Green synthesis of nanoparticles is considered a better alternative for the chemical and physical synthesis of nanomaterials. The biologically synthesized nanoparticles have immense potential to adsorb, degrade, catalyze and filter the major pollutants from wastewater. These green nanoparticles can easily be separated and reused because of their specific nature and high stability and hence are more suitable and economical to the low income countries. As these are biological in origin these are considered to be eco-friendly in nature. Yet their potential toxicity to living organisms still needs to be considered before implementation at large scale. Further, studies are required on different aspects like cost–benefit analysis, commercialization impact, environment constraints, challenges, safer application, the profitability, and feasibility, etc.

References

Trouillas P, Marsal P, Siri D, Lazzaroni R, Duroux J-L (2006) A DFT study of the reactivity of OH groups in quercetin and taxifolin antioxidants: the specificity of the 3-OH site. Food Chem 97:679–688. https://doi.org/10.1016/j.foodchem.2005.05.042

Abbasi T, Anuradha J, Ganaie SU, Abbasi SA (2015) Gainful utilization of the highly intransigent weed ipomoea in the synthesis of gold nanoparticles. Journal of King Saud University—Science 27:15–22. https://doi.org/10.1016/j.jksus.2014.04.001

Aboyewa JA, Sibuyi NRS, Meyer M, Oguntibeju OO (2021) Green synthesis of metallic nanoparticles using some selected medicinal plants from Southern Africa and their biological applications. Plants 10. https://doi.org/10.3390/plants10091929

Agency EP (2011) Water treatment manual: disinfection. Wexford, Ireland

Akuma P, Okagu OD, Udenigwe CC (2019) Naturally occurring exosome vesicles as potential delivery vehicle for bioactive compounds. Front Sustain Food Syst 3:23

Al-khattaf FS (2021) Gold and silver nanoparticles: green synthesis, microbes, mechanism, factors, plant disease management and environmental risks. Saudi J Biol Sci 28:3624–3631. https://doi.org/10.1016/j.sjbs.2021.03.078

Al-Qahtani KM (2017) Cadmium removal from aqueous solution by green synthesis zero valent silver nanoparticles with Benjamina leaves extract. Egypt J Aquat Res 43:269–274. https://doi.org/10.1016/j.ejar.2017.10.003

Ali I, Ding T, Peng C, Naz I, Sun H, Li J, Liu J (2021) Micro- and nanoplastics in wastewater treatment plants: occurrence, removal, fate, impacts and remediation technologies. A CritAl Rev Chem Eng J 423:130205. https://doi.org/10.1016/j.cej.2021.130205

Arabatzis I, Stergiopoulos T, Andreeva D, Kitova S, Neophytides S, Falaras P (2003) Characterization and photocatalytic activity of Au/TiO_2 thin films for azo-dye degradation. J Catal 220:127–135. https://doi.org/10.1016/S0021-9517(03)00241-0

Ariza-Tarazona MC, Villarreal-Chiu JF, Hernández-López JM, Rivera De la Rosa J, Barbieri V, Siligardi C, Cedillo-González EI (2020) Microplastic pollution reduction by a carbon and nitrogen-doped TiO_2: effect of pH and temperature in the photocatalytic degradation process. J Hazard Mater 395:122632. https://doi.org/10.1016/j.jhazmat.2020.122632
Awasthi U, Tiwari S (2004) Seasonal trends in abiotic factors of a lentic habitat (Govindgarh lake) Rewa, M.P., India. Ecol, Environ Conserv 10:165–170
Bagheri S, TermehYousefi A, Do T-O (2017) Photocatalytic pathway toward degradation of environmental pharmaceutical pollutants: structure, kinetics and mechanism approach. Catal Sci & Technol 7:4548–4569. https://doi.org/10.1039/C7CY00468K
Banerjee P, Satapathy M, Mukhopahayay A, Das P (2014) Leaf extract mediated green synthesis of silver nanoparticles from widely available Indian plants: synthesis, characterization, antimicrobial property and toxicity analysis. Bioresour Bioprocess 1:3. https://doi.org/10.1186/s40643-014-0003-y
Bar H, Bhui DK, Sahoo GP, Sarkar P, Pyne S, Chattopadhyay D, Misra A (2012) Synthesis of gold nanoparticles of variable morphologies using aqueous leaf extracts of Cocculus hirsutus. J Exp Nanosci 7:109–119. https://doi.org/10.1080/17458080.2010.509875
Begum NA, Mondal S, Basu S, Laskar RA, Mandal D (2009) Biogenic synthesis of Au and Ag nanoparticles using aqueous solutions of Black Tea leaf extracts. Colloids Surf B: Biointerfaces 71:113–118. https://doi.org/10.1016/j.colsurfb.2009.01.012
Bharathi D, Diviya Josebin M, Vasantharaj S, Bhuvaneshwari V (2018) Biosynthesis of silver nanoparticles using stem bark extracts of Diospyros montana and their antioxidant and antibacterial activities. J Nanostructure Chem 8:83–92. https://doi.org/10.1007/s40097-018-0256-7
Bhattacharya M et al (2012) Nanofibrillar cellulose hydrogel promotes three-dimensional liver cell culture. J Control Release 164:291–298. https://doi.org/10.1016/j.jconrel.2012.06.039
Bhuiyan MSH et al (2020) Green synthesis of iron oxide nanoparticle using Carica papaya leaf extract: application for photocatalytic degradation of remazol yellow RR dye and antibacterial activity. Heliyon 6. https://doi.org/10.1016/j.heliyon.2020.e04603
Cavalcante FTT et al. (2021) Designing of nanomaterials-based enzymatic biosensors: synthesis, properties, and applications. Electrochem 2. https://doi.org/10.3390/electrochem2010012
Chen X et al (2019) Plant-based nanoparticles prepared from proteins and phospholipids consisting of a core–multilayer-shell structure: fabrication, stability, and foamability. J Agric Food Chem 67:6574–6584. https://doi.org/10.1021/acs.jafc.9b02028
Choudhary M, Kataria J, Sharma S (2018) Evaluation of the kinetic and catalytic properties of biogenically synthesized silver nanoparticles. J Clean Prod. https://doi.org/10.1016/j.jclepro.2018.09.015
Clem Gruen L (1975) Interaction of amino acids with silver(I) ions. Biochim Biophys Acta (BBA)—Protein Struct 386:270–274. https://doi.org/10.1016/0005-2795(75)90268-8
Contado C (2015) Nanomaterials in consumer products: a challenging analytical problem. Front Chem 3:48
Corradini E (2010) A preliminary study of the incorporation of NPK fertilizer into chitosan nanoparticles. Express Polym Lett 4:509–515. https://doi.org/10.3144/expresspolymlett.2010.64
Dada AO et al. (2018) Effect of operational parameters, characterization and antibacterial studies of green synthesis of silver nanoparticles using Tithonia diversifolia. PeerJ 6:e5865. https://doi.org/10.7717/peerj.5865
Daisy P, Saipriya K (2012) Biochemical analysis of Cassia fistula aqueous extract and phytochemically synthesized gold nanoparticles as hypoglycemic treatment for diabetes mellitus. Int J Nanomedicine 7:1189–1202. https://doi.org/10.2147/IJN.S26650
Dangi S, Gupta A, Gupta DK, Singh S, Parajuli N (2020) Green synthesis of silver nanoparticles using aqueous root extract of Berberis asiatica and evaluation of their antibacterial activity. Chem Data Collect 28:100411. https://doi.org/10.1016/j.cdc.2020.100411

Dawson A, Kamat PV (2001) Semiconductor−metal nanocomposites. Photoinduced fusion and photocatalysis of gold-capped TiO_2 (TiO_2/gold) nanoparticles. J Phys Chem B 105:960–966. https://doi.org/10.1021/jp0033263
De Canha MN et al (2021) The activity of gold nanoparticles synthesized using Helichrysum odoratissimum against Cutibacterium acnes biofilms. Front Cell Dev Biol 9:2288
Devanesan S, AlSalhi MS (2021) Green synthesis of silver nanoparticles using the flower extract of Abelmoschus esculentus for cytotoxicity and antimicrobial studies. Int J Nanomedicine 16:3343–3356. https://doi.org/10.2147/IJN.S307676
Devatha CP, Thalla AK, Katte SY (2016) Green synthesis of iron nanoparticles using different leaf extracts for treatment of domestic waste water. J Clean Prod 139:1425–1435. https://doi.org/10.1016/j.jclepro.2016.09.019
Dharmatti R, Phadke C, Mewada A, Thakur M, Pandey S, Sharon M (2014) Biogenic gold nanotriangles: cargos for anticancer drug delivery. Mater Sci Eng C Mater Biol Appl 44:92–98. https://doi.org/10.1016/j.msec.2014.08.006
Di Gioia S, Hossain MN, Conese M (2020) Biological properties and therapeutic effects of plant-derived nanovesicles. Open Med 15:1096–1122. https://doi.org/10.1515/med-2020-0160
Dimapilis EAS, Hsu C-S, Mendoza RMO, Lu M-C (2018) Zinc oxide nanoparticles for water disinfection. Sustain Environ Res 28:47–56. https://doi.org/10.1016/j.serj.2017.10.001
Dobrucka R (2017) Synthesis of titanium dioxide nanoparticles using Echinacea purpurea Herba. Iran J Pharm Res 16:756–762
Duclairoir C, Orecchioni AM, Depraetere P, Nakache E (2002) Alpha-tocopherol encapsulation and in vitro release from wheat gliadin nanoparticles. J Microencapsul 19:53–60. https://doi.org/10.1080/02652040110055207
Ehrampoush MH, Miria M, Salmani MH, Mahvi AH (2015) Cadmium removal from aqueous solution by green synthesis iron oxide nanoparticles with tangerine peel extract. J Environ Health Sci Eng 13
El-Kassas HY, Aly-Eldeen MA, Gharib SM (2016) Green synthesis of iron oxide (Fe_3O_4) nanoparticles using two selected brown seaweeds: characterization and application for lead bioremediation. Acta Ocean Sin 35:89–98. https://doi.org/10.1007/s13131-016-0880-3
Fanoro OT, Parani S, Maluleke R, Lebepe TC, Varghese JR, Mavumengwana V, Oluwafemi OS (2021) Facile green, room-temperature synthesis of gold nanoparticles using Combretum erythrophyllum leaf extract: antibacterial and cell viability studies against normal and cancerous cells. Antibiotics 10. https://doi.org/10.3390/antibiotics10080893
Fernandes F, Dias-Teixeira M, Delerue-Matos C, Grosso C (2021) Critical review of lipid-based nanoparticles as carriers of neuroprotective drugs and extracts. Nanomaterials 11. https://doi.org/10.3390/nano11030563
Fincheira P, Quiroz A, Medina C, Tortella G, Hermosilla E, Diez MC, Rubilar O (2020) Plant growth induction by volatile organic compound released from solid lipid nanoparticles and nanostructured lipid carriers. Colloids Surf A: Physicochemical Eng Asp 596:124739. https://doi.org/10.1016/j.colsurfa.2020.124739
Gawande MB et al. (2011) Synthesis and characterization of versatile MgO–ZrO_2 mixed metal oxide nanoparticles and their applications. Catal Sci Technol 1:1653–1664. https://doi.org/10.1039/C1CY00259G
Goutam SP, Saxena G, Singh V, Yadav AK, Bharagava RN, Thapa KB (2018) Green synthesis of TiO_2 nanoparticles using leaf extract of Jatropha curcas L. for photocatalytic degradation of tannery wastewater. Chem Eng J 336:386–396. https://doi.org/10.1016/j.cej.2017.12.029
Gulfam M, Kim JE, Lee JM, Ku B, Chung BH, Chung BG (2012) Anticancer drug-loaded gliadin nanoparticles induce apoptosis in breast cancer cells. Langmuir 28:8216–8223. https://doi.org/10.1021/la300691n
Harman W, Starr R, Carter M, Tweedy K, Clemmons M, Suggs K, Miller C (2012) A function-based framework for stream assessment and restoration projects, vol EPA 843-K-12-006. Washington, DC, USA

Hasan N et al (2020) Recent advances of nanocellulose in drug delivery systems. J Pharm Investig 50:553–572. https://doi.org/10.1007/s40005-020-00499-4

Hassanisaadi M, Bonjar GH, Rahdar A, Pandey S, Hosseinipour A, Abdolshahi R (2021) Environmentally safe biosynthesis of gold nanoparticles using plant water extracts. Nanomaterials 11. https://doi.org/10.3390/nano11082033

Hemlata, Meena PR, Singh AP, Tejavath KK (2020) Biosynthesis of silver nanoparticles using cucumis prophetarum aqueous leaf extract and their antibacterial and antiproliferative activity against cancer cell lines. ACS Omega 5:5520–5528. https://doi.org/10.1021/acsomega.0c00155

Hong S, Choi DW, Kim HN, Park CG, Lee W, Park HH (2020) Protein-based nanoparticles as drug delivery systems. Pharmaceutics 12. https://doi.org/10.3390/pharmaceutics12070604

Hristozov D, Malsch I (2009) Hazards and risks of engineered nanoparticles for the environment and human health. Sustainability 1:1161–1194

Hu Y, Gaillard PJ, de Lange ECM, Hammarlund-Udenaes M (2019) Targeted brain delivery of methotrexate by glutathione PEGylated liposomes: how can the formulation make a difference? Eur J Pharm Biopharm 139:197–204. https://doi.org/10.1016/j.ejpb.2019.04.004

Huang J et al (2011) Biogenic silver nanoparticles by Cacumen platycladi extract: synthesis, formation mechanism, and antibacterial activity. Ind Eng Chem Res 50:9095–9106. https://doi.org/10.1021/ie200858y

Hutchison JE (2008) Greener nanoscience: a proactive approach to advancing applications and reducing implications of nanotechnology. ACS Nano 2:395–402. https://doi.org/10.1021/nn800131j

Ilahi I et al (2021) Synthesis of silver nanoparticles using root extract of Duchesnea indica and assessment of its biological activities. Arab J Chem 14:103110. https://doi.org/10.1016/j.arabjc.2021.103110

Iravani S, Varma RS (2019) Plant-derived edible nanoparticles and miRNAs: emerging frontier for therapeutics and targeted drug-delivery. ACS Sustain Chem Eng 7:8055–8069. https://doi.org/10.1021/acssuschemeng.9b00954

Jain S, Mehata MS (2017) Medicinal plant leaf extract and pure flavonoid mediated green synthesis of silver nanoparticles and their enhanced antibacterial property. Scientific Reports 7:15867. https://doi.org/10.1038/s41598-017-15724-8

Jeevanandam J, Chan YS, Danquah MK (2020) Effect of pH variations on morphological transformation of biosynthesized MgO nanoparticles. Part Sci Technol 38:573–586. https://doi.org/10.1080/02726351.2019.1566938

Joye IJ, Nelis VA, McClements DJ (2015) Gliadin-based nanoparticles: Fabrication and stability of food-grade colloidal delivery systems. Food Hydrocoll 44:86–93. https://doi.org/10.1016/j.foodhyd.2014.09.008

Ju S et al (2013) Grape exosome-like nanoparticles induce intestinal stem cells and protect mice from DSS-induced colitis. Mol Ther 21:1345–1357. https://doi.org/10.1038/mt.2013.64

Kajal H, Misra A (2011) Preparation of tetanus toxoid and ovalbumin loaded gliadin nanoparticles for oral immunization. J Biomed Nanotechnol 7:211–212. https://doi.org/10.1166/jbn.2011.1273

Karthiga P (2018) Preparation of silver nanoparticles by garcinia mangostana stem extract and investigation of the antimicrobial properties. Biotechnol Res Innov 2:30–36. https://doi.org/10.1016/j.biori.2017.11.001

Karthikeyan K, Krishnaswamy VR, Lakra R, Kiran MS, Korrapati PS (2015) Fabrication of electrospun zein nanofibers for the sustained delivery of siRNA. J Mater Sci Mater Med 26:101. https://doi.org/10.1007/s10856-015-5439-x

Khan I, Saeed K, Khan I (2019) nanoparticles: properties, applications and toxicities. Arab J Chem 12:908–931. https://doi.org/10.1016/j.arabjc.2017.05.011

Kim J, Li S, Zhang S, Wang J (2021) Plant-derived exosome-like nanoparticles and their therapeutic activities. Asian J Pharm Sci. https://doi.org/10.1016/j.ajps.2021.05.006

Kowalczyk P et al. (2021) All that glitters is not silver—a new look at microbiological and medical applications of silver nanoparticles. Int J Mol Sci 22:854

Kredy H (2018) The effect of pH, temperature on the green synthesis and biochemical activities of silver nanoparticles from Lawsonia inermis extract. J Pharm Sci Res 10:2022–2026

Krishnaraj C, Jagan EG, Rajasekar S, Selvakumar P, Kalaichelvan PT, Mohan N (2010a) Synthesis of silver nanoparticles using Acalypha indica leaf extracts and its antibacterial activity against water borne pathogens. Colloids Surf B: Biointerfaces 76:50–56. https://doi.org/10.1016/j.colsurfb.2009.10.008

Krishnaraj C, Jagan EG, Rajasekar S, Selvakumar P, Kalaichelvan PT, Mohan N (2010b) Synthesis of silver nanoparticles using Acalypha indica leaf extracts and its antibacterial activity against water borne pathogens. Colloids Surf B: Biointerfaces 76:50–56. https://doi.org/10.1016/j.colsurfb.2009.10.008

Krithiga N, Rajalakshmi A, Jayachitra A (2015) Green synthesis of silver nanoparticles using leaf extracts of *Clitoria ternatea* and *Solanum nigrum* and study of its antibacterial effect against common nosocomial pathogens. J Nanosci 2015:928204. https://doi.org/10.1155/2015/928204

Kunduru KR, Nazarkovsky M, Farah S, Pawar RP, Basu A, Domb AJ (2017) 2—Nanotechnology for water purification: applications of nanotechnology methods in wastewater treatment. In: Grumezescu AM (ed) Water purification. Academic Press, pp 33–74. https://doi.org/10.1016/B978-0-12-804300-4.00002-2

Kupnik K, Primožič M, Kokol V, Leitgeb M (2020) Nanocellulose in drug delivery and antimicrobially active materials. Polymers 12. https://doi.org/10.3390/polym12122825

Lakshmanan G, Sathiyaseelan A, Kalaichelvan PT, Murugesan K (2018) Plant-mediated synthesis of silver nanoparticles using fruit extract of Cleome viscosa L.: Assessment of their antibacterial and anticancer activity. Karbala Int J Mod Sci 4:61–68. https://doi.org/10.1016/j.kijoms.2017.10.007

Lee MW, An S, Latthe SS, Lee C, Hong S, Yoon SS (2013a) Electrospun polystyrene nanofiber membrane with superhydrophobicity and superoleophilicity for selective separation of water and low viscous oil. ACS Appl Mater Interfaces 5:10597–10604. https://doi.org/10.1021/am404156k

Gardea-Torresdey JL, Parsons JG, Gomez E, Peralta-Videa J, Troiani HE, Santiago P, Yacaman MJ (2002) Formation and growth of Au nanoparticles inside live alfalfa plants. Nano Lett 2:397–401. https://doi.org/10.1021/nl015673+

Lee S, Alwahab NSA, Moazzam ZM (2013b) Zein-based oral drug delivery system targeting activated macrophages. Int J Pharm 454:388–393. https://doi.org/10.1016/j.ijpharm.2013.07.026

Li G, Zhao H, Hong J, Quan K, Yuan Q, Wang X (2017) Antifungal graphene oxide-borneol composite. Colloids Surf B: Biointerfaces 160:220–227. https://doi.org/10.1016/j.colsurfb.2017.09.023

Lin N, Dufresne A (2014) Nanocellulose in biomedicine: current status and future prospect. Eur Polymer J 59:302–325. https://doi.org/10.1016/j.eurpolymj.2014.07.025

Liu X, Ma R, Zhuang L, Hu B, Chen J, Liu X, Wang X (2021) Recent developments of doped g-C3N4 photocatalysts for the degradation of organic pollutants. Crital Rev Environ Sci Technol 51:751–790. https://doi.org/10.1080/10643389.2020.1734433

Lizundia E, Sipponen MH, Greca LG, Balakshin M, Tardy BL, Rojas OJ, Puglia D (2021) Multifunctional lignin-based nanocomposites and nanohybrids. Green Chem 23:6698–6760. https://doi.org/10.1039/D1GC01684A

Lohcharoenkal W, Wang L, Chen YC, Rojanasakul Y (2014) Protein nanoparticles as drug delivery carriers for cancer therapy. Biomed Res Int 2014:180549. https://doi.org/10.1155/2014/180549

Low LE et al (2021) Lignin nanoparticles: the next green nanoreinforcer with wide opportunity. Environ Nanotechnol Monit Manag 15:100398. https://doi.org/10.1016/j.enmm.2020.100398

Magnabosco G et al (2020) A plant bioreactor for the synthesis of carbon nanotube bionic nanocomposites. Front Bioeng Biotechnol 8:1287

Magne TM et al. (2021) Graphene and its derivatives: understanding the main chemical and medicinal chemistry roles for biomedical applications. J Nanostructure Chem:1–35. https://doi.org/10.1007/s40097-021-00444-3

Mahmoodi Esfanddarani H, Abbasi Kajani A, Bordbar AK (2018) Green synthesis of silver nanoparticles using flower extract of Malva sylvestris and investigation of their antibacterial activity. IET Nanobiotechnol 12:412–416. https://doi.org/10.1049/iet-nbt.2017.0166

Makarov VV, Love AJ, Sinitsyna OV, Makarova SS, Yaminsky IV, Taliansky ME, Kalinina NO (2014) "Green" nanotechnologies: synthesis of metal nanoparticles using plants. Acta Naturae 6:35–44

Mandal S, Selvakannan PR, Phadtare S, Pasricha R, Sastry M (2002) Synthesis of a stable gold hydrosol by the reduction of chloroaurate ions by the amino acid, aspartic acid. J Chem Sci 114:513–520. https://doi.org/10.1007/BF02704195

Masum MMI et al. (2019) Biogenic synthesis of silver nanoparticles using Phyllanthus emblica fruit extract and its inhibitory action against the pathogen Acidovorax oryzae strain RS-2 of rice bacterial brown stripe. Front Microbiol 10. https://doi.org/10.3389/fmicb.2019.00820

Mathew S, Victório CP, Sidhi MSJ, Baby Thanzeela BT (2020) Biosynthesis of silver nanoparticle using flowers of Calotropis gigantea (L.) W.T. Aiton and activity against pathogenic bacteria. Arab J Chem 13:9139–9144. https://doi.org/10.1016/j.arabjc.2020.10.038

Min L-L, Zhong L-B, Zheng Y-M, Liu Q, Yuan Z-H, Yang L-M (2016) Functionalized chitosan electrospun nanofiber for effective removal of trace arsenate from water. Sci Rep 6:32480. https://doi.org/10.1038/srep32480

Mirshahi T, Irache JM, Gueguen J, Orecchioni AM (1996) Development of drug delivery systems from vegetal proteins: legumin nanoparticles. Drug Dev Ind Pharm 22:841–846. https://doi.org/10.3109/03639049609065914

Mishra S, Kumar A, Shukla P (2016) Study of water quality in Hindon River using pollution index and environmetrics, India. Desalination and Water Treatment 57:19121–19130. https://doi.org/10.1080/19443994.2015.1098570

Mishra S, Kumar D, Yadav S, Singhal M (2015) Assessment of heavy metal contamination in Kali river, Uttar Pradesh, India. J Appl Nat Sci 7:1016–1020. https://doi.org/10.31018/jans.v7i2.724

Mittal AK, Chisti Y, Banerjee UC (2013) Synthesis of metallic nanoparticles using plant extracts. Biotechnol Adv 31:346–356. https://doi.org/10.1016/j.biotechadv.2013.01.003

Mohammadinejad R, Karimi S, Iravani S, Varma RS (2016) Plant-derived nanostructures: types and applications. Green Chem 18:20–52. https://doi.org/10.1039/C5GC01403D

Morozova OV (2021) Silver nanostructures: limited sensitivity of detection, toxicity and anti-inflammation effects. Int J Mol Sci 22:9928

Mostafa HS (2021) Banana plant as a source of valuable antimicrobial compounds and its current applications in the food sector. J Food Sci 86:3778–3797. https://doi.org/10.1111/1750-3841.15854

Muramatsu H et al (2014) Rice husk-derived graphene with nano-sized domains and clean edges. Small 10:2766–2770. https://doi.org/10.1002/smll.201400017

Murthy HCA, Desalegn T, Kassa M, Abebe B, Assefa T (2020) Synthesis of green copper nanoparticles using medicinal plant Hagenia abyssinica (Brace) JF. Gmel. Leaf extract: antimicrobial properties. J Nanomater 2020:3924081. https://doi.org/10.1155/2020/3924081

Nair RR, Wu HA, Jayaram PN, Grigorieva IV, Geim AK (2012) Unimpeded permeation of water through helium-leak–tight graphene-based membranes. Science 335:442–444. https://doi.org/10.1126/science.1211694

Naseer M, Aslam U, Khalid B, Chen B (2020) Green route to synthesize zinc oxide nanoparticles using leaf extracts of Cassia fistula and Melia azedarach and their antibacterial potential. Sci Rep 10:9055. https://doi.org/10.1038/s41598-020-65949-3

Nasrollahzadeh M, Sajjadi M, Iravani S, Varma RS (2021) Green-synthesized nanocatalysts and nanomaterials for water treatment: Current challenges and future perspectives. J Hazard Mater 401:123401. https://doi.org/10.1016/j.jhazmat.2020.123401

Nasrollahzadeh M, Sajjadi M, Sajadi SM, Issaabadi Z (2019) Chapter 5—Green nanotechnology. In: Nasrollahzadeh M, Sajadi SM, Sajjadi M, Issaabadi Z, Atarod M (eds) Interface science and technology, vol 28. Elsevier, pp 145–198. https://doi.org/10.1016/B978-0-12-813586-0.00005-5

Niluxsshun MCD, Masilamani K, Mathiventhan U (2021) Green synthesis of silver nanoparticles from the extracts of fruit peel of Citrus tangerina, Citrus sinensis, and Citrus limon for antibacterial activities. Bioinorg Chem Appl 2021:6695734. https://doi.org/10.1155/2021/6695734

Orlov A, Jefferson D, Macleod N, Lambert R (2004) Photocatalytic properties of TiO_2 modified with gold nanoparticles in the degradation of 4-chlorophenol in aqueous solution. Catal Lett 92:41–47. https://doi.org/10.1023/B:CATL.0000011084.43007.80

Ozkan ZY, Cakirgoz M, Kaymak ES, Erdim E (2018) Rapid decolorization of textile wastewater by green synthesized iron nanoparticles. Water Sci Technol 77:511–517

Paço A et al (2017) Biodegradation of polyethylene microplastics by the marine fungus Zalerion maritimum. Sci Total Environ 586:10–15. https://doi.org/10.1016/j.scitotenv.2017.02.017

Padalia H, Moteriya P, Chanda S (2015) Green synthesis of silver nanoparticles from marigold flower and its synergistic antimicrobial potential. Arab J Chem 8:732–741. https://doi.org/10.1016/j.arabjc.2014.11.015

Paliwal R, Palakurthi S (2014) Zein in controlled drug delivery and tissue engineering. J Control Release 189:108–122. https://doi.org/10.1016/j.jconrel.2014.06.036

Pandey S, Thakur M, Shah R, Oza G, Mewada A, Sharon M (2013) A comparative study of economical separation and aggregation properties of biologically capped and thiol functionalized gold nanoparticles: selecting the eco-friendly trojan horses for biological applications. Colloids Surf B: Biointerfaces 109:25–31. https://doi.org/10.1016/j.colsurfb.2013.03.011

Panigrahi S, Kundu S, Ghosh S, Nath S, Pal T (2004) General method of synthesis for metal nanoparticles. J Nanopart Res 6:411–414. https://doi.org/10.1007/s11051-004-6575-2

Parham H, Bates S, Xia Y, Zhu Y (2013) A highly efficient and versatile carbon nanotube/ceramic composite filter. Carbon 54:215–223. https://doi.org/10.1016/j.carbon.2012.11.032

Patel A, Hu Y, Tiwari JK, Velikov KP (2010) Synthesis and characterisation of zein–curcumin colloidal particles. Soft Matter 6:6192–6199. https://doi.org/10.1039/C0SM00800A

Patra JK et al (2018) Nano based drug delivery systems: recent developments and future prospects. J Nanobiotechnology 16:71–71. https://doi.org/10.1186/s12951-018-0392-8

Pereira DR, Silva-Correia J, Oliveira JM, Reis RL, Pandit A, Biggs MJ (2018) Nanocellulose reinforced gellan-gum hydrogels as potential biological substitutes for annulus fibrosus tissue regeneration. Nanomedicine 14:897–908. https://doi.org/10.1016/j.nano.2017.11.011

Pham T-H, Do H-T, Phan Thi L-A, Singh P, Raizada P, Chi-Sheng WuJ, Nguyen V-H (2021) Global challenges in microplastics: from fundamental understanding to advanced degradations toward sustainable strategies. Chemosphere 267:129275. https://doi.org/10.1016/j.chemosphere.2020.129275

Qu X, Alvarez PJJ, Li Q (2013) Applications of nanotechnology in water and wastewater treatment. Water Res 47:3931–3946. https://doi.org/10.1016/j.watres.2012.09.058

Rahman TU, Khan H, Liaqat W, Zeb MA (2021) Phytochemical screening, green synthesis of gold nanoparticles, and antibacterial activity using seeds extract of Ricinus communis L. Microsc Res Tech 85:202–208. https://doi.org/10.1002/jemt.23896

Rao Y, Inwati GK, Singh M (2017) Green synthesis of capped gold nanoparticles and their effect on Gram-positive and Gram-negative bacteria. Futur Sci OA 3:FSO239. https://doi.org/10.4155/fsoa-2017-0062

Ravikumar KVG et al (2020) Batch and column study on tetracycline removal using green synthesized NiFe nanoparticles immobilized alginate beads. Environ Technol Innov 17:100520. https://doi.org/10.1016/j.eti.2019.100520

Karny A, Zinger A, Kajal A, Shainsky-Roitman J, Schroeder A (2018) Therapeutic nanoparticles penetrate leaves and deliver nutrients to agricultural crops. Sci Rep 8:7589. https://doi.org/10.1038/s41598-018-25197-y

Ray PC, Yu H, Fu PP (2009) Toxicity and environmental risks of nanomaterials: challenges and future needs. J Environ Sci Health, Part C 27:1–35. https://doi.org/10.1080/10590500802708267

Rojas J (2015) Current trends in the production of cellulose nanoparticles and nanocomposites for biomedical applications. In: Cellulose—fundamental aspects and current trends. InTech

Saha I, Bhattacharya S, Mukhopadhyay A, Chattopadhyay D, Ghosh U, Chatterjee D (2013) Role of nanotechnology in water treatment and purification: potential applications and implications. Int J Chem Sci Technol 3(3):59–64

Salimi S, Sotudeh-Gharebagh R, Zarghami R, Chan SY, Yuen KH (2019) Production of nanocellulose and its applications in drug delivery: a critical review. ACS Sustain Chem Eng 7:15800–15827. https://doi.org/10.1021/acssuschemeng.9b02744

Shak KPY, Pang YL, Mah SK (2018) Nanocellulose: recent advances and its prospects in environmental remediation. Beilstein J Nanotechnol 9:2479–2498. https://doi.org/10.3762/bjnano.9.232

Shamaila S, Sajjad AKL, Ryma N-u-A, Farooqi SA, Jabeen N, Majeed S, Farooq I (2016) Advancements in nanoparticle fabrication by hazard free eco-friendly green routes. Appl Mater Today 5:150–199. https://doi.org/10.1016/j.apmt.2016.09.009

Shang Y, Hasan MK, Ahammed GJ, Li M, Yin H, Zhou J (2019) Applications of nanotechnology in plant growth and crop protection: a review. Molecules 24 https://doi.org/10.3390/molecules24142558

Shankar S, Jaiswal L, Aparna RSL, Prasad RGSV (2014) Synthesis, characterization, in vitro biocompatibility, and antimicrobial activity of gold, silver and gold silver alloy nanoparticles prepared from Lansium domesticum fruit peel extract. Materials Letters 137:75–78. https://doi.org/10.1016/j.matlet.2014.08.122

Shankar SS, Ahmad A, Pasricha R, Sastry M (2003) Bioreduction of chloroaurate ions by geranium leaves and its endophytic fungus yields gold nanoparticles of different shapes. J Mater Chem 13:1822–1826. https://doi.org/10.1039/B303808B

Shanker U, Jassal V, Rani M, Kaith BS (2016) Towards green synthesis of nanoparticles: from bio-assisted sources to benign solvents. A Rev Int J Environ Anal Chem 96:801–835

Sharma S, Basu S, Shetti NP, Nadagouda MN, Aminabhavi TM (2021) Microplastics in the environment: occurrence, perils, and eradication. Chem Eng J 408:127317. https://doi.org/10.1016/j.cej.2020.127317

Shittu O, Ihebunna O (2017) Purification of simulated waste water using green synthesized silver nanoparticles of Piliostigma thonningii aqueous leave extract. Adv Nat Sci: Nanosci Nanotechnol 8:045003. https://doi.org/10.1088/2043-6254/aa8536

Singh AK, Talat M, Singh DP, Srivastava ON (2010) Biosynthesis of gold and silver nanoparticles by natural precursor clove and their functionalization with amine group. J Nanopart Res 12:1667–1675. https://doi.org/10.1007/s11051-009-9835-3

Singh S, Kotia A, Lodha P, Hassan M (2020) Green synthesis of silver nanoparticles using root extract of Aerva tomentosa Forsk. And analyse antibacterial and antioxidant property. Int J Pharm Sci Res 13:6456–6462. https://doi.org/10.13040/IJPSR.0975-8232.11(12).6456-62

Singh S, Kumar V, Romero R, Sharma K, Singh J (2019) Applications of nanoparticles in wastewater treatment. In: Prasad R, Kumar V, Kumar M, Choudhary D (eds) Nanobiotechnology in bioformulations. Springer International, Cham, pp 395–418. https://doi.org/10.1007/978-3-030-17061-5_17

Sinha T, Ahmaruzzaman M (2015) Biogenic synthesis of Cu nanoparticles and its degradation behavior for methyl red. Mater Lett 159:168–171. https://doi.org/10.1016/j.matlet.2015.06.099

Solano RA, Herrera AP, Maestre D, Cremades A (2019) Fe-TiO_2 nanoparticles synthesized by green chemistry for potential application in waste water photocatalytic treatment. J Nanotechnol 2019:4571848. https://doi.org/10.1155/2019/4571848

Stan M et al (2017) Removal of antibiotics from aqueous solutions by green synthesized magnetite nanoparticles with selected agro-waste extracts. Process Saf Environ Prot 107:357–372. https://doi.org/10.1016/j.psep.2017.03.003

Subha V, Ravindran RSE, Preethi R, Cyrril J, Kirubanandan S, Renganathan S (2018) Silver nanoparticles—green synthesis with Aq. extract of stems Ipomoea pes-caprae, characterization, antimicrobial and anti-cancer potential. Int J Med Nano Res 5:024. https://doi.org/10.23937/2378-3664/1410024

Subramanian V, Wolf EE, Kamat PV (2004) Catalysis with TiO2/gold nanocomposites Effect of metal particle size on the fermi level equilibration. J Am Chem Soc 126:4943–4950. https://doi.org/10.1021/ja0315199
Suharta S et al. (2021) Plant-derived exosome-like nanoparticles: a concise review on its extraction methods, content, bioactivities, and potential as functional food ingredient. J Food Sci 86:2838–2850. https://doi.org/10.1111/1750-3841.15787
Sujitha MV, Kannan S (2013) Green synthesis of gold nanoparticles using Citrus fruits (Citrus limon, Citrus reticulata and Citrus sinensis) aqueous extract and its characterization. Spectrochim Acta Part A: Mol Biomol Spectrosc 102:15–23. https://doi.org/10.1016/j.saa.2012.09.042
Sushma D, Richa S (2015) Use of nanoparticles in water treatment: a review. Int Res J Environ Sci 4:103–106
Tan YN, Lee JY, Wang DIC (2010) Uncovering the design rules for peptide synthesis of metal nanoparticles. J Am Chem Soc 132:5677–5686. https://doi.org/10.1021/ja907454f
Tenchov R, Bird R, Curtze AE, Zhou Q (2021) Lipid nanoparticles—from liposomes to mRNA vaccine delivery, a landscape of research diversity and advancement. ACS Nano. https://doi.org/10.1021/acsnano.1c04996
Théry C, Ostrowski M, Segura E (2009) Membrane vesicles as conveyors of immune responses. Nat Rev Immunol 9:581–593. https://doi.org/10.1038/nri2567
Tofa TS, Kunjali KL, Paul S, Dutta J (2019a) Visible light photocatalytic degradation of microplastic residues with zinc oxide nanorods. Environ Chem Lett 17:1341–1346. https://doi.org/10.1007/s10311-019-00859-z
Tofa TS, Ye F, Kunjali KL, Dutta J (2019b) Enhanced visible light photodegradation of microplastic fragments with plasmonic platinum/zinc oxide nanorod photocatalysts. Catalysts 9. https://doi.org/10.3390/catal9100819
Traiwatcharanon P, Timsorn K, Wongchoosuk C (2015) Effect of pH on the green synthesis of silver nanoparticles through reduction with Pistiastratiotes L extract. Adv Mater Res 1131:223–226. https://doi.org/10.4028/www.scientific.net/AMR.1131.223
Tshikovhi A, Mishra SB, Mishra AK (2020) Nanocellulose-based composites for the removal of contaminants from wastewater. Int J Biol Macromol 152:616–632. https://doi.org/10.1016/j.ijbiomac.2020.02.221
Velusamy P, Kumar GV, Jeyanthi V, Das J, Pachaiappan R (2016) Bio-inspired green nanoparticles: synthesis, mechanism, and antibacterial application. Toxicol Res 32:95–102. https://doi.org/10.5487/TR.2016.32.2.095
Venkateswarlu S, Lee D, Yoon M (2016) Bioinspired 2D-carbon flakes and Fe_3O_4 nanoparticles composite for arsenite removal. ACS Appl Mater Interfaces 8:23876–23885. https://doi.org/10.1021/acsami.6b03583
Verma ML, Dhanya BS, Thakur M, Jeslin J, Jana AK (2021) Chapter 8—Plant derived nanoparticles and their biotechnological applications. In: Verma SK, Das AK (eds) Comprehensive analytical chemistry, vol 94. Elsevier, pp 331–362. https://doi.org/10.1016/bs.coac.2021.01.011
Waghmare RD et al (2021) Carbon nanodots derived from kitchen waste biomass as a growth accelerator for fenugreek plant. J Nanosci Nanotechnol 21:2234–2245. https://doi.org/10.1166/jnn.2021.19028
Wang T, Jin X, Chen Z, Megharaj M, Naidu R (2014) Green synthesis of Fe nanoparticles using eucalyptus leaf extracts for treatment of eutrophic wastewater. Sci Total Environ 466–467:210–213. https://doi.org/10.1016/j.scitotenv.2013.07.022
Wu Z, Su X, Lin Z, Owens G, Chen Z (2019) Mechanism of As(V) removal by green synthesized iron nanoparticles. J Hazard Mater 379:120811
Wang H et al. (2018) Visible-light-driven removal of tetracycline antibiotics and reclamation of hydrogen energy from natural water matrices and wastewater by polymeric carbon nitride foam. Water Res 144:215–225
Xu H, Jiang Q, Reddy N, Yang Y (2011) Hollow nanoparticles from zein for potential medical applications. J Mater Chem 21:18227–18235. https://doi.org/10.1039/C1JM11163A

Xu H, Shen L, Xu L, Yang Y (2015) Controlled delivery of hollow corn protein nanoparticles via non-toxic crosslinking: in vivo and drug loading study. Biomed Microdevice 17:8. https://doi.org/10.1007/s10544-014-9926-5

Yang C, Zhang M, Merlin D (2018) Advances in plant-derived edible nanoparticle-based lipid nano-drug delivery systems as therapeutic nanomedicines. J Mater Chem B 6:1312–1321. https://doi.org/10.1039/C7TB03207B

Yao L, Yang H, Chen Z, Qiu M, Hu B, Wang X (2021) Bismuth oxychloride-based materials for the removal of organic pollutants in wastewater. Chemosphere 273:128576. https://doi.org/10.1016/j.chemosphere.2020.128576

Yu Y, Yang Y, Yu L, Koh KY, Chen JP (2021) Modification of polyvinylidene fluoride membrane by silver nanoparticles-graphene oxide hybrid nanosheet for effective membrane biofouling mitigation. Chemosphere 268:129187. https://doi.org/10.1016/j.chemosphere.2020.129187

Yuan C-G, Huo C, Gui B, Cao W-P (2017) Green synthesis of gold nanoparticles using Citrus maxima peel extract and their catalytic/antibacterial activities. IET Nanobiotechnol 11:523–530. https://doi.org/10.1049/iet-nbt.2016.0183

Zhang M, Viennois E, Xu C, Merlin D (2016) Plant derived edible nanoparticles as a new therapeutic approach against diseases. Tissue Barriers 4:e1134415. https://doi.org/10.1080/21688370.2015.1134415

Chapter 5
Microbial Synthesis of Nanoparticles for Wastewater Remediation

Shailja Singh, Shiv Shankar, and Shikha

Abstract The wastewater originating from the industrial and domestic sector with potential hazardous organic and inorganic pollutants when discharged into the water bodies causes undesirable effects on the aquatic environment and human health. It is therefore desirable to treat wastewater prior to its disposal. Recently, the application of nanostructured materials for water remediation has gained attention due to their unique size-dependent properties like large surface area, stability, remarkable reusability and recyclability. Nanoparticles represent a promising new technology for wastewater remediation, not only because of their high treatment efficiency but also for their cost-effectiveness, as they have the flexibility for in-situ and ex-situ applications. Several conventional physical and chemical procedures have been reported for the synthesis of metallic and non-metallic nanoparticles. But the environmental hazards associated with the conventional methods restrict the large-scale production and application in water remediation. It is therefore desirable to synthesize nanoparticles using environmentally safer, rapid and inexpensive biogenic approaches based on microbes. Carbohydrates, proteins, polyphenols, vitamins, polymeric substances and several antioxidants obtained from bacteria, fungi and algae have proven their effectiveness as capping and stabilizing agents during greener synthesis of nanomaterials. Application of microbially synthesized nanomaterials for wastewater treatment is a relatively newer but rapidly escalating area of research. The present book chapter outlines the information on recent advances on the microbial synthesis of different types of metallic and non-metallic nanoparticles, their characterization and role in the removal of different types of organic and inorganic pollutants from wastewater.

S. Singh · Shikha (✉)
Department of Environmental Science, School of Earth and Environmental Sciences, Babasaheb Bhimrao Ambedkar University (A Central University), Vidya Vihar, Raebareli Road, Lucknow, Uttar Pradesh 226025, India
e-mail: dr_shikha2003@yahoo.co.in

S. Shankar
Department of Environmental Science, School of Vocational Studies and Applied Sciences, Gautam Buddha University (A State University), Yamuna Expressway, Gautam Buddha Nagar, Greater Noida, Uttar Pradesh 201312, India

J. P. N. Rai and S. Saraswat (eds.), *Nano-biotechnology for Waste Water Treatment*, Water Science and Technology Library 111,
https://doi.org/10.1007/978-3-031-00812-2_5

Keywords Microbial synthesis · Nanoparticles · Wastewater · Remediation · Recyclability · Cost-effectiveness · Non-metallic nanoparticles

5.1 Introduction

The present situation of pollution in aqueous streams has become a major problem for the society (Lu et al. 2016; Moustafa 2017). Water is very essential for the survival of the living beings as well as for the industrial and economic growth of the country. It has been estimated that around 1.1 billion people do not have access to adequate drinking water (Cassivi et al. 2021; Chen et al. 2021; Price et al. 2021). This is due to the increasing cost of potable water, rapidly growing populations and various climatic and environmental concerns. Many harmful chemicals such as pesticides, fertilizers, insecticides, organic compounds, heavy metals, industrial effluents and sewage wastewater are the major environmental contaminants (El-Gendy and Nassar 2021; Mandeep 2020). These contaminants are disposed of in fresh water bodies such as water streams and rivers which leads to water pollution and affects the aquatic ecosystem. The effluent coming out of industrial units are disposed directly in the rivers without treatment due to inefficient wastewater treatment plant. The use of synthetic dyes in excess amount for various industrial activities is also responsible for water as well as soil pollution.

Modern ways of industrialization has increased the tremendous pressure for water usage due to its application in production process. High production resulted in generation of enormous amount of industrial effluents and it requires sustainable and cost-effective technique for achieving sustainable development in case of wastewater treatment. Many electrochemical, valourization techniques and advanced oxidation processes have been utilized to decrease the toxicity of effluents from wastewater and making it fit for drinking and household purpose (Lu et al. 2016). But the techniques mentioned here are not very cost-effective for many industries. Moreover, these conventional methods of wastewater treatment has certain other limitations like energy intensive, excessive cost, requirement of high temperature and pressure, generation of toxic chemicals that are hazardous in nature (Khin et al. 2012; Shankar and Nill 2015). These are uneconomical method because they do not completely purify the wastewater, and it do not give any option to reuse the material. The biological method of wastewater treatment is also applied but they are normally slow, and sometime they cause toxicity to the microorganism due to some hazardous toxic contaminants (Aghalari et al. 2020; Bhat et al. 2020). The physical processes like filtration has tendency to remove the contaminants by changing one phase to another phase but it produces highly concentrated sludge, which is poisonous and it is difficult to dispose in environment.

So, in the aforesaid mentioned context, the wastewater treatment requires more efficient and powerful technology to treat municipal and industrial wastewaters. This can be done either by developing new methods or by improving the techniques that exists. Among several technologies, nanotechnology has emerged as an incredible way for remediating wastewater and certain other environmental related problems. The nanotechnological ways are more effective than their conventional counterparts owing to their small size, excellent chemical properties, high reactivity and large surface area (Aragaw et al. 2021; Kumari et al. 2019; Magalhães-Ghiotto et al. 2021). Synthesizing green nanomaterials from the microorganisms and their extracts have opened the doors towards the environmentally friendly treatment of wastewater pollutants. Nanomaterials possess exceptional surface chemistry in comparison to other conventional methods, due to which they are able to target contaminants with the help of their functional groups for remediation. The properties of nanomaterials such as absorptive capacity, shape, size and chemical components can be altered to enhance the performance of nanomaterial for treatment of contaminants. Application of nanomaterial will help in promoting green technology because there will be less production of a sludge, and it can be the safer option for the remediation of environmental pollutants (Gehrke et al. 2015; Lu et al. 2016; Yaqoob et al. 2020). Iron nanoparticles are the well-known green nanoparticles that are used in remediation of pollutants due to their nontoxic nature, magnetic susceptibility and redox potential. Another type of nanomaterial involved in effective removal of effluent are membrane associated nanomaterial. Nanomaterial are known to improve the permeability of membrane, mechanical and thermal strength for pollutant degradation.

The size of nanoparticles range between 1 and 100 nm. Production of nanoparticles is generally done through top-down and bottom-up methods. In case of top-down method, large structures are converted into small ones by means of physical methods whereas in bottom-up method small atoms, and molecules are utilized to produce nanoparticles by supra-molecular chemistry or self-assembly. The commercial synthesis of nanoparticles is generally carried out by various chemical and physical routes. Among the chemical approaches, the most common methods are sol-gel, coprecipitation, hydrothermal and solvothermal (Ganachari et al. 2017).

The biological method of synthesizing nanoparticle through bottom-up approach has gained popularity as a novel green strategy. The synthesis of nanoparticles biologically is less toxic, biocompatibe, eco-friendly and energy efficient process which produces less sludge. The biological synthesis of nanoparticles is the best alternative method which will overcome the drawback of the abovementioned problems. Among the biological methods, microbial-based synthesis of nanoparticles is of utmost importance due to its cost-effectiveness, ease of synthesis and eco-friendly nature (Singh et al. 2016; Prasad 2016). Microorganisms which are found in contaminated environment have tendency to modify and adapt themselves for degrading xenobiotic compounds and exhibit immense catabolic activity towards polluted environment. Number of studies have been done that shows the efficiency of many newly developed nanomaterials. So, the present chapter focuses upon microbial synthesis of nanomaterials for the application of wastewater treatment.

5.2 Water Pollution: Sources and Environmental Concerns

Water is the most important material for the support of life process in organism. It is the basic need of humans and all the living beings on earth. It is the most precious reserve among all natural resources. Earth is considered as a blue planet as it contains plenty of water on its surface. It is the most valuable asset among all form of natural resources. Freshwater is very important for sustenance as it is needed for household purposes and also for industrial as well as farming purpose. The water used by humans for their day to day activity should be safe i.e. colourless, odourless and free from harmful chemicals and pathogen. The quality of water is a very important issue because decrease in water quality reduces its application for human beings and other living organisms.

Any alteration in the chemical, physical and biological features of water which causes a destructive result on human health and living beings can be called as water pollution. Water plays a vital role in range of human activities like drinking, bathing and other household use. Plenty of water is required for irrigation purpose, and it is a crucial part of agriculture without which the crop production may be affected. Water also plays a major role in biogeochemical reaction that occur below the surface. In the last few decades, scarcity of freshwater has been observed due to rapidly growing population, increased urbanization and farming activities. The industrial revolution has been responsible for rapid industrialization among fast developing countries which has been a major issue as large amount of water is required in industrial processes. Presently, the problem of water scarcity is being observed all over the world among which less developed countries are likely to be more affected (Anderson et al. 2007; Madhura et al. 2019; Shan et al. 2009; Weng 2021).

Many industrial units like thermal power plants, fertilizer, distilleries, oil refineries, cement industries, pulp and paper industries, tanneries, sugar mills, pharmaceuticals, pesticides and textiles along with dye and dye-intermediates generate problem of water pollution through their effluents (Chatterjee et al. 2013; Khan et al. 2005; Sharma et al. 2021; Singh 2001; Solomon 2005). These manufacturing units dispose their effluent containing wastewater into the nearby water bodies or sewage system that create the environmental crisis.

5.2.1 *Sources of Water Pollution*

Sources of water pollution are broadly classified under two categories:

i. Point source: In point, the harmful substances are directly emitted into the aquatic body e.g. water pollution created by industrial unit discharging its effluent directly into a river. Point source pollution can be monitored and regulated. The strict policies designed and efforts by the government of countries can tackle the problems related to point source water pollution in efficient manner.

ii. Nonpoint source: In nonpoint source, the water pollution occurs through diffuse sources. It negatively affects water bodies. Its example includes runoff from agricultural land draining into a river. Nonpoint source pollution have many sources which makes it difficult to manage. It is very difficult to control nonpoint source water pollution it comes from daily activities of different people, using pesticide, or constructing activity. The major sources of nonpoint source wastewater include nutrient loss from grassland, crop fields, road surfaces and forest. The two major nutrients phosphorus and nitrogen are of great concern. Phosphorus is main part of fertilizers used in agriculture. It is transported to aquatic water bodies through soil erosion. Nitrogen is the other crucial component of fertilizers, and it is a major pollutant of saltwater or brackish estuarine systems where nitrogen is considered as a limiting nutrient. Large amounts of nitrogen causes eutrophication and algal blooms

The major sources of water pollution are described below.

5.2.1.1 Sewage (Wastewater)

Sewage refers to the wastewater generated from domestic and industrial processes. Sewage disposal is a major concern in developing countries because many people in such countries don't have access to hygienic sanitary conditions and clean water (Kiguchi et al. 2016; Nazeer et al. 2016; Shrivastava et al. 2020). Untreated sewage water can contaminate the environment thus, causing diseases like diarrhoea. Sewage treatment is done in water treatment plants and is often disposed in the sea. Sewage is biodegradable, and it is easily broken down in the environment.

5.2.1.2 Agricultural Pollution

The agriculture sector contribute a major portion of water pollution. Agricultural processes like spreading of manure and slurries, tillage, ploughing on land, pesticide and fertilizers use on land causes water pollution (Evans et al. 2019; Logan 1993).

5.2.1.3 Oil Pollution

Oil spillages severely affects the quality of water in many ways. Oil makes drinking water unsafe for drinking. Large amount of oil released into the oceans and seas can destroy aquatic life and ecosystems that sustain them (Hernández Ruiz et al. 2021; Ukhurebor et al. 2021). Oil spills reduces oxygen in the water environment. The main reason for oil related water pollution are:

- Leakage from storage facilities
- Spillage during oil delivery
- Disposal of waste oil into drainage systems

5.2.1.4 Radioactive Substances

Radioactive waste is also among one of the source of water pollution. Radioactive substances are mostly used in nuclear power plants. They are also used medical, industrial and other scientific processes. These substances are also used in making watches, x-ray machinery and television sets. If the radioactive substance are not properly disposed of, they could result in severe water pollution incidents (Radioactive Pollution 2014).

5.2.1.5 River Dumping

Many people dump their household waste such as discarded bicycles, garden waste and electronic waste into the nearby river. River dumping is not only responsible for water pollution but it also affects wildlife and lead to occurrence of severe floods.

5.2.1.6 Marine Dumping

The Worldwide Fund for Nature (WWF) estimated that substantial amount of waste is dumped into the sea every year. The waste dumped into the sea generally include plastics and other non-degradable materials that is blown or washed from the land. These plastics get fragmented into microplastics and engulfed by the aquatic fauna which affects their metabolism severely.

5.2.2 Concerns of Water Pollution

Infectious diseases of humans are one of the most serious effect of water pollution. Developing countries are much affected by such infectious disease of water pollution because of poor sanitation practices followed by the people. Waterborne diseases affects human beings when parasites or other microorganisms responsible for causing disease are transmitted through contaminated water, mainly the water contaminated by pathogens from excreta (Some et al. 2021). Waterborne diseases include typhoid, intestinal disease, enteric and diarrheal diseases that are caused by various bacteria, viruses, and, parasites. The most severe parasitic diseases are ascariasis, amoebiasis and giardiasis. Water-based diseases are similar to waterborne diseases but they are not directly an effect of water pollution. They occurs from infectious agents that occur in water and naturally spend a part of their life cycle in water. Humans become infected when intake such water or come in contact with it (Abia et al. 2015; Akita et al. 2021; Seurinck et al. 2005; Teklehaimanot et al. 2014). Major water based disease is schistosomiasis, which currently infects many people around the world. Water pollution causes nutrient pollution which is the major, chronic environmental problem of aquatic ecosystem. Nutrients like nitrogen and phosphorus comes from agriculture, waste disposal and fossil fuel use. Once these nutrient reaches into the aquatic water bodies, they stimulates overgrowths of algae, and causes harmful algal bloom, which is very harmful and have direct toxic effects on aquatic flora and fauna. It causes depletion of oxygen in the water body. Zooplankton engulfs the toxic algae and passes these toxins up the food chain, thus, affecting seabirds, marine animals and humans. This results in illness and sometimes proves to be fatal.

Some other sources of pollution like sewage, industrial waste, chemicals, oil spills destroys the marine as well as terrestrial biodiversity. A new emerging threat is hormone-disrupting chemicals. The effects of hormone-disrupting chemicals include thyroid system disorders; reduced immune response; inability to breed and abnormal mating and parenting behaviour. In humans, endocrine disruptors lead to weaken immune function, mental impairment, decreased fertility and increases in some types of cancers.

5.3 Wastewater Remediation: Existing Technologies and Recent Advances

The choice of technology for the treatment of wastewater depends upon its nature. One of the primary reason behind the development of new technology for wastewater treatment is heavy fines which are levied on the industries and factories for the disposal of wastewater that does not meet the standard limits. This has negative impact on the financial aspect of industries that fuelled the need for new and improved wastewater treatment technologies. Some of the commonly used technologies are explained below.

5.3.1 Membrane Technology

Membrane technology (MT) utilizes the scientific as well as engineering approach for the transport of components, through the membranes. This technique is basically adopted for the mechanical separation of gaseous or liquid streams. Membranes are the thin layer barrier for the separation of substances having different size (Santoro et al. 2021). This technology is then combined with chemical and biological treatment processes. In a typical membrane technology, the driving force is a semi-permeable barrier that controls the movement rate of substances by fractional permeation and rejection through pores of various sizes. The permeation and selective rejection depends upon the membrane pore size and chemical affinity that allows for a product stream containing no target components. The major advantages for the usage of membrane technology in wastewater treatment are as follows:

1. It is energy saving and clean technology.
2. It has tendency to replace conventional process of treatment.
3. It has flexible system design

4. It is easier to use

This technology is used in many industries, pharmaceutical process, metallurgical process including treatment of water for domestic use, biological, chemical and in various separation processes.

5.3.2 Microalgal Wastewater Treatment (MWWT)

Phytoremediation is an eco-friendly technique to remove residual pollutants from wastewater and make it potent for re-introduction in the water supply system. The microalgae-based wastewater treatment system has been used for domestic, agricultural, commercial and industrial wastewater. The advantages of microalgae-based wastewater treatment system are as follows:

1. reduces pollutants and pathogens
2. recover nutrients as biomass
3. energy savings
4. mitigate CO_2 gas emissions
5. recovery of metabolites

Microalgae-based wastewater treatment technology has many advantages but it also have some challenges. These include requirement of land, environmental and operational condition influence, harvesting of biomass and valourization. Some other limitations are algae biomass separation of algae biomass from water, and less ability of the algae biomass to reduce and remove micropollutant content in wastewater discourages the use of technology at large scale (Abdel-Raouf et al. 2012; Alcántara et al. 2015; Molazadeh et al. 2019).

5.3.3 Microbial Fuel Cells for Wastewater Treatment

Microbial fuel cells are considered as the sustainable way of managing the increasing energy demands for wastewater treatment. It is the promising technique to deal with environmental pollution (Guo et al. 2020). An MFC is basically a device which converts organic matter into electricity by utilizing microbes as biocatalyst. Generally, MFCs consists of three main components: electrodes, electrogens and separator. MFCs have two electrodes, which, can be separated either into one or two chambers. These chambers are known to operate as mixed reactors. Below, each electrode a proton exchange membrane (PEM) or the cation exchange membrane (CEM) is placed. The anode faces towards the chamber that contains the liquid phase, whereas cathode faces towards the chamber that is contained with air. Cation exchange membrane or a salt bridge is used as a separator to keep the chamber. The potential difference which develops between the two chambers make electrons move through the circuit and the microbial degradation of wastewater acts as the substrate to produce bioelectricity. Use of MFCs in treatment of wastewater include a number of advantages like use of renewable resources, long-term sustainability, bio-hydrogen production, degradation of organic and inorganic waste and removal of certain compounds like nitrates (Gude 2016).

5.4 Nanoparticles as a Promising Tool of Remediation of Wastewater

Clean water is a basic need for human and for multi-dimensional development of society and emerging economy. Rise of population, expansion of industries, extensive agriculture practices and urbanization resulted in wastewater generation. This has made water polluted and deadly for drinking purpose. Many people die due to diseases through consumption of contaminated water. Even though several methods are available for wastewater treatment but their use is limited due to number of limitations due to harmful chemicals used, formation of disinfection by-products (DBPs), consumption of time and costly approach (Khin et al. 2012; Shankar and Nill 2015). Nanotechnology is a science of manipulating matter at molecular or atomic level to design new structures and devices that possess excellent electronic, conductive, optical, magnetic and mechanical properties. It is gaining importance as a promising technology, that has shown remarkable results in various areas including wastewater treatment. Nanomaterials possess high surface to volume ratio, a high reactivity and sensitivity, high capacity of adsorption and easy functionalization that makes them appropriate for usage in wastewater treatment (Santoro et al. 2021).

5.5 Nanoparticles: Types, Properties and Synthesis

Nanoparticles are materials which have size ranging from 1 to 100 nm. They are classified into various categories based on their properties, sizes or shapes. The different groups of nanoparticles include fullerenes, metallic nanoparticles, ceramic N nanoparticles and polymeric nanoparticles. Nanoparticles have unique chemical and physical properties due to large surface area and its nanoscale size (Kshtriya et al. 2021; Shukla et al. 2021). The optical properties of nanoparticle are dependent on size, which gives different colour because of absorption in visible region. Their properties like reactivity, toughness are dependent on their size, shape and structure. Due to such character, they are suitable for many commercial and domestic usage, that include catalysis, imaging, energy-based research, medical-based applications and in environmental applications. Heavy metal nanoparticles of lead and mercury have been reported to be very rigid and stable. Their degradation is not easily achieved, and it can lead to several environmental toxicities.

5.5.1 *Types of Nanoparticles*

Nanoparticles are divided into many categories based on their size, morphology and chemical properties. On the basis of physical and chemical character, some of the popularly known classes of NPs are as follows:

5.5.1.1 Carbon-Based Nanoparticles

Fullerenes and carbon nanotubes are the two main classes of carbon-based NPs. Fullerenes consists of nanomaterial made up of globular hollow cage of allotropic forms of carbon. They are used frequently in various applications due to good electrical conductivity, structure, electron affinity, high strength and versatility (Kokorina et al. 2020; Maiti et al. 2019; Patel et al. 2019). These nanoparticles are elongated and tubular in structure, of 1–2 nm in diameter. They are synthesized by the deposition of carbon precursors mainly the atomic carbons, which are vapourized from graphite by laser or by an electric arc on metal particles. Due to their unique properties, they are not only used in pristine form but also they are used in nanocomposites for number of commercial applications like fillers, as an efficient gas adsorbents for remediation of environment, and as a support medium for many inorganic and organic catalysts.

5.5.1.2 Metal Nanoparticles

Metal nanoparticles are made up of metals precursors. Due to its localized surface plasmon resonance (LSPR) character, such nanoparticles shows unique optoelectrical properties. Nanoparticles of the alkali and noble metals like copper, silver and gold shows broad absorption band in visible zone of electromagnetic solar spectrum. Due to their excellent optical properties, metal nanoparticles have many applications in research areas (Luong et al. 2011; Shnoudeh et al. 2019; Vanden Bout 2002). Gold nanoparticles coating is used for the sampling in scanning electron microscopy (SEM), to increase the electronic stream, that helps in getting high quality SEM images.

5.5.1.3 Ceramics Nanoparticles

Ceramics nanoparticles are the inorganic non-metallic solids, that are synthesized via heat and successive cooling. They are commonly found in polycrystalline, amorphous, dense, hollow forms or porous. These NPs are used frequently in several applications such as catalysis, photodegradation of dyes, photocatalysis and imaging applications (De Guire et al. 2006; Thomas et al. 2015).

5.5.1.4 Semiconductor Nanoparticles

Semiconductor materials shows properties between metals and nonmetals due to which they are used in various applications. Semiconductor nanoparticles have wide bandgaps and upon bandgap tuning they show significant alterations in properties. They are considered as an important materials for photo optics, photocatalysis and electronic devices (Nayak et al. 2017; Uchida and Matsui 2001; Wang et al. 2003). Many semiconductor nanoparticles are efficient in water splitting applications, due to suitable bandgap and band edge positions.

5.5.1.5 Polymeric Nanoparticles

These are organic based nanoparticles. They are nanospheres or nano-capsular shaped. In nanosphere shaped polymeric nanoparticles, matrix particles are generally solid and the other molecules are adsorbed over the outer boundary of the spherical surface (Zielińska et al. 2020). In nano-capsular shaped polymeric nanoparticles, the solid mass is encapsulated completely within the particle. The polymeric nanoparticles are used in numerous applications.

5.5.1.6 Lipid-Based Nanoparticles

These nanoparticles contains lipid moieties. They are used effectively in many biomedical applications. Generally, a lipid nanoparticles is spherical having diameter that range from 10 to 1000 nm. Lipid nanoparticles contains a solid core made up of lipid and a matrix that possess soluble lipophilic molecules (García-Pinel et al. 2019).

5.5.2 Synthesis of Nanoparticles

Various methods have been used for the synthesis of nanoparticles (Fig. 5.1). These methods are classified into two main classes.

1. Bottom-up approach, and
2. Top-down approach

These approaches are divide into various subclasses on the basis of operation, reaction condition and protocols adopted.

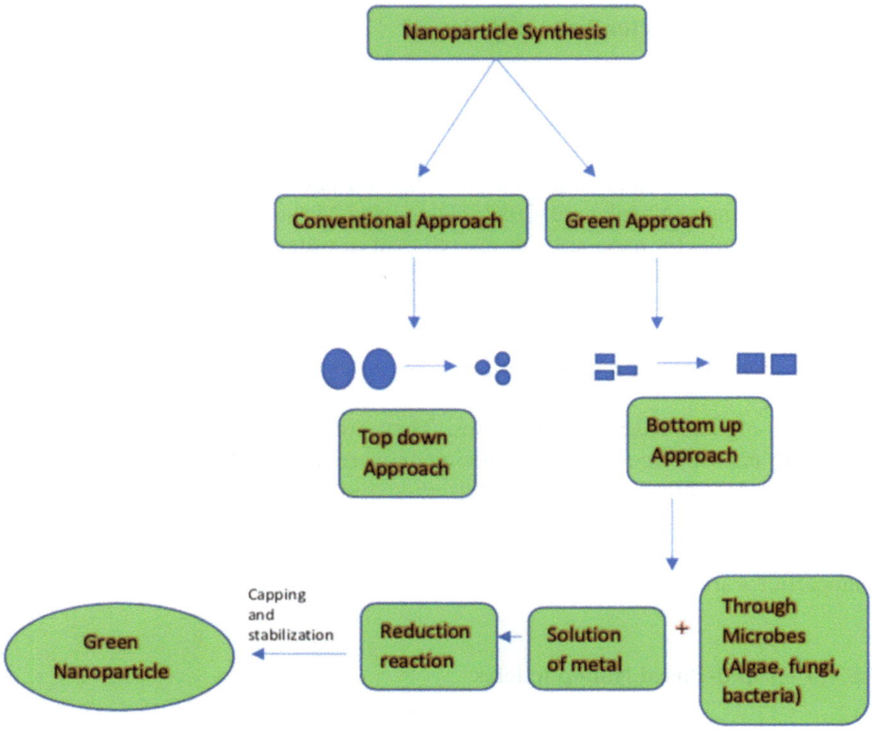

Fig. 5.1 Schematic diagram of synthesis of nanoparticle

5.5.2.1 Top-Down Syntheses

This method utilizes destructive approach. A larger molecule is decomposed into smaller molecules, and then these molecules are converted into nanoparticles. Examples for this method include grinding/milling, physical vapour deposition (PVD) (Aryal et al. 2019; Baig et al. 2021). This approach is also used in synthesizing coconut shell nanoparticles. The colloidal carbon spherical particles with control size are also synthesized from simple top-down approach. The technique of synthesis is based on the continuous adsorption of polyoxometalates (POM) chemically on the carbon interfacial surface. Adsorption transformed the carbon black aggregates into smaller spherical particles. Micrographs revealed that the size of carbon particles reduced with sonication time. Transition-metal dichalcogenide nanodots (TMD-NDs) are also synthesized by top-down techniques using their bulk crystals.

5.5.2.2 Bottom-Up Approach

This approach is in reverse of top-down syntheses as nanoparticles are formed of simpler substances. So, this approach is also known as building up approach. Examples of bottom-up approach are reduction and sedimentation techniques. It includes sol-gel, green synthesis, spinning and biochemical synthesis (Ramanathan et al. 2021). Through this method, TiO_2 anatase nanoparticles are synthesized with graphene domains. In TiO_2 nanoparticles, titanium isopropoxide and alizarin precursors are used to synthesize the photoactive composite for the degradation of methylene blue photocatalytically. Alizarin is used because it provide strong binding capacity with TiO_2 nanoparticles through the axial hydroxyl terminal groups.

5.5.3 Physicochemical Properties of Nanoparticles

Unique properties of nanoparticles such as large surface area, optically active, mechanically strong and chemically reactive make makes it a suitable material for various applications. Some of the important properties of nanoparticles are discussed below.

5.5.3.1 Magnetic Properties

Magnetic property of nanoparticles are of great importance as they are used in number of disciplines. It is used in heterogeneous and homogenous catalysis, magnetic fluids, biomedicine, magnetic resonance imaging (MRI) and in environmental remediation for water decontamination (Akbarzadeh et al. 2012; Issa et al. 2013).

5.5.3.2 Electronic and Optical Properties

The electronic and optical properties of nanoparticles are interdependent. For example, noble metals nanoparticles have optical properties dependent on size and shows a strong UV–visible extinction band which is not found in the spectrum of bulk metal (Asha and Narain 2020; McConnell et al. 2000). This excitation band is seen when incident frequency of photon is constant with the excitation of the conduction electrons and is called as localized surface plasma resonance (LSPR) (Kumbhakar et al. 2014; Zhang 2009). It is well-known that peak wavelength of LSPR depends upon the shape, size, interparticle spacing of nanoparticles, its own dielectric properties and its local environment which includes adsorbates, substrate and solvents. Gold colloidal nanoparticles are responsible for the rusty colours seen in blemished glass door and windows, while silver nanoparticles are yellow. These free electrons over the surface nanoparticles (d electrons in silver and gold) are freely transportable through nanomaterial.

5.5.3.3 Mechanical Properties

Due to distinct mechanical properties nanoparticles, they are used in many important fields like surface engineering, nanomanufacturing and nanofabrication (Reghunadhan et al., 2018; Wu et al. 2020). Different mechanical parameters studied to know the mechanical nature of nanoparticle. Beside such parameters coagulation, surface coating and lubrication are also the mechanical properties of nanoparticle. Nanoparticles show different mechanical properties when compared to microparticles. Good command over the mechanical features of nanoparticles and their interactions with any type of surface are important for enhancing the surface quality and increasing material removal. Fruitful outcomes in these fields generally need a deep insight into the basics of the Some of the mechanical properties of nanoparticles like elastic modulus, hardness, friction and interfacial adhesion, movement law and their size-dependent characteristics (Guo et al. 2014; Wu et al. 2020).

5.5.3.4 Thermal Properties

Metals nanoparticles have thermal conductivities higher than fluids in solid form, for e.g. thermal conductivity of copper is about 700 times higher than that of water at room temperature and about 3000 times higher than that of engine oil (Das et al. 2020; Savage and Rao 2004). Oxides like alumina (Al_2O_3) have thermal conductivity greater than that of water. Thus, the fluids that contain suspended solid particles display significantly higher thermal conductivities compared to conventional heat transfer fluids. Nanofluids are known to exhibit excellent properties. Nanofluids are formed by dispersing the nanometric solid particles in the liquid like water, oils or ethylene glycol. Recently, it has been examined that the nanofluids containing CuO or Al_2O_3 nanoparticles in ethylene or water exhibit greater thermal conductivity.

5.6 Microbial Synthesis of Nanoparticle

Number of bacteria, fungi and algae are well-known to produce nanoparticles (Fig 5.2). These nanoparticles of microbial origin are then utilized in number of applications. Some of the important bacteria, fungi and algae for the synthesis of nanoparticles are discussed below.

5.6.1 Synthesis of Nanoparticle by Bacteria

The bacteria are very well-known as for the generation of nanoparticles like gold, platinum, silver, titanium, palladium, magnetite cadmium sulphide and titanium dioxide. Bacteria are the important microbes for synthesis of nanoparticles due

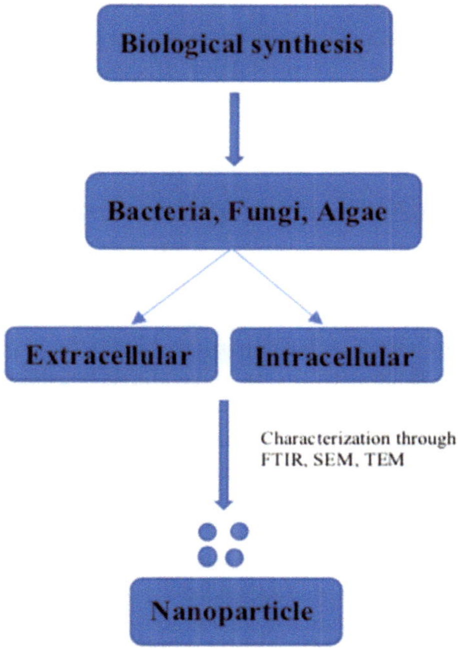

Fig. 5.2 Schematic diagram of biological synthesis of nanoparticle

to their characteristic feature of getting adapted to adverse environmental conditions. It has been reported in some study that some bacteria are known to reduce or precipitate the soluble inorganic toxic ions into nontoxic metal nanoparticles which are insoluble. Bacteria forms nanoparticles either intracellularly or extracellularly under certain physicochemical conditions such as exposure period, temperature, pH, metal salts and concentration of bacteria. The biomolecules which are present in medium have tendency to reduce the metal ions in extracellular process. But, in intracellular process, the functional groups found on the cell wall attracts the metal, metalloids. Metal ions interacts with proteins that are present inside the cells for producing nanoparticles. Extracellular reduction is more favourable as compared to the intracellular reduction because the extraction procedure is easy and efficiency is high. Fang et al. (2019) revealed that dead bacteria can also be used to synthesize the nanoparticles same as living bacteria. Bacteria can also be used as biocatalyst because they act as biological platform for the mineralization process (Iqtedar et al. 2019). The bacteria mobilizes or immobilizes the metals and reduces or precipitate the metal ions. Bacteria catalyses different reactions due to the presence of various enzymes and produces inorganic nanoparticles. The large amount of nanoparticles can be produced via extracellular secretion of bacterial enzymes in a pure form. The bacterial cells possess metal binding capacity and the presence of S-layer makes them useful for their use in bioremediation. The bacterial cell wall plays important function as metals could percolate in the cytoplasm through cell wall, and it is transferred back to wall for the extracellular secretion. The cell wall of bacteria with metal binding sites could be changed by chemical reactions for some specific groups, like amines and carboxyl groups, that converts positive charge into negative charge. Use of bacteria for production of nanoparticles is profitable process as it do not require any costly and toxic chemicals for the synthesis and stabilization processes.

5.6.2 Synthesis of Nanoparticle by Fungi

The filamentous fungi are used as a potent source for synthesizing nanoparticles. The fungus mycelium has high surface area which is known to secrete large amount of proteins which can participate directly in production of nanoparticles. Nanoparticle production by filamentous fungi is considered to be better due to the capacity of fungi to secrete, enzymes, proteins and metabolites, economic feasibility, simple scaling up and downstream handling, higher surface area due to occurrence of mycelia and low-cost input for production procedures. Some filamentous fungi species are known to grow fast and their maintenance is easy at laboratory conditions (Fouda et al. 2018). The nanoparticles production having nanoscale dimension via fungi shows more monodispersity when compared to those synthesized from bacteria. *Fusarium oxysporum* produces gold nanoparticles in presence of aqueous $AuCl4^-$ ions with NADH-enzyme-mediated reaction that releases reducing agents in the solution and forms gold nanoparticles. The nanoparticles synthesized show stability for long-term due to protein binding capacity by association of cysteine and lysine residues (Das et al. 2017). The filamentous fungi possess greater regenerative ability for synthesis of metal nanoparticles in good amount with its commercial feasibility. *Aureobasidium pullulans, Cladosporium resinae, Aspergillus nigerPenicillium* species, *Ganoderma lucidum, Funalia trogii, Trametes versicolor* and *Rhizopus arrhizus* have been studied to absorb heavy metals from polluted sites which was the used for production of nanoparticles. Salvadori et al. (2013) have studied the uptake of Cu(II) by *Hypocrea lixii* dead biomass and then produced copper nanoparticles. The same fungus was also able to produce NiO nanoparticles by extra and intracellularly both. *Fusarium oxysporum* showed extracellular synthesis of gold and silver nanoparticles when treated with the equimolar mixture of silver nitrate and tetrachloroaurate ion (Senapati et al. 2005). In the presence of hexachloroplatinic acid, it produces platinum nanoparticles (Riddin et al. 2006). *Aspergillus flavus* produced silver nanoparticles of 9 nm size because it reduces silver ions due to the presence of sil genes in its plasmid (Vigneshwaran et al. 2007). Salvadori et al. (2014) revealed that *Aspergillus aculeatus* dead biomass produced NiO nanoparticles of 5.89 nm size which were organized in the form of film. Due to the occurrence of metabolites, fungi are known as the better resource for synthesizing nanoparticles as compared to bacteria (Singh et al. 2016). Zhang et al. (2011) found that gold nanoparticles have been biosynthesised in vacuoles of filamentous fungi. They also described the functions of fungal proteins for the capping of gold nanoparticles. Filamentous fungi are the better candidate for the synthesis of metallic nanoparticles due to the occurrence of different enzymes in their cells and simple handling procsess (Khandel and Shahi

2018). Filamentous fungi also shows metal uptake capacities. It can easily be cultured in large amount by using solid substrate fermentation. *Verticillium* species are known to produce gold nanoparticles intracellularly when it is exposed to chloroauric acid solution. Gericke and Pinches (2006) discovered the synthesis of gold nanoparticles by fungus *Verticillium luteoalbum.* The age do not show any effect on the shape of gold nanoparticles but the number of nanoparticles decreased significantly when old cells are used. The biomass of *Fusarium oxysporum* is also used for production of silver nanoparticles (Karbasian et al. 2008). Saxena et al. (2014) studied that genetic modification procedures can be used to increase the properties of nanoparticles.

5.6.3 Synthesis of Nanoparticle by Algae

The algae are the aquatic oxygenic photoautotrophs which are used for production of nanoparticles (Castro et al., 2013). *Chlorella vulgaris*, *Nannochloropsis oculata* and *Dunaliella salina* show tendency to produce silver nanoparticles of size less than 15 nm inside the cells in 48 h (Mohseniazar et al. 2011). The bio reduction character of algae showed significant potential in synthesis of various metal oxide nanoparticles like gold, silver, palladium, platinum, zinc oxide and copper oxide (Momeni and Nabipour 2015). Algae produces complex inorganic nanomaterials by using both intracellular and extracellular process (Sau and Murphy 2004). The gold nanoparticles are synthesized intracellularly in *Rhizoclonium fontinale* and by extracellular process in *Lyngbya majuscula* and *Spirulina subsalsa* (Chakraborty et al. 2009). The metal ions can attach to the cell surface through electrostatic interactions between ions and carboxylate groups. Afterwards, ions are reduced by enzymes which cause nuclei formation and growth with reduction of metal ions (Parial et al., 2012). Accumulation of gold nanoparticles of size 9–20 nm was stated in *Chlorella vulgaris* dried cell suspension (Hosea et al., 1986). Cyanobacteria are also known to reduce gold (III)–chloride to metallic gold. The intermediates formed are gold (I)–sulphide and Au (I) (Lengke et al. 2006).

5.7 Mechanism of Microbial Synthesis of Nanoparticles

Microorganisms produces nanoparticles by extracellular and intracellular enzymes as defined below.

5.7.1 Extracellular Enzymes

The extracellular microbial enzymes play crucial role in production of metallic nanoparticles and work as a reducing agent in their production. (Subbaiya et al.

2017). The extracellular enzymes like acetyl xylan esterase, glucosidase and cellobiohydrolase D found in fungi play a great role in synthesizing metallic nanoparticles (Ovais et al. 2018). *Rhodopseudomonas capsulata* have been studied to produce gold nanoparticles through extracellular process by electron transfer to NADH-reliant reductase enzymes from NADH. After acceptance of electrons, gold ions get reduced to produce gold nanoparticles (He et al. 2007). *Fusarium oxysporum* was also utilized as a reducing agent for synthesizing gold and silver nanoparticles. *Fusarium* species are known to produce nitrate-reliant reductase and shuttle quinone which are then used for the production of nanoparticles through extracellular process (Senapati et al. 2005). Other two species of *Fusarium* which are *Fusarium solani* and *Fusarium semitectum* also produce silver nanoparticles through extracellular process by using their enzymes (Ingle et al. 2009). Silver nanoparticles are also synthesized through extracellular process by *Coriolus versicolor* and *Cladosporium cladosporioides* (Balaji et al., 2009). *Aspergillus fumigatus* produced silver nanoparticles extracellularly in in a time span of 10 min in comparison to chemical and physical techniques (Bhainsa and D'Souza 2006). *Sargassum wightii* has been used to reduce Au^{3+} ions to gold nanoparticles (Singaravelu et al. 2007). *Chlorella vulgaris* has been used to synthesize gold nanoparticles (Lengke et al. 2006).

5.7.2 Intracellular Enzymes

Actinomycetes like *Rhodococcus* and *Thermomonospora,* which are alkalo-thermophilic and alkalo-tolerant, are used for production of gold nanoparticles through intracellular process (Ahmad et al. 2003). *Verticillium* species produced silver nanoparticles when it is exposed to Ag^+ ion solution through intracellular reduction. Same method was used for fabrication of gold nanoparticles using *Verticillium* as a good source of reducing enzymes (Ovais et al. 2018). Salvadori et al. (2017) suggested a natural method for the synthesis of metal nanoparticles intracellularly using yeasts. Possible mechanism for the production of nanoparticles intracellularly is the electrostatic interaction between amide groups and metal cations which are present in enzymes of yeast cell wall and reduction of ions by the enzymes which then results in metal ions accumulation and nanoparticles formation.

5.8 Characterization of Nanoparticles

Different techniques of characterization have been employed for the analysis of physicochemical properties of nanoparticles. It includes techniques like X-ray diffraction (XRD), infrared (IR), X-ray photoelectron spectroscopy (XPS), Brunauer–Emmett–Teller (BET), Scanning electron microscopy (SEM), transmission electron microscopy (TEM) and particle size analysis.

5.8.1 *Morphological Characterizations*

The morphological attributes of nanoparticles have always gained great interest as morphology always affects most of the properties of nanoparticles. There are many characterization techniques for the morphological studies, but some microscopic techniques like SEM, polarized optical microscopy (POM), and TEM are the important one. SEM is based on the principle of electron scanning, and it provides all the information available about the nanoparticles at nanoscale level. Numerous literature is available, where researchers have used this instrumentation technique to study the morphology of nanomaterials, as well as the dispersion of nanoparticles in the bulk or matrix. Similar to SEM, TEM is based on the principle of electron transmittance, so it provides information about the bulk material from lesser to higher magnification. The morphologies of gold nanoparticles have been studied through this technique. TEM provides important basic information about the layer of material, like, quadrupolar hollow shell structure of Co3O4 nanoparticle which was observed through TEM. These nanoparticles are exceptionally active as anode in Li-ion batteries (Wang et al. 2013).

5.8.2 *Structural Characterizations*

The structural features are of great importance in order to study the nature of bonding and composition of nanomaterials. Some of the common techniques which are utilized to study the structural properties of nanoparticles are X-Ray diffraction (XRD), XPS, energy dispersive X-ray (EDX), BET, Raman, IR and Zieta size analyzer. XRD is most important instrumentation technique that reveal the structural properties of nanoparticles. It provides enough information about crystallinity and phase of nanoparticles (Ullah et al. 2017). In case of nanoparticles having size less than hundred of atoms, the correct measurement of structure and may be difficult. Moreover, nanoparticles which have amorphous character different inter atomic lengths can affect the XRD diffractogram. EDX, which is usually attached with field emission scanning electron miscopy (FE-SEM) or TEM instrument, is used widely to know the elemental composition, and it gives a rough idea of weight percent. XPS is also considered as a most sensitive instrumentation technique to determine the exact ratio of element and exact nature of bond of the elements in nanoparticle materials. It is a surface sensitive instrumentation technique and is used for depth profiling studies to determine the overall composition and compositional variation alongwith depth. XPS is another instrumentation technique which is based on basic principles of spectroscopy and a XPS spectrum shows the binding energy (eV) of the electrons on versus X-axis number of electrons on Y-axis plot. FT-IR and Raman spectroscopies give information about the vibrational character of nanoparticles (Dablemont et al. 2008).

5.8.3 Particle Size and Surface Area Characterization

There are many techniques to determine the size of the nanoparticle which include SEM, XRD, AFM, TEM and dynamic light scattering (DLS). SEM, AFM TEM and XRD can give better idea about the particle size (Kestens et al. 2016). Zeta potential size analyser and DLS are used to find the size of nanoparticle at very low level. But, in case of hydrophilicity and agglomeration, DLS may be incapable of correct measurement, so in this case other high-resolution technique like differential centrifugal sedimentation (DCS) should be used (Sikora et al. 2016). Nanoparticle tracking analysis (NTA) is a newer technique, which helpful for dealing with biological systems like proteins, and DNA. In NTA technique, nanoparticles can be visualized and analyzed in liquids media. This technique helps us to find the size distribution profile of nanoparticles with diameter ranging between 10 and 1000 nm in a liquid medium. BET is one of the best technique to estimate the surface area of nanoparticle materials. This instrumentation technique works on adsorption and desorption principle and according to Brunauer–Emme–Teller (BET) theorem.

5.8.4 Optical Characterizations

Optical properties are important in photocatalytic applications and therefore, these characterizations works on the famous beer-lambert law and the basic principles of light. These techniques provide information about the reflectance, absorption, phosphorescence and luminescence properties of nanoparticle. It is very well-known that nanoparticle, especially metallic and semiconductor nanoparticle show different colours and thus, best used for photo-related applications (Kumbhakar et al. 2014; Uchida and Matsui 2001).

5.9 Application of Nanoparticles in Removal of Organic Pollutants from Wastewater

Natural organic matter (NOM) forms various group of hydrophobic and hydrophilic organic compounds. Its contribution is significant towards water contamination (Farrokhi et al., 2013). Number of carbon-based adsorbents are used for the removal of NOM from the wastewater and many factors affect the sorption of NOM (Farrokhi et al. 2013; Ye et al. 2019). Some of the nanomaterials used for the removal of such organic compounds are discussed below.

5.9.1 CNTs

Number of nanomaterial like nano-sorbents like carbon nanotubes (CNTs), polymeric materials, such as dendrimers, and zeolites have shown exceptional properties of adsorption and are used for removal of organic compound from wastewater (Ye et al. 2019; Zhang 2009). CNTs have gained special importance because of their exceptional water treatment capabilities. CNTs have been studied widely. Removal of NOM by using CNTs is higher when compared to carbon-based adsorbents because of large surface areas and some other factors. CNTs have also proved to be effective in removal of polycyclic aromatic organic compounds and atrazine.

5.9.2 TiO_2 Nanoparticles

TiO_2 and CeO_2 are the common nanomaterial of metal oxides that are used as a catalysts for quick and complete degradation of organic contaminant (Farrokhi et al. 2013). TiO_2 nanoparticles are used in the treatment of water which is contaminated with organic contaminants like polychlorinated biphenyls (PCBs), chlorinated alkanes and benzenes (Upadhyayula et al. 2009). The addition of TiO_2 nanoparticles in the wastewater enhanced the Removal of total organic carbon. TiO_2 nanoparticles have also been used in a 'falling film' reactor for degrading microcystins in water. Functionalization of multiwalled CNTs with Fe nanoparticles proved to be an effective sorbents for removal of aromatic compounds such as benzene and toluene.

5.9.3 Zero-Valent Iron

Catalysts such as zero-valence metal, bimetallic nanoparticles and semiconductor materials have shown their usage in degradation of environmental pollutants like PCBs, azo dyes and pesticides, due to their large surface area alongwith its shape dependent properties (Kang and Choi 2009; Li et al. 2021; Liu et al. 2021). Magnetic nano-sorbents have also been used effectively in removal of organic contaminants. Nanomaterial of iron oxide showed better capabilities of removal of organic pollutants in comparison to bulk materials. They have also been utilized for removal of coloured humic acids from wastewater.

5.9.4 *Other Nanomaterials*

Nanostructured ZnO semiconductor films have been used successfully for the degradation of organic contaminants like 4-chlorocatechol (Yu et al. 2015). The nanocatalyst made up of silver and amidoxime fibres have been utilized effectively for the degradation of some organic dyes. Nanocomposite of Pd- Cu/γ-alumina have also been utilized for the reduction of nitrate. Hydrogen and the palladium-based nanoparticles have been investigated to biodegrade traces of halogenated organic compounds. Palladium nanoparticles and bimetallic Pd/Au (gold) nanoparticles have been used for hydrodechlorination of trichloroethylene (TCE) (Arsiya et al. 2017). Mineralization of was accelerated in one study by using films of Hydrogen peroxide and manganese oxide (MnO_2) nanoparticles has also been investigated for mineralization of organic dyes.

5.10 Application of Nanoparticles in Removal of Inorganic Pollutants from Wastewater

Nanomaterials of different types have been developed for the heavy metals removal from wastewater. Nano-sorbents like CNTs, dendrimers and zeolites have shown exceptional adsorption properties (Upadhyayula et al. 2009). The CNTs ability to adsorb heavy metals has been studied by several researchers. Composites of CNTs with iron and cerium oxide (CeO_2) have been reported for the removal of heavy metal ions (Amin et al. 2014). Nanoparticles of Cerium oxide supported on CNTs have been used in effective manner to adsorb arsenic of CNTs are known to have fast adsorption kinetics due to short intraparticle diffusion distance and highly accessible adsorption sites (Fausey et al. 2019).

The nanomaterials that are metal based showed better removal of heavy metals in comparison to activated carbon. TiO_2 nanoparticles and nanosized magnetite have been used for adsorption of arsenic. The usage of TiO_2 nanoparticles has been examined to reduce the toxic metal ions in water. Nanocrystalline TiO_2 has been effective in removal of different forms of arsenic it has been proven to be more effective photocatalyst as compared to commercially available TiO_2 nanoparticles that shows maximum removal of arsenic at neutral pH. Nanocomposite of TiO_2 nanoparticles attached on graphene sheet reduce Cr(VI) to Cr(III) in presence of sunlight. Chromium ions were removed using palladium nanoparticles in a study. Removal of arsenic was investigated by utilizing high specific surface area of Fe_3O_4 nanocrystals. Polymer-grafted Fe_2O_3 nanocomposite has been used effectively for the removal of divalent heavy metal ions of copper, cobalt and nickel at a pH range of 3 to 7.

For the removal of radioactive metal toxins like uranium dioxide ($UO_2^{2+)}$ from water, bisphosphonate-modified magnetite nanoparticles has been used efficiently. Zero-valent iron or iron nanoparticles (nZVI or Fe^0) have shown to be effective in transformation of heavy metal ions like As(III), As(V), Cu(II), Ni(II), Pb(II) and Cr(VI) (Jézéquel and Chu, 2006).

As(V) and Cr(VI) have been shown to be successfully adsorbed by novel self-assembled 3D flower-like iron oxide nanostructures (Fausey et al., 2019). The 3D nanostructures of CeO_2 also proved to be a good adsorbents for arsenic and chromium. The efficacy of NaP1 zeolites was examined for removing heavy metals like (Cr(III), Zn(II), Cu(II), Ni(II), and Cd(II)) from the wastewater stream. Dendritic polymers have also shown their capability for treatment of toxic metal ions. Biopolymers were used for remediating heavy metal from aqueous wastes. Nanoparticles of chitosan for sorption of Pb(II) were also reported. (Arsiya et al. 2017)

Nanofilters has been examined for removal of arsenic and radioactive waste from ground and surface waters (Darab et al. 2007). It has also shown its potential to remove uranium (VI) from the sea water. Novel nanofilter membranes have been prepared by assembling) negative poly (styrene sulfonate) and positive poly (allylamine hydrochloride on porous alumina. They displayed high retention of Ca^{2+} and Mg^{2+}ions. A dendrimer-UF system showed the removal of Cu^{2+} ion from water. The addition of iron (hydr)oxide nanoparticles in the porous carbon materials has successfully removed inorganics and organics. Therefore, these filters can be used as point-of-use applications.

5.11 Application of Nanoparticles in Removal of Pathogenic Microbes from Wastewater

Biological contaminants are classified in three categories, such as, natural organic matter (NOM), microorganisms and biological toxins. Microbial contaminants contain free living microbes and human pathogens. Removal of cyanobacterial toxins is a major issue in conventional water treatment methods. Many adsorbents have shown good removal efficiencies but number of factors affects the removal process. Despite several development in disinfection technology, waterborne infections are prevalent in many areas. Therefore, advanced disinfection techniques should be used to eliminate the pathogens. There are several kind of nanomaterials like silver, zinc and titanium which are able to disinfect waterborne disease-causing microorganism. TiO_2 photocatalysts and metal oxide nanoparticles are considered among the most effective nanomaterials which showed antimicrobial properties (Azizi-Lalabadi et al. 2019; Darab et al. 2007; Othman et al. 2014). The potential of metal ions for disinfection in water has been investigated by many researchers. Some of the nanoparticle used for removing microbial contamination has been discussed below.

5.11.1 Silver Nanoparticles

Silver is widely used metal because of its microbial inactivation property in water and low toxicity (Pandey et al. 2019). Silver nanoparticles are obtained from its salts like silver chloride and silver nitrate. Although, the antibacterial effect is dependent on size, smaller silver nanoparticles (8 nm) have been most effective, while larger particle size (11–23 nm) showed lower bactericidal activity. Truncated triangular silver nanoplates showed better antibacterial property in comparison to rod-shaped and spherical nanoparticles which indicate their dependency on shape. The mechanisms which is involved during bactericidal effects of silver nanoparticles include, formation of free radicals that damage the bacterial membranes, damaging DNA, alteration in the properties of the membrane by adhering to the cell surface and by damaging enzyme.

5.11.2 TiO_2 Nanoparticles

TiO_2 nanoparticles are the emerging and promising photocatalysts for purification of water. The basic mechanism of photocatalysts such as low-cost TiO_2 that have good photoactivity and less toxicity involves production of highly reactive oxidants, like OH radicals, for disinfecting microorganisms (Desa and Kowshik 2009; Senarathna et al. 2017). TiO_2 after simulated solar exposure of 8 hours has been shown to decrease the viability of many waterborne pathogens. Full inactivation of faecal coliforms in the presence of sunlight has been reported, expressing the photocatalytic disinfection potential of TiO_2.

5.11.3 CNTs and Others

CNTs have shown their potential in removal of bacterial pathogens. CNTs have received special attention for their ability of removing biological contaminants from the water (Al-Jumaili et al. 2017; Kang and Choi 2009; Liu et al. 2018; Teixeira-Santos et al. 2021). CNTs have shown antimicrobial characteristics against wide range of microbes including bacteria and viruses. Adsorption of cyanobacterial toxins over CNTs is also high when compared to the carbon-based adsorbents due to large specific surface area, large composition of mesoporous volume and external diameter of CNTs.

5.12 Mechanism of Nano-Remediation of Wastewater

Nanotechnology-based pathways, which are being employed for wastewater remediation, are adsorption and biosorption, nanofiltration, photocatalysis, disinfection and pathological control, sensing and monitoring. The mechanism of the nano-remediation of wastewater is discussed below.

5.12.1 Adsorption and Biosorption

Adsorption a surface phenomenon and exothermic process, which involves process of transfer of a phase called adsorbate, on a solid surface called adsorbent to form a monomolecular layer over the surface through physicochemical or the chemical interactions under some specific conditions (Sadegh et al. 2017). Biosorption is an adsorption process in which biological substances like bacteria, fungi and algae act as an adsorbents. Owing to their intrinsic property they bind up heavy metals, from a very dilute aqueous solution or by making use of ATP or through spontaneous physicochemical pathways of uptake. The process of biosorption mainly involves microprecipitation, cell surface and complexation ion exchange.

5.12.2 Nanofiltration

Water filtration is a process of reducing the amount of particulate matter, like suspended particles, microbes and other dangerous biological and chemical pollutant from polluted water to make it safe and clean for drinking, medical applications and pharmaceutical (Abdel-Fatah 2018; Shon et al. 2013). Membrane technology has been used widely and the most important development in membrane technology is the nanofiltration (NF) membrane. NF membranes possess molecular weight cut-off (MWCO) for the uncharged particles in nanometer range. NF membranes are the recent and most preferably used technology for wastewater treatment and drinking water. Nanofilters works on pressure-driven membrane process which lies between ultrafiltration and reverse osmosis, with a pressure in the range of 5–20 bars and pore size between 0.5 and 2.0 nm. NF is widely employed techniques for wastewater treatment due to its exclusive filtration mechanism and availability of several types of membranes. NF is are the suitable technique to filter out all the organic and inorganic contaminants, including harmful microbes from the wastewater.

5.12.3 Photocatalysis

Photocatalysis is a light-induced reaction accelerated by a catalyst. Photocatalysis involves a solid material which is a photocatalyst that absorbs radiation and induces chemical reaction. Photocatalysis is considered as one of the Advanced Oxidation Processes (AOPs) which comprises of in-situ production of potent chemical oxidants with the help of Fenton's reagent, UV light, hydrogen peroxide (H_2O_2), ozone or a catalyst. Photocatalysis is basically a surface phenomenon, and its mechanism involves five steps discussed below:

- Diffusion of pollutants on the surface of photocatalyst
- Adsorption of pollutants over the surface of photocatalyst
- Reaction of the adsorbed pollutants
- Desorption of the products formed, from the surface
- Removal or diffusion of products from the interface Some of the widely used nanostructured photocatalysts are Fe_2O_3, TiO_2, ZnO, zinc sulphide (ZnS), zirconium dioxide (ZrO_2), cadmium sulphide (CdS) and tungsten trioxide (WO_3) (Chimupala et al. 2020; Mustapha et al. 2020; Sagir et al. 2020).

5.12.4 Disinfection and Pathological Control

Nanomaterials possess tremendous ability to inactivate pathogens present in water, due to its large surface area and the specific reactivity. The mechanism of inactivation for the nanoparticles involves surface-based electrostatic interaction and the photochemical reactions. It induces the generation of reactive oxygen species (ROS), disruption of the cell walls and the targeted delivery of disinfecting agents. Due to excellent surface properties and reactivity of nanomaterials, pathogens inhibition in water can be done easily. Nanomaterials which are based on CuO, TiO_2, silver, polymeric nanoparticles and CNTs have been examined for the disinfection of wastewater (Deshmukh et al. 2019; Teixeira-Santos et al. 2021).

5.12.5 Sensing and Monitoring

Monitoring of quality water, on large scales, is a challenging work because of low concentrations of contaminants, variability and complexity of wastewater matrices. To solve these issues, fast and effective techniques are needed to be developed. In recent past, the researchers have shown their interest towards development of sensors based on nanomaterial to monitor quality of water. Due to their excellent properties, like excellent recognition of trace pollutant, and fast analysis, nanosensors are defined as a device or material that are sensitive towards alterations in surrounding stimuli, like heat, mechanical and chemical stress, changes in volume, gravitational, concentration and the magnetic, as well as electrical forces. They are used to provide physical, chemical and biological information about behaviour and character of nanoparticle from the nanoscale level to macroscopic level (Deshmukh et al. 2019; Lee and El-Sayed 2006; Luo et al. 2019; Segev-Bar and Haick 2013; Willner and Vikesland 2018). Nanosensors contains three main parts specifically, a recognizing component like nanometals, nanowires, nanotubes, nanoparticles, etc. which is connected to a transducer (such as amperometric, spectrophotometric, conductometric) and a display for real time monitoring.

5.13 Constraints in Application of Nanoparticles in Water Remediation

Commercialization of nanomaterials for wastewater and water technology depends on their impact on aquatic ecosystem. Several studies like toxicity tests, technology assessment, life cycle analysis and dispersal of nanomaterial in aquatic water bodies have been done to evaluate the health risks caused by nanomaterials (Ray et al. 2009; Subramaniam et al. 2019). The findings of such studies have given a better understanding of behaviour of nanoparticles like CNTs, silver nanoparticles and TiO_2 in aquatic system. Therefore, stakeholders from all fields have supported to form a new laws and regulations or the modification of present ones. Several studies have produced contradictory results, because no general standards for experimental procedures and measurements have been formulated, which slows the necessary process of decision. Nanomaterials that goes into water, do not affect the humans directly, but there is a possibility to uptake nanomaterials through consumption of fish. So the impact of nanoparticles on aquatic organisms should be taken seriously. Nanoparticles are emitted into the environment through various point sources like landfill sites, production units and wastewater treatment plants. Nonpoint sources, for the emission of nanoparticles include washing machines, clothes or other substances that contain nanoparticles, for example, nanosilver nanoparticles are released from clothes after washing. Nanosocks that are commercially available leach 25% of its total silver content at pH 10 within minutes. So, it is necessary to perform the research on the toxicity of nanomaterials in air, water and soil. Moreover, the impact of nanoparticles in the food chain should also be studied thoroughly as it will affect many organism.

5.14 Conclusion and Future Prospectus

The biological resources like microorganisms and their enzymes, when applied successfully, could help in synthesis of nanoparticles biologically which can be termed as a potential strategy used in an eco-friendly and eco-incentive manner.

Microbial possess a significant potential for the synthesis of metal nanoparticles because they show less toxic nature alongwith high degradation capacity. It has been found that the mechanism of biologically synthesized nanoparticles are not clear, despite knowing that the stable nanoparticles are generated by selecting appropriate microorganisms and optimizing conditions. Therefore, there is a requirement of choosing appropriate microorganism or microbial consortia for sustainable production of nanoparticles on large-scale or commercial scale. Synthesis of nanoparticles by microorganism can be obtained without the use of pressure, high temperature, stabilizers, energy and toxic chemicals. More research is required to synthesize nanoparticles having wide range of organic functional groups by using microbial enzymes for selective and multi-pollutants removal from the wastewater. The synthesis of nanoparticles through microbes like bacteria, fungi, actinomycetes, yeast and algae has number of benefits like easy production, high efficiency, non-expensive and eco-friendly approach. The nanoparticles obtained from microorganism can be utilized at contaminated sites for the treatment of various pollutants. The residues which are left after degradation of pollutants by nanoparticles produced microbially are biocompatible in nature and can be separated by filtration or by using precipitation technique. The value-added products like construction material can be prepared using left residues by adding biochar, thus, there will be no waste at the end. Synthesis of nanoparticle opens a new channels for number of biotechnological applications using greener path. Bio synthesis of metallic nanoparticles has been done at laboratory scale but its production at industrial scale still needs to be investigated for the mass production. The use of efficient microorganism for nanotechnology can enhance the industrial economy at commercial level but unfortunately only 1% of the nanotechnological material have been commercialized. The cost-incentive and cost-effective production of nanoparticles from microbes is needed to make this procedure economically viable and sustainable. The cost-benefit analysis also needs to be conducted for the commercial exploitation of the usage of microbial nanoparticles because no cost related data is available yet. Utilization of salts, chemicals, stabilizing and reducing agents in chemical synthesis method are costly whereas in microbial process of synthesizing nanoparticles, the usage of metal salts as well as media for growth of microorganism is also expensive. The waste biomass that can be recycled can be prove to be an alternative for producing nanoparticles to reduce the expenditure. Research investigations about biosynthetic pathways of microorganism and research in genetic engineering may pave the way for breakthrough development of nanomaterials for their industrial exploitation as efficient and sustainable strategies in the field of bioremediation. Advanced tools on computation is required to use the omics derived data for good understanding of microbial processes. Thus, green chemistry can be applied successfully for the production of nanoparticles by microorganism and research efforts in this direction can be a giant step for the adoption of green nanotechnology.

References

Abdel-Fatah MA (2018) Nanofiltration systems and applications in wastewater treatment: Review article. Ain Shams Eng J 9:3077–3092. https://doi.org/10.1016/j.asej.2018.08.001

Abdel-Raouf N, Al-Homaidan AA, Ibraheem IBM (2012) Microalgae and wastewater treatment. Saudi J Biol Sci 19:257–275. https://doi.org/10.1016/j.sjbs.2012.04.005

Abia ALK, Ubomba-Jaswa E, du Preez M, Momba MNB (2015) Riverbed sediments in the Apies River, South Africa: recommending the use of both Clostridium perfringens and Escherichia coli as indicators of faecal pollution. J Soils Sed 15:2412–2424. https://doi.org/10.1007/s11368-015-1209-0

Aghalari Z, Dahms H-U, Sillanpää M, Sosa-Hernandez JE, Parra-Saldívar R (2020) Effectiveness of wastewater treatment systems in removing microbial agents: a systematic review. Glob Health 16:13. https://doi.org/10.1186/s12992-020-0546-y

Ahmad A, Senapati S, Kumar R, Ramani R, Srinivas V, Sastry M (2003) Intracellular synthesis of gold nanoparticles by a novel alkalotolerant actinomycetes, *Rhodococcus* Species. Nanotechnology 14:824

Akbarzadeh A, Davaran SM (2012) Magnetic nanoparticles: preparation, physical properties, and applications in biomedicine. Nanoscale Res Lett 7:144. https://doi.org/10.1186/1556-276X-7-144

Akita LG, Laudien J, Biney C, Akrong MO (2021) A baseline study of spatial variability of bacteria (total coliform, E. coli, and Enterococcus spp.) as biomarkers of pollution in ten tropical Atlantic beaches: concern for environmental and public health. Environ Sci Pollut Res https://doi.org/10.1007/s11356-021-15432-x

Al-Jumaili A, Alancherry S, Bazaka K, Jacob M (2017) Review on the antimicrobial properties of carbon nanostructures. Materials 10:1066. https://doi.org/10.3390/ma10091066

Alcántara C, Posadas E, Guieysse B, Muñoz (2015) Microalgae-based wastewater treatment. In: Handbook of marine microalgae. Elsevier, pp 439–455. https://doi.org/10.1016/B978-0-12-800776-1.00029-7

Amin MT, Alazba AA, Manzoor U (2014) A Review of Removal of pollutants from water/wastewater using different types of nanomaterials. Adv Mater Sci Eng 2014:1–24. https://doi.org/10.1155/2014/825910

Anderson BA, Romani JH, Phillips H, Wentzel M, Tlabela K (2007) Exploring environmental perceptions, behaviors and awareness: waterland water pollution in South Africa. Popul Environ 28:133–161. https://doi.org/10.1007/s11111-007-0038-5

Aragaw TA, Bogale FM, Aragaw BA (2021) Iron-based nanoparticles in wastewater treatment: a review on synthesis methods, applications, and removal mechanisms. J Saudi Chem Soc 25:101280. https://doi.org/10.1016/j.jscs.2021.101280

Arsiya F, Sayadi M, Sobhani S (2017) Arsenic (III) adsorption using palladium nanoparticles from aqueous solution. J Water Environ Nanotechnol 2. https://doi.org/10.22090/jwent.2017.03.004

Aryal S, Park H, Leary JF, Key J (2019) Top-down fabrication-based nano/microparticles for molecular imaging and drug delivery. Int J Nanomed 14:6631–6644. https://doi.org/10.2147/IJN.S212037

Asha, AB, Narain R (2020) Nanomaterials properties. In: Polymer science and nanotechnology. Elsevier, pp 343–359. https://doi.org/10.1016/B978-0-12-816806-6.00015-7

Azizi-Lalabadi M, Ehsani A, Divband B, Alizadeh-Sani M (2019) Antimicrobial activity of Titanium dioxide and Zinc oxide nanoparticles supported in 4A zeolite and evaluation the morphological characteristic. Sci Rep 9:17439. https://doi.org/10.1038/s41598-019-54025-0

Baig N, Kammakakam I, Falath W (2021) Nanomaterials: a review of synthesis methods, properties, recent progress, and challenges. Mater Adv 2:1821–1871. https://doi.org/10.1039/D0MA00807A

Balaji D, Basavaraja S, Deshpande R, Mahesh DB, Prabhakar B, Venkataraman A (2009) Extracellular biosynthesis of functionalized silver nanoparticles by strains of *Cladosporium cladosporioides* fungus. Colloids Surf B 68:88–92

Bhainsa KC, D'Souza S (2006) Extracellular biosynthesis of silver nanoparticles using the fungus *Aspergillus fumigatus*. Colloids Surf B 47:160–164

Bhat SA, Cui G, Li W. Wei Y, Li F (2020) Effect of heavy metals on the performance and bacterial profiles of activated sludge in a semi-continuous reactor. Chemosphere 241:125035. https://doi.org/10.1016/j.chemosphere.2019.125035

Cassivi A, Tilley E, Waygood EOD, Dorea C (2021) Evaluating self-reported measures and alternatives to monitor access to drinking water: a case study in Malawi. Sci Total Environ 750:141516. https://doi.org/10.1016/j.scitotenv.2020.141516

Castro L, Blazquez ML, Munoz JA, Gonzalez F, Ballester A (2013) Biological synthesis of metallic nanoparticles using algae. Nanobiotechnol. IET 7:109–116

Chakraborty N, Banerjee A, Lahiri S, Panda A, Ghosh AN, Pal R (2009) Biorecovery of gold using cyanobacteria and an eukaryotic alga with special reference to nanogold formation-a novel phenomenon. J Appl Phycol 21:145–152

Chatterjee S, Mitra A, Datta S, Veer V (2013) Phytoremediation protocols: an overview. In: Gupta DK (ed) Plant-based remediation processes, soil biology. Springer Berlin Heidelberg, Berlin, Heidelberg, pp 1–18. https://doi.org/10.1007/978-3-642-35564-6_1

Chen L, Deng Y, Dong S, Wang H, Li P, Zhang H, Chu W (2021) The occurrence and control of waterborne viruses in drinking water treatment: a review. Chemosphere 281:130728. https://doi.org/10.1016/j.chemosphere.2021.130728

Chimupala Y, Phromma C, Yimklan S, Semakul N, Ruankham P (2020) Dye wastewater treatment enabled by piezo-enhanced photocatalysis of single-component ZnO nanoparticles. RSC Adv 10:28567–28575. https://doi.org/10.1039/D0RA04746E

Dablemont C, Lang P, Mangeney C, Piquemal J-Y, Petkov V, Herbst F, Viau G (2008) FTIR and XPS study of Pt nanoparticle functionalization and interaction with alumina. Langmuir 24:5832–5841

Darab JG, Amonette AB, Burke DSD, Orr RD, Ponder SM, Schrick B, Mallouk TE, Lukens WW, Caulder DL, Shuh DK (2007) Removal of pertechnetate from simulated nuclear waste streams using supported zerovalent iron. Chem Mater 19:5703–5713. https://doi.org/10.1021/cm0607379

Das RK, Pachapur VL, Lonappan L (2017) Biological synthesis of metallic nanoparticles: plants, animals and microbial aspects. Nanotechnol Environ Eng 2:18. https://doi.org/10.1007/s41204-017-0029-4

Das L, Habib K, Saidur R, Aslfattahi N, Yahya SM, Rubbi F (2020) Improved thermophysical properties and energy efficiency of aqueous ionic liquid/MXene nanofluid in a hybrid PV/T solar system. Nanomaterials 10:1372. https://doi.org/10.3390/nano10071372

De Guire MR, Hu MZ, Gogotsi Y, Lu SW (eds) (2006) Ceramic nanomaterials and nanotechnology II: De Guire/Ceramic. Wiley, Hoboken, NJ, USA. https://doi.org/10.1002/9781118406083

Desa VS, Kowshik M (2009) Antimicrobial activity of titanium dioxide nanoparticles synthesized by sol-gel technique. Res J Microbio 4:97–103. https://doi.org/10.3923/jm.2009.97.103

Deshmukh SP, Patil SM Mullani SB, Delekar SD (2019) Silver nanoparticles as an effective disinfectant: a review. Mater Sci Eng C 97:954–965. https://doi.org/10.1016/j.msec.2018.12.102

El-Gendy NSh, Nassar HN (2021) Biosynthesized magnetite nanoparticles as an environmental opulence and sustainable wastewater treatment. Sci Total Environ 774:145610. https://doi.org/10.1016/j.scitotenv.2021.145610

Evans AE, Mateo-Sagasta J, Qadir M, Boelee E, Ippolito A (2019) Agricultural water pollution: key knowledge gaps and research needs. Curr Opin Environ Sustain 36:20–27. https://doi.org/10.1016/j.cosust.2018.10.003

Fang X, Wang Y, Wang Z, Jiang Z, Dong M (2019) Microorganism assisted synthesized nanoparticles for catalytic applications. Energies 12:190

Farrokhi M, Yang J-K, Lee S-M, Shirzad-Siboni M (2013) Effect of organic matter on cyanide removal by illuminated titanium dioxide or zinc oxide nanoparticles. J Environ Health Sci Eng 11:23. https://doi.org/10.1186/2052-336X-11-23

Fausey CL, Zucker I, Shaulsky E, Zimmerman JB, Elimelech M (2019) Removal of arsenic with reduced graphene oxide-TiO2-enabled nanofibrous mats. Chem Eng J 375:122040. https://doi.org/10.1016/j.cej.2019.122040

Fouda A, Saad E, Salem SS, Shaheen TI (2018) *In-vitro* cytotoxicity, antibacterial, and UV protection properties of the biosynthesized zinc oxide nanoparticles for medical textile applications. Microb Pathog 125:252–261

Ganachari SV, Banapurmath NR, Salimath B, Yaradoddi JS, Shettar AS, Hunashyal AM et al (2017) Synthesis techniques for preparation of nanomaterials. In: Martínez LMT, Kharissova OV, Kharisov BI (eds) Handbook of ecomaterials. Springer, Cham, pp 1–21

García-Pinel B, Porras-Alcalá C, Ortega-Rodríguez A, Sarabia F, Prados J, Melguizo C, López-Romero JM (2019) Lipid-based nanoparticles: application and recent advances in cancer treatment. Nanomaterials 9:638. https://doi.org/10.3390/nano9040638

Gehrke I, Geiser A, Somborn-Schulz A (2015) Innovations in nanotechnology for water treatment. Nanotechnol. Sci Appl l. https://doi.org/10.2147/NSA.S43773

Gude VG (2016) Microbial fuel cells for wastewater treatment and energy generation. In: Microbial electrochemical and fuel cells. Elsevier, pp 247–285. https://doi.org/10.1016/B978-1-78242-375-1.00008-3

Guo D, Xie G, Luo J (2014) Mechanical properties of nanoparticles: basics and applications. J Phys Appl Phys 47:013001. https://doi.org/10.1088/0022-3727/47/1/013001

Guo Y, Wang J, Shinde S, Wang X, Li Y, Dai Y, Ren J, Zhang P, Liu X (2020) Simultaneous wastewater treatment and energy harvesting in microbial fuel cells: an update on the biocatalysts. RSC Adv 10:25874–25887. https://doi.org/10.1039/D0RA05234E

He S, Guo Z, Zhang S, Wang J, Gu N (2007) Bio-synthesis of gold nanoparticles using the bacteria Rhodo*pseudomonas* capsulata. Mater Lett 61:3984–3987

Hernández Ruiz L, Ekumah B, Asiedu DA, Albani G, Acheampong E, Jónasdóttir SH, Koski M, Nielsen TG (2021) Climate change and oil pollution: a dangerous cocktail for tropical zooplankton. Aquat Toxicol 231:105718. https://doi.org/10.1016/j.aquatox.2020.105718

Hosea M, Greene B, McPherson R, Henzl M, Dale Alexander M, Darnall DW (1986) Accumulation of elemental gold on the alga *Chlorella vulgaris*. Inorg Chim Acta 123:161–165

Ingle A, Rai M, Gade A, Bawaskar M (2009) Fusarium solani: a novel biological agent for the extracellular synthesis of silver nanoparticles. J Nanopart Res 11:2079

Iqtedar M, Aslam M, Akhyar M, Shehzaad A, Abdullah R, Kaleem A (2019) Extracellular biosynthesis, characterization, optimization of silver nanoparticles using *Bacillus mojavensis* BTCB15 and its antimicrobial activity against multidrug resistant pathogens. Prep Biochem Biotechnol 49:136–142

Issa B, Obaidat I, Albiss B, Haik Y (2013) Magnetic nanoparticles: surface effects and properties related to biomedicine applications. Int J Mol Sci 14:21266–21305. https://doi.org/10.3390/ijms141121266

Jézéquel H, Chu KH (2006) Removal of arsenate from aqueous solution by adsorption onto titanium dioxide nanoparticles. J Environ Sci Health Part A 41:1519–1528. https://doi.org/10.1080/10934520600754201

Kang SH, Choi W (2009) Oxidative degradation of organic compounds using zero-valent iron in the presence of natural organic matter serving as an electron shuttle. Environ Sci Technol 43:878–883. https://doi.org/10.1021/es801705f

Karbasian M, Atyabi S, Siadat S, Momen S, Norouzian D (2008) Optimizing nano-silver formation by *Fusarium oxysporum* PTCC 5115 employing response surface methodology. Am J Agric Biol Sci 3:433–437

Kestens V, Roebben G, Herrmann J, Jämting Å, Coleman V, Minelli C, Clifford C, De Temmerman P-J, Mast J, Junjie L, Babick F, Cölfen H, Emons H (2016) Challenges in the size analysis of a silica nanoparticle mixture as candidate certified reference material. J Nanopart Res 18:171

Khandel P, Shahi SK (2018) Mycogenic nanoparticles and their bio-prospective applications: current status and future challenges. J Nanostruct Chem 8:369–391

Khan R, Israili SH, Ahmad H, Mohan A (2005) Heavy metal pollution assessment in surface water bodies and its suitability for irrigation around the Neyveli Lignite Mines and associated industrial complex, Tamil Nadu, India. Mine Water Environ 24:155–161. https://doi.org/10.1007/s10230-005-0087-x

Khin MM, Nair AS, Babu VJ, Murugan R, Ramakrishna S (2012) A review on nanomaterials for environmental remediation. Energy Environ Sci 5:8075. https://doi.org/10.1039/c2ee21818f

Kiguchi O, Sato G, Kobayashi T (2016) Source-specific sewage pollution detection in urban river waters using pharmaceuticals and personal care products as molecular indicators. Environ Sci Pollut Res 23:22513–22529. https://doi.org/10.1007/s11356-016-7437-z

Kokorina AA, Ermakov AV, Abramova AM, Goryacheva IYu, Sukhorukov GB (2020) Carbon nanoparticles and materials on their basis. Colloids Interfaces 4:42https://doi.org/10.3390/colloids4040042

Kshtriya V, Koshti B, Gour N (2021) Green synthesized nanoparticles: classification, synthesis, characterization, and applications. In: Comprehensive analytical chemistry. Elsevier, pp 173–222. https://doi.org/10.1016/bs.coac.2020.12.009

Kumari P, Alam M, Siddiqi WA (2019) Usage of nanoparticles as adsorbents for waste water treatment: an emerging trend. Sustain Mater Technol 22:00128. https://doi.org/10.1016/j.susmat.2019.e00128

Kumbhakar P, Ray SS, Stepanov AL (2014) Optical properties of nanoparticles and nanocomposites. J Nanomater 2014:1–2. https://doi.org/10.1155/2014/181365

Lee K-S, El-Sayed MA (2006) Gold and silver nanoparticles in sensing and imaging: sensitivity of plasmon response to size, shape, and metal composition. J. Phys Che B 110:19220–19225. https://doi.org/10.1021/jp062536y

Lengke MF, Fleet ME, Southam G (2006) Synthesis of platinum nanoparticles by reaction of filamentous cyanobacteria with platinum (IV)- chloride complex. Langmuir 22:7318–7323

Li Q, Chen Z, Wang H, Yang H, Wen T, Wang S, Hu B, Wang X (2021) Removal of organic compounds by nanoscale zero-valent iron and its composites. Sci Total Environ 792:148546. https://doi.org/10.1016/j.scitotenv.2021.148546

Liu D, Mao Y, Ding L (2018) Carbon nanotubes as antimicrobial agents for water disinfection and pathogen control. J Water Health 16:171–180. https://doi.org/10.2166/wh.2018.228

Liu J, Zhang L, Lu G, Jiang R, Yan Z, Li Y (2021) Occurrence, toxicity and ecological risk of Bisphenol A analogues in aquatic environment—a review. Ecotoxicol Environ Saf 208:111481. https://doi.org/10.1016/j.ecoenv.2020.111481

Logan TJ (1993) Agricultural best management practices for water pollution control: current issues. In: Agriculture and the environment. Elsevier, pp 223–231. https://doi.org/10.1016/B978-0-444-89800-5.50020-8

Lu H, Wang J, Stoller M, Wang T, Ba Y, Hao H (2016) An Overview of nanomaterials for water and wastewater treatment. Adv Mater Sci Eng 2016:1–10. https://doi.org/10.1155/2016/4964828

Luo Z, Tu Y, Li H, Qiu B, Liu Y, Yang Z (2019) Endocrine-disrupting compounds in the Xiangjiang River of China: spatio-temporal distribution, source apportionment, and risk assessment. Ecotoxicol Environ Saf 167:476–484. https://doi.org/10.1016/j.ecoenv.2018.10.053

Luong NH, Long NN, Vu LV, Hai NH, Nghia PT, Anh NTV (2011) Metallic nanoparticles: synthesis, characterisation and application. Int J Nanotechnol 8:227. https://doi.org/10.1504/IJNT.2011.038201

Madhura L, Singh S, Kanchi S, Sabela M, Bisetty K, Inamuddin (2019) Nanotechnology-based water quality management for wastewater treatment. Environ Chem Lett 17:65–121. https://doi.org/10.1007/s10311-018-0778-8

Magalhães-Ghiotto GAV, de Oliveira AM, Natal JPS, Bergamasco R, Gomes RG (2021) Green nanoparticles in water treatment: a review of research trends, applications, environmental aspects and large-scale production. Environ Nanotechnol Monit. Manag 16:100526. https://doi.org/10.1016/j.enmm.2021.100526

Maiti D, Tong X, Mou X, Yang K (2019) Carbon-based nanomaterials for biomedical applications: a recent study. Front Pharmacol 9:1401. https://doi.org/10.3389/fphar.2018.01401

Mandeep SP (2020) Microbial nanotechnology for bioremediation of industrial wastewater. Front Microbiol 11:590631. https://doi.org/10.3389/fmicb.2020.590631

McConnell WP, Novak JP, Brousseau LC, Fuierer RR, Tenent RC, Feldheim DL (2000) Electronic and optical properties of chemically modified metal nanoparticles and molecularly bridged nanoparticle arrays. J Phys Chem B 104:8925–8930. https://doi.org/10.1021/jp000926t

Mohseniazar M, Barin M, Zarredar H, Alizadeh, Shanehbandi D (2011) Potential of microalgae and Lactobacilli in biosynthesis of silver nanoparticles. Bioimpacts 1:149–152

Molazadeh M, Ahmadzadeh H, Pourianfar HR, Lyon S, Rampelotto PH (2019) The use of microalgae for coupling wastewater treatment with CO_2 biofixation. Front Bioeng Biotechnol 7:42. https://doi.org/10.3389/fbioe.2019.00042

Momeni S, Nabipour I (2015) A simple green synthesis of palladium nanoparticles with Sargassum alga and their electrocatalytic activities towards hydrogen peroxide. Appl Biochem Biotechnol 176:1–13

Moustafa MT (2017) Removal of pathogenic bacteria from wastewater using silver nanoparticles synthesized by two fungal species. Water Sci 31:164–176. https://doi.org/10.1016/j.wsj.2017.11.001

Mustapha S, Ndamitso MM, Abdulkareem AS, Tijani JO, Shuaib DT, Ajala AO, Mohammed AK (2020) Application of TiO_2 and ZnO nanoparticles immobilized on clay in wastewater treatment: a review. Appl Water Sci 10:49. https://doi.org/10.1007/s13201-019-1138-y

Nayak MK, Singh J, Singh B, Soni S, Pandey VS, Tyagi S (2017) Introduction to semiconductor nanomaterial and its optical and electronics properties. In: Metal semiconductor core-shell nanostructures for energy and environmental applications. Elsevier, pp 1–33. https://doi.org/10.1016/B978-0-323-44922-9.00001-6

Nazeer S, Ali Z, Malik RN (2016) Water quality assessment of River Soan (Pakistan) and source apportionment of pollution sources through receptor modeling. Arch Environ Contam Toxicol 71:97–112. https://doi.org/10.1007/s00244-016-0272-x

Othman SH, Abd Salam NR, Zainal N, Kadir Basha R, Talib RA (2014) Antimicrobial activity of TiO_2 nanoparticle-coated film for potential food packaging applications. Int J Photoenergy 2014:1–6.

Ovais M, Khalil AT, Islam NU, Ahmad I, Ayaz M, Saravanan M, et al. (2018) Role of plant phytochemicals and microbial enzymes in biosynthesis of metallic nanoparticles. Appl Microbiol Biotechno 102:6799–6814

Pandey A, Shankar S, Shikha ANK (2019) Amylase-assisted green synthesis of silver nanocubes for antibacterial applications. Bioinspired Biomim Nanobiomater 8:161–170. https://doi.org/10.1680/jbibn.17.00031

Parial D, Patra HK, Dasgupta AKR, Pal R (2012) Screening of different algae for green synthesis of gold nanoparticles. Eur J Phycol 47:22–29

Patel KD, Singh RK, Kim H-W (2019) Carbon-based nanomaterials as an emerging platform for theranostics. Mate Horiz 6:434–469. https://doi.org/10.1039/C8MH00966J

Prasad R (2016) Advances and applications through fungal nanobiotechnology. Springer International Publishing, Cham. ISBN 978-3-319-42989-2. http://www.springer.com/us/book/9783319429892

Price HD, Adams EA, Nkwanda PD, Mkandawire W, Quilliam RS (2021) Daily changes in household water access and quality in urban slums undermine global safe water monitoring programmes. Int J Hyg Environ Health 231:113632. https://doi.org/10.1016/j.ijheh.2020.113632

Radioactive Pollution (2014) Methods of measuring environmental parameters. Wiley, Hoboken, NJ, USA, pp 380–384. https://doi.org/10.1002/9781118914236.ch36

Ramanathan S, Gopinath SCB, Arshad MKM, Poopala P, Perumal V (2021) Nanoparticle synthetic methods: strength and limitations. In: Nanoparticles in analytical and medical devices. Elsevier, pp 31–43. https://doi.org/10.1016/B978-0-12-821163-2.00002-9

Ray PC, Yu H, Fu PP (2009) Toxicity and environmental risks of nanomaterials: challenges and future needs. J Environ Sci Health Part C 27:1–35. https://doi.org/10.1080/10590500802708267

Reghunadhan A, Kalarikkal N, Thomas S (2018) Mechanical property analysis of nanomaterials. In: Characterization of nanomaterials. Elsevier, pp 191–212. https://doi.org/10.1016/B978-0-08-101973-3.00007-9

Riddin T, Gericke M, Whiteley CG (2006) Analysis of the inter and extracellular formation of platinum nanoparticles by *Fusarium oxysporum* f. sp. lycopersici using response surface methodology. Nanotechnology 17:3482

Sadegh H, Ali GAM, Gupta VK, Makhlouf ASH, Shahryari-ghoshekandi R, Nadagouda MN, Sillanpää M, Megiel E (2017) The role of nanomaterials as effective adsorbents and their applications in wastewater treatment. J Nanostructure Chem 7:1–14. https://doi.org/10.1007/s40097-017-0219-4

Sagir M, Tahir MB, Akram J, Tahir MS, Waheed U (2020) Nanoparticles and significance of photocatalytic nanoparticles in wastewater treatment: a review. Curr Anal Chem 17:38–48. https://doi.org/10.2174/1573411016999200607172553

Salvadori MR, Ando RA, Nascimento CA, Corrêa B (2014) Bioremediation from wastewater and extracellular synthesis of copper nanoparticles by the fungus Trichoderma koningiopsis. J Environ Sci Health A Tox Hazard Subst Environ Eng 49(11):1286–1295

Salvadori MR, Ando RA, Nascimento CAO, Corrêa B (2017) Dead biomass of Amazon yeast: a new insight into bioremediation and recovery of silver by intracellular synthesis of nanoparticles. J Environ Sci Health A Tox Hazard Subst Environ Eng 52:1112–1120

Salvadori MR, Lepre LF, Ando RA, Oller do Nascimento CA, Corrêa B (2013) Biosynthesis and uptake of copper nanoparticles by dead biomass of *Hypocrea lixii* isolated from the metal mines in the Brazilian Amazon region. PLoS ONE 8:80519

Sau TK, Murphy CJ (2004) Room temperature, high-yield synthesis of multiple shapes of gold nanoparticles in aqueous solution. J Am Chem Soc 126:8648–8649

Santoro S, Estay H, Avci AH, Puglies L, Ruby-Figueroa R, Garcia A, Aquino M, Nasirov S, Straface S, Curcio E (2021) Membrane technology for a sustainable copper mining industry: The Chilean paradigm. Clean Eng Technol 2:100091. https://doi.org/10.1016/j.clet.2021.100091

Savage T, Rao AM (2004) Thermal properties of nanomaterials and nanocomposites. In: Tritt TM (ed) Thermal conductivity, physics of solids and liquids. Springer US, pp 261–284. https://doi.org/10.1007/0-387-26017-X_12

Saxena J, Sharma MM, Gupta S, Singh A (2014) Emerging role of fungi in nanoparticles synthesis and their applications. World J Pharmacy Pharmaceutical Sci 3:1586–1613

Segev-Bar M, Haick H (2013) Flexible sensors based on nanoparticles. ACS Nano 7:8366–8378. https://doi.org/10.1021/nn402728g

Senapati S, Ahmad A, Khan MI, Sastry M, Kumar R (2005) Extracellular biosynthesis of bimetallic Au-Ag alloy nanoparticles. Small 1:517–520

Senarathna ULNH, Fernando SSN, Gunasekara TDCP, Weerasekera MM, Hewageegana HGSP, Arachchi NDH, Siriwardena HD, Jayaweera PM (2017) Enhanced antibacterial activity of TiO2 nanoparticle surface modified with Garcinia zeylanica extract. Chem Cen J 11:7. https://doi.org/10.1186/s13065-017-0236-x

Seurinck S, Verstraete W, Siciliano SD (2005) Microbial source tracking for identification of fecal pollution. Rev Environ Sc Biotechnol 4:19–37. https://doi.org/10.1007/s11157-005-4997-7

Sikora A, Shard AG, Minelli C (2016) Size and f-potential measurement of silica nanoparticles in serum using tunable resistive pulse sensing. Langmuir 32:2216–2224

Singaravelu G, Arockiamary J, Kumar VG, Govindaraju K (2007) A novel extracellular synthesis of monodisperse gold nanoparticles using marine alga, Sargassum wightii Greville. Colloids Surf B 57:97–101

Singh RP (2001) Effect of wastewater disposal and extent of industrial pollution in and around Kanpur, Uttar Pradesh, India. Bull Eng Geol Environ 60:31–35. https://doi.org/10.1007/s10064 0000079

Singh P, Kim YJ, Zhang D, Yang DC (2016) Biological synthesis of nanoparticles from plants and microorganisms. Trends Biotechnol 34:588–599

Shan G, Surampalli R, Tyagi RD, Zhang TC (2009) Nanomaterials for environmental burden reduction, waste treatment, and nonpoint source pollution control: a review. Front Environ Sci Eng China 3:249–264. https://doi.org/10.1007/s11783-009-0029-0

Shankar S, Nill S (2015) Effect of metal ions and redox mediators on decolorization of synthetic dyes by Crude Laccase from a novel white rot Fungus Peniophora sp. (NFCCI-2131). Appl Biochem Biotechnol 175:635–647. https://doi.org/10.1007/s12010-014-1279-2
Sharma R, Kumar SDK, Sarkar M, Mishra BK, Puri V, Priyadarshini I, Thong PH, Ngo PTT, Nhu V-H (2021) Water pollution examination through quality analysis of different rivers: a case study in India. Environ Dev Sustain. https://doi.org/10.1007/s10668-021-01777-3
Shnoudeh AJ, Hamad I, Abdo RW, Qadumii L, Jaber AY, Surchi HS, Alkelany SZ (2019) Synthesis, characterization, and applications of metal nanoparticles. In: Biomaterials and bionanotechnology. Elsevier, pp 527–612. https://doi.org/10.1016/B978-0-12-814427-5.00015-9
Shon HK, Phuntsho S, Chaudhary DS, Vigneswaran S, Cho J (2013) Nanofiltration for water and wastewater treatment—a mini review. Drink Water Eng Sci 6:47–53. https://doi.org/10.5194/dwes-6-47-2013
Shrivastava KK, Shrivastava RK, Jain SV (2020) Pollution abatement strategies in a stretch of River Kahn at Indore, India. J Inst Eng India Ser A 101:829–834. https://doi.org/10.1007/s40030-020-00479-9
Shukla S, Khan R, Daverey A (2021) Synthesis and characterization of magnetic nanoparticles, and their applications in wastewater treatment: a review. Environ Technol Innov 24:101924. https://doi.org/10.1016/j.eti.2021.101924
Solomon SK (2005) Environmental pollution and its management in sugar industry in India: an appraisal. Sugar Tech 7:77–81. https://doi.org/10.1007/BF02942422
Some S, Mondal R, Mitra D, Jain D, Verma D, Das S (2021) Microbial pollution of water with special reference to coliform bacteria and their nexus with environment. Energy Nexus 1:100008. https://doi.org/10.1016/j.nexus.2021.100008
Subbaiya R, Saravanan M, Priya AR, Shankar KR, Selvam M, Ovais M et al (2017) Biomimetic synthesis of silver nanoparticles from *Streptomyces atrovirens* and their potential anticancer activity against human breast cancer cells. IET Nanobiotechnol 11:965–972
Subramaniam VD, Prasad SV, Banerjee A, Gopinath M, Murugesan R, Marotta F, Sun X-F, Pathak S (2019) Health hazards of nanoparticles: understanding the toxicity mechanism of nanosized ZnO in cosmetic products. Drug Chem Toxicol 42:84–93. https://doi.org/10.1080/01480545.2018.1491987
Teixeira-Santos R, Gomes M, Gomes LC, Mergulhão FJ (2021) Antimicrobial and anti-adhesive properties of carbon nanotube-based surfaces for medical applications: a systematic review. iScience 24:102001. https://doi.org/10.1016/j.isci.2020.102001
Teklehaimanot GZ, Coetzee MAA, Momba MNB (2014) Faecal pollution loads in the wastewater effluents and receiving water bodies: a potential threat to the health of Sedibeng and Soshanguve communities, South Africa. Environ Sci Pollut Res 21:9589–9603. https://doi.org/10.1007/s11356-014-2980-y
Thomas S, Harshita BSP, Mishra P, Talegaonkar S (2015) Ceramic nanoparticles: fabrication methods and applications in drug delivery. Curr Pharm Des 21:6165–6188. https://doi.org/10.2174/1381612821666151027153246
Uchida Y, Matsui K (2001) Optical properties of ZnS:Mn nanoparticles in sol-gel glasses. In: Studies in surface science and catalysis. Elsevier, pp 331–334. https://doi.org/10.1016/S0167-2991(01)82099-1
Ukhurebor KE, Athar H, Adetunji CO, Aigbe UO, Onyancha RB, Abifarin O (2021) Environmental implications of petroleum spillages in the Niger Delta region of Nigeria: a review. J Environ Manage 293:112872. https://doi.org/10.1016/j.jenvman.2021.112872
Ullah H, Khan I, Yamani ZH, Qurashi A (2017) Sonochemical-driven ultrafast facile synthesis of SnO2 nanoparticles: growth mechanism structural electrical and hydrogen gas sensing properties. Ultrason Sonochem 34:484–490
Upadhyayula VKK, Deng S, Mitchell MC, Smith GB (2009) Application of carbon nanotube technology for removal of contaminants in drinking water: a review. Sci Total Environ 408:1–13. https://doi.org/10.1016/j.scitotenv.2009.09.027

Vanden Bout DA (2002) Metal nanoparticles: synthesis, characterization, and applications (ed: Feldheim DL, Foss, Jr, CA). Marcel Dekker, Inc., New York and Basel. x+ 338 pp $150.00. ISBN: 0-8247-0604-8. J Am Chem So 124:7874–7875. https://doi.org/10.1021/ja015381a

Vigneshwaran N, Ashtaputre NM, Varadarajan PV, Nachane RP, Paralikar KM, Balasubramanya RH (2007) Biological synthesis of silver nanoparticles using the fungus *Aspergillus flavus*. Mater Lett 61:1413–1418

Wang ZL, Liu Y, Zhang Z (eds) (2003) Semiconductor nanoparticles. In: Handbook of nanophase and nanostructured materials. Springer US, Boston, MA, pp 813–848. https://doi.org/10.1007/0-387-23814-X_23

Wang J, Yang N, Tang H, Dong Z, Jin Q, Yang M, Kisailus D, Zhao H, Tang Z, Wang D (2013) Accurate control of multishelled Co3O4 hollow microspheres as high-performance anode materials in lithium-ion batteries. Angew Chemie Int Ed 52:6417–6420

Weng C-H (2021) Environmental concerns and pollution control in the context of developing countries. Environ Sci Pollut Res 28:46085–46088. https://doi.org/10.1007/s11356-021-15004-z

Willner MR, Vikesland PJ (2018) Nanomaterial enabled sensors for environmental contaminants. J Nanobiotechnol 16:95. https://doi.org/10.1186/s12951-018-0419-1

Wu Q, Miao W, Zhang Y, Gao H, Hui D (2020) Mechanical properties of nanomaterials: a review. Nanotechnol Rev 9:259–273. https://doi.org/10.1515/ntrev-2020-0021

Yaqoob AA, Parveen T, Umar K, Mohamad Ibrahim MN (2020) Role of nanomaterials in the treatment of wastewater: a review. Water 12:495. https://doi.org/10.3390/w12020495

Ye Y, Bruning H, Liu W, Rijnaarts H, Yntema D (2019) Effect of dissolved natural organic matter on the photocatalytic micropollutant removal performance of TiO_2 nanotube array. J Photochem Photobiol Chem 371:216–222. https://doi.org/10.1016/j.jphotochem.2018.11.012

Yu X, Qin A, Liao L, Du R, Tian N, Huang S, Wei C (2015) Removal of organic dyes by nanostructure ZnO-bamboo charcoal composites with photocatalysis function. Adv Mater Sci Eng 2015:1–6. https://doi.org/10.1155/2015/252951

Zhang JZ (2009) Optical properties and spectroscopy of nanomaterials. World Sci. https://doi.org/10.1142/7093

Zhang X, He X, Wang K, Yang X (2011) Different active biomolecules involved in biosynthesis of gold nanoparticles by three fungus species. J Biomed Nanotechnol 7:245–254

Zielińska A, Carreiró F, Oliveira AM, Neves A, Pires B, Venkatesh DN, Durazzo A, Lucarini M, Eder P, Silva AM, Santini A, Souto EB (2020) Polymeric nanoparticles: production, characterization. Toxicology and Ecotoxicology. Molecules 25:3731. https://doi.org/10.3390/molecules25163731

Chapter 6
Recent Trends in Synthesis and Applicability of GO-Based Nanomaterials in Environmental Remediation

Rimmy Singh, Sunil Kumar, and Rachna Bhateria

Abstract Graphene oxide (GO) has been discovered as the most attractive material for multidisciplinary research. Due to unique physico-chemical properties and abundance of functional groups, graphene oxide nanomaterials have achieved high end development for pollution treatment. Go-based nanomaterials have been coupled time to time with different elements to produce highly advanced GO-based materials for environmental remediation. GO has a potential of dispersing and solubilizing in aqueous media and other solvents which are polar and organic in nature. This enables its wet processing. Despite having various defects, GO-based nanomaterials can be easily functionalized, cut into pieces, doped, and punched holes. Apart from this, GO can be assembled into macroscopic materials with a property of living building block. Sheets of GO can be considered as individual molecule/particle or soft polymer material. Hence, it is important to have an overview of GO as 2D molecular structure which will enable us to study its various intrinsic properties and will also help in the development of this subject. Hence, the chapter is framed to provide insight about synthesis, structure, and application of graphene oxide nanomaterials in environmental remediation.

Keywords Graphene oxide · Exfoliation · Luminescent · Remediation · Sensing

6.1 Introduction

Graphene and graphene oxide (GO)-based nanomaterials are often utilized due to their unique properties. Their various physico-chemical properties include their high adsorption potential, thermal conductivity, electrical conductivity, and high specific

R. Singh
DPG Institute of Technology and Management, Gurugram, India

S. Kumar · R. Bhateria (✉)
Maharshi Dayanand University, Rohtak, India
e-mail: rachna.env@mdurohtak.ac.in

J. P. N. Rai and S. Saraswat (eds.), *Nano-biotechnology for Waste Water Treatment*,
Water Science and Technology Library 111,
https://doi.org/10.1007/978-3-031-00812-2_6

area. They possess 2D structure and several atomic layers and bear excellent properties. Graphene can be easily enclosed into fullerenes (0-dimensional), enveloped as carbon nanotubes (1-dimensional), or can be heaped as 3-dimesional graphite. It is 2D nanostructures which contribute to its basic structure and responsible for various allotropic forms which are highly crystalline in nature. These unique and beneficial properties possessed by graphene have attracted the attention of researchers and enable them to explore its applications in various fields. Graphene is often used individually, but nowadays it is used as graphene oxide (GO) due to its better oxidant properties. Hydroxyl (eOH), carboxyl (eCOOH), and epoxy (CeOeC) are present in GO which enhances their application and makes their synthesis and processing easy. GO spreads easily in comparison with graphene thus making its synthesis, processing, and wide applicability. GO has found its applicability in purification of drinking water due to its imperishable hydrophilicity. GO is also used to make various nanocomposites having varied applications owing to its vast surface area. However, high dispersibility of GO compounds in aqueous media often makes their separation difficult.

Sulfonated graphene sheets in nanosize were synthesized by Zhao et al. (2011a) for removal of 1-naphthol and naphthalene. Nowadays the highest potential of sorption is ~2.3–2.4 mmol/g in case of naphthalene and 1-naphthol. In carbon nanotubes, there is a need for oxidation processes of introduction of hydrophilic groups for improvement in sorption of metal ion, but for making graphene oxide nanosheets using "Hummer's method" from graphite, functional groups like COOH, −CdO, and −OH are introduced on graphene oxide nanosheet's surface. The introduced functional groups help in heavy metal (HMs) sorption. However, oxidized graphene such as graphene oxides has oxygen (O) in their functional groups on their surfaces. Due to this, graphene oxide has high surface area (2620 m^2 g^{-1}) which enhances sorption capacity of GO nanosheets. But for removal of HMs, the use of GO nanosheets is limited as a sorbent from aqueous solution (Zhao et al. 2011b), specifically in the presence of humic substances. Humic substances are prevalently available in environment and are having the potential to form strong complexation with the metal ions as they have functional group having attached O. So, to study GO nanosheets and their sorption mechanism with metal ions is significant in the presence of humic substances (Zhao et al. 2011c).

Currently, excessively engineered and modified GO has been proposed as a sorbent for water purification (Perreault et al. 2015a) from metallic and organic contaminants. Certainly, the advancement in adsorption of HMs such as cadmium (Cd) and lead (Pb) in addition to azo dyes, e.g., methylene blue, in wastewater has been reported using amino-modified GO, EDTA, and sulfonate (Wei et al. 2018).

In current years, GO is one of the carbon nanomaterials as it has strong adsorption potential with high surface area. GO is a thick sheet of carbon atoms (thickness 0.334 nm, surface area 2630 m^2 g^{-1}) as hexagonally arrayed sp^2 structures (Schedin et al. 2007). Graphene oxide is a significant graphene derivative; it is a potential substance for preconcentrating HMs ions as it has large specific area, various functional groups, as well as high mechanical strength (Chandra et al. 2010). GO is proved to be much beneficial in removal of HMs solution (like Pb (II)) (Weng et al. 2019). However, graphene sheets have sturdy inter-functional bond which results in reduced surface area, less dispersion as well as high agglomeration, and reduced adsorption potential (Ramesha et al. 2011). Materials utilizing GO involved chemically modified GO, fabricating metal/GO oxide, organic/GO compound, and photocatalytic/GO composites were produced systemically to enhance and remove different HMs ions from wastewater as outstanding adsorbents of high efficiency. Although these amazing advances have been made, GO-based composites still face several important hurdles: (i) The GO-based composite interaction mechanisms with HMs ions remain unknown due to varied surface characteristics of several GO-based composites; (ii) great endeavors are required for reuse and recovery of GO-based composites, since the aqueous solution's dispersion is high and metal ions strongly interact with functional groups; (iii) actually familiar of minimal cost, the large volume along with green synthesis methods, cheap costs continue to be a demanding challenge.

6.2 Synthesis

Graphene was discovered in 2004 (Novoselov et al. 2004). After that, its exclusive characteristics attracted considerable interest from several disciplines of research. GO is a graphene derivative that has gained recognition in the remediation of environmental pollution. Till date, GO techniques of preparation include mainly the Brodie method (Brodie 1859), Hummers' method (Hummers and Offeman 1958), and Staudenmaier method (Staudenmaier 1898). Synthesis of GO was initially showed by Brodie (1859) at "University of Oxford." Some graphite was strongly blended with potassium chlorate in synthesis (ratio 1:3), in a distiller; this prepared mixture was cast up. Then, suitable quantities of fuming nitric acid were provided. Distillery was kept in a water bath and temperature was maintained at 60 °C for some time until there were no yellow fumes. The material was then put into enough distilled water and cleaned to get free from acid and salt by decanting. It was dried and oxidized with an equal amount of nitric acid and potassium chlorate until there were no more changes. Ultimately, the GO obtained was vacuum dried at 100 °C. The process was repeated with similar quantity of nitric acid and potassium chlorate till no change occurs. Then, prepared GO was dried in vacuum at 100 °C.

A natural mineral, flake graphite is purified to remove heteroatomic contamination (Fig. 6.1). For chemical reactions, this is the source of graphite (Wissler 2006). This flake graphite is used to prepare GO so that it can be easily dissolved in water. Besides, this can be used on broader scale to make large graphitic films to bind carbon products as well as a part of cathode (lithium batteries) (Ishikawa and Nagaoki 1986; Maire et al. 1968; Ishikawa et al. 1978; Touzain et al. 1986). Furthermore, the hydrophilic nature of the prepared GO permits its uniform deposition on the substrates as thin film, which is essential for electronic application (Mkhoyan et al. 2009). Also, it is most important to transform GO back into its conductive graphitic material, i.e., in bulk or thin films. However, chemical reduction can be used to partially restore the

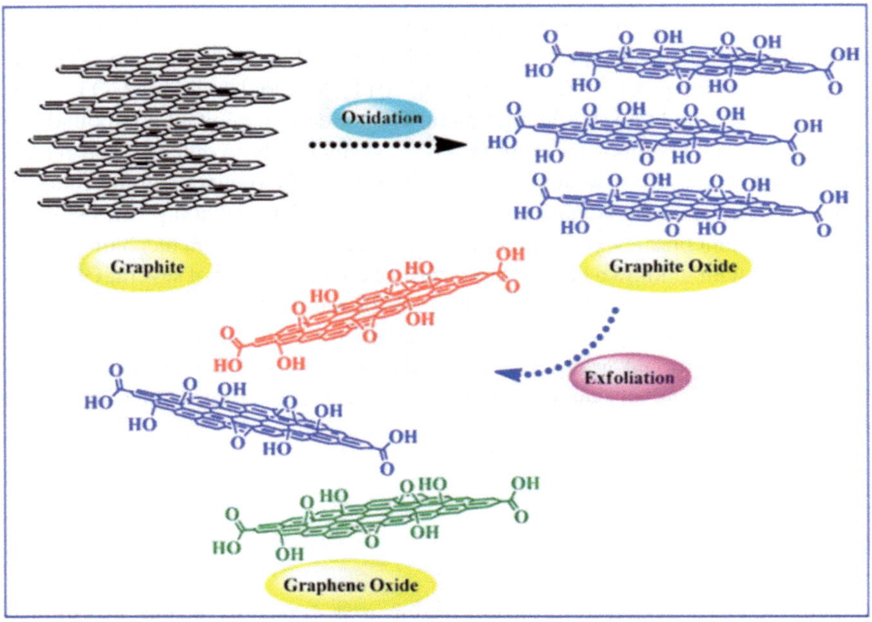

Fig. 6.1 Preparation of graphene oxide (Yu et al. 2020)

graphitic structure (Li et al. 2008) to chemically converted graphene (CCG). By using these conditions, graphitic structure cannot be restored and numerous deficiencies have been introduced (Gomez-Navarro et al. 2010).

Graphite is a "stack of many sheets of graphene" bind together by week van der Waals force. However, if these bonds are broken possibly, graphene can be produced from highly pure sheets of graphite. In the process, van der Waals forces are broken mechanically or chemically by exfoliation and single sheets of graphene can be obtained. An attempt was made in this field and potassium metal was applied to introduce a "pure graphite sheet" followed by flaking using ethanol to separate C sheets (Viculis et al. 2003).

A process of repeating peeling is basically called exfoliation. Novoselov et al. (2005) subjected a pyrolytic graphite sheet (thickness 1 mm) to dry etching in "oxygen plasma" for preparation of 5 μm deep mesas having an area of 0.4 to 4 mm^2. Further, placed on a photoresist, baked to get them adhered to mesas. Scotch tape was used to peel off graphene sheets (layers). Flakes of graphene sheets which were adhered to photoresist got liberate in acetone and then channelized to silicone substrate. Graphene sheets ranging from "single to few layer" could be seen in them.

Single- to few-layered graphene sheets were prepared by exfoliation and dispersion of pure graphite in N-methyl-pyrrolidone (Hernandez et al. 2008). The monolayer yield from 1 wt% to 12 wt% could be enhanced with processing. The energy needed for exfoliating graphite into one-layered graphene was anticipated by the

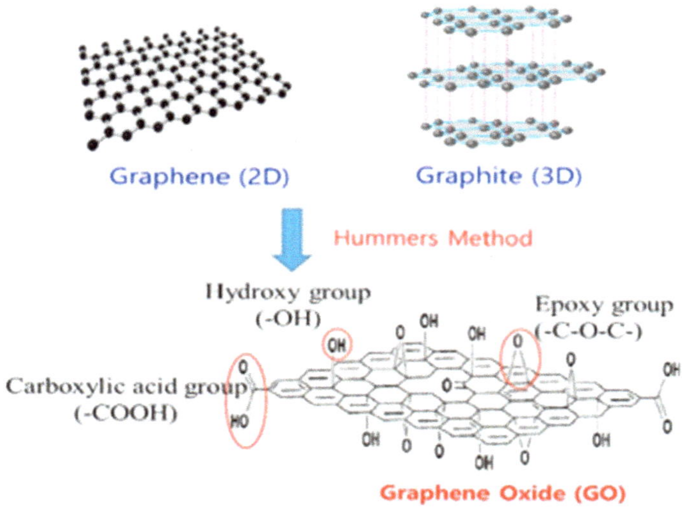

Fig. 6.2 Graphene oxide produced from graphite (Lingamdinne et al. 2019)

solvent-graphene inter-relation, with the solvent having surface energy similar to graphene. Graphite's exfoliation in organic solvent (a liquid-based exfoliation) has assured graphene production at larger scale. Similarly, single- to few-layer graphene sheets were fabricated in a same way, through the dispersion of graphite powder in sodium dodecylbenzene sulfonation, and accompanied by sonication, for exfoliation of graphite into graphene (Lotya et al. 2009). In Fig. 6.2, different images of graphene can be seen prepared through different methods. Different efforts have been done to produce the graphene from graphite as well as from graphite oxide powder applying solvents (Si and Samulski, 2008; Tung et al. 2008; Tung et al. 2009).

Organic molecules are required for hydrothermal methods to occur (Fig. 6.3a) and alkaline media (Fig. 6.3b). Different temperatures from 160 and 180 °C are being used in hydrothermal method (Teflon-line autoclave) (Zhou et al. 2014) also known by the name of "solvo-chemical method." Mu et al. (2017) stated usage of in situ or one-pot production magnetic graphene/ polyaniline nanocomposite by "intercalation polymerization" and hydrothermal co-precipitation method for dyes elimination.

6.3 Morphology and Type of Graphene Oxide-Based Nanomaterials

Spectroscopic methods such as FT-IR, XRD, and XPS can be used for the synthesis and knowing structural functionality of the prepared MGOs. Microscopic techniques such as SEM and AFM were used to know the size, dimensions, porosity, and surface morphology of the fabricated MGOs.

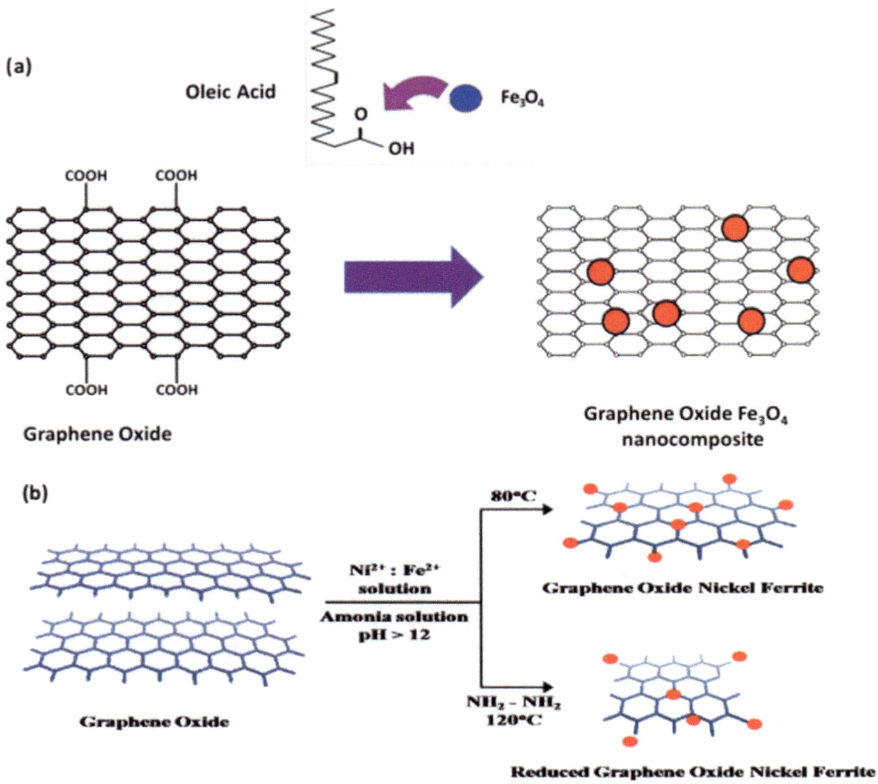

Fig. 6.3 **a** Synthesis of Go-Fe_3O_4 nanoparticle hybrid **b** porous magnetic nickel ferrite-Go-rGO-based nanocomposites preparation

Major adsorption properties of the surface area and adsorption on surface were assessed by BET analysis. The magnetometer was used to measure the magnetic property of MGOs. The XRD is used mostly for the identification of MGO production, structure, and crystalline character. For GO, the peak occurs in a range of 8–12° as revealed by XRD. However, by magnetization, GO reduces its crystallinity as mesoporous carbonate nature is increased with a change in the actual site of GO peak (Maaz et al. 2009). Also, XRD peaks are used for ferrite identification in MGOs. MGO's size decreases as porosity increases and hence, diffraction peaks shift toward low range (Marcano et al. 2010). FT-IR shows the functional groups present and MGO's formation along with peaks in between 1100–1300 cm^{-1} and 1400–1600 cm^{-1} range of wavelengths.

Peaks in a wavelength range of 1400–1600 cm^{-1} determines the presence of "stretching vibrations of epoxy groups" of GO in MGOs. Peaks in between 1650 and 1750 cm^{-1} range and a broad peak around 3000–3400 cm^{-1} determine the presence of carboxylic acids in MGOs (Senthilkumar et al. 2011). Primarily, FT-IR indicates ferrites in addition to magnetic material in synthesized MGOs. In octahedral and tetrahedral M–O bonds, spinel magnetic composites are indicated by the peaks in the range of 500–600 cm^{-1} (Nejati and Zabihi 2012).

6.3.1 *Wrinkled Reduced GO Sheets*

Carbon materials including carbon fibers, carbon nanotubes, and graphene (Sawangphruk et al. 2013), conducting polymers including polypyrrole (Davoglio et al. 2013), polyaniline (Chen et al. 2014b), and polythiophene and transition metal oxides were explored to synthesize fiber-shaped electrodes for flexible supercapacitors.

The two-dimensional materials like graphene are attracting research interests owing to large specific surface area (Booth et al. 2008), room-temperature electron mobility (Bolotin et al. 2008), and good mechanical flexibility (Lee et al. 2008), which are widely used for the preparation of composite materials for supercapacitor electrodes (Wang et al. 2009). As the graphene oxide gets reduced with time, restacking of reduced graphene oxide (rGO) sheets often contributes to reduction in functional groups present at the surface, reduced specific surface area, and restricted charge mobility. The PEDOT: PSS can act like isolated layers within rGO nanosheets because of the "oxygen rich functional groups of GO" and hydrophilic features offered by PEDOT: PSS. Hence, graphene nanosheet's restacking is prevented to a certain extent if rGO and PEDOT:PSS are well distributed with each other.

The wet spinning method is used for preparation of rGO-PEDOT: PSS composite fibers with wrinkled structure at the surface and pores on inner side. Based on composite fibers, FSCs and ISCs show outstanding electrochemical properties and good flexibility with high specific capacitance and energy density.

The main issue which needs to be considered while introducing the material into devices is "formation of wrinkles" (Zhang et al. 2015; Kim et al. 2014). There are theoretical (Zhang and Arroyo 2014) as well as research studies based on the "effect of wrinkles on electrical conduction of different types of graphene"—GO (Cote et al. 2010), rGO (Ladak et al. 2013), and pristine, with varied results (Xu et al. 2009). According to theoretical predictions, wrinkles create additional scattering sites which reduce electrical conductivity (Guinea et al. 2009).

TEM can be used to characterize the structure of rGO sheets. It is also used for characterization of electronic properties along with the influence of wrinkles. It was observed that predominant factor in the measured values was contact. Large amount of wrinkles allows the large contact areas with the probe (Fig. 6.4).

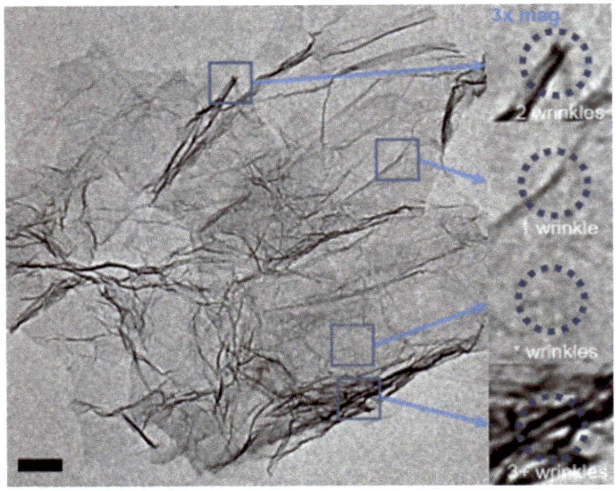

Fig. 6.4 TEM Image showing rGO on SiN membrane (Nilsson et al. 2017)

6.3.2 Luminescent2D GO Sheets

Graphene is hexagonally arranged, individual thick sheet of carbon atom with sp2 hybridization. It has exclusive characteristics such as flexibility, optical transparency, high mobility, environmental stability, and robustness which make graphene a capable material to use in the field of nanoelectronics. The conduction and valence bands are cone-shaped in the band structure of graphene and meet at the K point of the Brillouin zone and a bandgap up to zero (Castro et al. 2007). Graphene has been researched to produce an adaptable bandgap with PL behavior despite the absence of any bandgap (Castro et al. 2007; Partoens and Peeters, 2006; Elias et al. 2009). All of the methods involve generation and misuse of the faults. These faults of graphene and its derivatives are considered significant to generate a compatible bandgap and light emissions. GO share covalent bond with independent fluorophores, for example hydroxyl, epoxy and carboxyl groups (Eda and Chhowalla 2010). Above-mentioned fluorophores are "aromatic π-conjugated sp^2 -hybridized subsystems of carbon atoms" and a sp^3 zone surrounds it. For GO, PL is pH-dependent due to the occurrence of different functional groups having oxygen atom in GO sheets. Through the reduction of GO, reduced GO (rGO) is formed (Bagri et al. 2010). GO and rGO have different behaviors primarily due to reduced atom of oxygen present in them. rGO and GO have unique PL characteristics and thoroughly studied and modified over a wide spectral range. In general, the luminescent samples are the solutions or solids of rGO or GO (Pal 2015; Quan et al. 2017).

In recent times, GO Quantum Dots (GOQD) are gaining researchers' attention because of its exclusive properties like robust chemical inertness, high luminescent, low toxicity, and simple synthesis. It is a type of luminescent nanomaterials prepared from carbon having outstanding optical-electronic properties and higher biocompatibility and also has a "single atom layered structure with lateral dimension" which ranges from 2 to 20 nm. The functional groups containing oxygen atom attached at the edges and in the basal plane make thin sheet of GOQD (Ponomarenko et al. 2008). More significantly, they also lead other materials into a small number of single layers and create a variety of inestimable photophysical structures and luminous features that open up a brand-new realm as a source of 2D materials.

6.3.3 GO-Based Membranes

In recent times, materials based on carbon, viz. graphene, its derivatives like graphene oxide (GO), carbon nanotubes (CNTs), are considered best in membrane-based separation areas due to their high resistance to organic solvents, strong mechanical strength, and availability (Kalra et al. 2003). For the fabrication of novel separation membrane, GO was opted as one of the rising nano-building materials along with them. The mesoporous GO sheets are successfully prepared by reoxidizing GO using $KMnO_4$ and assemble them into laminar GO membranes subsequently to separate molecules. Due to the introduction of in-plane pores, the amount of effective channels for transportation of water is increased and it also remarkably diminishes the path of transport (Fig. 6.5a, b). As compared with original GO membranes, mesoporous GO membranes exhibit around twofold to threefold increase in permeability having similar rejection rate for small molecules (3 nm). Additionally, excellent structural stability is shown by mesoporous GO membranes as verified by the process of pressure loading and releasing.

Composite (with many layers) (Ganesh et al. 2013) or free-standing membranes, GO membranes have been developed by researchers. There are different types of techniques found in the literature, like simple filtration (Park and Ruoff 2009), drop-casting (Nair et al. 2012), and vacuum filtration (Huang et al. 2015). On a supporting membrane surface, GO layers are deposited as a "coating layer" using layer-by-layer (LBL) (Gao et al. 2014; Goh et al. 2015), pressure-assisted self-assembly (Hung et al. 2014), and spray coating (Moon et al. 2013) techniques. GO-based polymeric membranes are used in water purification however the underlying polymeric membrane is poorly adhesive to GO layers. It mechanically renders GO layers

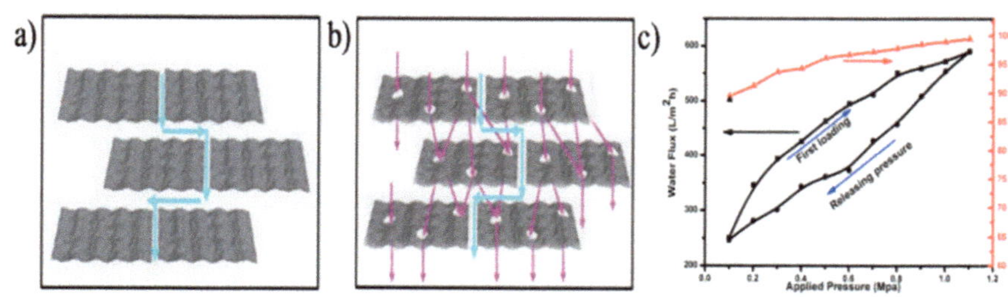

Fig. 6.5 Diagrams showing transport route of original GO (**a**); mesoporous membrane (**b**); separation property offered by mesoporous GO membrane (Ma et al. 2017)

fragile, so at the time of water treatment they may readily be removed. Therefore, it is highly crucial to enhance adhesion within the polymer membrane and deposited GO layers as a prerequisite of high-performance GO membrane. Several strategies are used to enhance the stability of GO layers on the membrane surface. Few approaches involve GO flakes functioning with several functional groups. Minář et al. (2019) described that by the establishment of covalent connection between GO and polyamide surface chains, the usage of functional GO with poly(β-caprolactone) might enhance adhesivity.

6.4 Structural Models

The actual GO chemical structure has been the subject of significant dispute throughout the years beyond the operational oxidative processes, and still, no clear model today is present. Various explanations are there, but it is the intricacy of the content that makes the fundamental contribution (involving variability from sample to sample) because of its berthollide character and amorphous and less accurate characterization techniques. Despite these limitations, considerable attempts have been made to comprehend the structure of GO, and most of it successfully applied.

Earlier, the GO structural model consisted of discrete "repeat units." Hofmann and Holst's structure (Fig. 6.6) consists of epoxy groups (molecular formula C_2O) which are distributed across the basal planes of graphite (Hofmann and Holst 1939). In 1946, Ruess proposed a variation in this model, which incorporates hydroxyl groups into the basal plane that accounts for the "hydrogen content" of GO (Ruess and Monatsh 1946). Instead of sp^2 hybridization, Ruess's model changed the basal plane structure into sp^3 hybridized model of Hofmann and Holst. Still, this model is based on the assumption of repeated unit. But, one-fourth of cyclohexanes at 1, 3 positions consisted of epoxides whereas hydroxylated at 4th position, through formation of regular lattice structure. Also, this gets the support from Mermoux based on the observation of structural similarities to poly(carbon monofluoride), $(CF)_n$ (Mermoux et al. 1991). The structure requires "the formation of C–F bonds through the complete rehybridization of the *sp*^{2} planes in graphite to *sp*^{3} cyclohexyl structures" (Dubois et al. 2006). In 1969, a model was suggested that eliminated epoxy and ether entirely from corrugated backbone replacing normal quinoid species (Scholz and Boehm 1969). The formation mechanism of GO is the best contribution to understand the chemical nature of GO 3 oxidation protocols (Nakajima and Matsuo 1994).

Fig. 6.6 Different structural models of GO (Szabo et al. 2006; Gao et al. 2015)

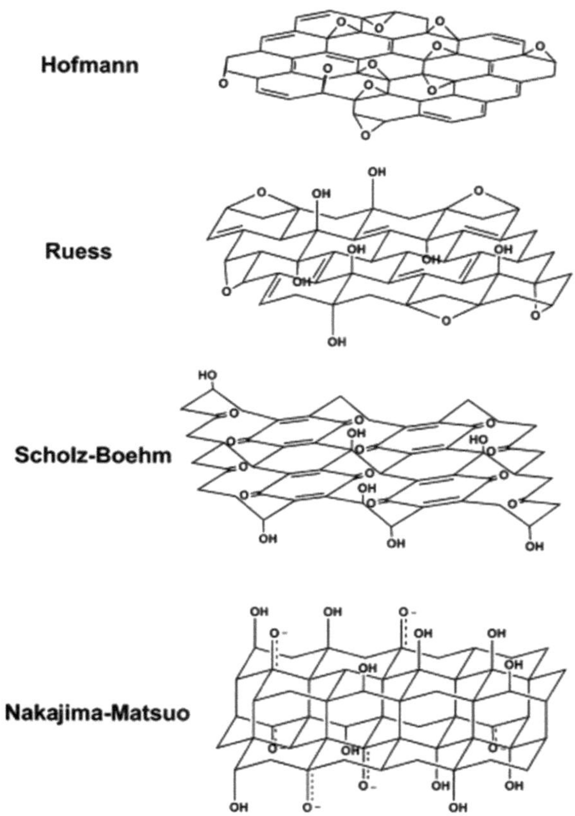

6.5 Applications

6.5.1 Adsorbents

Heavy metals are major pollutants in waste water that possess serious threat to the health of human beings. They can be removed from the from the waste water using various techniques including adsorption. However, adsorption as a method for the purification of water is known to be the efficient, economical, and facile in operation in comparison with conventional procedures (Oraby and Eksteen 2015; Koduru et al. 2014). Controversially, inconvenience of regeneration and filtration limits the use of adsorbents. However, these difficulties can be coped up by using magnetic materials for regeneration and filtration purposes. Hence, various researchers have adopted GO-based magnetic nanostructured materials for the abatement of various contaminants. Inverse spinal ferrites and their derivatives have dominated the water purification sector as they have chemical and magnetic stability and porous surface structure (Li et al. 2011a, 2011b). Nanometal ferrites were shown to have lower stability

(Maaz et al. 2009). Hence by adopting hybrid materials synthesized using magnetic ferrites and GO, these limitations can be overcome. In fact, various researchers have utilized magnetic organic composites (Lv et al. 2009) and functionalized magnetic carbon composites for the recovery of metals. Magnetic graphene oxide (MGO) composites having an approximate size of 10 nm have been synthesized and found to have maximum performance for water remediation (Xu et al. 2009). The developed superparamagnetic rMGO composites possess a maximum adsorption efficiency of 99.9% for As(III) as well as As (V) metal ions.

6.5.2 Photocatalyst

In recent era, remediation of global environmental contamination is encouraged by the "semiconductor-based photocatalytic decomposition of industrial organic byproducts" (Fox and Dulay 1993; Herrmann 1999; Hoffmann et al. 1995; Ravelli et al. 2009). The degradation and conversion of toxic contaminants into harmless products are referred to as "photocatalytic decomposition." Heterogeneous photocatalysis "involves organic synthesis, water splitting, photoreduction, hydrogen transfer, disinfection, water detoxification, and gaseous pollutant eradication, etc." In semiconductor photocatalysis, the photocatalytic reaction initiates with the exciton generation which adsorbs the energy, i.e., similar to the "band gap of semiconductor" (Bard 1979). Two-dimensional (2D) GO is the promising candidate for semiconductor-based photocatalysis due to its physico-chemical features as follows: (Krishnamoorthy et al. 2011) (1) An enhanced surface area is provided by the 2D GO that leads to the enormous number of active sites for organic dyes; (2) in fact, to start the redox reaction, 2D structure provides frequent charge transfer to the surface; (3) for an aqueous photocatalytic system, hydrophilic oxygen functionalities are promptly significant; (4) the band gap of GO composites via "controlled ratio of sp^2 and sp^3 hybridized carbon can probably be tuned from an insulator to a semiconductor" (Yan et al. 2009); (5) besides, "GO has been found to have a band gap between 1.7 and 4.3 eV which is tunable up to the range 2.6 to 1.9 eV to become active over the range of the visible spectrum for the efficient photocatalysis" (Jeong et al. 2009; Mathkar et al. 2012).

6.5.3 Antibacterial

In 2010, Akhavan and Ghaderi firstly reported the toxicity due to GO nanowalls in case of two Gram-positive bacterial strains "(*Staphylococcus aureus* (*S. aureus*)" and one Gram-negative (*Esherichia coli* (*E. coli*)) strain of bacteria (Fig. 6.7). They revealed that through uninterrupted association GO nanosheets can dominantly interact with bacteria at their pointed edges. Toxicity of GO nanowalls was investigated by "quantifying the intracellular products (concentration of RNA) in the

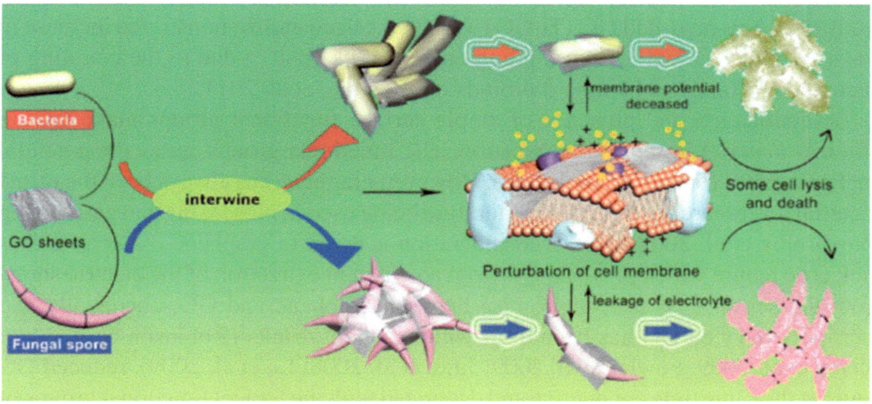

Fig. 6.7 A representation of GO and pathogen interaction mechanism. Fungal spores and bacterial cells are intertwisted with nested GO sheets (Chen et al. 2014a; Sajjad et al. 2017)

phosphate buffer solution (PBS) of the bacteria that were kept under the influence of GO nanowalls, due to the damage of bacterial cell membrane with the pointed peripheries of GO nanowalls." In the presence of GO nanowalls, the concentration of RNA was considerably higher in the solution in comparison with the concentration in control (in the absence of GO). However, antibacterial activity of GO was significantly higher in *S.aureus* than for *E.coli* bacterial strains. It was recognized that *E.coli* has strong resistance toward direct contact in the GO nanowalls as the Gram-negative *E.coli* has exterior covering in its structure. Thus, "according to the proposed mechanism by Akhavan and Ghaderi (2010) for the antibacterial activity of GO, edges of GO nanosheets can act as the cutters to damage the cell membrane that eventually results in bacterial decay."

In cell suspension and surface coatings, Perreault et al. (2015b) estimated the antimicrobial behavior of GO on the basis of their size. They also reported that by reducing GO sheet area, antimicrobial performance can be intensified on GO surface coatings.

6.5.4 Sensing

GO composited with other metal oxides enhances the sensing ability of GO. During gas molecule adsorption, the metal oxides are observed as attaching sites on the surface of GO. It is observed that metal-composited GO facilitates the adsorption of acidic gas molecules with high binding energy. The reason behind such behavior is the intensified amount of electron transfer from the metals to the adsorbed gaseous molecule. Consequently, the interaction strength among gas and metal increases due to the decrease in distance between metals and gases (Chen et al. 2014b). In a study, titanium-modified GO (Ti-GO) was synthesized and demonstrated increase in adsorption of CO with a high binding energy (70 $kJmol^{-1}$) due to the coupling of occupied p orbitals and empty d orbitals of Ti (Wang et al. 2011).

Diversification and presence of ample surface functional groups make GO an efficient material in sensing application. The functional groups act as the potential anchoring sites for various molecules. However, graphene has always been the prime material for research as reported in literature being used in biosensing and gas sensing (Novoselov et al. 2004). Fewer consideration is given to graphene oxide as it "has dielectric nature which prevents GO from an efficient extraction of the electric signal from the sensor." As a "compromise between graphene and GO, chemically and thermally reduced GO has been used with excellent results (Robinson et al. 2008; Zhang et al. 2009; Fowler et al. 2009; Jung et al. 2008; Lu et al. 2009), reduced GO being at the same time conductive and decorated with the typical functional groups of GO, which represent the adsorption sites. In order to increase selectivity, molecules, and nanoparticles (NPs) have been also anchored to the functional groups" (Cheng et al. 2012; Mohanty and Berry 2008; Li et al. 2011a, 2011b).

6.5.5 *Catalysis*

Basically, catalysis is categorized as homogeneous catalysis and heterogeneous catalysis. Homogeneous catalysis is "a catalytic process where catalysts and reactants coexist in the same phase, usually providing a higher reactivity and selectivity since the reactions are not limited by mass transfer" (Marras 2010). However, homogeneous catalysis is the practical challenge of efficiently separating unreacted reactants or spent catalysts from products after completion of a reaction. Contrarily, homogenous catalysis is a mechanism where both reactants and catalysts exist in different phase. In fact, this kind of configuration results into smaller contact area having active sites (Gomes et al. 2008); moreover, due to easy handling, ease of separation, and facile maintenance, heterogeneous catalysis is intensively used at industrial processes.

Nowadays, due to wide area of contact, low cost, and oxygen-containing groups, the use of GO has gained potential hike as Fe carrier "as the oxygen-containing group" enhances metal interaction. The facile reduction of formed metal oxides on the surface of GO has been facilitated by the presence of high electron density and extended π-system. For well-dispersed Fe nanoparticles, the defect sites arranged in piled layers act as excellent anchoring sites (Wei et al. 2018) and thus help in providing remarkable stability and catalytic properties (Moussa et al. 2014). "The oxygen-containing groups on the surface result like the acidic surface of GO and to increase the conversion of synthesis gas the high temperature treatment is usually applied to reduce the GO and further enhance the FTS reaction" (Zhao et al. 2013).

Sun et al. (2012) reported that for the activation of PMS, reduced graphene can act as metal-free catalyst by comparison with cobalt catalyst (Co_3O_4). The results concluded that huge variation was observed in the recovery of phenol from GO and rGO, suggesting that number of active sites is the determining factor for graphene, and there were zigzag edges.

6.6 Conclusion

This chapter summarizes the development and synthesis of GO-based nanocomposites and their potential environmental utility, especially in remediation. Due to the strong affinity and high efficiency for different metal ions, GO-based nanocomposites have been comprehensively studied as competitive adsorbents. For the environmental remediation of contaminants such as dyes, metal ions, and radionuclides, GO-based nanocomposites and their derivatives were reviewed in this chapter. Graphene oxides possess excellent efficiency for remediation of contaminants and other applications. In fact, GO has salient features of high adsorption performance, high surface area, nanosize, magnetic ability at different pH, chemical stability, robust structure, and thermal stability as well that makes it potential candidate in wide range of applications. Moreover, GO exhibits outstanding thermal, electronic, mechanical, magnetic, and optical properties that allow scientists to harvest fundamental research on it and have huge potential for intensive industrial applications. Therefore, it was recommended by researchers that fundamental parameters like defect-free engineering, crystallite graphitic domain size, and compactness of graphene layers have to be further improved to make more efficient and defect-free graphene oxide-based materials.

References

Akhavan O, Ghaderi E (2010) Toxicity of graphene and graphene oxide nanowalls against bacteria. ACS Nano 4:5731–5736

Bagri A, Mattevi C, Acik M, Chabal YJ, Chhowalla M, Shenoy VB (2010) Structural evolution during the reduction of chemically derived graphene oxide. Nat Chem 2:581

Bard AJ (1979) Photoelectrochemistry and heterogeneous photocatalysis at semiconductors. J Photochem 10:59–75

Bolotin KI, Sikes KJ, Jiang Z, Klima M, Fudenberg G, Hone J, Kim P, Stormer HL (2008) Ultrahigh electron mobility in suspended graphene. Solid State Comm 146:351–355

Booth TJ, Blake P, Nair RR, Jiang D, Hill EW, Bangert U, Bleloch A, Gass M, Novoselov KS, Katsnelson MI (2008) Macroscopic graphene membranes and their extraordinary stiffness. Nano Lett 8:2442–2446

Brodie BC (1859). On the atomic weight of graphite. Philos Trans Roy. Soc Lond 149:249e259.

Castro E, Novoselov K, Morozov S, Peres N, dos Santos J, Nilsson J, Guinea F, Geim A, Neto A (2007) Biased Bilayer Graphene: semiconductor with a gap tunable by the electric field effect. Phys Rev Lett 99:216802

Chandra V, Park J, Chun Y, Lee JW, Hwang IC, Kim KS (2010) Water-dispersible magnetite-reduced graphene oxide composites for arsenic removal. ACS Nano 4(7):3979–3986

Chen J, Peng H, Wang X, Shao F, Yuan Z, Han H (2014a) Graphene oxide exhibits broad-spectrum antimicrobial activity against bacterial phytopathogens and fungal conidia by intertwining and membrane perturbation. Nanoscale 6:1879–1889

Chen C, Xu K, Ji X, Miao L, Jiang J (2014b) Enhanced adsorption of acidic gases (CO_2, NO_2 and SO_2) on light metal decorated graphene oxide. Phys Chem Chem Phys 16:11031–11036

Cheng M, Yang R, Zhang L, Shi Z, Yang W, Wang D, Xie G, Shi D, Zhang G (2012) Restoration of graphene from graphene oxide by defect repair. Carbon 50:2581–2587

Cote LJ, Kim J, Zhang Z, Sun C, Huang J (2010) Tunable assembly of graphene oxide surfactant sheets: wrinkles, overlaps and impacts on thin film properties. SoftMatter 6(24):6096–6101.

Dubois M, Giraudet J, Guerin K, Hamwi A, Fawal Z, Pirotte P, Masin F (2006) Solid-state nmr studies of poly(dicarbon monofluoride) $(C2F)_n$. J Phys Chem B 110:11800–11808

Eda G, Chhowalla M (2010) Chemically Derived Graphene Oxide: Towards Large-Area Thin-Film Electronics and Optoelectronics. Adv Mater 22:2392

Elias DC, Nair RR, Mohiuddin TMG, Morozov SV, Blake P, Halsall MP, Ferrari AC, Boukhvalov DW, Katsnelson MI, Geim AK, Novoselov KS (2009) Control of graphene's properties by reversible hydrogenation: evidence for graphane. Science 323:610

Fowler JD, Allen MJ, Tung VC, Yang Y, Kaner RB, Weiller BH (2009) Practical chemical sensors from chemically derived graphene. ACS Nano 3:301–306

Fox MA, Dulay MT (1993) Heterogeneous photocatalysis. Chem Rev 93:341–357

Ganesh BM, Isloor AM, Ismail AF (2013) Enhanced hydrophilicity and salt rejection study of graphene oxide-polysulfone mixed matrix membrane. Desalination 313:199–207

Gao Y, Hu M, Mi B (2014) Membrane surface modification with TiO_2–graphene oxide for enhanced photocatalytic performance. J Membr Sci 455:349–356

Goh K, Setiawan L, Wei L, Si R, Fane AG, Wang R, Chen Y (2015) Graphene oxide as effective selective barriers on a hollow fiber membrane for water treatment process. J Membr Sci 474:244–253

Gomes JF et al (2008) Development of heterogeneous catalysts for transesterifcation of triglycerides. React Kinet Catal Lett 95(2):273–279

Gomez-Navarro C, Meyers JC, Sundaram RS, Chuvilin A, Kurash S, Burghard M, Kern K, Kaizer U (2010) Atomic structure of reduced graphene oxide. Nano Lett 10:1144–1148

Guinea F, Horovitz B, Le Doussal P (2009) Gauge fields, ripples and wrinkles in graphene layers. Solid State Commun 149(27–28):1140–1143

Hernandez Y, Nicolosi V, Lotya M, Blighe FM, Sun Z, De S, Mcgovern IT et al (2008) High-yield production of graphene by liquid-phase exfoliation of graphite. Nature Nanotechnol 3:563

Herrmann J-M (1999) Heterogeneous photocatalysis: fundamentals and applications to the removal of various types of aqueous pollutants. Catal Catal Today 53:115–129

Hoffmann MR, Martin ST, Choi W, Bahnemann DW (1995) Environmental applications of semiconductor photocatalysis. Chem Rev 95:69–96

Hofmann U, Holst R (1939) Über die Säurenatur und die Methylierung von Graphitoxyd. Ber Dtsch Chem Ges B 72:754–771

Huang T, Zheng B, Liu Z, Kou L, Gao C (2015) High rate capability supercapacitors assembled from wet-spun graphene films with a $CaCO_3$ Template. J Mater Chem A 3:1890–1895

Hummers WS, Offeman RE (1958) Preparation of graphitic oxide. J Am Chem Soc 80:1339

Hung W-S, An Q-F, De Guzman M, Lin H-Y, Huang S-H, Liu W-R, Hu C-C, Lee K-R, Lai J-Y (2014) Pressure-assisted self-assembly technique for fabricating composite membranes consisting of highly ordered selective laminate layers of amphiphilic graphene oxide. Carbon 68:670–677

Ishikawa T, Nagaoki T (1986) Shin tanso kogyo (New carbon industry), 2nd ed. Kindai hensyusya, Tokyo, pp 125–136.

Ishikawa T, Kanemaru T, Teranishi H, Onishi K (1978) Composites of oxidized graphite material and expanded graphite material. U.S. Patent 4094951, June 13.

Jeong H, Jin M, So K, Lim S, Lee Y (2009) Tailoring the characteristics of graphite oxides by different oxidation times. J Phys D: Appl Phys 42:065418

Jung I, Dikin D, Park S, Cai W, Mielke SL, Ruoff RS (2008) Effect of water vapor on electrical properties of individual reduced graphene oxide sheets. J Phys Chem C 112:20264–20268
Kalra A, Garde S, Hummer G (2003) Osmotic water transport through carbon nanotube membranes. Proc Natl Acad Sci USA 100:10175–10180
Kim NH, Ko Y, Cho S-R, Chang ST (2014) Formation and characterization of wrinkle structures of chemically-derived graphene thin films and micro-patterns. J Nanosci Nanotechnol 14(5):3774–3777.
Koduru JR, Chang Y, Kim I (2014) Low-cost Schizandra chinesis fruit peel for Co (II) removal from aqueous environment: adsorption properties and mechanism. Asian J Chem 26(1):289–297
Krishnamoorthy K, Mohan R, Kim S-J (2011) Graphene oxide as a photocatalytic material. Appl Phys Lett 98:244101–244101
Ladak S, Ball JM, Moseley BD, Eda G, Branford WR, Chhowalla M, Anthopoulos TD, Cohen LF (2013) Observation of wrinkle induced potential drops in biased chemically derived graphene thinfilm networks. Carbon 64:35–44
Lee C, Wei X, Kysar JW, Hone J (2008) Measurement of the elastic properties and intrinsic strength of monolayer graphene. Science 321:385–388
Li J, Guo S, Zhai Y, Wang E (2008) Nafion–graphene nanocomposite film as enhanced sensing platform for ultrasensitive determination of cadmium. Electrochem Commun 11(5):1085–1088
Li Z, Mi Y, Liu X, Liu S, Yang S, Wang J (2011a) Flexible graphene/MnO2 composite papers for supercapacitor electrodes. J Mater Chem 21:14706–14711
Li W, Geng X, Guo Y, Rong J, Gong Y, Wu L, Zhang X, Li P, Xu J, Cheng G (2011b) Reduced graphene oxide electrically contacted graphene sensor for highly sensitive nitric oxide detection. ACS Nano 5:6955–6961
Lingamdinne LP, Koduru JR, Karri RR (2019) A comprehensive review of applications of magnetic graphene oxide based nanocomposites for sustainable water purification. J Environ Manage 231:622–634
Lotya M, Hernandez Y, King PJ, Smith RJ, Nicolosi V, Karlsson LS, Blighe FM et al (2009) Liquid phase production of graphene by exfoliation of graphite in surfactant/water solutions. J Am Chem Soc 131:3611
Lu G, Ocola LE, Chen J (2009) Gas detection using low-temperature reduced graphene oxide sheets. Appl Phys Lett 94:083111
Lv S, Chen X, Ye Y, Yin S, Cheng J, Xia M (2009) Rice hull/MnFe2O_4 composite: preparation, characterization and its rapid microwave-assisted COD removal for organic wastewater. J Hazard Mater 171(1–3):634–639
Ma J, Ping D, Dong X (2017) Recent developments of graphene oxide-based membranes: a review. Membranes 7(3):52
Maaz K, Karim S, Mumtaz A, Hasanain SK, Liu J, Duan JL (2009) Synthesis and magnetic characterization of nickel ferrite nanoparticles prepared by co-precipitation route. J Magn Magn Mater 321(12):1838–1842
Maire J, Colas H, Maillard P (1968) Membranes de Carbone et de Graphite et Leurs Proprieties. Carbon 6:555–560
Marcano DC, Kosynkin DV, Berlin JM, Sinitskii A, Sun Z, Slesarev A, Alemany LB, Lu W, Tour JM (2010). Improved synthesis of graphene oxide. ACS Nano 4(8):4806–4814.
Marras F (2010) Recovery and recycling of homogeneous catalysts: silica as temporary or permanent support. In: Faculty of Science (FNWI). University of Amsterdam, p 138
Mathkar A, Tozier D, Cox P, Ong P, Galande C, Balakrishnan K, Leela Mohana Reddy, A, Ajayan PM (2012) Controlled, stepwise reduction and band gap manipulation of graphene oxide. J Phys Chem Lett 3:986–991
Mermoux M, Chabre Y, Rousseau A (1991) FTIR and 13C NMR study of graphite oxide. Carbon 29:469–474
Minář J, Brožek J, Michalcová A, Hadravová R, Slepička P (2019) Functionalization of graphene oxide with poly(ε-caprolactone) for enhanced interfacial adhesion in polyamide 6 nanocomposites. Compos Part B Eng 174:107019

Mkhoyan K, Contryman A, Silcox J, Stewart D, Eda G, Mattevi C, Miller S, Chhowalla M (2009) Atomic and electronic structure of graphene-oxide. Nano Lett 9:1058–1063

Mohanty N, Berry V (2008) Graphene-Based Single-bacterium resolution biodevice and DNA transistor: interfacing graphene derivatives with nanoscale and microscale biocomponents. Nano Lett 8:4469–4476

Moon IK, Kim JI, Lee H, Hur K, Kim WC, Lee H (2013) 2D graphene oxide nanosheets as an adhesive over-coating layer for flexible transparent conductive electrodes. Sci Rep 3:2673–2684

Moussa SO et al (2014) Graphene-supported, iron-based nanoparticles for catalytic production of liquid hydrocarbons from synthesis gas: the role of the graphene support in comparison with carbon nanotubes. ACS Catal 4(2):535–545

Mu B, Tang J, Zhang L, Wang A (2017) Facile fabrication of superparamagnetic graphene/polyaniline/Fe 3 O 4 nanocomposites for fast magnetic separation and efficient removal of dye. Sci Rep 7(1):1–12

Nair RR, Wu HA, Jayaram PN, Grigorieva IV, Geim AK (2012) Unimpeded permeation of water through helium-leak-tight graphene-based membranes. Science 335:442–444

Nakajima T, Matsuo Y (1994) Formation process and structure of graphite oxide. Carbon 32:469–475

Nejati K, Zabihi R (2012) Preparation and magnetic properties of nano size nickel ferrite particles using hydrothermal method. Chem Cent J 6(1):1–6

Nilsson HM, de Knoop L, Cumings J, Olsson E (2017) Localized resistance measurements of wrinkled reduced graphene oxide using in-situ transmission electron microscopy. Carbon 113:340–345

Novoselov KS, Geim AK, Morozov SV, Jiang D, Zhang Y, Dubonos SV, Grigorieva IV, Firsov AA (2004) Electric field effect in atomically thin carbon films. Science 306:666–669

Novoselov KS, Jiang D, Schedin F, Booth TJ, Khotkevich VV, Morozov SV, Geim AK (2005) Two-dimensional atomic crystals. PNAS 102:10451

Oraby EA, Eksteen JJ (2015) The leaching of gold, silver and their alloys in alkaline glycine–peroxide solutions and their adsorption on carbon. Hydrometallurgy 152:199–203

Pal SK (2015) Versatile photoluminescence from graphene and its derivatives. Carbon 88:86

Park S, Ruoff RS (2009) Chemical methods for the production of graphenes. Nat Nanotechnol 4:217

Partoens B, Peeters FM (2006) From graphene to graphite: electronic structure around the K point. Phys Rev B 74:1

Perreault F, Fonseca de Faria A, Elimelech M (2015a) Environmental applications of graphene-based nanomaterials. Chem Soc Rev 44:5861–5896.

Perreault F, De Faria AF, Nejati S, Elimelech M (2015b) Antimicrobial properties of graphene oxide nanosheets: why size matters. ACS Nano 9:7226–7236

Ponomarenko LA, Schedin F, Katsnelson MI, Yang R, Hill EW, Novoselov KS et al (2008) Chaotic dirac billiard in grapheme quantum dots. Science 320(5874):356–358

Quan Q, Lin X, Zhang N, Xu Y-J (2017) Graphene and its derivatives as versatile templates for materials synthesis and functional applications. Nanoscale 9:2398

Ramesha GK, Kumara AV, Muralidhara HB, Sampath S (2011) Graphene and graphene oxide as effective adsorbents toward anionic and cationic dyes. J Colloid Interface Sci 361(1):270–277

Ravelli D, Dondi D, Fagnoni M, Albini A (1999) Photocatalysis. A multi-faceted concept for green chemistry. Chem Soc Rev 38:1999–2011

Robinson JT, Keith Perkins F, Snow ES, Wei Z, Sheehan PE (2008) Reduced graphene oxide molecular sensors. Nano Lett 8:3137–3140

Ruess G (1946) Über das graphitoxyhydroxyd (graphitoxyd). Monatsh Chem 76:381–417

Sajjad S, Khan Leghari SA, Iqbal A (2017) Study of graphene oxide structural features for catalytic, antibacterial, gas sensing, and metals decontamination environmental applications. ACS Appl Mater Interfaces 9(50):43393–43414

Sawangphruk M, Srimuk P, Chiochan P, Krittayavathananon A, Luanwuthi S, Limtrakul J (2013) High-performance supercapacitor of manganese oxide/reduced graphene oxide nanocomposite coated on flexible carbon fiber paper. Carbon 60:109–116

Schedin F, Geim AK, Morozov SV, Hill EW, Blake P, Katsnelson MI, Novoselov KS (2007) Detection of individual gas molecules adsorbed on graphene. Nat Mater 6(9):652–655
Scholz W, Boehm HP, Anorg Z (1969) Untersuchungen am Graphitoxid. VI. Betrachtungen zur Struktur des Graphitoxids. Allg Chem 369:327–340
Senthilkumar B, Selvan RK, Vinothbabu P, Perelshtein I, Gedanken A (2011) Structural, magnetic, electrical and electrochemical properties of NiFe2O4 synthesized by the molten salt technique. Mater Chem Phys 130(1–2):285–292
Si Y, Samulski ET (2008) Synthesis of water-soluble graphene. Nano Lett 8:1679
Staudenmaier L (1898) Verfahren zur darstellung der graphitsaure. Ber Deutsche € Chem. Ges 31:1481–1487
Sun H et al (2012) Reduced graphene oxide for catalytic oxidation of aqueous organic pollutants. ACS Appl Mater Interfaces 4(10):5466–5471
Szabo T, Berkesi O, Forgo P, Josepovits K, Sanakis Y, Petridis D, Dekany I (2006) Evolution of surface functional groups in a series of progressively oxidized graphite oxides. Chem Mater 18:2740–2749
Touzain P, Yazumi R, Maire J (1986) Insertion compounds of graphite with improved performances and electrochemical applications of those compounds. U.S. Patent 4584252, April 22
Tung VC, Allen MJ, Yang Y, Kaner RB (2008) High throughput solution processing of large-scale graphene. Nature Nanotechnol 4:25
Tung VC, Chen L, Allen MJ, Wassei JK, Nelson K, Kaner RB, Yang Y (2009) Low-temperature solution processing of graphene-carbon nanotube hybrid materials for high-performance transparent conductors. Nano Lett 9:1949
Viculis LM, Mack JJ, Kaner RB (2003) A chemical route to carbon nanoscrolls. Science 299:1361
Wang DW, Li F, Zhao J, Ren W, Chen ZG, Tan J, Wu ZS, Gentle I, Lu GQ, Cheng HM (2009) Fabrication of graphene/polyaniline composite paper via in situ anodic electropolymerization for high-performance flexible electrode. ACS Nano 3:1745–1752
Wang L, Zhao J, Wang L, Yan T, Sun Y-Y, Zhang SB (2011) Titanium-decorated graphene oxide for carbon monoxide capture and separation. Phys Chem Chem Phys 13:21126–21131
Wei Y, et al (2018) Enhanced Fischer–Tropsch performances of graphene oxide-supported iron catalysts via argon pretreatment. Catal Sci Technol 8(4):1113–1125, 147.
Weng X, Wu J, Ma L, Owens G, Chen Z (2019) Impact of synthesis conditions on Pb (II) removal efficiency from aqueous solution by green tea extract reduced graphene oxide. Chem Eng J 359:976–981
Wissler M (2006) Graphite and carbon powders for electrochemical applications. J Power Sources 156:142–150
Xu K, Cao P, Heath JR (2009) Scanning tunneling microscopy characterization of the electrical properties of wrinkles in exfoliated graphene monolayers. Nano Lett 9(12):4446–4451
Yan J-A, Xian L, Chou MY (2009) Structural and electronic properties of oxidized graphene. Phys Rev Lett 103:086802
Yu W, Sisi L, Haiyan Y, Jie L (2020) Progress in the functional modification of graphene/graphene oxide: a review. RSC Adv 10(26):15328–15345
Zhang YH, Chen YB, Zhou KG, Liu CH, Zeng J, Zhang HL, Peng Y (2009) Improving gas sensing properties of graphene by introducing dopants and defects: a first-principles study. Nanotechnology 20:185504
Zhang Z, Zhang Q, Guo K, Li Y, Li X, Wang L, Luo Y, Li H, Zhang Y, Guan G, Wei B, Zhu X, Peng HS (2015) Flexible electroluminescent fiber fabricated from coaxially wound carbon nanotube sheets. J Mater Chem C 3:5621–5624
Zhang K, Arroyo M (2014) Understanding and strain-engineering wrinkle networks in supported graphene through simulations. J Mech Phys Solids 72:61–74.
Zhao GX, Jiang L, He YD, Li JX, Dong HL, Wang XK, Hu WP (2011a) Sulfonated graphene for persistent aromatic pollutant management. Adv Mater 23:3959–3963

Zhao GX, Ren XM, Gao X, Tan XL, Li JX, Chen CL, Huang YY, Wang XK (2011b) Removal of Pb(II) ions from aqueous solutions on few-layered graphene oxide nanosheets. Dalton Trans 40:10945–40952

Zhao G, Li J, Ren X, Chen C, Wang X (2011c). Few-layered graphene oxide nanosheets as superior sorbents for heavy metal ion pollution management. Environ Sci Technol 45(24):10454–10462

Zhao HB et al (2013) Iron oxide nanoparticles supported on pyrolytic graphene oxide as model catalysts for Fischer Tropsch synthesis. Appl Catal A 456:233–239

Zhou, L., Yang, X., Yang, B., Zuo, X., Li, G., Feng, A., Tang H, Zhang H, Wu M, Ma Y, Jin S, Sun Z, Chen, X. (2014). Controlled synthesis of CuInS2/reduced graphene oxide nanocomposites for efficient dye-sensitized solar cells. J Power Sour 272:639–646

Chapter 7
Nanomaterials for the Removal of Inorganic Contaminants from Industrial Wastewater

Rashmi Paliwal and Jai Prakash Narain Rai

Abstract Water pollution has become a critical issue for the present world population. A variety of industrial operations release toxic organic and inorganic substances/contaminants into the water bodies. The continued accumulation of these toxic compounds in natural water resources may put living organisms at risk. Different conventional methods are well explored and have been applied to the treatment of contaminated water. Current research focuses on the development of low-cost and more efficient technologies, with the potential advantage of reusability in continuously operating wastewater treatment plants. In this context, nanotechnology as an emerging field can provide immense opportunities for removing contaminants from wastewater, thus improving water quality. Applications of nanomaterials have been successfully discovered and documented in different fields of biomedical science. Certain unique properties of nanomaterial toward the contaminants offer additional benefits for the application of these nano-biotechnological tools for the treatment of wastewater. This chapter is an attempt to explore the role of nanomaterials in the treatment of wastewater focusing primarily on the removal of inorganic contaminants. The advantages, limitations, and future perspectives of nanomaterials applications are also discussed.

Keywords Nanotechnology · Nanomaterial · Inorganic contaminants · Wastewater

7.1 Introduction

Scanty freshwater resources together with water pollution due to various industrial activities have become a challenge for the global population. The pressure on demand for water is likely to increase with the growing human population and climate change.

R. Paliwal (✉) · J. P. N. Rai
Department of Environmental Sciences, Govind Ballaph Pant University of Agriculture and Technology, Pantnagar, India
e-mail: paliwal39561@gmail.com

J. P. N. Rai
e-mail: jamesraionline@gmail.com

J. P. N. Rai and S. Saraswat (eds.), *Nano-biotechnology for Waste Water Treatment*,
Water Science and Technology Library 111,
https://doi.org/10.1007/978-3-031-00812-2_7

The industrial revolution and rapid population growth have not only put immense pressure on existing freshwater resources but also deteriorating their water quality by adding organic and inorganic contaminants on a large scale. Most developing countries discharged ≈70% of their industrial wastewater into the environment without proper treatment. More than 2 billion people are living in the countries experiencing high water stress (UNESCO 2019). According to an estimation, approx. 600 million of the world's children will be in extreme water stress regions by 2040 (UNICEF 2017) and 700 million people will be displaced worldwide due to extreme water scarcity by 2030 (Hameeteman 2013). Industries contribute toxic contaminants by dumping their untreated or partially treated wastewater into the local surface water bodies and jeopardize environmental health. Wastewater released from industrial activities contributes a variety of organic and inorganic toxic pollutants to the environment. The organic class of contaminants includes phenolic compounds, chlorinated biphenyls, dyes, pharmaceuticals, pesticides, humic substances, etc. Inorganic contaminants include heavy metals of toxic nature i.e., cadmium (Cd), chromium (Cr), arsenic (As), mercury (Hg), nickel (Ni), lead (Pb), etc., (Chowdhry et al. 2016; Kim et al. 2018).

Heavy metals such as lead, cadmium, chromium, mercury, and arsenic are known to have deleterious effects on human health and can disrupt the metabolic functions of the body in various ways. Heavy metals may accumulate in vital body organs, disrupting the normal biological functions (Rehman et al. 2018). Conventional treatment processes such as activated carbon-based adsorption, membrane filtration, ion exchange, solvent extraction, coagulation and flocculation, reverse osmosis, flotation and extraction, electrochemical treatment, advanced oxidation processes, etc., are available and being employed for the treatment of industrial wastewater (Rao et al. 2010; Thines et al. 2017; Mohammad et al. 2017; Ma et al. 2018; Werkneh & Rene 2019; Jain et al. 2020). However, certain issues still exist related to the applicability of conventional methods, which include high energy demands, unsustainable, process efficiency, and economic benefits (Werkneh & Rene 2019). Therefore, the real challenge is to make the treatment process more efficient and economically acceptable.

In recent years, nanotechnology has received enormous attention for its application in wastewater treatment systems. Nanomaterials (≈1–100 nm size) possess unique features such as, their nano-size, high surface-area-to-volume ratio, high mobility in solution, high adsorption capacity, and reactivity that make them a suitable candidate for their application in wastewater treatment (Kunduru et al. 2017; Kiri et al. 2019; Yu et al. 2021). Recently, successful application of nanomaterials for the treatment of wastewater of different sources has been reviewed and verified by several studies (Kumar et al. 2014; Bora & Dutta 2014; Yadav et al. 2020). Present work aims to review the recent development of nano-technological tools i.e., nanomaterials, and their role in removing heavy metal contaminants from industrial wastewater. Additionally, we discuss the environmental limitations and risks associated with the application of nanomaterials and future research prospects of the promising area.

7.2 Inorganic Contaminants

Major inorganic contaminants released from industrial activities are heavy metals, which by nature can persist in the environment and can also bio-accumulate in living organisms e.g., mercury (Hg) in fishes. Metals with an elemental density of 5 g/cm^3 or more are considered as heavy metals (Ali et al. 2019). Wastewater release from electronics, smelting, battery, dyes, and alloys industries is the anthropogenic source of heavy metals. Heavy metals are highly reactive and can cause severe health issues in life forms when exceeding the standards prescribed by WHO (Srivastav & Ranjan 2020). Toxic heavy metals along with their sources, prescribed standards, and health effects are listed in Table 7.1 (Srivastav & Ranjan 2020).

Table 7.1 Major heavy metal contaminants, prescribed standards, and their human health effects

Heavy metal	Industrial sources	Maximum concentration levels (mcl) (mg/l) as prescribed by USEPA[a]	Tolerance limits (mg/l) IS:10,500, 1992[b]	Potential health effects
Arsenic (As)	Pesticides, Mining, Smelting	0.05	0.05	Skin lesions and carcinogenic effects
Cadmium (Cd)	Paints and pigments, Phosphate Fertilizer, Electroplating	0.005	0.01	Kidney damage
Chromium (Cr)	Tanneries, Steel industries	0.1	0.05	Allergic dermatitis
Copper (Cu)	Pesticides, Fertilizers, Mining, and Smelting	1.3	0.05	Gastrointestinal distress (short term exposure) Liver and Kidney damage (long-term exposure)
Mercury (Hg)	Mining,	0.002	0.001	Kidney and spinal-cord disorder
Lead (Pb)	Batteries, Insecticides, and herbicides	0.015	0.05	Delayed physical and mental growth, Central nervous system (CNS) problems, Kidney issues, and high blood pressure

[a] US Environmental Protection Agency; [b] Indian Standard: 10,500, 1992

7.3 Nanomaterials

Successful application of nanotechnology in the diverse field of the environment including industrial wastewater treatment depends upon the innovation of different nanomaterials. Nanomaterials possess certain unique attributes that facilitate them to be used in the wastewater remediation processes (Sánchez et al. 2011; Khin et al. 2012). These characteristics are:

- Small size—Small size on nano-scale provides nanomaterials' high specific surface area which increases with size shrink.
- High surface-area-to-volume ratio—This property intends to improve the adsorptive capacity of nanomaterials.
- High mobility—The small size of nanomaterials allows them to move freely in the aqueous solution, which results in the fast treatment or scanning of a large volume of solution by the small amounts of nanomaterials. A small amount of nanomaterial thus can scan the total volume of solution in a very short time (Sánchez et al. 2011).
- High adsorption capacity and reactivity—The nanomaterials have a highly reactive surface due to their reduced size and large radii of curvature.

The field of nanomaterials is broad and has generated a large number of defined terms. A variety of terms such as nanoparticles, nanomaterials, nanocomposites, nanocrystals, nano-sized mediums, and nano-phases are used to define the materials of nano-scales sized. Nanomaterials can be classified based on dimension and composition. However, the requirement of precise classification is still needed with the innovations in the nanotechnology field. Dimensionality based nanomaterials' classification comprised zero (eg., C_{60} fullerene, core–shell nanomaterial, etc.), one (eg., nano-wires, nano-rods, carbon nanotubes, etc.), two (graphene sheet, carbon-coated nanomaterial, etc.) and three dimensional (eg., nano-diamond, future 3D electrode) (Prajitha et al. 2019). Among the two approaches, classification on a composition basis is reliable after dimensionality (Prajitha et al. 2019). Nanomaterial based on chemical composition can be grouped into carbon-based, metal and metal oxides-based, noble metal nanomaterials, and nanocomposites (Fig. 7.1).

7.4 Carbon-Based Nanomaterials

Nanomaterials (eg., fullerenes, graphene, and carbon nanotubes) entirely composed of carbon are grouped under the carbon-based category of nanomaterials. Carbon-based nanomaterials have unique properties of strength and reactivity, hence are widely accepted for their use in industrial wastewater treatment technologies.

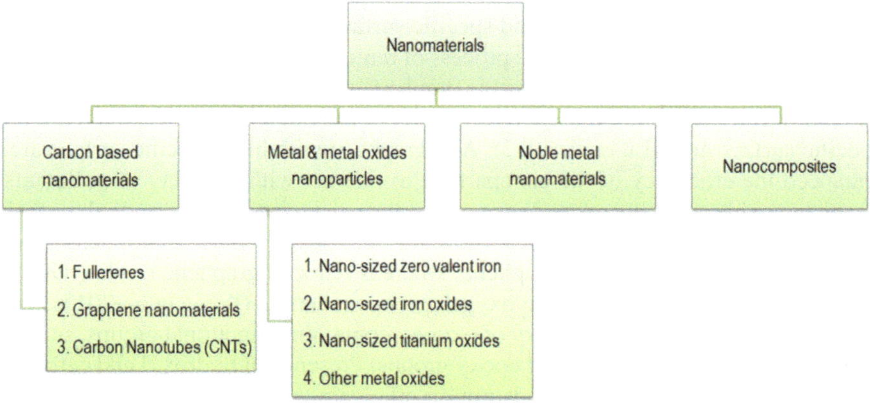

Fig. 7.1 Classification of nanomaterials based on chemical composition

7.4.1 Fullerenes

Fullerenes are purely composed of carbon defined as synthetic carbon allotropes (SCAs) and are also known as carbon nanomaterials forming different shapes such as the hollow sphere, cylindrical, ellipsoid, etc. Fullerenes are large, closed-cage, carbon bundles that are naturally non-toxic and soluble in a variety of organic solvents (Kumar and Kumbhat 2016). Fullerenes have widely been explored for their applications in biomedical science. However, due to their unique properties fullerenes have been proposed to use as a base material for nanomaterial-based wastewater treatment technology development (Bottero et al. 2006; Chae et al. 2009). Membrane fabricated with fullerene nanomaterial has been assessed for removing the salts, organic pollutants, and pathogens by several workers (Hummer et al. 2001; Naguib et al. 2004; Majumder et al. 2005; Holt et al. 2006). Fullerene-polymer composites have been reported for successful application in pressure-driven membranes due to their strength thereby resisting the breakage (Polotskaya et al. 2005). The biomedical applications of fullerenes reveal the antimicrobial properties of these nanomaterials. The fullerenes (C_{60} and C_{70}) are being widely used in medical treatments such as inactive microbial pathogens (bacteria and viruses), to treat the tumor and to cleave DNA, strongly suggesting the application of these nanomaterials for disinfecting the water (Chae et al. 2009). However, research to explore the versatility of fullerenes is further needed.

7.4.2 Graphene Nanomaterials

Graphene is monolayer carbon allotrope, exhibiting a tightly bound hexagonal honeycomb structure. The unique properties of graphene are remarkable thermal, electrical conductivity, chemical stability, and specific surface area. Some graphene nanomaterials playing important role in the process of remediation are graphene oxide (GO), reduced graphene oxide (rGO), pristine graphene (pGr), few-layer graphene (FLG), and multilayer graphene (MLG). Modification of pristine graphene can increase the specific surface area (Lü et al. 2012). And graphene with high specific surface area enhanced the efficiency of the nanoparticle to interact with a variety of pollutants. Graphene oxide and reduced graphene oxides have been extensively studied for their heavy metal removal efficiency in wastewater treatment processes. Graphene oxide (GO) is the oxidative form of graphene, whereas reduced graphene oxide (rGO) is produced by eliminating the ionic oxygen groups it carries (Wang et al. 2013). The graphene oxide comprises a variety of oxygen-containing functional groups, such as hydroxyl, carboxyl, carbonyl, and epoxy groups. Presence of hydroxyl and carboxyl functional groups enhances the adsorption efficiency of graphene oxide for heavy metals (Li et al. 2009; Lü et al. 2012). GO and rGO have been reported to possess great adsorption capacity by many studies (Wang et al. 2013; Madadrang et al. 2012; Yang et al. 2013; Ding et al. 2014). Ding et al. (2014), studied the adsorption capacity of GO-layered fixed bed sand column for the treatment of heavy metals (Cu(II) and Pb(II)) containing aqueous solution.

7.4.3 Carbon Nanotubes (CNTs)

Carbon nanotubes are one-dimensional chemical structures with unique mechanical, thermal, and electrical properties. CNTs can be further classified into single-walled (SWCNTs), multi-walled (MWCNTs), and CNTs composites (Zhang et al. 2016). Carbon nanotubes structurally can be classified as armchair, chiral, and zigzag nanotubes. Similar to GO, CNTs interact with the contaminants through hydrophobic interactions, electrostatic interaction, hydrogen-bonding, pi–pi interaction, Van der Waals force, etc. (Gupta et al. 2013; Zhang et al. 2016; Mashkoor et al. 2020). The adsorptive capacity of CNTs can be influenced by certain structural factors as discussed by Ren et al. (2011):

- Open-end nanotubes offer the adsorption sites, hence their presence and amount can influence the adsorption.
- Presence of impurities can hinder the adsorption process.
- Presence of functional groups (hydroxyl, carboxyl, and carbonyl), can improve the performances.

Madima et al. (2020), explained the mechanism of enhanced adsorption efficiency of carbon nanotubes through surface oxidation (Fig. 7.2). Adsorption of pollutants on oxidative carbon nanotubes takes place at both the surface of nanomaterial and the functional groups present on the surface of the nanomaterials. Whereas, in the case of untreated carbon nanotubes the adsorption occurs on the surface of the nanomaterial only (Madima et al. 2020). The functional groups added to the surface of

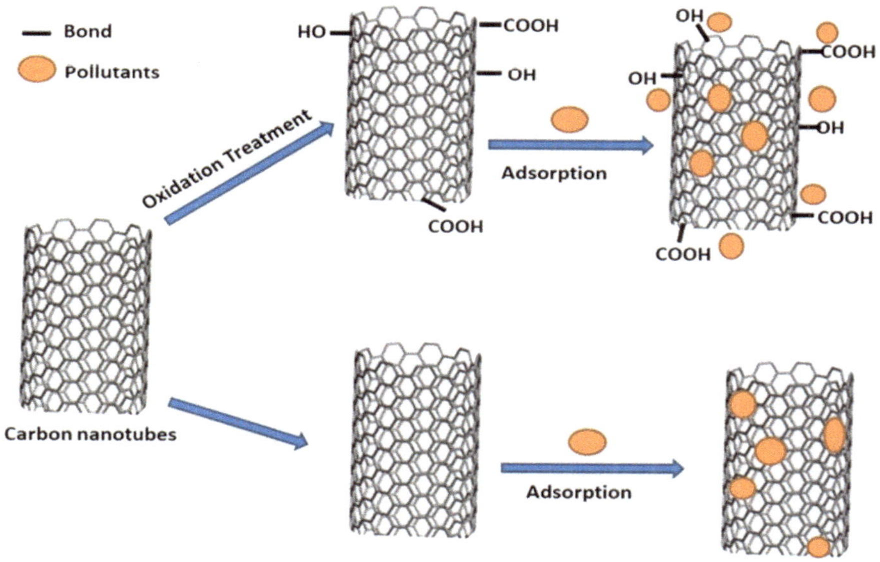

Fig. 7.2 Adsorption mechanism of pollutants by treated and untreated carbon nanotube. Reproduced with permission from Madima et al. (2020). Copyright 2020, Springer Nature

CNTs following oxidation treatment act as the primary adsorption site for inorganic contaminants.

7.5 Metal & Metal Oxides Nanomaterials

The metal-based nanomaterials include nano-sized metals (eg. titanium, gold, silver, etc.) and metal oxides. The metal and metal oxide-based nanomaterials include zero-valent iron, iron oxides, titanium oxides, aluminum oxides, magnesium oxides, and cerium oxides (Zhang et al. 2016). These nanomaterials possess a strong affinity toward the adsorption of heavy metal contaminants from the aqueous solution (Table 7.2). Nano-sized metals and metal oxides have been reported to show remarkable affinity toward the inorganic contaminants such as Cr(VI), Cd(II), Pb(II), Ni, Se(IV), etc. removal from solution (Hu et al. 2006; Shang et al. 2007; Engates and Shipley 2011; Wei et al. 2012).

Table 7.2 Metal and metal oxide-based nanomaterial applied for the heavy metal removal

Metal and metal oxides Nano-adsorbent	Heavy metal contaminant	Reference
Titanium oxide (TiO_2)	Pb(II), Cd(II), Cu(II), Ni(II), Zn(II)	Engates and Shipley (2011)
Al_2O_3	Pb(II), Cd(II), Cr(III), Co(II), Ni(II), Mn(II)	Afkhami et al. (2010)
γ-Al_2O_3	Ni(II)	Ghaedi et al. (2008)
ZnO	Pb(II) Zn(II), Cd(II), Hg(II)	Shang et al. (2007) Sheela et al. (2012)
Magnetite (Fe_3O_4)	Se(IV)	Wei et al. (2012)
Hematite (α-Fe_2O_3)	Pb(II), Cd(II), Cu(II), Zn(II)	Shipley et al. (2013)
Maghemite (γ-Fe_2O_3)	Cr(VI) Cr(VI), Cu(II), Ni(II) Cu (II), Ni (II), Mn (II), Cd (II), and Cr (VI)	Hu et al. (2005) Hu et al. (2006) Akhbarizadeh et al. (2014)
nZVI	As(V) Cd Cr As(III), U(VI), Se(VI) Cu(II)	Kanel et al. (2006); Ramos et al. (2009) Boparai et al. (2011) Yu et al. (2014) Crane et al. (2011) Li et al. (2014)

7.5.1 Nano-sized Zero-valent Iron (nZVI)

Iron is efficiently responsive to both air (oxygen) and water. The reactivity of iron can be significantly increased toward the various organic and inorganic compounds as the particle size of iron is reduced. Thus, nano-sized zero-valent iron (nZVI) becomes a potential candidate studied extensively for the treatment of various organic and inorganic contaminants (Kanel et al. 2006; Tosco et al. 2014). Zero-valent iron (Fe^0) has a standard reduction potential (E^0) i.e., −0.440 V, which means that Fe^0 is an effective electron donor and can reduce any chemical species that has a higher reduction potential than −0.440 V (Zhang et al. 2010; Adusei-Gyamfi & Acha 2016). Nano-sized zero-valent iron is often tended to aggregate, which affects their efficiency of the reaction. The efficiency of nZVI particles can be improved by immobilization using different supports such as sodium carboxy-methyl cellulose (CMC), guar gum, zeolite, bentonite/iron, and kaolin (He & Zhao 2005; Xu & Zhao 2007; Li et al. 2007; Tiraferri et al. 2008; Yuan et al. 2008; Shahwan et al. 2010).

The immobilization of nZVI improves efficiency by reducing the aggregation property of nanoparticles and improves the dispersion also. Several studies have reported the remarkable remediation properties of nZVI for the treatment of heavy metals contaminated environments (Boparai et al. 2011; Yang et al. 2018; Shang et al. 2017). Boparai et al. (2011), studied the nZVI for their efficient adsorbent capacity to remove Cd(II) from the contaminated water and observed the maximum adsorption capacity 769.2 mg g^{-1}, achieved at 297 K temperature. Yang et al. (2018), observed the adsorption capacity (195.1, 161.9, and 109.7 mg g^{-1}) of corn stalk-derived, bio-char supported nZVI for the removal of heavy metal ions (Pb(II), Cu(II), and Zn(II), respectively) from the water after 6 h. Shang et al. (2017), applied the nZVI supported on herb-residue bio-char for the treatment of Cr(VI) contaminated water. The results confirm the reduction and co-precipitation reactions occurred during the treatment (Shang et al. 2017). Applications of chemically stable support materials for nZVI particles immobilization have been reported to stabilize and improve the efficiency of nanoparticles (Yuan et al. 2008).

7.5.2 *Nano-sized Iron Oxides*

Iron oxide is polymorphic and consisting about sixteen compounds between oxides, hydroxides, and oxy-hydroxides (Macera et al. 2020). The nano-sized iron oxides such as non-magnetic goethite (α-FeOOH) and hematite (α-Fe_2O_3), magnetic magnetite (Fe_3O_4) and maghemite (γ-Fe_2O_3), and hydrous ferric oxides (HFO) have been extensively studied for their application in wastewater treatment (Hu et al. 2005; Chen and Li 2010). Kefeni et al. (2018), synthesized and assessed the efficiency of superparamagnetic hematite nanoparticles for the removal of Al (III), Mg (II), Mn (II), Zn (II), Ni (II), etc., from acid mined drainage. The magnetic property of nano-sized iron oxide makes the separation and recovery of these nanoparticles easy from aqueous solution (Hua et al. 2012).

7.5.3 *Nano-sized Titanium Oxides*

Titanium oxide (TiO_2) nanomaterials due to their photocatalytic property have been reported to remove several organic contaminants from the wastewater. The TiO_2 activated by light, producing activated oxidant species like hydroxyl (OH) radicals, superoxide anions (O_2^-), and peroxides (O_2^{--}) thus disinfect the water from pathogenic microbial organisms (Cho et al. 2005; Zan et al. 2007; Guerra et al. 2018).

7.5.4 *Other Metal Oxides*

Metal oxides nanoparticles are composed of metals such as aluminum (Al), Magnesium (Mg), Manganese (Mn), Zinc (Zn), Cerium (Ce), etc., and oxygen elements (Sarma et al. 2019). Metal oxides in nano-sized can be synthesized in different shapes such as wires, rods, cubes, cross-linked, etc. (Gangwar et al. 2016). The adsorption capacity of nano-sized alumina (γ-Al_2O_3) for heavy metals in aqueous solution has been studied by many researchers (Sharma et al. 2008; Türker 2007; Zhang et al. 2008). γ-Al_2O_3 nanoparticles were reported as an effective adsorbent for the removal of Pb(II), Ni(II), and Zn(II) ions from the aqueous solution (Rahmani et al. 2010).

7.6 Noble Metal Nanomaterials

Certain transition metals include Gold (Au), Silver (Ag), Platinum (Pt), and Palladium (Pd) are considered noble metals (Pradeep and Anshup 2009). Synthesized noble metal nanoparticles are studied to treat organic contaminants due to their photocatalytic property. Noble metal nanomaterials have also been observed to possess anti-bacterial properties, hence, also studied for their water disinfectant efficiency (Morones et al. 2005; Xiu et al. 2011).

7.7 Nanocomposites

Nanocomposites are multiphase materials, where one or more phases are of nanoscale dimension, ie., 100 nm or less (Cammarata 2004; Prajitha et al. 2019). The nanocomposite shows some advantageous characteristics over other nanomaterials including, reusability, regeneration, stability, dispersion, etc. (Zhang et al. 2016). Various studies are being done on synthesizing the nanocomposites for industrial wastewater treatment processes. Nanocomposites can be classified based on matrix material such as organic and inorganic supported. Organic supported nanocomposites include polymer-based nanocomposites whereas; inorganic supports for nanocomposites include activated carbon, carbon nanotubes, and natural minerals (zeolite, bio-char, and clay) (Zhang et al. 2016).

The idea behind constructing nanocomposites is to improve the basic properties of nanoparticles by fixing them to some organic or inorganic supports. The polymer-based nanocomposite, nano-sized hydrous iron-oxide (HFO-001) has been reported to remove heavy metals—Pb(II), Cd(II), and Cu(II) efficiently from aqueous solution (Pan et al. 2010). The nano-sized HFO were confined within cation-exchange resin D-001 (R-SO_3-Na) and were tested for the removal of Pb(II), Cd(II), and Cu(II) (Pan et al. 2010). Another class of nanocomposites includes magnetic nanocomposites and nanocomposites membranes. Magnetic nanocomposites are made up of metal oxides such as magnetite (Fe_3O_4), maghemite (Fe_2O_3), jacobsite ($MnFe_2O_4$), etc. nanoparticles supported on the polymeric matrix or some functional nanoparticles (such as Ag, TiO_2, CNTs, GO, Pd, SiO_2, etc.) can be immobilized on the metal oxides core (Gómez-Pastora et al. 2017; Beyene and Ambaye 2019). Magnetic nanocomposites are easy to separate from the aqueous solutions after the treatment process (Gómez-Pastora et al. 2017). Nanocomposites membranes are nanomaterials imbedded in the polymeric matrix and also considered as advanced membranes for the application in wastewater treatment processes. The advanced regeneration technologies can make the use of nanocomposites more suitable for large-scale applications.

7.8 Application of Nanomaterial for Wastewater Treatment

Nanotechnology is advanced with the development of nanomaterials used for different wastewater treatment approaches. These approaches can be classified into adsorption, separation, catalysis, photo-catalysis, and disinfection (Zhang et al. 2016; Guerra et al. 2018).

7.8.1 Adsorption

Adsorption of contaminants from wastewater is a surface property of an adsorbent. The process of adsorption is regarded as technically viable, simple, and environmentally sound. Nanotechnology-based adsorption processes have various advantages over conventional adsorbents including high reactivity, large surface area for adsorption, specific affinity toward the various contaminants, and cost-effectiveness (Ai & Jiang 2012; Zhang et al. 2016). Efforts are being made to develop new adsorbents for the treatment of the heavy metal contaminated environment. Several nanomaterials-based adsorbents such as carbon-based nanomaterial, metal oxides, and magnetic metal–organic nanomaterial have been verified for removing the inorganic as well as organic pollutants from wastewater (Jiang et al. 2014; Xiong et al. 2015; Kiri et al. 2019). The adsorptive capacity of nanomaterial adsorbents can be enhanced or modified by some structural improvements. The most common interaction of graphene nanoparticles with organic and inorganic pollutants is pi–pi interaction and availability of functional groups on the surface of nanomaterial, respectively. Electrostatic attraction is another important interaction in the adsorption mechanism between modified graphene nanoparticles and inorganic pollutants. The electrostatic attractions occur between positively charged ions, and the negatively charged functional group available on the surface of changed graphene nanomaterials (Kemp et al. 2013; Madima et al. 2020). The general mechanism of adsorption–desorption of pollutants by graphene-based nanomaterial is illustrated in Fig. 7.3.

Wang et al. (2013) reported the remarkable adsorptive capacity of graphene oxide toward the Zn(II) at pH 7.0. Madadrang et al. (2012) observed the enhanced adsorptive capacity of GO when the heavy metal chelating groups i.e., N-(trimethoxysilylpropyl) ethylenediamine tri-acetic acid (EDTA-silane) and hydroxyl groups linked on GO surface through the process of silanization. The modified EDTA-GO showed an increase in adsorptive capacity toward Pb(II). The study also reported the regeneration and reusability of EDTA-GO nanomaterial, suggesting the potential use of the same in remediation processes. Similarly, the surface-oxidized CNTs were observed to have high adsorption capacity for the removal of Cd^{2+} from the

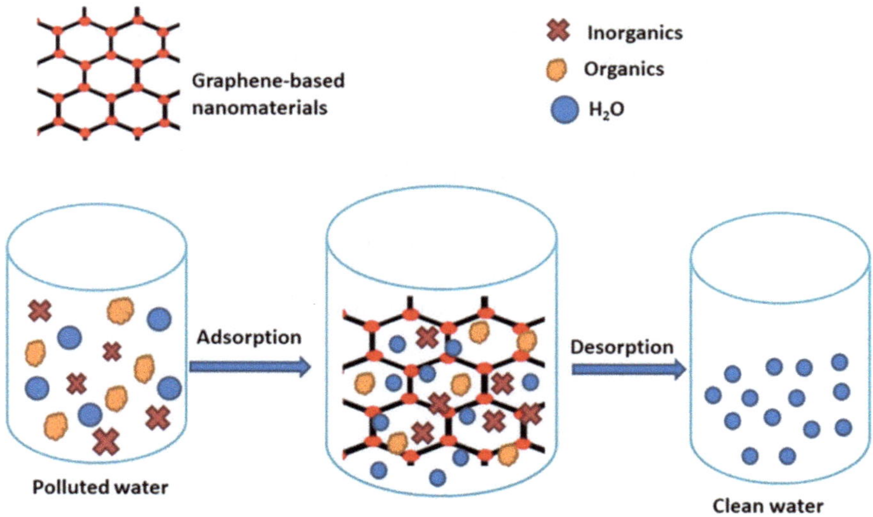

Fig. 7.3 Adsorption–desorption mechanism of organic and inorganic pollutants in aqueous solution by graphene-based nanoparticles. The desorption process is generally performed in the presence of acid or base. Reproduced with permission from Madima et al. (2020). Copyright 2020, Springer Nature

solution (Li et al. 2003). The functional groups such as carboxylic acid, carbonyl, and hydroxyl enhanced the adsorption capacity of the modified CNTs nanomaterial (Li et al. 2003; Vuković et al. 2010). Application of metal and metal oxides based nanomaterials for the treatment of inorganic contaminants such as arsenic, cadmium, chromium, uranium, etc. have been verified by several researchers (Lakshmipathiraj et al. 2006; Engates and Shipley 2011; Yu et al. 2014; Crane et al. 2011; Wen et al. 2014).

7.8.2 Separation

Toxic contaminants removal by separation method includes the phenomenon such as filtration or ultra-filtration, size-exclusion, reverse osmosis, etc. Nano-filtration membranes came into recognition in 1980s (Tul et al. 2017). These membranes are specifically designed to remove organic molecules and inorganic salts. The application of nanofiltration membranes offers some advantages such as high flux, low maintenance, and operational cost, low operation pressure, and high retention of anionic salts (Tul et al. 2017). However, certain limitations resist the application of the nano-separation method on large-scale treatment facilities. These limitations include nano-membrane fouling, stability, blocking, and discharge of nanomaterial (Tul et al.

2017; Bhat et al. 2018; Yaqoob et al. 2020). With the advancement in nanotechnology new nano-composite membranes have been developed by introducing functional nanoparticles into the membrane. The new nano-composite membranes exhibit advanced properties such as enhanced stability (mechanical or thermal), porosity and hydrophilicity, enhanced permeability, antimicrobial capabilities to address the limitations like stability, fouling, and blocking (Yan et al. 2005; Pendergast & Hoek 2011; Yin & Deng 2015).

7.8.3 *Catalytic and Photocatalytic Activity*

The term photo-catalysis is used when any reaction for example decomposition or degradation of the compound is catalyzed by the activity of photons/light. The process of photo-catalysis is also a type of advanced oxidation process involving the oxidative removal of micro-pollutants and pathogens from wastewater (Yadav et al. 2020). Photocatalysts initiate or modify the reaction rate of a chemical transformation (Lens et al. 2013; Yaqoob et al. 2020). Nano-photocatalyst improves the chemical reactivity such as the oxidation process by enhancing the production of oxidative species at the material surface, thereby; increasing the degradation of contaminants in wastewater (Gómez-Pastora et al. 2017; Ong et al. 2018; Yaqoob et al. 2020). Metal oxides and nanoparticles are known to possess strong catalytic properties and catalyze oxidation reactions very effectively (Zekić et al. 2018). Zero-valance-based metal, semiconductor, and bimetallic type nanoparticles are commonly used to clean organic pollutants such as dyes, pesticides, etc. in wastewater (Samanta et al. 2016).

7.9 Limitations and Future Perspectives

The application of nanomaterials in wastewater treatment has become widely acceptable and economically viable with the recent improvements in technology. However, some limitations such as loss of activity (Lofrano et al. 2016), difficulty in separation from the system (Al-Hamadani et al. 2015), the fate of nanomaterials in the environment, and their further impact on human health, etc. (Varma 2012; Dale et al. 2015) still exist and limit their applications in a wastewater treatment process. The Cost-effectiveness of the nanotechnology-based treatment process can be achieved by reusing the nanoparticles (Yadav et al. 2020). Management of recovered exhausted nanomaterials and development of strategies for their safe disposal is crucial as the nanomaterials could emerge as secondary pollutants (Lata et al. 2015). Researches in the field of enhancing the reusability of nanoparticles in a treatment process are needed. Bystrzejewska-Piotrowska et al. (2009), highlight the problem associated with the uncontrolled discharge of nanoparticles in the environment and stress that a comprehensive understanding of the physical, chemical, and biological properties of nanoparticle waste is a prerequisite before dealing with the waste generated from the application of nanotechnology. Researchers have studied the application of nanoparticles to clean up the environment while avoiding the short- and long-term effects of nanoparticles on the environment. Therefore, studies related to the risk assessment (environmental risk factors of exposure to nanoparticles) and management of the recovered and exhausted nanoparticles for a better understanding of their fate in the environment are also essential. Sánchez et al. (2011), have well assessed the toxic effects of nanoparticles available in the literature and concluded the need for more research efforts to support the stakeholders in promoting the scientific regulation of nanotechnology and for the betterment of society.

7.10 Conclusion

Nanotechnology offers a wide scope for the development of pollution abatement technologies. Nanomaterials have been studied for the treatment of various inorganic (heavy metals) contaminants released from industries. Different mechanisms such as adsorption, photo-catalysis, and separation are offered by nanomaterials for the removal of inorganic pollutants from wastewater. Among the various characteristics of nanomaterials, the magnetic property of metal and metal oxide-based nanomaterials makes them economically feasible for their large-scale applications. Similarly, nanocomposites being multiphase materials possess a great advantage over the other nanomaterials due to some properties like reusability, regeneration, stability, and dispersion. Therefore, the multiphase property of nanocomposites makes them a potential candidate for the development of advanced techniques and also for their large-scale application. However, studies on industrial applications, environmental fates, and the cost of nanomaterials are still needed.

References

Adusei-Gyamfi J, Acha V (2016) Carriers for nano zerovalent iron (nZVI): Synthesis, application and efficiency. RSC Adv 6(93):91025–91044

Afkhami A, Saber-Tehrani M, Bagheri H (2010) Simultaneous removal of heavy-metal ions in wastewater samples using nano-alumina modified with 2, 4-dinitrophenylhydrazine. J Hazard Mater 181(1–3):836–844

Ai L, Jiang J (2012) Removal of methylene blue from aqueous solution with self-assembled cylindrical graphene–carbon nanotube hybrid. Chem Eng J 192:156–163

Akhbarizadeh R, Shayestefar MR, Darezereshki E (2014) Competitive removal of metals from wastewater by maghemite nanoparticles: A comparison between simulated wastewater and AMD. Mine Water Environ 33:89–96

Al-Hamadani YA, Chu KH, Son A, Heo J, Her N, Jang M, … Yoon Y (2015) Stabilization and dispersion of carbon nanomaterials in aqueous solutions: A review. Sep Purif Technol 156:861–874

Ali H, Khan E, Ilahi I (2019) Environmental chemistry and ecotoxicology of hazardous heavy metals: environmental persistence, toxicity, and bioaccumulation. J Chem

Bereket G, Arog AZ, Özel MZ (1997) Removal of Pb (II), Cd (II), Cu (II), and Zn (II) from aqueous solutions by adsorption on bentonite. Journal of Colloïd and Interface Science 187(2):338–343

Beyene HD, Ambaye TG (2019) Application of sustainable nanocomposites for water purification process. In Sustainable polymer composites and nanocomposites, Springer, Cham, pp 387–412

Bhat AH, Rehman WU, Khan IU, Khan I, Ahmad S, Ayoub M, Usmani MA (2018). Nanocomposite membrane for environmental remediation. In Polymer-based nanocomposites for energy and environmental applications, Woodhead Publishing, Cambridge, UK, pp 407–440

Boparai HK, Joseph M, O'Carroll DM (2011) Kinetics and thermodynamics of cadmium ion removal by adsorption onto nano zerovalent iron particles. J Hazard Mater 186(1):458–465

Bora T, Dutta J (2014) Applications of nanotechnology in wastewater treatment—A review. J Nanosci Nanotechnol 14(1):613–626

Bottero JY, Rose J, Wiesner MR (2006) Nanotechnologies: tools for sustainability in a new wave of water treatment processes. Integrated Environmental Assessment and Management: an International Journal 2(4):391–395

Bystrzejewska-Piotrowska G, Golimowski J, Urban PL (2009) Nanoparticles: Their potential toxicity, waste and environmental management. Waste Manage 29(9):2587–2595

Cammarata R (2004) Nanocomposites. In: Di Ventra M, Evoy S, Heflin JR (eds) Introduction to nano scale science and technology, Springer Publishers, USA

Chae SR, Hotze EM, Wiesner MR (2009) Possible applications of fullerene nanomaterials in water treatment and reuse. In Nanotechnology applications for clean water, William Andrew Publishing, pp 167–177

Chen YH, Li FA (2010) Kinetic study on removal of copper (II) using goethite and hematite nano-photocatalysts. J Colloid Interface Sci 347(2):277–281

Cho M, Chung H, Choi W, Yoon J (2005) Different inactivation behaviors of MS-2 phage and Escherichia coli in TiO2 photocatalytic disinfection. Appl Environ Microbiol 71(1):270–275

Chowdhury S, Khan N, Kim GH, Harris J, Longhurst P, Bolan NS (2016) Zeolite for nutrient stripping from farm effluents. In Environmental materials and waste, Academic Press, pp 569–589

Crane RA, Dickinson M, Popescu IC, Scott TB (2011) Magnetite and zero-valent iron nanoparticles for the remediation of uranium contaminated environmental water. Water Res 45(9):2931–2942

Dale AL, Casman EA, Lowry GV, Lead JR, Viparelli E, Baalousha M (2015) Modeling nanomaterial environmental fate in aquatic systems. Environ Sci Technol 49(5):2587–2593

Ding Z, Hu X, Morales VL, Gao B (2014) Filtration and transport of heavy metals in graphene oxide enabled sand columns. Chem Eng J 257:248–252

Engates KE, Shipley HJ (2011) Adsorption of Pb, Cd, Cu, Zn, and Ni to titanium dioxide nanoparticles: effect of particle size, solid concentration, and exhaustion. Environ Sci Pollut Res 18(3):386–395

Gangwar J, Gupta BK, Srivastava AK (2016) Prospects of emerging engineered Oxide nanomaterials and their applications. Defence Sci J 66(4)

Ghaedi M, Niknam K, Shokrollahi A, Niknam E, Rajabi HR, Soylak M (2008) Flame atomic absorption spectrometric determination of trace amounts of heavy metal ions after solid phase extraction using modified sodium dodecyl sulfate coated on alumina. J Hazard Mater 155(1–2):121–127

Gómez-Pastora J, Dominguez S, Bringas E, Rivero MJ, Ortiz I, Dionysiou DD (2017) Review and perspectives on the use of magnetic nanophotocatalysts (MNPCs) in water treatment. Chem Eng J 310:407–427

Guerra FD, Attia MF, Whitehead DC, Alexis F (2018) Nanotechnology for environmental remediation: Materials and applications. Molecules 23(7):1760

Gupta VK, Kumar R, Nayak A, Saleh TA, Barakat MA (2013) Adsorptive removal of dyes from aqueous solution onto carbon nanotubes: A review. Adv Coll Interface Sci 193:24–34

Hameeteman E (2013) Future water (In) security: Facts, figures, and predictions, vol 16, Global Water Institute

He F, Zhao D (2005) Preparation and characterization of a new class of starch-stabilized bimetallic nanoparticles for degradation of chlorinated hydrocarbons in water. Environ Sci Technol 39(9):3314–3320

Holt JK, Park HG, Wang Y, Stadermann M, Artyukhin AB, Grigoropoulos CP, … Bakajin O (2006) Fast mass transport through sub-2-nanometer carbon nanotubes. Science 312(5776):1034–1037

Hu J, Chen G, Lo IM (2005) Removal and recovery of Cr (VI) from wastewater by maghemite nanoparticles. Water Res 39(18):4528–4536

Hu J, Chen G, Lo IM (2006) Selective removal of heavy metals from industrial wastewater using maghemite nanoparticle: Performance and mechanisms. J Environ Eng 132(7):709–715

Hua M, Zhang S, Pan B, Zhang W, Lv L, Zhang Q (2012) Heavy metal removal from water/wastewater by nanosized metal oxides: A review. J Hazard Mater 211:317–331

Hummer G, Rasaiah JC, Noworyta JP (2001) Water conduction through the hydrophobic channel of a carbon nanotube. Nature 414(6860):188–190

Jain M, Garg VK, Paliwal R, Kadirvelu K, Chaudhry S (2020). Optimization of cadmium (II) removal from water using sunflower waste carbon—A statistical approach. Toxin reviews, 1–10

Jiang W, Cai Q, Xu W, Yang M, Cai Y, Dionysiou DD, O'Shea KE (2014) Cr (VI) adsorption and reduction by humic acid coated on magnetite. Environ Sci Technol 48(14):8078–8085

Kanel SR, Greneche JM, Choi H (2006) Arsenic (V) removal from groundwater using nano scale zero-valent iron as a colloidal reactive barrier material. Environ Sci Technol 40(6):2045–2050

Karri RR, Shams S, Sahu JN (2019) Overview of potential applications of nano-biotechnology in wastewater and effluent treatment. In Nanotechnology in water and wastewater treatment, Elsevier, pp 87–100

Kefeni KK, Msagati TA, Nkambule TT, Mamba BB (2018) Synthesis and application of hematite nanoparticles for acid mine drainage treatment. J Environ Chem Eng 6(2):1865–1874

Kemp KC, Seema H, Saleh M, Le NH, Mahesh K, Chandra V, Kim KS (2013) Environmental applications using graphene composites: Water remediation and gas adsorption. Nanoscale 5(8):3149–3171

Khin MM, Nair AS, Babu VJ, Murugan R, Ramakrishna S (2012) A review on nanomaterials for environmental remediation. Energy Environ Sci 5(8):8075–8109

Kim S, Park CM, Jang M, Son A, Her N, Yu M, … Yoon Y (2018) Aqueous removal of inorganic and organic contaminants by graphene-based nanoadsorbents: a review. Chemosphere 212:1104–1124

Kumar N, Kumbhat S (2016) Carbon-based nanomaterials, In Essentials in Nanoscience and nanotechnology. New York, US, Wiley, pp 189–236. https://doi.org/10.1002/9781119096122.ch5

Kumar S, Ahlawat W, Bhanjana G, Heydarifard S, Nazhad MM, Dilbaghi N (2014) Nanotechnology-based water treatment strategies. J Nanosci Nanotechnol 14(2):1838–1858

Kunduru KR, Nazarkovsky M, Farah S, Pawar RP, Basu A, Domb AJ (2017) Nanotechnology for water purification: Applications of nanotechnology methods in wastewater treatment. In Water purification, Academic Press, pp 33–74

Lakshmipathiraj P, Narasimhan BRV, Prabhakar S, Raju GB (2006) Adsorption of arsenate on synthetic goethite from aqueous solutions. J Hazard Mater 136(2):281–287

Lata S, Singh PK, Samadder SR (2015) Regeneration of adsorbents and recovery of heavy metals: a review. Int J Environ Sci Technol 12(4):1461–1478

Lens P, Virkutyte J, Jegatheesan V, Al-Abed S (Eds.) (2013) Nanotechnology for water and wastewater treatment. Iwa Publishing

Li J, Guo S, Zhai Y, Wang E (2009) Nafion–graphene nanocomposite film as enhanced sensing platform for ultrasensitive determination of cadmium. Electrochem Commun 11(5):1085–1088

Li S, Wang W, Yan W, Zhang WX (2014) Nanoscale zero-valent iron (nZVI) for the treatment of concentrated Cu (II) wastewater: A field demonstration. Environ Sci Process Impacts 16(3):524–533

Li YH, Ding J, Luan Z, Di Z, Zhu Y, Xu C, … Wei B (2003) Competitive adsorption of Pb2+, Cu2+ and Cd2+ ions from aqueous solutions by multiwalled carbon nanotubes. Carbon 41(14):2787–2792

Li Z, Jones HK, Zhang P, Bowman RS (2007) Chromate transport through columns packed with surfactant-modified zeolite/zero valent iron pellets. Chemosphere 68(10):1861–1866
Lofrano G, Carotenuto M, Libralato G, Domingos RF, Markus A, Dini L, ... Chattopadhyaya MC (2016) Polymer functionalized nanocomposites for metals removal from water and wastewater: an overview. Water Res 92:22–37
Lü K, Zhao G, Wang X (2012) A brief review of graphene-based material synthesis and its application in environmental pollution management. Chin Sci Bull 57(11):1223–1234
Ma L, Wei Q, Chen Y, Song Q, Sun C, Wang Z, Wu G (2018) Removal of cadmium from aqueous solutions using industrial coal fly ash-nZVI. Royal Society Open Science 5(2):171051
Macera L, Taglieri G, Daniele V, Passacantando M, D'Orazio F (2020) Nano-sized Fe (III) oxide particles starting from an innovative and eco-friendly synthesis method. Nanomaterials 10(2):323
Madadrang CJ, Kim HY, Gao G, Wang N, Zhu J, Feng H, ... Hou S (2012) Adsorption behavior of EDTA-graphene oxide for Pb (II) removal. ACS Appl Mater Interfaces 4(3):1186–1193
Madima N, Mishra SB, Inamuddin I, Mishra AK (2020) Carbon-based nanomaterials for remediation of organic and inorganic pollutants from wastewater. A Review. Environmental Chemistry Letters 18(4):1169–1191
Majumder M, Chopra N, Andrews R, Hinds B (2005) Erratum: Nanoscale hydrodynamics: Enhanced flow in carbon nanotubes. Nature 438(7070):930–930
Martynková GS, Valášková M (2014) Antimicrobial nanocomposites based on natural modified materials: A review of carbons and clays. J Nanosci Nanotechnol 14(1):673–693
Mashkoor F, Nasar A (2020) Carbon nanotube-based adsorbents for the removal of dyes from waters: A review. Environ Chem Lett. https://doi.org/10.1007/s10311-020-00970-6
Mohammad R, Akram H, Rim H (2017) Removal of cadmium (II) Ions from waste water by adsorption onto the powder of lebanese anacyclus nigllifolius boiss: a comparative study. American J Phytomedicine and Clin Ther 5(1):1–6
Morones JR, Elechiguerra JL, Camacho A, Holt K, Kouri JB, Ramírez JT, Yacaman MJ (2005) The bactericidal effect of silver nanoparticles. Nanotechnology 16(10):2346
Naguib N, Ye H, Gogotsi Y, Yazicioglu AG, Megaridis CM, Yoshimura M (2004) Observation of water confined in nanometer channels of closed carbon nanotubes. Nano Lett 4(11):2237–2243
Ong CB, Ng LY, Mohammad AW (2018) A review of ZnO nanoparticles as solar photocatalysts: Synthesis, mechanisms and applications. Renew Sustain Energy Rev 81:536–551
Pan B, Qiu H, Pan B, Nie G, Xiao L, Lv L, ... Zheng S (2010) Highly efficient removal of heavy metals by polymer-supported nanosized hydrated Fe (III) oxides: Behavior and XPS study. Water Res 44(3):815–824
Pendergast MM, Hoek EM (2011) A review of water treatment membrane nanotechnologies. Energy Environ Sci 4(6):1946–1971
Polotskaya G, Biryulin Y, Rozanov V (2005) Asymmetric membranes based on Fullerene-containing Polyphenylene Oxide. Fullerenes, Nanotubes, Carbon Nanostruct 12(1–2):371–376
Pradeep T (2009) Noble metal nanoparticles for water purification: A critical review. Thin Solid Films 517(24):6441–6478
Prajitha N, Athira SS, Mohanan PV (2019) Bio-interactions and risks of engineered nanoparticles. Environ Res 172:98–108
Rahmani A, Mousavi HZ, Fazli M (2010) Effect of nanostructure alumina on adsorption of heavy metals. Desalination 253(1–3):94–100
Ramos MA, Yan W, Li XQ, Koel BE, Zhang WX (2009) Simultaneous oxidation and reduction of arsenic by zero-valent iron nanoparticles: Understanding the significance of the core– shell structure. The J Phys Chem C 113(33):14591–14594
Rao KS, Mohapatra M, Anand S, Venkateswarlu P (2010) Review on cadmium removal from aqueous solutions. Int J Eng Sci Tech 2(7)
Rehman K, Fatima F, Waheed I, Akash MSH (2018) Prevalence of exposure of heavy metals and their impact on health consequences. J Cell Biochem 119(1):157–184
Ren X, Chen C, Nagatsu M, Wang X (2011) Carbon nanotubes as adsorbents in environmental pollution management: A review. Chem Eng J 170(2–3):395–410

Samanta HS, Das R, Bhattachajee C (2016) Influence of nanoparticles for wastewater treatment—A short review. Austin Chem Eng 3(3):1036

Sánchez A, Recillas S, Font X, Casals E, González E, Puntes V (2011) Ecotoxicity of, and remediation with, engineered inorganic nanoparticles in the environment. TrAC, Trends Anal Chem 30(3):507–516

Sarma GK, Gupta SS, Bhattacharyya KG (2019) Nanomaterials as versatile adsorbents for heavy metal ions in water: A review. Environ Sci Pollut Res 26(7):6245–6278

Shahwan T, Üzüm Ç, Eroğlu AE, Lieberwirth I (2010) Synthesis and characterization of bentonite/iron nanoparticles and their application as adsorbent of cobalt ions. Appl Clay Sci 47(3–4):257–262

Shang J, Zong M, Yu Y, Kong X, Du Q, Liao Q (2017) Removal of chromium (VI) from water using nanoscale zerovalent iron particles supported on herb-residue biochar. J Environ Manage 197:331–337

Shang TM, Sun JH, Zhou QF, Guan MY (2007) Controlled synthesis of various morphologies of nanostructured zinc oxide: Flower, nanoplate, and urchin. Crystal Research and Technology: Journal of Experimental and Industrial Crystallography 42(10):1002–1006

Sharma YC, Srivastava V, Upadhyay SN, Weng CH (2008) Alumina nanoparticles for the removal of Ni (II) from aqueous solutions. Ind Eng Chem Res 47(21):8095–8100

Sheela T, Nayaka YA, Viswanatha R, Basavanna S, Venkatesha TG (2012) Kinetics and thermodynamics studies on the adsorption of Zn (II), Cd (II) and Hg (II) from aqueous solution using zinc oxide nanoparticles. Powder Technol 217:163–170

Srivastav AL, Ranjan M (2020) Inorganic water pollutants. In Inorganic pollutants in water, Elsevier, pp 1–15

Thines RK, Mubarak NM, Nizamuddin S, Sahu JN, Abdullah EC, Ganesan P (2017) Application potential of carbon nanomaterials in water and wastewater treatment: a review. J Taiwan Inst Chem Eng 72:116–133

Tiraferri A, Chen KL, Sethi R, Elimelech M (2008) Reduced aggregation and sedimentation of zero-valent iron nanoparticles in the presence of guar gum. J Colloid Interface Sci 324(1–2):71–79

Tosco T, Papini MP, Viggi CC, Sethi R (2014) Nanoscale zerovalent iron particles for groundwater remediation: a review. J Clean Prod 77:10–21

Tul Muntha S, Kausar A, Siddiq M (2017) Advances in polymeric nanofiltration membrane: A review. Polym-Plast Technol Eng 56(8):841–856

Türker AR (2007) New sorbents for solid-phase extraction for metal enrichment. Clean-Soil, Air, Water 35(6):548–557

UNESCO W (2019) The United Nations world water development report 2019: Leaving no one behind

UNICEF (2017) Thirsting for a future: water and children in a changing climate. UNICEF

Varma RS (2012) Greener approach to nanomaterials and their sustainable applications. Curr Opin Chem Eng 1(2):123–128

Vuković GD, Marinković AD, Čolić M, Ristić MĐ, Aleksić R, Perić-Grujić AA, Uskoković PS (2010) Removal of cadmium from aqueous solutions by oxidized and ethylenediamine-functionalized multi-walled carbon nanotubes. Chem Eng J 157(1):238–248

Wang H, Yuan X, Wu Y, Huang H, Zeng G, Liu Y, … Qi Y (2013) Adsorption characteristics and behaviors of graphene oxide for Zn (II) removal from aqueous solution. Appl Surf Sci 279:432–440

Wei X, Bhojappa S, Lin LS, Viadero RC Jr (2012) Performance of nano-magnetite for removal of selenium from aqueous solutions. Environ Eng Sci 29(6):526–532

Wen Z, Zhang Y, Dai C (2014) Removal of phosphate from aqueous solution using nanoscale zerovalent iron (nZVI). Colloids Surf, A 457:433–440

Werkneh AA, Rene ER (2019) Applications of nanotechnology and biotechnology for sustainable water and wastewater treatment. In Water and wastewater treatment technologies, Springer, Singapore, pp 405–430

Xiong C, Wang W, Tan F, Luo F, Chen J, Qiao X (2015) Investigation on the efficiency and mechanism of Cd (II) and Pb (II) removal from aqueous solutions using MgO nanoparticles. J Hazard Mater 299:664–674
Xiu ZM, Ma J, Alvarez PJ (2011) Differential effect of common ligands and molecular oxygen on antimicrobial activity of silver nanoparticles versus silver ions. Environ Sci Technol 45(20):9003–9008
Xu Y, Zhao D (2007) Reductive immobilization of chromate in water and soil using stabilized iron nanoparticles. Water Res 41(10):2101–2108
Yadav D, Singh, P, Kumar P (2020) Application of nanoparticles for inorganic water purification. In Inorganic Pollutants in Water, Elsevier, pp 221–243
Yan L, Li YS, Xiang CB (2005) Preparation of poly (vinylidene fluoride) (pvdf) ultrafiltration membrane modified by nano-sized alumina (Al_2O_3) and its antifouling research. Polymer 46(18):7701–7706
Yang F, Zhang S, Sun Y, Cheng K, Li J, Tsang DC (2018) Fabrication and characterization of hydrophilic corn stalk biochar-supported nanoscale zero-valent iron composites for efficient metal removal. Biores Technol 265:490–497
Yang Y, Xie Y, Pang L, Li M, Song X, Wen J, Zhao H (2013) Preparation of reduced graphene oxide/poly(acrylamide) nanocomposite and its adsorption of Pb(II) and methylene blue. Langmuir 29(34):10727–10736
Yaqoob AA, Parveen T, Umar K, Mohamad Ibrahim MN (2020) Role of nanomaterials in the treatment of wastewater: A review. Water 12(2):495
Yin J, Deng B (2015) Polymer-matrix nanocomposite membranes for water treatment. J Membr Sci 479:256–275
Yu G, Wang X, Liu J, Jiang P, You S, Ding N, … Lin F (2021) Applications of nanomaterials for heavy metal removal from water and soil: A review. Sustainability 13(2):713
Yu RF, Chi FH, Cheng WP, Chang JC (2014) Application of pH, ORP, and DO monitoring to evaluate chromium (VI) removal from wastewater by the nanoscale zero-valent iron (nZVI) process. Chem Eng J 255:568–576
Yuan P, Annabi-Bergaya F, Tao Q, Fan M, Liu Z, Zhu J, … Chen T (2008) A combined study by XRD, FTIR, TG and HRTEM on the structure of delaminated Fe-intercalated/pillared clay. J Colloid Interface Sci 324(1–2):142–149
Zan L, Fa W, Peng T, Gong ZK (2007) Photocatalysis effect of nanomaterial TiO2 and TiO2-coated ceramic plate on Hepatitis B virus. J Photochem Photobiol, B 86(2):165–169
Zekić E, Vuković Ž, Halkijević I (2018) Application of nanotechnology in wastewater treatment. Građevinar 70(04):315–323
Zhang L, Huang T, Zhang M, Guo X, Yuan Z (2008) Studies on the capability and behavior of adsorption of thallium on nano-Al_2O_3. J Hazard Mater 157(2–3):352–357
Zhang X, Lin S, Lu XQ, Chen ZL (2010) Removal of Pb (II) from water using synthesized kaolin supported nanoscale zero-valent iron. Chem Eng J 163(3):243–248
Zhang Y, Wu B, Xu H, Liu H, Wang M, He Y, Pan B (2016) Nanomaterials-enabled water and wastewater treatment. Nanoimpact 3:22–39

Chapter 8
Application of Nanobiotechnology for Heavy Metal Remediation

Arjita Punetha and Amir Khan

Abstract Societies across the globe are facing a major challenge of the ill effects of heavy metals. However, there is no decline in the production of these toxic metals due to the ever increasing pace of industrialisation and other anthropogenic activities. In order to solve the problem of heavy metal contamination so as to avoid its intake, we need to search for a technology that can act as a silver bullet for its removal from the environment. Nanobiotechnology is a new field of study with the potential to solve such environmental problems by combining cutting-edge information technology and nanotechnology applications. Nanobiotechnology, which is the consequence of combining nanotechnology and biology, involves using biological organisms to create nanomaterials that can be used for heavy metal remediation. The spatial arrangements of atoms present in nanomaterials that emerge in diverse geometries increase the efficiency and specificity of nanostructures. Several types of nanostructures such as nanobiosensors, nanomembranes, nanocomposites and nanoparticles have recently proved quite effective in the removal of heavy metals. This technology expected to be a good substitute for traditional treatments, but the risks of environmental nano-toxicity are still unknown. Extensive studies in the field are required to completely understand the benefits and negative impacts of the technology.

Keywords Heavy metals · Nanobiosensors · Nanocomposites · Remediation

8.1 Introduction

Increased anthropogenic activities have led to increase in the release of environmental pollutants, which has become one of the major issues that societies are facing

---

A. Punetha (✉)
Research Fellow CSIR-Central Institute of Medicinal and Aromatic Plants, Pantnagar, India
e-mail: arjitap@gmail.com

A. Khan
Research Scholar, Department of Microbiology, College of Basic Science and Humanities, Govind Ballabh Pant University Agriculture and Technology, Pantnagar, India

J. P. N. Rai and S. Saraswat (eds.), *Nano-biotechnology for Waste Water Treatment*, Water Science and Technology Library 111,
https://doi.org/10.1007/978-3-031-00812-2_8

globally. Heavy metals and their derivatives are among the most harmful environmental contaminants, with several negative consequences on the biosphere. Metals and metalloids that have an atomic mass greater than 20 and a specific gravity greater than 5 are termed as heavy metals. Examples of heavy metals include arsenic (As), cadmium (Cd), chromium (Cr), lead (Pb), mercury (Hg), nickel (Ni) and zinc (Zn). From a biological standpoint, 'heavy' refers to a group of metals or metalloids that can be hazardous to plants and animals even at low concentrations (Rascio and Navariizzo 2011; Li et al. 2019). A substantial amount of heavy metals in the environment is released from anthropogenic activities such as metallurgical operations, petrochemical refining and mining. Heavy metal contamination in soil has emerged as a global environmental problem that has piqued public interest, owing to growing concerns about the safety of agricultural goods (Hu et al. 2017). Total number of sites impacted with heavy metal and other major pollutants pollution worldwide counts to 5 million, covering a total of 500 million hectares of land (Li et al. 2019). Because of their toxicity, non-biodegradable nature, and ability to bioaccumulate, heavy metals and their derivatives are among the most dangerous environmental pollutants. For the treatment of heavy metal contaminated environments, a variety of physical and chemical approaches have been used. However, there technologies are not accepted widely due to several issues that arise with their use. These technologies are quite expensive, result in alteration in the quality of soil, affect native microorganisms, as well as can result in secondary pollution. In recent years, biological treatment approaches such as microbial and phytoremediation have been introduced and have proven to be quite efficient. The technique of bioremediation is cost-effective, eco-friendly and can be used for a number of pollutants effectively. However, the technique faces a few limitations. The process is slow and is limited by factors like bioavailability of pollutants, environmental conditions, etc. For advancing bioremediation beyond its current constraints, nanotechnology and the integration of nanomaterials with biotechnology can prove to be very effective. Nanomaterials are extremely small, measuring between 1 and 100 nm in diameter, and have a unique property that allows them to be used in several novel applications (Qu et al. 2013; Werkneh et al. 2018a, b). They have a high surface area-to-volume ratio, which is important for producing 'highly reactive surface area' compared to their bulk counterparts (Bethi et al. 2018). Biotechnology, on the other hand, is concerned with the physiological and other metabolic processes of living creatures, including microbes. Environmental biotechnology is a rapidly expanding discipline that aims to solve complicated environmental problems in the ecosystem and natural environment. Nano-technology and Biotechnology on merging together give rise to Nanobiotechnology. Integration of these emerging technologies i.e. nanobiotechnology has the potential to revolutionise bioremediation (Rizwan et al. 2014). This integrated strategy can cover a broader array of probable applications with lower expenses and fewer harmful environmental implications. Nanobiotechnology combines the principles of nanotechnology and characteristics of nanomaterials with biological processes to create new products, and it offers a promising opportunity to improve our fundamental understanding of interaction of nanomaterials with intracellular structures, activities and the environment. The majority of available research in this area has focused on biofabrication of Nanoparticles (NPs) by using biological materials rather than traditional physico-chemical methods. This results in prevention of the usage of potentially toxic solvents and reagents, preventing the generation of toxic by-products and the reduction of energy consumption. Nanobiosensors are other examples of nanobiotechnology that can be used to detect the presence of pollutants in the environment. Other approaches include the use of nanomembranes and nanoadsorbents derived from living organisms for heavy metal clean-up, most commonly from water. This chapter summarises the sources and consequences of heavy metals in the environment, as well as the use of nanobiotechnology to mitigate their impacts.

8.2 Sources of Heavy Metals and Their Impacts on Human Health

Heavy metals enter the ecosystem through both natural sources and anthropogenic activities. Several studies have found that in comparison to anthropogenic activities, natural sources of heavy metals in the environment are usually insignificant (Dixit et al. 2015). The parent material from which heavy metals are derived is the principal natural source of heavy metals in soils. Heavy metals chiefly Cd, Co, Cu and Ni are abundant in basaltic igneous rocks, whereas Mn, Pb, and Zn are abundant in shales (Sarwar et al. 2016). Natural activities such as wind and water erosion, leaching and volcanic processes introduce heavy metals into the soil environment. Anthropogenic processes disrupt the natural gradual cycling of heavy metals in the environment, due to which, the rate of their accumulation in the soil has accelerated (Dixit et al. 2015). Industrialisation, urbanisation and current agricultural practices have contributed significantly to rising levels of heavy metal contamination in the soil. Increase in activities like fossil fuel combustion, municipal waste disposal (Khan et al. 2016), mining and smelting (Chen et al. 2015), use of pesticides in fields (Ogunlade and Agbeniyi 2011), fertiliser use (Atafar et al. 2010) and irrigation of lands using sewage water (Sun et al. 2013), all contribute to increased heavy metal concentrations. Terrestrial and aquatic water bodies are also frequently found to be contaminated with high concentrations of heavy metals, which can be associated to the untreated dumping of liquid or solid waste products resulting from electroplating, smelting, mining, agrochemicals runoff, bio-solids, municipal solid waste and sludge, paint and textile industries (Barkat 2011; Mani and Bharagava 2016; Yadav et al. 2017;). Mining activities result in extraction of heavy metals like As, Cd, Fe due to the very common practices of mining, these metals are found in abundance in the soil. Metal contamination into the environment occurs with the transportation of metals from mining sites to other businesses for use in a wide range of products. Because of the substantial use of Cr salts in the chrome tanning process, leather industries are principally responsible for the release of Cr in the soil and water bodies (Mishra and Bharagava 2016). The paint and electroplating industries pollute the environment with Pb-based pollution. Metal pollution into soils is also a result of the increasing use of fertilisers, insecticides and fungicides in agricultural activities. Chemical-based fertilisers, which are widely used in agricultural activities to improve soil fertility and increase yields, have grown to be one of the majorly significant causes of metal contamination (Yadav et al. 2016).

Metals are very toxic and strong contaminants that cause severe soil contamination and have an impact on soil fertility and productivity (Prajapati and Meravi 2014). These metals have a tendency to bioaccumulate and therefore harm the plants and aquatic life and wreak havoc on ecological equilibrium (Chen 2002). Long-term exposure to heavy metals poses a serious threat to plant diversity as well as aquatic and terrestrial creatures. Heavy metals have a deleterious impact on plant and microbial variety and species richness (Arao et al. 2010; Zhang et al. 2011). When heavy metals are consumed in excess of their recommended dietary intake, they cause a variety of toxicological effects that have been thoroughly researched (Dong et al. 2010). Surfeit metal concentration in farmed soils influence the quality of food by accumulating in it thereby increasing the danger of kidney and liver failure, reproductive diseases even infertility, neurological breakdown, leukaemia, etc. (Khan et al. 2011). Presence of heavy metals in urban soils degrades the environmental quality by polluting the food, air and water causing direct harm to the population, particularly children and adolescents, through skin contact and inhalation (Nagajyoti et al. 2010). Cd is the sixth most dangerous material according to the ATSDR Committee because it disrupts or alters calcium metabolism, causing calcium insufficiency and resulting in cartilage degeneration, bone fractures and other health problems (ATSDR 2008). Pb enters the human body primarily through the gastrointestinal and respiratory tracts, and then its circulation in the blood in the form of soluble salts, protein complexes, ions and other ions begins. 95 per cent of the Pb accumulates in the bones in the insoluble form of phosphates. It leads to damage of kidney, liver, reproductive system, neurological system, urinary system and immunological system, as well as the basic physiological processes and genetic expressions (Chao et al. 2014). Some essential trace metals, such as Ni, Cu, Zn and Mo, are required for enzymatic and physiological functions, but excessive amounts of these metals can harm human health if consumed in large quantities. Cu and Ni are carcinogenic in nature, which is a major issue worldwide. Lung cancer and nasopharyngeal carcinoma are caused by direct exposure to Ni-manufacturing products by industrial employees (Chen 2011; Chao et al. 2014). Cr (VI) is extremely dangerous and poisonous. If inhaled, it causes severe symptoms of itching and irritation of the nose and skin, as well as damage to the nasal septum and ulcers, whereas high Cr (VI) dose absorption and ingestion can result in kidney and liver damage, nausea, gastrointestinal irritation, ulcers in stomach and even death (Mishra and Bharagava 2016). The sources of heavy metals in the environment, their acceptable limitations, and their effects on the human body are summarised in Table 8.1.

Table 8.1 Heavy metals, their sources, effects and permissible limits

S.No	Heavy metal	Source	Effect on humans	Permissible limit(mg/L)
1	Lead	Paint industries, pesticides, coal burning, automobile emission, battery manufacturing, mining	Mental retardation, learning disability, development delay in children, chronic damage to nervous system, liver and kidney damage, gastrointestinal damage	0.1
2	Arsenic	Textile, wood and wood products, electrical, paper manufacturing units, pesticides, fungicides, metal smelters	Skin diseases, visceral cancers, vascular disease, bronchitis, dermatitis	0.05
3	Nickel	Stainless steel manufacturing industries, electroplating factory discharge	Neurotoxic, Carcinogenic and genotoxic, Nickel Dermatitis	0.20
4	Cadmium	Electroplating factory, Cd-Ni batteries making, TV phosphors, welding, nuclear fission plant, pesticides	Kidney and liver diseases, Renal Dysfunction, Gastrointestinal damage, bronchitis, bone marrow damage, cancer	0.01
5	Chromium	Mines, electroplating, leather tanning, industrial coolants	Gastrointestinal, Hepatic, Renal, Neuronal damage	0.05
6	Mercury	Pharmaceuticals and electronics, thermal power plants, paper industry, batteries	Rheumatoid Arthiritis, kidney problems, diseases of circulatory and nervous system, protoplasm poisoning	0.003

8.3 Heavy Metal Remediation Techniques

In order to make the environment safer for humans, contaminated water bodies and land must be cleaned to remove heavy metals and trace elements. Several methods devised for the treatment of heavy metals (Fig. 8.1) are discussed as follows:

8.3.1 Extraction via Electrokinetics

Electrokinetic extraction uses electrical adsorption for the removal of heavy metals from polluted soils. This process works on the principle of establishing an appropriate electric field gradient on both sides of an electrolytic tank having contaminated soil

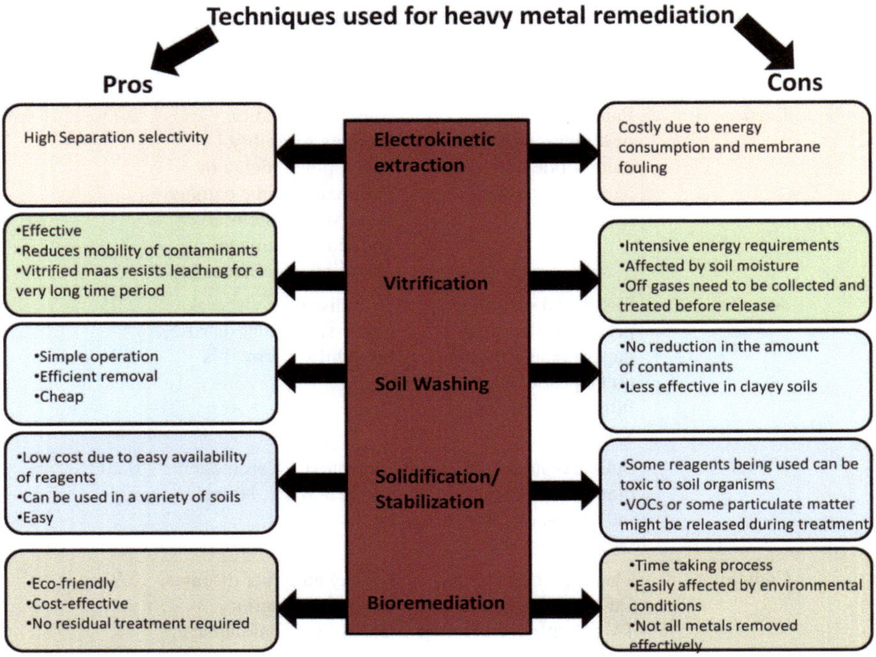

Fig. 8.1 Pros and Cons of technologies being used for heavy metal remediation

in saturated form. On application of low-density direct current (DC) to the ground via electrodes, cations in the solution phase of contaminated soil move to the cathode, while anions migrate to the anode due to the attractive forces of electrical field. Electroplating, precipitation, solution pumping or ion exchange resin complexation are used to remove metal impurities that have accumulated at the polarised electrodes in the process of electrokinetic extraction (FRTR 2012). Preliminary dissolving is required to remove heavy metals with low conductivity (e.g. sulphides) or those present in the form of metals (e.g. Hg). In these circumstances, the use of a suitable electrolyte (distilled water, organic acids or synthetic chelates) may improve the electrokinetic remediation method's removal efficiency (Iannelli et al. 2015). The removal efficiency, however, varies with the chemical utilised and the metal to be remediated.

8.3.2 *Vitrification*

This method makes the use high temperature to reduce the mobility of heavy metals inside soil of the polluted site (Mallampati et al. 2015), which results in the development of vitreous substances. Some heavy metals like Hg may get volatilised under high temperatures during this process and need to be collected for disposal or treatment. Electric current is applied through the soil during in situ vitrification by vertically inserting an array of electrodes into the polluted area. Dellisanti (2016) used Joule heating to vitrify tonnes of Zn and Pb-rich ceramic trash in the field, reaching temperatures of around 1850 °C. They claimed that the vitrification technology was extremely effective in cleaning tonnes of heavy metals-contaminated waste materials and that it could be used to treat large amounts of soil. The temperature at which heavy metals in soil samples are immobilised is influenced by the temperature at which they are vitrified. Navarro et al. (2013), for example, employed solar technology to vitrify waste from Ag-Pb mines in Spain and discovered that at 1350 °C, vitrification rendered Cu, Fe, Mn, Ni and Zn immobile, but at 1050 °C, these heavy metals remained immobilised.

8.3.3 Soil Washing

This process removes heavy metals from soil with the use of reagents and extractants that can leach heavy metals from the soil (Park and Son 2016; Guo et al. 2016). Depending on the type of metal and soil, the contaminated soil is scrapped out and blended with an appropriate extractant solution during soil washing. The solution used as extractant and dirt is combined for a required amount of time. As a result of this heavy metals present in solid phase in the soil are transferred to the liquid phase and are then separated from the leachate (Ferraro et al. 2015). The separated soil that meets regulatory requirements can then be returned to its original location. Chelating agents like EDTA, EDDS or organic acids, humic compounds, surfactants, cyclodextrins, etc. have all been employed to mobilise and remove heavy metals from soil (Shahid et al. 2014; Kulikowska et al. 2015). The ability of the extractant to dissolve heavy metals in soils determines the efficacy of soil washing. Therefore, extractants capable of dissolving high quantities of metals would be suitable for soil cleaning.

8.3.4 Solidification/Stabilisation

This approach is typically used to reduce the mobility of heavy metals in soil by the addition of chemicals. The method of physical encapsulation of contaminants in a solid matrix using asphalt, bitumen, cement and thermoplastic binders is known as solidification, while chemical processes to restrict contaminant mobility are known as stabilisation. Calcined oyster and cockle shells, lime-based agents, eggshells, waste mussel shells are just a few examples of eco-friendly and cost-effective resources that have been reported to immobilize heavy metals and also improve the soil quality (Islam et al. 2017).

8.3.5 *Bioremediation*

It is a process of removing, destroying or immobilising dangerous pollutants from a polluted environment through the use of biological mechanisms found in plants and microbes (Ayangbenro and Babalola 2017). In comparison to traditional chemical and physical procedures for heavy metal removal, which are typically very expensive and unsuccessful, and produce substantial amounts of toxic sludge, bioremediation is an environmentally safe and cost-effective technology (Ojuederie and Babalola 2017). Heavy metal bioremediation can be accomplished using microbes, plants or both.

Although bioremediation provides a good strategy for clean-up of heavy metals, however, it may not provide a feasible strategy at sites with very high concentrations of heavy metals that are toxic and kill most of the microorganisms responsible to degrade them. Integration of nanomaterials with the bioremediation technology provides additional benefits to the removal of heavy metals.

8.3.6 *Nanomaterials for Heavy Metal Remediation*

NMs have been shown to be effective for bioremediation of environmental pollutants for the following reasons:

1. Because NPs have a high surface-to-volume ratio per unit mass of material, they considerably accelerate surface-dependent bioremediation processes.
2. Because of the quantum effect, the material requires very little activation energy at the nanoscale dimension, resulting in requirement of very low activation energy for a physical or chemical process using NP.
3. Another phenomena displayed by NPs that can be exploited to detect harmful substances is surface plasmon resonance.
4. Because of their submicroscopic size, NPs can access harmful chemicals that are hidden from microorganisms.

However, the most serious disadvantage of these NPs is that they are all hazardous when used for a long-term. Metals or their derivatives are used to make them in the majority of cases. Recent advances in synthesis of nanomaterials, on the other hand, have led to the production of biologically generated nanomaterials that are resistant to the hazardous effects of metal-based NPs.

8.4 Biotechnological Approaches to Nanoparticle Production

Several biotechnological approaches such as green nanoparticles, nanosensors, nanbiomembranes, nanobioadsorbents, etc. (Fig. 8.2) are being used for heavy metal remediation. Green synthesis of nanomaterials is the most widely used technique. Pollutants can be transformed into molecules with lower toxicity, solubility and mobility using nanoparticles made from biological components (Wasi et al. 2008). Biologically manufactured metallic nanoparticles are more stable at room temperature for longer periods of time than chemically created metallic nanoparticles (Balakrishnan et al. 2017). The biosynthetic methods are more stable because natural proteins are capped over the metallic nanoparticle surface. By using correct methodologies, the cost of producing nanoparticles can be reduced to 1/10th of the cost of chemical synthesis processes. A significant quantity of pollutants may be eliminated with lesser amounts of biogenic nanoparticles. Table 8.2 presents few examples of nanoparticles synthesised using biological organisms.

8.4.1 Plant-Based Nanomaterials

The presence of biomolecules in plant extracts enables the bio-reduction of NPs. Although the specific mechanism of this process is still unknown, amino acids, citric acid, phenolics, sugars, membrane proteins, tartaric acid and functional groups (alcohols, aldehydes, amines, carboxylic acid ketones) all operate as reducing and capping agents (Al-Shnani et al. 2017; Selvan et al. 2018; Weng et al. 2013). Several

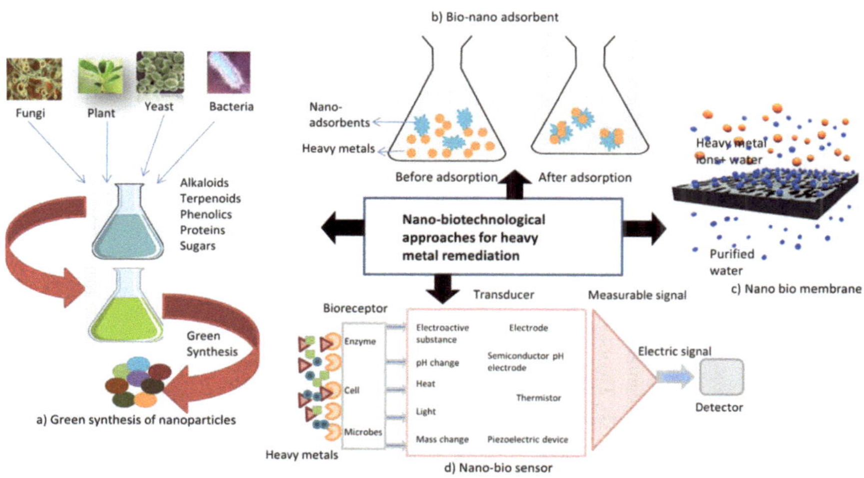

Fig. 8.2 Nanobiotechnological approaches for heavy metal remediation

Table 8.2 Green synthesis of naoparticles for heavy metal remediation

S.No	Biological organism	Salt/solution used	Nanoparticles synthesised	Use	Reference
1	Lemon juice	Copper acetate monohydrat	Cupric oxide	Removal of Cr(VI)	Mohan et al. 2015
2	Tangerine peel extract	$FeCl_3$ and $FeCl_2.4H_2O$	Iron oxide	Adsorbent for cadmium ions removal from Contaminated solution	Ehrampoushet al. 2015
3	Rosa damascene (RD), Thymus vulgaris (TV), and Urtica dioica (UD)	Polyphenols, proteins and organic acids	nZVI	Decontamination of aqueous solution containing hexavalent chromium	Fazlzadeh et al. 2016
4	Aloe vera leaf	Anhydrous ferric chloride ($FeCl_3$)	Fe_2O_3	Arsenic(V) remediation	Mukherjee et al. 2016
5	Eucalyptus leaf extracts	0.05 m $FeSO_4$solution	Fe nanoparticles	Removemixed Cr (VI) and Cu(II)	Weng et al. 2016
6	Coconut husk	10 mm $FeCl_2$ solution	Iron oxide	Treat metal contaminated water and heavy metal stress in *oryza* *Sativa*	Sebastian et al. 2018
7	*Murrayakoenigii* leaves	1 mm iron (III) chloride	Fe_3O_4 magnetic Nanoparticles	Removal of Pb (II) from aqueous solution	Prasad et al. 2017
8	Sawdust	Fe $(NO_3)3.9H_2O$	Fe_3O_4	Cadmium (II) Removal from water	Kataria and Garg 2018
9	*Vitex agnus-castus* fruit	SNO_2 solution	SNO_2	Degradation of organic dye and removal of heavy metal ions	Ebrahimian et al. 2020
10	*Syzygium cumini* leaf extract	5 mm titanium tetra isopropoxide	TiO_2 nanoparticles	Removal of lead (Pb) in explosive industrial wastewater	Sethy et al. 2020

studies have demonstrated the production of nanoparticles from plant parts that have been used for heavy metal remediation. Fruit from the plant *Terminalia Chebula* has been utilised to make Ag, Au, Fe and Pd nanoparticles for efficient removal of Cr(VI) (Gopinath et al. 2014). During the synthesis of Au NPs, chemicals like flavonoids, glycosides, polyphenolic, tannin, terpenoid and saponins act as reducing and stabilising agents, according to Gopinath et al. (2014). Plants such as aloe vera, oat, alfalfa, tulsi, lemon, neem, coriander, mustard and lemon grass can be used for the synthesis of nanoparticles like Zn, Ni, Co and Cu (Marchiol 2012). ZnO nanoparticles have also been made from several plant leaf extracts, including crown flower and copper leaf (Gnanasangeetha and Saralathambavani 2014), China rose (Hibiscus rosa-sinensis) (Devi and Singh 2014), Green Tea (Camellia sinensis) (Maensiri et al. 2008; Gunalan et al. 2012).

8.4.2 Bacterial Nanoparticles

Bacteria are considerable microorganisms for nanoparticle manufacture because of their capacity to adapt to harsh environments (Wang et al. 2017). *Bacillus cereus, Bacillus amyloliquefaciens, Bacillus indicus, E. coli, Lactobacillus casei, Pseudomonas proteolytica, Phaeocystis antarctica, Enterobacter cloacae, Arthrobacter gangotriensis*, etc. are few examples of bacteria that have been employed for the synthesis of nanoparticles. Few bacteria have the ability to convert harmful inorganic ions into harmless metal nanoparticles (Garole et al. 2018). Under various physico-chemical parameters such as duration of exposure, temperature, pH, concentration of metal salt and bacterial abundance, bacterial nanoparticles can be synthesised either intracellularly or extracellularly (Fang et al. 2019). In an extracellular process, biomolecules present in the medium or components of cell wall lead to decrease in metal ions. However, in the intracellular process, metal and metalloids are attracted to functional groups on the cell wall by electrostatic interactions, and metal ions interact with proteins inside the cells to produce nanoparticles. Various bacterial species such as *Bacillus megaterium* D01, *Bacillus subtilis* 168, *E. coli* DH5a, *Shewanella alga, Desulfovibrio desulfuricans, Plectonemaboryanum UTEX 485* and *Rhodopseudomonas capsulate* have been extensively employed in the synthesis of gold nanoparticles.

8.4.3 Fungal Nanoparticles

Synthesis of nanoparticles by fungi is also quite efficient method for the production of nanoparticles having well-defined morphologies. Because fungi contain a range of intracellular enzymes, they are superior biological agents for the purpose of metal and metal oxide nanoparticle's synthesis (Chen et al. 2009). When compared to bacteria, competent fungi can produce more nanoparticles (Mohanpuria et al. 2008). Furthermore, due to the presence of enzymes, proteins and reducing components on their cell surfaces, fungi have numerous advantages over other organisms (Narayanan and Sakthivel 2011). Enzymatic reduction by the action of reductase in the fungal cell is the most likely mechanism for synthesizing metallic nanoparticles. Nanoparticles such as Ag, Au, titanium dioxide (TiO_2) and ZnO are synthesised using a variety of fungal species. Because of their ability to secrete multiple enzymes, metabolites and proteins, as well as their economic feasibility, bigger surface area due to the presence of mycelia and lower rate of production, filamentous fungi are preferred for nanoparticle production (Fouda et al. 2018; Spagnoletti et al. 2019). In the presence of aqueous $AuCl_4$ ions, *Fusarium oxysporum* releases some reducing agents into the solution via a NADH-enzyme-mediated process, resulting in the production of gold nanoparticles. Heavy metals are known to be absorbed by *Funaliatrogii*, c *Aspergillus niger*, *Cladosporium resinae*, *Trametes versicolor*, *Penicillium* species, *Ganoderma lucidum, Aureobasidium pullulans and Rhizopus arrhizus* which were employed to make nanoparticles (Say et al. 2003). Cu(II) absorption by dead biomass of *Hypocrealixii* and synthesis of Cu nanoparticles has been reported by Salvadori et al. (2013). Both extracellular and internal NiO nanoparticles have also been reported to be produced by the same bacterium (Salvadori et al. 2015). Au–Ag nanoparticles were produced extracellularly by *Fusarium oxysporum* when it was treated with a mixture of tetrachloroaurate ion and silver nitrate in equimolar concentrations. It can also produce platinum nanoparticles in the presence of hexachloroplatinic acid (Senapati et al. 2005). Due to the inclusion of sil genes in their plasmid, *Aspergillus flavus* generated silver nanoparticles (9 nm size) to decrease silver ions (Vigneshwaran et al. 2007).

8.4.4 Nanomaterials Made from Yeast

The majority of yeast has the ability to accumulate considerable amounts of heavy metals. Glutathione, metallothioneins and phytochelatins are involved in the detoxification process in yeast cells (Kapoor et al. 2021). Several research groups have reported successful production of nanoparticles/ nanomaterials using yeast. For the production of many metallic nanoparticles, a wide range of species is used. *Yarrowia lipolytica* produced gold nanoparticles both extracellularly and intracellularly (Pimprikar et al. 2009). *Rhodosporidium diobovatum* has been utilised to produce stable lead sulphide nanoparticles intracellularly (Seshadri et al. 2011). *Candida utilis*, *Candida albicans* and *Saccharomyces boulardii* were found to synthesise silver nanoparticles, according to Soliman et al. (2018). Lian et al. (2019) reported that *Magnusiomyces ingens* LHF1 could produce stable selenium nanoparticles.

8.5 Other Nanobiotechnological Approaches for Heavy Metal Remediation

8.5.1 *Nanobiosensors*

These are naturally occurring or engineered microorganisms that provide a detectable signal in response to environmental stimuli (van der Meer and Belkin 2010). A chemical of interest triggers the promoter which induces the expression of a reporter gene, resulting in an output signal in form of colorimetry, luminometry or fluorimetry in the majority of the sensors. Organic xenobiotics such as naphthalene, polychlorinated biphenyls, benzene, toluene, xylene, alkylsulphonates, heavy metals and metalloids can be detected by biosensors. Contact between pollutants and bacteria is an important prerequisite for a biosensor to work effectively. Microbes modify the functioning of their cells in reaction to the immediate environmental conditions. As a result of this contact, they subsequently express the correct genes as needed. To detect certain heavy metals, bacteria use two component regulatory system (TCRSs). Since heavy metals serve as both poisonous and crucial elements, bacterial cells need TCRSs to maintain their state of homeostasis. For instance, *Escherichia coli* uses HydHG TCRS (also called ZraSR) to control the expression of the zraP gene, which codes for zinc efflux protein, under high concentrations of Zn^{2+} and Pb^{2+} in aerobic conditions (Leonhartsberger et al. 2001). Similarly, the CusRS TCRS has also been discovered in *E. coli* K-12 in response to Cu^{2+} ions. This TCRS is necessary to induce the production of pcoE, which is part of the pco operon. Genes induced in this operon activate the copper efflux system, that enables the departure of excess Cu^{2+} from the cell (Munson et al. 2000).

8.5.2 *Nano-adsorbents*

Adsorption happens as a result of chemical or physical forces and defined as the attachment of molecules to the adsorbent's surfaces. Through the active sites on the sorbent surface, the contaminants in the solution are absorbed into the sorbent surface (Barakat 2011; Varjani et al. 2019). Nanoadsorbent is a substance composed of nanoscale particles that may adsorb contaminants from waste water. An example of nanoadsorbents is nanocellulose composite. Although the extraction and production of nanoscale cellulose material adsorbents have been extensively studied over the last decade, their use as reinforcing agents in the preparation of composite materials for novel applications is a relatively new research field that is attracting increasing interest among nanobiotechnology researchers. This is because nanocellulose materials are lower in weight, have a larger surface area-to-volume ratio and have stronger strength and stiffness than cellulose. Nanobiochar has also proven to be a good adsorbent due to its superior physical, chemical, and structural capabilities above regular biochar. Its enhanced surface area due to its nanosize and abundance of functional groups with superior mechanical and thermal stability make it distinctive and useful (Noreen et al. 2021). Furthermore, larger negative zeta potentials, a smaller hydrodynamic radius, and a higher number of functional groups with oxygen and carbon defects may form reactive organic species (ROS), resulting in better adsorption than macrobiochar equivalents (Ramanayaka et al. 2020). When compared to ordinary biochar, nanobiochar and biochar enhanced with nanominerals displayed exceptional adsorption capability for the remediation of diverse pollutants due to its unique properties (Ma et al. 2019, Samsudin et al. 2019). Furthermore, biochar can remediate pollution by adsorbing hazardous substances such as pesticides and immobilising metals, which can cause serious environmental and health problems. As a result, nanobiochar has been hailed as a promising bioremediation agent for a wide range of pollutants.

8.5.3 Nano-membranes

Biological nanomembranes are widespread and essential; without them, life as we know it would be impossible. They permit extraordinarily sophisticated and selective interchange of materials between the cell and its environment because to these protective activities. This comprises sophisticated and balanced import and export of various materials, ranging from ions to very complex organic molecules, cell signalling via receptors and a variety of other tasks. It seems to reason that the synthesis and usage of artificial nanomembranes, particularly those with imparted biomimetic functions, should be one of the primary goals of biomimetics, particularly within the context of nanoscience, nanotechnologies and micro/nanosystems. Biomimetic nanomembranes with several functions can be used in a variety of applications. The property of selective transmembrane transport, for example, has previously demonstrated its utility in wastewater treatment (Rambabu et al. 2019), desalination (Fuwad et al. 2019; Giwa et al. 2017) and the elimination of dangerous substances such as heavy metals (Singh et al. 2011). Choia et al. (2020) examined the capacity of a thiol-functionalised cellulose nanofibre membrane to adsorb Cu (II), Cd (II) and Pb (II) ions in their research. Electro spinning was used to create the cellulose nanofibre membrane, which was then deacetylated to add thiol groups to the surface. They found that this adsorbent can remove 49 mg/g of Cu, 45.9 mg/g of Cd and 22 mg/g of Pb.

8.6 Future Prospects and Challenges

Nanotechnologies applied in bioremediation processes are projected to promote technical growth in established and emerging countries for the enhancement of environmental quality (Bartke et al. 2018; Medina-Pérez et al. 2019). In addition to the aforementioned nanobiotechnological used in heavy metal remediation, advancements in other disciplines of nanobiotechnology may be developed that further improve the effectiveness of the process of remediation. Despite the fact that a substantial studies have been done to establish the mechanisms of heavy metal remediation (Kumar et al. 2009), pilot scale investigations are still needed. The use of biomolecules to functionalize nanostructures is an issue worth investigating. This method has been put to the test using cutting-edge experimental designs based on natural molecular processes. Bolisetty et al. (2016), for example, created membranes with activated carbon having pores and amyloid proteins for heavy metal removal. The importance of discovering a low-cost source of biomolecules to develop structures useful in heavy metal remediation was underlined in this study. Naturally found proteins also provide a number of advantages. They can, for example, be produced in a cost-effective manner using well-established recombinant technologies and can contain up to 20 different types of amino acids, giving them a huge ability to interact with other molecules and produce new catalytic structures (Ljubetic et al. 2017). Large-scale applications of molecular biotechnology, such as water desalination, have also been proposed. Kumar et al. (2007) studied the integration of the bacterial protein namely aquaporin Z with polymeric membranes and established that it could restrict the movement salt and produce clean water. This approach was modified using modern protein modelling algorithms and molecular biology approaches to develop porin proteins with improved exclusion activity for both organic and inorganic water solutes (Chowdhury et al. 2018). Enzymes may be used in the above-mentioned configurations in the future to monitor more complicated contaminants or mixtures of pollutants. Bionanoparticles have a number of advantages over metallic nanoparticles, including biodegradability, which means they have a lower environmental impact. Current nanotechnologies could be applied in air, water or soil remediation procedures, but new cost-effective production methods should emerge. Although batch experiments have demonstrated the synergistic relation between nanoparticles and microrobes for the remediation of few pollutants, however, understanding about the synergy between nanomaterials and biotechnologies during the nanobioremediation process and the most effective nanomaterial for heavy metal remediation is still lacking. A better knowledge of the interactions between nanomaterials and bio-based technologies during remediation procedures under varying environmental circumstances, could lead to improved regulation. Besides, a knowledge gap about the safety regarding the long-term use of NPs with microbes also exists. The regulatory framework is an important consideration when using these materials.

8.7 Conclusion

The field of nanobiotechnology is still in its infancy. Nanobiotechnology, a multidisciplinary discipline, is bringing the science of the almost incomprehensibly small device closer to reality. These breakthroughs will have such far-reaching consequences that they will almost certainly influence every branch of science and technology at some point. Nanobiotechnology research innovation opens up a new option for developing nano-remediation techniques. Through nanobiotechnology, a far-reaching sustainable process development for tailoring nanoparticles on size and shape control has been achieved. Nano-remediation adaptation has the potential to minimise the overall cost of environmental clean-up on a broad scale. In-situ site nano-remediation could even bring the contamination level down to near zero. Ex-situ nano-remediation, on the other hand, is still in its early stages but has demonstrated to be effective in removing microorganisms, organic and inorganic pollutants. Nanobioremediation, which is still in its early stages of development, is seen as a possible technique for reducing pollution. These technical advancements are expected to be a good substitute for traditional treatments, but the risks of environmental nano-toxicity are still unknown. Nanobioremediation will undoubtedly be a promising technique for achieving environmental sustainability once these research gaps have been revealed.

References

Al-Shnani F, Al-Haddad T, Karabet F, Allaf AW (2017) Chitosan loaded with silver nanoparticles, CS-AgNPs, using thymus syriacus, wild mint, and rosemary essential oil extracts as reducing and capping agents. Journal of Physical Organic Chemistry. 30, e3680.AMM.295–298.1594

Arao T, Ishikawa S, Murakam IM, Abe K, Maejima Y, Makino T (2010) Heavy metal contamination of agricultural soil and counter, measures in Japan. Paddy Water Environ, 8(3):247–257

Atafar Z, Mesdaghinia A, Nouri J, Homaee M, Yunesian M, Ahmadimoghaddam M, Mahvi AH (2010) Effect of fertilizer application on soil heavy metal concentration. Environ Monit Assess 160(1–4):83–89. https://doi.org/10.1007/s10661-008-0659-x

ATSDR (2008) Agency for toxic substances and disease registry. Atlanta: U.S. Department of Health and Human Services, Public Health Service. Toxicological profile for chromium

Ayangbenro AS, Babalola OO (2017) A new strategy for heavy metal polluted environments: A review of microbial biosorbents. International Journal Environmental Research and Public Health. 14:1–16. https://doi.org/10.3390/ijerph14010094

Balakrishnan S, Mukherjee S, Das S, Bhat FA, Raja SP, Patra CR (2017) Gold nanoparticles-conjugated quercetin induces apoptosis via inhibition of EGFR/PI3K/Akt-mediated pathway in breast cancer cell lines (MCF-7 and MDA-MB-231). Cell Biochemistry and Function 35, 217–231. https://doi.org/10.1002/cbf.3266

Barakat MA (2011) New trends in removing heavy metals from industrial wastewater. Arab J Chem 4:361–377

Bartke S, Hagemann N, Harries N, Hauck J, Bardos P (2018) Market potential of nanoremediation in Europe—Market drivers and interventions identified in a deliberative scenario approach. Sci Total Environ 619:1040–1048

Bethi B, Sonawane SH, Bhanvase BA, Gumfekar SP (2016) Nanomaterials-based advanced oxidation processes for wastewater treatment: A review. Chem Eng Process 109:178–189

Bolisetty S, Mezzenga R (2016) Amyloid-carbon hybrid membranes for universal water purification. Nat Nanotechnol 11:365

Chao S, LiQin J, WenJun Z (2014) A review on heavy metal contamination in the soil worldwide: Situation, impact and remediation techniques. Environmental Skeptics and Critics 3(2):24–38

Chen HM (2002) Behaviours and environmental quality of chemical substances in the soil. Science Press, Beijing

Chen YF (2011) Review of the research on heavy metal contamination of China's city soil and its treatment method China population. Res Environ 21(3):536–539

Chen YL, Tuan HY, Tien CW (2009) Augmented biosynthesis of cadmium sulfide nanoparticles by genetically engineered *Escherichia coli*. Biotechnol Prog 25:1260–1266

Chen ZF, Zhao Y, Fan LD, Xing LT, Yang YJ (2015) Cadmium (Cd) localization in tissues of Cotton (Gossypium hirsutum L.) and Its phytoremediation potential for Cd-contaminated soils bull. Arch Environ Contam Toxicol 95:784–789. https://doi.org/10.1007/s00128-015-1662-x

Choia HY, Bae JH, Hasegawad Y, Ane S, Kimd IS, Leea H, Kim M (2020) Thiol-functionalized cellulose nanofiber membranes for the effective adsorption of heavy metal ions in water. Carbohyd Polym 234:115881. https://doi.org/10.1016/j.carbpol.2020.115881

Chowdhury R, Ren T, Shankla M, Decker K, Grisewood M, Prabhakar J, Baker C, Golbeck JH, Aksimentiev A, Kumar M (2018) PoreDesigner for tuning solute selectivity in a robust and highly permeable outer membrane pore. Nat Commun 9:1–10

Dellisanti, (2016) In-field remediation of tons of heavy metal-rich waste by Joule heating vitrification. Int J Miner Process 93:239–245

Devi HS, Singh TD (2014) Synthesis of copper oxide nanoparticles by a novel method and its application in the degradation of methyl orange. Advances in Elect Electrical Eng 4:83–88

Dixit R, Wasiullah MD, Pandiyan K, Singh UB, Sahu A, Shukla R, Singh BP, Rai JP, Sharma PK, Lake H, Paul D (2015) Bioremediation of heavy metals from soil and aquatic environment: An overview of principles and criteria of fundamental processes. Sustainability 7:2189–2212. https://doi.org/10.3390/su7022189

Dong X, Li C, Li J, Wang J, Liu S, Ye B (2010) A novel approach for soil contamination assessment from heavy metal pollution: A linkage between discharge and adsorption. J Hazard Mater 175:1022–1030

Ebrahimian J, Mohsennia M, Khayatkashani M (2020) Photocatalytic-degradation of organic dye and removal of heavy metal ions using synthesized SnO2 nanoparticles by *Vitex agnus-castus* fruit via a green route. Mater Lett 263:127255

Ehrampoush MH, Miria M, Salmani MH, Mahvi AH (2015) Cadmium removal from aqueous solution by green synthesis iron oxide nanoparticles with tangerine peel extract. J Environ Health Sci Eng 13(1):1–7

Fang X, Wang Y, Wang Z, Jiang Z, Dong M (2019) Microorganism assisted synthesized nanoparticles for catalytic applications. Energies 12:190. https://doi.org/10.3390/en12010190

Fazlzadeh M, Rahmani K, Zarei A, Abdoallahzadeh H, Nasiri F, Khosravi R (2017) A novel green synthesis of zero valent iron nanoparticles (NZVI) using three plant extracts and their efficient application for removal of Cr (VI) from aqueous solutions. Adv Powder Technol 28(1):122–130

Ferraro A, vanHullebusch ED, Huguenot D, Fabbricino M, Esposito G (2015) Application of an electrochemical treatment for EDDS soil washing solution regeneration and reuse in a multistep soil washing process: Case of a Cu contaminated soil. J Environ Manage 163:62–69

Fouda A, Saad E, Salem SS, Shaheen TI (2018) In-vitro cytotoxicity, antibacterial, and UV protection properties of the biosynthesized zinc oxide nanoparticles for medical textile applications. Microb Pathog 125:252–261. https://doi.org/10.1016/j.micpath.2018.09.030

FRTR (2012) Remediation technologies screening matrix and reference guide, Version 4.0. Federal Remediation Technologies Roundtable, Washington, DC

Fuwad A, Ryu H, Malmstadt N, Kim SM, Jeon TJ (2019) Biomimetic membranes as potential tools for water purification: Preceding and future avenues desalination. 458:97–115

Garole DJ, Choudhary BC, Paul D, Borse AU (2018) Sorption and recovery of platinum from simulated spent catalyst solution and refinery wastewater using chemically modified biomass as a novel sorbent. Environ Sci Pollut Res 25:10911–10925. https://doi.org/10.1007/s11356-018-1351-5

Giwa A, Hasan S, Yousuf A, Chakraborty S, Johnson D, Hilal N (2017) Biomimetic membranes: A critical review of recent progress. Desalination 420:403–424

Gnanasangeetha D, Saralathambavani D (2014) Biogenic production of zinc oxide nanoparticles using *Acalypha indica*. J Chem Biol Phy Sci 4:238–246

Gopinath K, Gowri S, Karthika V, Arumugam A (2014) Green synthesis of gold nanoparticles from fruit extract of Terminalia arjuna, for the enhanced seed germination activity of Gloriosa superba. J Nanostruct Chem 4:115

Gunalan S, Sivaraj R, Rajendran V (2012) Green synthesized ZnO nanoparticles against bacterial and fungal pathogens. Progress in Natural Science: Materials International 22:693–700

Guo X, Wei Z, Wu Q, Li C, Qian T, Zheng W (2016) Effect of soil washing with only chelators or combining with ferric chloride on soil heavy metal removal and phyto availability: field experiments. Chemosphere 147:412–419

Hu BF, Chen SC, Hu J, Xia F, Xu JF, Li Y, Shi Z (2017) Application of portable XRF and VNIR sensors for rapid assessment of soil heavy metal pollution. PLoS ONE 12:e0172438. https://doi.org/10.1371/journal.pone.0172438

Iannelli R, Masi M, Ceccarini A, Ostuni MB, Lageman R, Muntoni A, Spiga D, Polettini A, Marini A, Pomi R (2015) Electro kinetic remediation of metal-polluted marine sediments: Experimental investigation for plant design. Electrochim Acta 181:146–159

Islam MN, Taki G, Xuan PN, Jo YT, Kim J, Park JH (2017) Heavy metal stabilization in contaminated soil by treatment with calcined cockle shell. Environmentall Science and Pollution Research 24:7177–7183. https://doi.org/10.1007/s11356-016-8330-5

Kapoor RT, Salvadori MR, Rafatullah M, Siddiqui MR, Khan MA, Alshareef SA (2021) Exploration of microbial factories for synthesis of nanoparticles—A sustainable approach for bioremediation of environmental contaminants. Frontiers in Microbiology 12:658294. https://doi.org/10.3389/fmicb.2021.658294

Kataria N, Garg VK (2018) Green synthesis of Fe3O4 nanoparticles loaded sawdust carbon for cadmium (II) removal from water: Regeneration and mechanism. Chemosphere 208:818–828

Khan I, Ghani A, Rehman AU, Awan SA, Noreen A, Khalid I (2016) Comparative analysis of heavy metal profile of Brassica campestris (L.) and Raphanus sativus (L.) irrigated with municipal waste water of sargodha city. Journal of Clinical Toxicology 6:1–4. https://doi.org/10.4172/2161-0495.1000307

Khan MN, Wasim AA, Sarwar A, Rasheed MF (2011) Assessment of heavy metal toxicants in the roadside soil along the N-5, National Highway, Pakistan. Environ Monit Assess 182:587–595

Kulikowska D, Gusiatin ZM, Bulkowska K, Klik B (2015) Feasibility of using humic substances from compost to remove heavy metals (Cd, Cu, Ni, Pb, Zn)from contaminated soilaged fordifferent periods of time. J Hazard Mater 300:882–891

Kumar M, Grzelakowski M, Zilles J, Clark M, Meier W (2007) Highly permeable polymeric membranes based on the incorporation of the functional water channel protein aquaporin Z. Proceedings of the National Academy of Sciences USA 104:20719–20724

Kumar SR, Gopinath P (2009) Nano-bioremediation: Applications of nanotechnology for bioremediation. In: Wang KL, Wang SM-H, Hung Y-T, Shammas NK, Chen JP (eds) Handbook of advanced industrial and hazardous wastes management, vol 1st edn. CRC Press, Boca Raton, FL, USA, pp 27–48

Leonhartsberger S, Huber A, Lottspeich F, Böck A (2001) The hydH/G genes from Escherichia coli code for a zinc and lead responsive two-component regulatory system. J Mol Biol 307:93–105

Li C, Kehai Z, Wenqiang Q, Changjiu T, Miao Q, Xiaoming Y, Wenbing H (2019) A review on heavy metals contamination in soil: Effects, sources, and remediation techniques. Soil and Sediment Contam: An Int J. https://doi.org/10.1080/15320383.2019.1592108

Lian S, Diko CS, Yan Y, Li Z, Zhang H, Ma Q (2019). Characterization of biogenic selenium nanoparticles derived from cell-free extracts of a novel yeast *Magnusiomyces ingens*. 3Biotech 9:221. https://doi.org/10.1007/s13205-019-1748-y

Lianwen L, Li, Wei, Weiping S, Mingxin G (2018) Remediation techniques for heavy metal-contaminated soils: Principles and applicability the science of the total environment. 633:206–219. https://doi.org/10.1016/j.scitotenv.2018.03.161

Ljubetič A, Lapenta F, Gradišar H, Drobnak I, Aupič J, Strmšek Ž, Lainšček D, Hafner-Bratkovič I, Majerle A, Krivec N (2017) Design of coiled-coil protein-origami cages that self-assemble in vitro and in vivo. Nat Biotechnol 35:1094–1101

Ma SQ, Jing FQ, Sohi SP, Chen JW (2019) New insights into contrasting mechanisms for PAE adsorption on millimeter, micron and nano-scale biochar. Environ Sci Pollut Res 26(2019):18636–18650

Maensiri S, Laokul P, Klinkaewnarong J (2008) Indium oxide (in 2O3) nanoparticles using aloe vera plant extract: Synthesis and optical properties. J Optoelectron Adv Mater 10:161–165

Mallampati SR, Mitoma Y, Okuda T, Simion C, Lee BK (2015) Dynamic immobilizationofsimulatedradionuclide133Csinsoilbythermaltreatment/vitrificationwith nanometallic Ca/CaO composites. J Environ Radioact 139:118–124

Mani S, Bharagava RN (2016) Exposure to Crystal Violet, its toxic, genotoxic and carcinogenic effects on environmental and its degradation and detoxification for environmental safety. Rev Environ Contam Toxicol 237:71–104

Marchiol L (2012) Synthesis of metal nanoparticles in living plants. Int J Agron 7:274–282

Medina-Pérez G, Fernández-Luqueño F, Vazquez-Nuñez E, López-Valdez F, Prieto-Mendez J, Madariaga-Navarrete A, Miranda-Arámbula M (2019) Remediating polluted soils using nanotechnologies: Environmental benefits and risks. Pol J Environ Stud 28:1–18

Mishra S, Bharagava RN (2016) Toxic and genotoxic effects of hexavalent chromium in environment and its bioremediation strategies. J Environ Sci Health C. https://doi.org/10.1080/10590501.2015.1096883

Mohan S, Singh Y, Verma DK, Hasan SH (2015) Synthesis of CuO nanoparticles through green route using Citrus limon juice and its application as nanosorbent for Cr (VI) remediation: Process optimization with RSM and ANN-GA based model. Process Saf Environ Prot 96:156–166

Mohanpuria P, Rana NK, Yadav SK (2008) Biosynthesis of nanoparticles: Technological concepts and future applications. J Nanopart Res 10:507–517

Mukherjee D, Ghosh S, Majumdar S, Annapurna K (2016) Green synthesis of α-Fe2O3 nanoparticles for arsenic (V) remediation with a novel aspect for sludge management. J Environ Chem Eng 4(1):639–650

Munson GP, Lam DL, Outten FW, O'Halloran TV (2000) Identification of a copperresponsive two-component system on the chromosome of *Escherichia coli* K-12 J Bacteriology 182:5864–71

Nagajyoti PC, Lee KD, Sreekanth TVM (2010) Heavy metals, occurrence and toxicity for plants: A review. Environ Chem Lett 8:199–216

Narayanan KB, Sakthivel N (2011) Synthesis and characterization of nanogold composite using *Cylindrocladium floridanum* and its heterogeneous catalysis in the degradation of 4-nitrophenol. J Hazard Mater 189:519–525

Navarro A, Cardellach E, Cañadas I, Rodríguez J (2013) Solar thermal vitrification of mining contaminated soils. Int J Miner Process 119:65–74

Noreen S, KA Abd-Elsalam (2021) Biochar-based nanocomposites: A sustainable tool in wastewater bioremediation. In: KA Abd-Elsalam M Zahid (eds) Applications of nanomaterials for water purification, micro and nano technologies, aquananotechnology, Elsevier, pp 185–200

Ogunlade MO, Agbeniyi SO (2011) Impact of pesticides use on heavy metals pollution in cocoa soils of Cross-River State, Nigeria. Africal Journal of Agricultural Research 6:3725–3728

Ojuederie OB, Babalola OO (2017) Microbial and plant–Assisted bioremediation of heavy metal polluted environments: A review. Int J Environ Res Pub Health 14, pii: E1504. https://doi.org/10.3390/ijerph14121504

Park B, Son Y, (2016) Ultrasonic and mechanical soil washing processes for the removal of heavy metals from soils. Ultrasonics Sonochemistry

Pimprikar PS, Joshi SS, Kumar AR, Zinjarde SS, Kulkarni SK (2009). Influence of biomass and gold salt concentration on nanoparticle synthesis by the tropical marine yeast *Yarrowia lipolytica* NCIM 3589 Colloids and Surfaces: B Biointerfaces 74, 309–316. https://doi.org/10.1016/j.colsurfb.2009.07.040

Prajapati SK, Meravi N (2014) Heavy metal speciation of soil and Calotropis procera from thermal power plant area. Proceedings of the International Academy of Ecology and Environmental Sciences 4(2):68–71

Prasad C, Murthy PK, Krishna RH, Rao RS, Suneetha V, Venkateswarlu P (2017) Bio-inspired green synthesis of RGO/Fe3O4 magnetic nanoparticles using Murrayakoenigii leaves extract and its application for removal of Pb (II) from aqueous solution. J Environ Chem Eng 5(5):4374–4380

Qu X, Alvarez PJ, Li Q (2013) Applications of nanotechnology in water and wastewater treatment. Water Resour 47(12):3931–3946

Cernansky R (2015) Agriculture: State–of–the–art soil. Nature 517:258–260

Ramanayaka SM, Vithanage D, Alessi WJ, Liu ACA, Ok JYS (2020) Nanobiochar: production, properties, and multifunctional applications. Environ Sci Nano. Journal 7:3279–3302

Rambabu K, Bharath G, Monash P, Velu S, Banat F, Naushad M, Arthanareeswaran G, Show PL (2019) Effective treatment of dye polluted wastewater using nanoporous CaCl2 modified polyethersulfone membrane. Process Saf Environ Prot 124:266–278

Rascio N, Navariizzo F (2011) Heavy metal hyper accumulating plants: How and why do they do it, and what makes them so interesting. Plant Sci 18:169–181. https://doi.org/10.1016/j.plantsci.2010.08.016

Rizwan MD, Singh M, Mitra CK, Morve RK (2014) Ecofriendly application of nanomaterials for nanobioremediation. J Nanoparticles 1:1–7

Salvadori MR, Lepre LF, Ando RA, Oller do Nascimento CA, Corrêa B, (2013) Biosynthesis and uptake of copper nanoparticles by dead biomass of Hypocrea lixii isolated from the metal mines in the Brazilian Amazon region. PLoS ONE 8:e80519. https://doi.org/10.1371/journal.pone.0080519

Salvadori MR, Ando RA, Nascimento CAO, Correa B (2015) Extra and intracellular synthesis of nickel oxide nanoparticles mediated by dead fungal biomass. PLoS ONE 10:e0129799. https://doi.org/10.1371/journal.pone.0129799

Samsudin MH (2019) MA Hassan. J Idris a one-step self-sustained low temperature carbonization of coconut shell biomass produced a high specific surface Area biocharderived nano-adsorbent waste management and research 37:551–555

Sarwar N, Imran M, Shaheen MR, Ishaque W, Kamran MA, Matloob A, Rehim A, Hussain S (2016) Phytoremediation strategies for soils contaminated with heavy metals: Modifications and future perspectives. Chemosphere 171:710–721. https://doi.org/10.1016/j.chemosphere.2016.12.116

Say R, Yimaz N, Denizli A (2003) Removal of heavy metal ions using the fungus Penicillium canescens. Adsorpt Sci Technol 21:643–650. https://doi.org/10.1260/026361703772776420

Sebastian A, Nangia A, Prasad MNV (2018) A green synthetic route to phenolics fabricated magnetite nanoparticles from coconut husk extract: Implications to treat metal contaminated water and heavy metal *stress in Oryza sativa L.* J Clean Prod 174:355–366

Selvan DA, Mahendiran D, Kumar RS, Rahiman AK (2018) Garlic, green tea and turmeric extracts-mediated green synthesis of silver nanoparticles: Phytochemical, antioxidant and in vitro cytotoxicity studies. J Photochem Photobiol, B 180:243

Senapati S, Ahmad A, Khan MI, Sastry M, Kumar R (2005) Extracellular biosynthesis of bimetallic Au-Ag alloy nanoparticles. Small 1:517–520. https://doi.org/10.1002/smll.200400053

Seshadri S, Saranya K, Kowshik M (2011) Green synthesis of lead sulphide nanoparticles by the lead resistant marine yeast, *Rhodosporidium diobovatum.* Biotechnol Prog 27:1464–1469. https://doi.org/10.1002/btpr.651

Sethy NK, Arif Z, Mishra PK, Kumar P (2020) Green synthesis of TiO2 nanoparticles from *Syzygium cumini* extract for photo-catalytic removal of lead (Pb) in explosive industrial wastewater. Green Processing and Synthesis 9(1):171–181

Shahid M, Pourrut B, Dumat C, Nadeem M, Aslam M, Pinelli E (2014) Heavy-metal- induced reactive oxygen species: Phytotoxicity and physic chemical changes in plants. Rev Environ Contam Toxicol 232:1–44

Singh B, Mahato TH, Srivastava AK, Prasad GK, Ganesan K, Vijayaraghavan R, Jain R (2011) Significance of porous structure on degradatin of 2 20 dichloro diethyl sulphide and 2 chloroethyl ethyl sulphide on the surface of vanadium oxide nanostructure. J Hazard Mater 190:1053–1057 oliman H, Elsayed A, Dyaa A, (2018) Antimicrobial activity of silver nanoparticles biosynthesized by Rhodotorula sp. strain ATL72. Egypt J Basic Appl Sci 5:228–233. https://doi.org/10.1016/j.ejbas.2018.05.005

Spagnoletti FN, Spedalieri C, Kronberg F, Giacometti R (2019) Extracellular biosynthesis of bactericidal Ag/AgCl nanoparticles for crop protection using the fungus Macrophomina phaseolina. J Environ Manage 231:457–466. https://doi.org/10.1016/j.jenvman.2018.10.081

Sun WH, Jiang YX, Li X (2013) Research of the evaluation on heavy-metal pollution in rice by sewage irrigation. Appl Mech Mater 295–298:1594–1599. https://doi.org/10.4028/www.scientific.net/

van der Meer JR, Belkin S (2010) Where microbiology meets microengineering: design and applications of reporter bacteria. Nature Reviews Microbiology 8(7):511–522. https://doi.org/10.1038/nrmicro2392

Varjani S, Joshi R, Srivastava VK, Ngo HH, Guo W (2019) Treatment of wastewater from petroleum industry: current practices and perspectives. Environ Sci Pollut Res 1–9. https://doi.org/10.1007/s11356-019-04725-x

Vigneshwaran N, Ashtaputre NM, Varadarajan PV, Nachane RP, Paralikar KM, Balasubramanya RH (2007) Biological synthesis of silver nanoparticles using the fungus *Aspergillus flavus*. Mater Lett 61:1413–1418. https://doi.org/10.1016/j.matlet.2006.07.042

Wang X, Zhang D, Pan X, Lee DJ, Al-Misned FA, Mortuza MG (2017) Aerobic and anaerobic biosynthesis of nano-selenium for remediation of mercury contaminated soil chemosphere 170:266–273. https://doi.org/10.1016/j.Chemosphere.2016.12.020

Wasi S, Jeelani G, Ahmad M (2008) Biochemical characterization of a multiple heavy metal, pesticides and phenol resistant Pseudomonas fluorescens strain Chemosphere 71, 1348–1355. https://doi.org/10.1016/j.chemosphere.2007.11.023

Weng X, Huang L, Chen Z, Megharaj M, Naidu R (2013) Synthesis of iron-based nanoparticles by green tea extract and their degradation of malachite. Ind Crops Prod 51:342–347

Weng X, Jin X, Lin J, Naidu R, Chen Z (2016) Removal of mixed contaminants Cr (VI) and Cu (II) by green synthesized iron based nanoparticles. Ecol Eng 97:32–39

Werkneh AA, Rene RR, Lens PNL (2018a) Application of nanomaterials in food, cosmetics and other process industries. In: Kumar V, Dasgupta N, Ranjan S (eds) Nanotoxicology: toxicity evaluation, risk assessment and management, vol 1. CRC Press Taylor and Francis Group, USA, pp 63–79

Werkneh AA, Rene RR, Lens PNL (2018b) Simultaneous removal of selenite and phenol from wastewater in an upflow fungal pellet bioreactor. J Chem Technol Biotechnol 93:1003–1011

Yadav A, Chowdhary P, Kaithwas G, Bharagava RN (2017) Toxic metals in environment, threats on ecosystem and bioremediation approaches. In: Das S, Singh (eds) Handbook of metal-microbe interactions and bioremediation. Boca Raton, CRC Press/Taylor & Francis Group

Yadav A, Raj A, Bharagava RN (2016) Detection and characterization of a multi-drug and multi-metal resistant Enterobacterium Pantoea sp. from tannery wastewater after secondary treatment process. Int J Plant Environ 2(1–2), 37–42

Zhang WJ, Jiang FB, Ou JF (2011) Global pesticide consumption and pollution: With China as a focus. Proceedings of the International Academy of Ecology and Environmental Sciences 1(2):125–144

Chapter 9
Applications of Nanoparticles for Microbial Contaminants and Pathogens Removal from Wastewater

Nandkishor More, Anjali Verma, Deepthi Wickramasinghe, Rajnish Gautam, Dimuth Navaratna, A. V. P. R. Jonnada, and Ram Naresh Bharagava

Abstract Because of their unique physical and chemical properties, nanomaterial's have drawn the attention of many scientists and researchers in various fields of environmental science, especially scientists and researchers in the field of ecological purity. Bioremediation offers a good cleaning strategy for certain types of waste. For example, bioremediation may not be the right strategy because of the high concentration of chemicals that are toxic to most microorganisms. These include heavy metals and salts. In addition, advances in science and technology have improved living standards and contributed directly or indirectly to more waste and toxic substances. Therefore, pollution remediation using current technology is neither effective nor effective in disinfecting the environment. For bioremediation, although the immobilization can enhance the wastewater treatment efficiency, the applications of immobilized pathogens are still confined to laboratory research. Usually, several pollutants may coexist in wastewater that will increase the difficulty of treatment; thereby obtaining more stable support for microbes is necessary. Moreover, various modified nanoparticles can be prepared and designed for improving the bioremediation. In addition,

---

N. More (✉) · A. Verma
Department of Environment Science, School of Environmental Sciences, Babasaheb Bhimrao Ambedkar University (A Central University), Lucknow 226025, Uttar Pradesh, India
e-mail: nkmore2010@gmail.com

D. Wickramasinghe
Department of Zoology and Environment Sciences, University of Colombo, Colombo, Sri Lanka

R. Gautam · D. Navaratna
Institute of Sustainable Industries and Liveable Cities, Victoria University International, Melbourne, VIC, Australia

A. V. P. R. Jonnada
Department of Biotechnology, D D U Gorakhpur University, Gorakhpur, Uttar Pradesh, India

R. N. Bharagava
Department of Microbiology, Babasaheb Bhimrao Ambedkar University (A Central University), Lucknow 226025, Uttar Pradesh, India
e-mail: ramnaresh_dem@bbau.ac.in; bharagavarnbbau11@gmail.com

J. P. N. Rai and S. Saraswat (eds.), *Nano-biotechnology for Waste Water Treatment*, Water Science and Technology Library 111,
https://doi.org/10.1007/978-3-031-00812-2_9

although the presence of nanoparticles can enhance the bioremediation performance of microorganisms at certain extent, the toxicity of nanoparticles can also be experimented. Therefore, nanomaterial's can be used in bioremediation, not only have less toxic effects on microorganisms, but also improve the microbial activity of some wastes and toxic substances and shorten the overall time. And the total costs of this chapter summarize the main types of nanomaterial's that have been used in the bioremediation of waste and toxins so far. Stabilization can improve wastewater treatment efficiency, but the use of inactive pathogens is limited to laboratory biological remediation studies. In general, several pollutants can be present in the wastewater at the same time, which makes their treatment difficult. That is why we need more stable support for microorganisms. In addition, modified or doped nanoparticles can be manufactured and developed to improve the biological process. The presence of nanoparticles can improve the bioremediation performance of microorganisms to a certain extent, but the toxicity of nanoparticles can also be felt. Thus, nanomaterial's can be used in natural wastewater treatment, which not only toxic microorganisms, but also enhance the microbial exertion of some wastes and poisons, and reduce consumption and time.

9.1 Introduction

The rapid growth of population, tired water resources and weather changes due to long droughts and floods are a competitive source in many parts of the world. The biggest problem in the world today is everything in liquid waste or water and wastewater from homes, hospitals and factories and any other use of water facilities in its facilities. The development of practical and stable materials to provide fresh water. When it comes to health and emerging pollutants, numerous operations, applied and high-cost technology costs are cost-effective solutions to treat water and sewage that depends on large infrastructure or central systems. The purpose of this chapter is to remove the potential applications of pollutant nanoparticles such as different nanomaterial's (package or fibre) to remove viruses, amazing abdomen, heavy metals, metal ions, complex organic compounds, nitrate and other pollution in the surface and in industrial water. The sewage infected with human use and the effect of family, industrial and commercial use, and so on constantly configuring it changes and determines. The hygienic composition contains 99.9% water and 0.1% remaining includes the problem of membership, microorganisms and vehicles. General species of water pollution, oxygen dissipation, vegetable food, organic organic chemistry, mineral chemicals, plastic flour, sediments, radioactive materials, oil and heat. This approx 0.1% in foods such as phosphorus, nitrogen, fat, oil etc. Among different pollutants in wastewater, microbial factors are increasingly important and efficient in different sewage treatment plants (Wang et al. 2017). Biological contaminants in sewage are different types of bacteria (Escherichia coli, Salmonella, parasites, viruses and fungi). Each of them can be dangerous for the environment and human health, depending on its nature and quantity (Suolale and Okoh 2017). Microbial contamination of water environments is one of the most important issues of water health used to provide drinking water, entertainment and marine cranes due to the pollution of possible diseases of diseases, protocols or viruses. Waste includes homes, shops, offices, factories, farms, fuel, warehouses, ships, mines and mines. The lesions of this case are not for good treatment, especially the disposal of liquid and powerful waste from families and communities, serious health risks and lead to the release of infectious diseases. Skin infections and blood are obtained against direct contact with polluted waste and wounds; eye infections and respiratory diseases affect the injured material, especially during landfill; Many diseases lead to animal bites that disappear; Glossy infections are transmitted through fed flies with waste; Some chemicals, if they are released without treatment (e.g. cyanide, mercury diphenyl and PCB), are very toxic and exposed to illness or death. Over the years, these efforts have helped and provided opportunities to understand the processes; technologies related to the development and formulation of international and national standards for controlling pollution in the world. Wastewater is a resource that requires effective management to ensure future water safety. Treatment of untreated and treated wastewater in urban areas and sewage for various other purposes. This helps to control pollution and minimize the impact on health. Given its economic and health importance, research into the treatment and reuse of wastewater will continue in the future. Microbial bioremediation has been used in one form or another for thousands of times and the most effective system with the use of prokaryotes (or microbial metabolism) to remove adulterants in sewage treatment shops. Bioremediation occurs naturally in the climate, but has also been used to remove agricultural chemicals similar as certain poisonous essence and oxides, similar as selenium, arsenic composites, soil in groundwater and sub ground. Bioremediation is a relatively new technology that has been the subject of more intense investigations focused on the destruction or immobilization of toxic wastes present in these water sources as rapid industrialization and urbanization release toxic compounds in natural water bodies, polluting both freshwater and seawater resources. The concept of Nano Bioremediation (NBR) was proposed, with the aim of integrating two recycling technologies to achieve effective, efficient and ultimately sustainable development. Nanotechnology is all about understanding and controlling matter at the nanoscale. At the nanoscale, unique phenomena make new applications possible. Nanotechnology includes the science, engineering and technology of nanoparticles.

9.2 Wastewater

Wastewater has been poisoned by natural use and affected by domestic, processed and corporate use, therefore constantly changing its composition and making it relatively delicate to define. Water composition is of about 99.9% and the remaining 0.1 contains organic matter, microorganisms and emulsion inorganic substances. Wastewater is a byproduct of domestic, artificial, marketable or farming conditioning.

Types of wastewater include domestic wastewater from homes, external collaborative wastewater (also known as sewage) and artificial wastewater which may contain physical, chemical and natural contaminations that it can produce from flushes, cesspools, dishwashers, washing machines, barrels and showers, etc. The treated wastewater or also called effluent is discharged into an entering water body. The terms "wastewater exercise" and "water recovery" apply if the treated waste is used for other purposes. In developing countries and sparsely populated pastoral areas, wastewater is frequently treated by colourful on-point sanitation systems, including septic tanks connected to drainage fields, on-point sewage systems (OSS), fungicide pollutants, etc. Marketable wastewater comes from non-domestic sources, similar as beauty salons, taxidermy, cabinetwork finishing, musical instrument cleaning or bus body form shops. This wastewater may contain dangerous accoutrements and bear special treatment or disposal.

9.2.1 Types of Wastewater

Wastewater entering a wastewater treatment factory is occasionally called raw wastewater or raw sewage.

Wastewater is divided into two orders, depending on the source.

1. Gray water—Gray water comes from showers, bathrooms, hot barrels, washing machines, dishwashers and sinks other than the kitchen Gomorrah.
2. Black water—Black water comes from toilets and kitchen cesspools.

9.2.2 Microbial Contaminants in Wastewater

Bacteria with in soil and water absorb organic waste from sewage and use it as food and energy to grow quickly. Aerobic bacteria consume organic waste and produce sludge from new bacterial cells and dissolved saline waste in a natural aquatic environment where a significant amount of oxygen dissolves in water. Anaerobic bacteria break decompose organic waste in undiluted wastewater and produce foul-smelling gases such hydrogen sulphide. Gases such as methane and carbon dioxide can also be generated in their natural state. This is a significant amount of water that has been wasted; all oxygen has been used, and anaerobic bacteria had taken over the water, contaminating it and represent a threat to fish and other oxygen-dependent species.

9.2.3 Sources of Generation of Wastewater

The composition of liquid waste, also called wastewater, is very varied and depends mainly on its origin. In cities, the three main sources are residential, commercial and industrial areas (Abdel-Shafy and Mansour 2018). Water used in toilets, showers, bathrooms, kitchen sinks and laundry rooms in homes and offices is domestic wastewater. Industrial operations such as food processing or metal refining are industrial or commercial waste. This includes liquid wastes from any process (for example, water used to cool machines or clean factories and equipment). Rainwater, a form of sewage, is the runoff that flows to and from urban areas such as rooftops, parks, gardens, roads, paths and gutters into storm sewers after the rain. Household wastewater comes from various activities such as washing, using the toilet, bathing, food preparation and laundry, etc. Commercial wastewater is discharged from non-domestic sources. The origin of industrial wastewater is industrial and its composition varies according to the type of industry. This type of wastewater is also released from some commercial manufacturing processes such as agriculture and is generally more difficult to treat than household waste.

1. **Liquid wastes from Residential areas**
   Domestic sewage is a term used in urban areas to describe liquid waste from residential areas. This wastewater is generated by our daily activities and includes wastewater from food preparation, washing, bathing and toilet use.

2. **Liquid wastes from Commercial areas**
   Wastewater from commercial areas, such as businesses, shops, outdoor markets, restaurants and cafes Restaurant and cafe effluents can contain high levels of cooking oil, but this can be avoided by installing a grease trap in the drain pipes.

3. **Liquid wastes from Industrial areas**
   Liquid waste that is generated in industrial areas by processing or manufacturing industries and also service industries such as auto repair shops. The composition of wastes which can be hazardous and therefore potentially harmful is defined by the type of industry. The presence of hazardous materials (e.g., food waste, packaged products, disposable electronics, used) and commercial and industrial waste (e.g., demolition debris, waste incineration, sludge refinery).

9.2.4 The Problem Caused by Wastewater

The degradation of the receiving surface water body is caused by the low quality of wastewater effluent. To eliminate health concerns to users of surface water resources and the aquatic ecology, wastewater effluents must be treated adequately. The environment and human health are both harmed when raw and poorly treated wastewater is discharged into waterways. As a result, proper enforcement of water and environmental laws is required to preserve the health of residents in both rural and urban areas. Waste can result in a variety of issues, all of which can have a severe influence on people's health and well-being, as well as the environment. These can be a dangerous source of information.

9.2.5 Environmental and Human Health Effects

Polluted water is one of the biggest problems in wastewater treatment caused by pathogens in the wastewater. If not managed properly, pathogens can contaminate surface and ground water and contaminated drinking water can cause many diseases (Choudri et al. 2018). It is about taking care of the environment and our health. There are numerous reasons why maintaining clean water is a key responsibility. Clean water is essential for aquatic plants and animals. It is critical for fishing, sport hunters and the future of the sport. Our rivers and oceans are teeming with life, and they rely on the coasts, beaches and salt marshes for survival. Hundreds of fish and other aquatic life rely on them because of survival. Migrants use these sites as rest and feeding areas. Water is a wonderful place for every one of us since it provides a tranquil environment and a high quality of life. Many individuals choose to reside here because of the beauty and recreational appeal of our waters. Water can spread disease if it is not properly controlled. Wastewater treatment has the potential to harm the environment as well as human health.

Plants are unlikely to absorb sufficient amounts of harmful substances. For consumers and are toxic to plants long before reaching levels toxic to humans. The negative effects are related to the presence of pathogens and toxic chemicals in the wastewater Irrigation with wastewater will lead to many potential risks to human health, as it will be threatened by ingestion or exposure to pathogenic microorganisms, heavy metals and organic chemicals.

9.3 Wastewater Management and Challenges

The degradation of the receiving surface water body is caused by the poor quality of wastewater effluent. To avoid health risks to users of surface water resources and the aquatic ecosystem, wastewater effluents must be effectively treated. The discharge of raw and poorly treated wastewater into waterways has both short- and long-term environmental and human health consequences. As a result, adequate enforcement of water and environmental laws is required to protect the health of residents of rural and urban communities. Waste can cause a variety of issues that have a negative impact on people's and the environment's health and well-being. Cities around the world are facing great challenges due to increasing urbanization and one of the biggest challenges is the increasing amount of garbage and waste generated due to the high demand for food and other products (Singh et al. 2017). Public trash cans are filling up faster than ever and inevitably many trash cans end up overflowing before collection, not only causing congested streets and bad smells, but also negative impacts on health and the environment.

9.3.1 Bioremediation and Control

Bioremediation is a process that uses environmental conditions to stimulate the growth of microorganisms in order to degrade target pollutants and treat contaminated media such as water, soil and subsoil materials. It is a biological treatment that is less expensive and more environmentally friendly than other sanitation alternatives for waste treatment, such as sewage, industrial waste and solid waste. The microbial degradation of organic substances into inorganic substances accounts for a large portion of water bioremediation. Mercury is used in industry in the production of batteries as an active ingredient in some pesticides. It is highly toxic because it accumulates in living tissue.

9.3.2 Nanobioremediation

Poor waste product effluent quality is to blame for the degradation of the receiving surface water body. Waste product effluents should be effectively treated to avoid health risks to users of surface water resources and, as a result, the aquatic ecosystem. The discharge of raw and poorly treated wastewater into waterways has both immediate and long-term effects on the environment and human health. As a result, adequate social control of water and environmental laws are required to protect the health of residents of rural and urban communities. Waste can cause a variety of problems that can have a negative impact on people's health and well-being, as well as the environment. Excretory product has long been harmful to our environment. Humans generate an excessive amount of waste and can't manage it during a property manner. Waste that's not perishable and cannot be properly recycled, they fill our oceans and landfills take waste plastic (Figs. 9.1, 9.2, 9.3) Table 9.1.

9.3.3 Nanobiotechnology

Regarding nanotechnology, physicists, chemists and biologists work in the field of nanobiotechnology (also known as bionanotechnology, biomedical nanotechnology or nanomedicine), which employs biological raw materials, biological design principles, or has biological or medical applications. This discipline may represent a synthesis of biological research and various fields of nanotechnology. Nanobiology

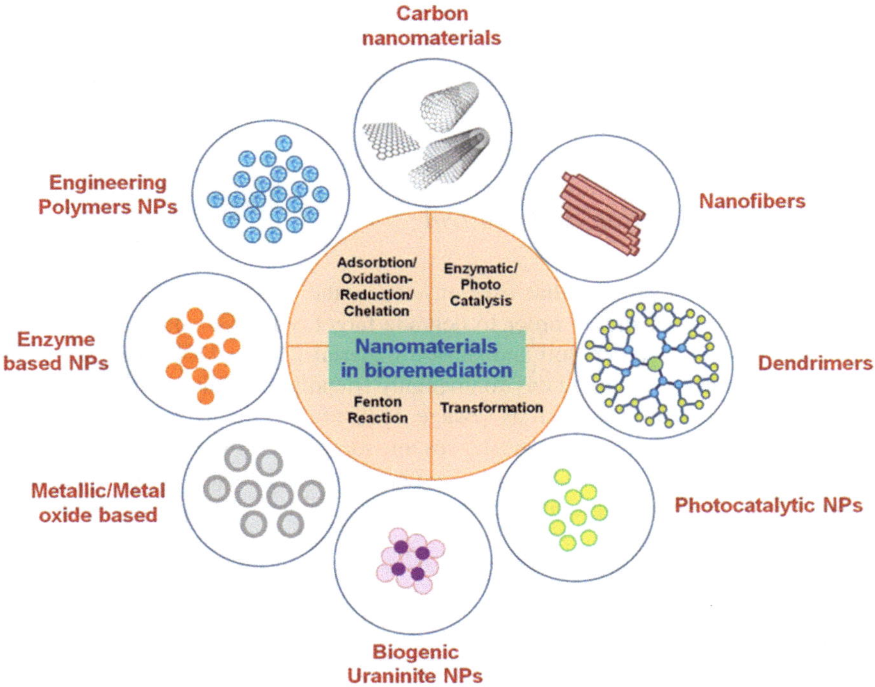

Fig. 9.1 Different types and classes of nanoparticles

has improved concepts such as nanodevices (such as biological machines), nanoparticles and nanoscale phenomena that occur in the discipline of nanotechnology. This technical approach to biology enables scientists to imagine and build usable biological research systems. Nanoscience is the study of materials found at the nanometric scale. The conversion of any material to the nanometric scale leads to the alteration of its physicochemical, biological, mechanical, optical, electronic properties, etc. These (new) material properties acquired through conversion to the nanoscale can be used for several useful activities. Therefore, it is an enabling technology, relevant to different sectors, such as chemicals, consumer products, health, energy, various other sectors and the environment.

9.3.4 *Nanotechnology*

Nanotechnology is the study, engineering and application of science, engineering and technology at the nanoscale, including imaging, measurement, modelling and manipulation of matter at this length scale. Nanotechnology is all about understanding and controlling matter at the nanoscale. At the nanoscale, unique phenomena can give

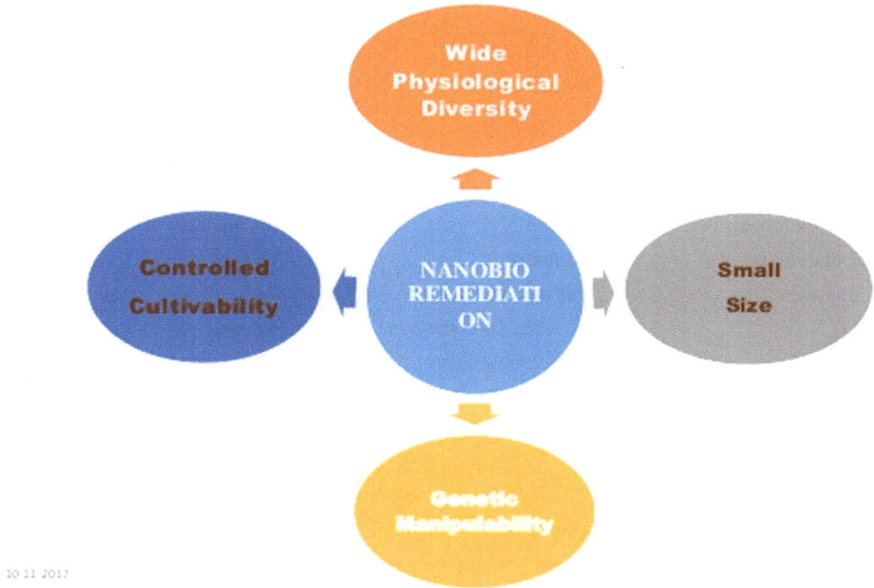

Fig. 9.2 Nanoparticle diversity and its manipulability as enumerated in the block diagram

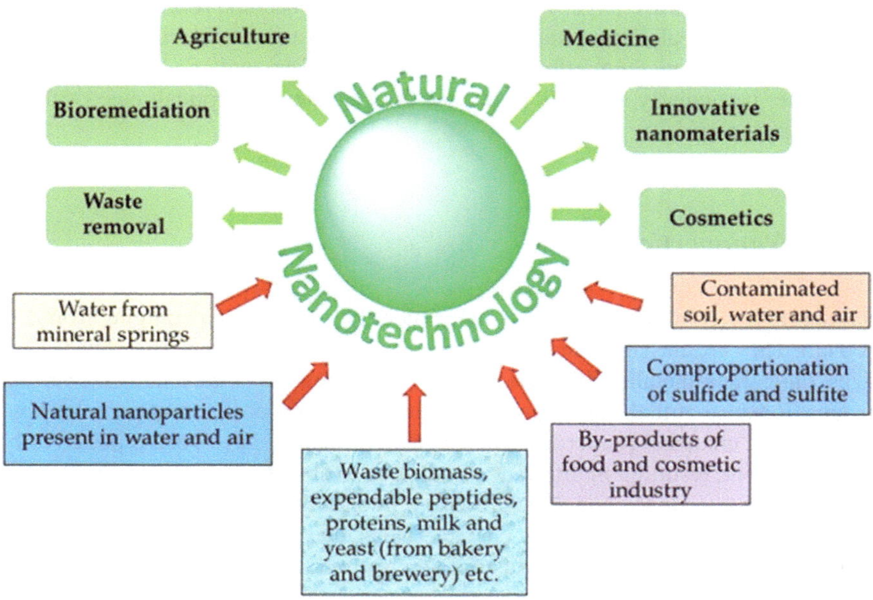

Fig. 9.3 Versatile applications of nanoparticles in agriculture and environmental clean up

Table 9.1 Various pollutants remediated by iron nanoparticles

Carbon tetrachloride	Chrysoidine	Cis-Dichloroethene
Chloroform	Tropacolin	Trans-Dichloroethene
Dichloromethane	Acid orange	1,1-Dichloroethane
Hexachlorobenzene	Mercury	PCBs
Pentachlorobenzene	Nickel	Dioxins
Tetrachlorobenzenes	Silver	Pentachlorophenol
Dichlorobenzens	Bromoform	TNT
Chlorobenzens	Dibromochloromethane	Dichromate
DDT	Dibromochloromethane	Arsenic
Lindane	Tetrachloroethene	Perchlorate
Orange II	Trichloroethane	Nitrate
Chloromethane	Acid red	Vinyl chloride
Trichlorobenzene	Cadmium	NDMA

rise to new applications. The fact that material properties can differ at the nanoscale is at the heart of nanotechnology. There are two main reasons for this: For starters, nanomaterials have a relatively large surface area when compared to materials of comparable quality produced in a larger form. Materials that are more chemically reactive (in some cases, larger forms of inert materials become reactive when transformed into nanoscale forms) and have an effect on resistance or electrical properties. Second, so-called quantum effects may begin to dominate matter's behaviour at the nanoscale, particularly at lower levels and affect materials' optical, electrical and magnetic properties.

9.3.5 Nanoparticles

Nanoparticles (NPs) are extremely small materials with diameters ranging from 1 to 100 nm. They are classified into various classes based on their properties, shapes and sizes. Fullerenes, metallic NPs, ceramic NPs and polymer NPs are among the various groups. Because of their large surface area and nanometric size, nanoparticles have physical and chemical properties. Their optical properties have been reported to be size dependent, resulting in different colours due to absorption in the visible region. Their unique size, shape and structure also influence their reactivity, toughness and other properties. Because of these characteristics, they are suitable candidates for a wide range of commercial and industrial applications.

NNs provide performance high in wastewater treatment with a primary application of adsorption (Li et al. 2005), membranes and membrane processes (Diagne et al. 2012), photocatalyst, disinfection and microbial control, detection and monitoring (Tanvir et al. 2012). Considering that the increase in nanoparticles will find a final path in wastewater treatment plants, some researchers have expressed their concern about the possible influence on the elimination process (Wang et al. 2016). For scientific recommendations, a quantitative and qualitative assessment of the role of NNs in wastewater treatment is required. Liu et al. (2018) investigated the impact of wastewater (WE) containing old agricultural nanoparticles (NPs). Nanotechnologies have the potential to effectively eliminate pollutants and germs in the field of water purification (Crane and Scott 2012). Metals (e.g. cadmium, copper, lead, mercury, nickel, zinc), nutrients (e.g. phosphate, ammonia, nitrates and nitrites), cyanide, organic substances, algae (e.g. cyanobacterial toxins), viruses and bacteria are all examples of nanoparticles, nanomembranes and nanopowders used for chemical and biological detection and removal.

9.3.6 Nanotechnology and the Future of Advanced Materials

The current and future development of a large number of nanomaterials is the foundation for future nanotechnology products. The development of a diverse range of nanomaterials will result in a completely new method of producing materials and devices. Essentially, it will have an impact on every aspect of our lives. Faster computers, advanced drugs, controlled drug delivery; biocompatible materials, nerve and tissue repair, anti-cracking surface coatings, improved skin care and protection, more efficient catalysts and smaller sensors are all on the horizon. Most current applications of nanomaterials, such as nanocomposites, nanoclays, nanocoatings, nanostructured surfaces and nanolubricants, are extensions of existing technology (Xiao et al. 2009).

9.4 Applications of Nanotechnology

Besides the application of the aforementioned nanotechnology to environmental biotechnology, innovations may also emerge in other areas where nanotechnology is developing rapidly, such as medical nanotechnology. The functionalization of nanostructures with biomolecules is a field of research. Biotechnology can also help provide environmental methods for the functionalization of nanoparticles. Nanotechnology applications typically combine industrial, medicinal and energy applications. More durable building materials, therapeutic drug delivery and environmentally friendly high density hydrogen fuel cells are among them (Khan et al. 2019). Nanoparticles and nanodevices have found applications in nanoscale electronics, cancer treatments, vaccines, hydrogen fuel cells and nanobatteries due to their versatility in modifying their properties. Nanotechnology advancements have transformed a variety of scientific and industrial fields, including the food industry (Rizwan et al. 2014). Nanotechnology applications have emerged in response to the growing demand for nanoparticles in various fields of food science and food microbiology.

The following is a list of some of the applications of nanoparticles in biology and medicine:

1. Tissue engineering is number one
2. Thermotherapy-assisted tumour elimination (hyperthermia)
3. Biological molecule and cell separation and purification
4. Enhancement of MRI contrast
5. Phagokinetic research
6. Biological markers that glow in the dark
7. Delivery of drugs and genes
8. Examining the structure of DNA
9. Pathogen bio detection
10. Protein Detection is number ten on the list.

9.5 Future Aspects of Nanotechnology Advancement and Its Scope

Nanotechnologies utilized in bioremediation procedures are projected to spur technological advancements in developed and developing countries, improving environmental quality. The mechanisms of decontamination and sanitation have been the subject of extensive research. Furthermore, the worldwide nanotechnology and bioremediation markets are predicted to expand and develop new niches in order to promote not only environmental but also human well-being (Grandclement et al. 2017). Nanotechnology can help to improve the efficiency with which contaminants are removed (Cecchin et al. 2017). Nanomaterials have both physical and chemical properties, and as a result, have gotten a lot of attention.

9.6 Discussion and Conclusion

Wastewater is wastewater that has had its physical, chemical or biological qualities altered as a result of the absorption of certain compounds, making it unsafe for specific purposes. Ingestion or physical contact with water contaminated with human, chemical or industrial wastes can induce a range of ailments, including dengue fever, filariasis, malaria, onchocerciasis, trypanosomiasis and yellow fever. The impact on development and on the community is greater. Most homes and businesses send their wastewater to a wastewater treatment plant where many pollutants are removed from the water. Wastewater contains nitrogen and phosphorus from human waste, food and some soaps and detergents, as well as federal authorities; it is typically discharged into a local body of water, where it can pollute the environment with nitrogen and phosphorus. Depending on their technology and how they treat wastewater, some treatment plants are able to remove more nitrogen and phosphorus from drains than others. When compared to plants employing traditional treatment methods, improved treatment systems allow some treatment plants to produce drainage with less nitrogen. Physics, chemical and biological processes and rules of economic management of wastewater from different sources. Wastewater comes from the processes of normal bathroom, toilet, laundry, kitchen utensils, etc. Wastewater generally contains a variety of pollutants and pollutants such as Chrome, zinc and lead) pathogens (such as viruses, bacteria, worms and protozoa) phytonutrients (such as nitrogen, potassium and phosphorus) organic pollutants (such as pesticides, PCBs, biodegradable and polycyclic aromatic hydrocarbons) micropollutants (such as cosmetics, pharmaceuticals and detergents). All of the above quality issues can cause environmental and health issues and ultimately has an economic impact on society. Wastewater treatment is an important part of water conservation. Although natural factors influence water quality, the term "pollution" often refers to human activity. Water pollution is mostly produced by the discharge of contaminated wastewater into surface or subsurface water and wastewater treatment is an important aspect of water pollution mitigation. Nanotechnology's functional aspect (NBT) strives to either improve the performance of materials or minimize the amount of materials utilized for this purpose. Treatment and cleanup, as well as pollution detection and prevention, are among its possible environmental attributes.

References

Abdel-Shafy HI, Mansour MSM (2018) Solid waste issue, sources, composition, disposal, recycling and valorisation. Egypt J Petrol 27(4):1275–1290. https://doi.org/10.1016/j.ejpe.2018.07.003

Cecchin I, Reddy KR, Thomé A, Tessaro EF, Schnaid F (2017) Nanobioremediation: integration of nanoparticles and bioremediation for sustainable remediation of chlorinated organic contaminants in soils. Int Biodeterior Biodegrad 119:419–428

Choudri BS, Charabi Y, Ahmed M (2018) Health effects associated with wastewater treatment, reuse and disposal. Water Environ Res 90(10):1759–1776. https://doi.org/10.2175/106143018x15289915807425

Crane RA, Scott TB (2012) Nanoscale zero-valent iron: future prospects for an emerging water treatment technology. J Hazard Mater 211:112–125

Diagne F, Malaisamy R, Boddie V, Holbrook RD, Eribo B, Jones KL (2012) Polyelectrolyte and silver nanoparticle modification of microfiltration membranes to mitigate organic and bacterial fouling. Environ Sci Technol 46:4025–4033

Grandclement C, Seyssiecq I, Piram A, Wong-Wah-Chung P, Vanot G, Tiliacos N, Roche N, Doumenq P (2017) From the conventional biological wastewater treatment to hybrid processes, the evaluation of organic micropollutant removal: a review. Water Res 111:297–317. https://doi.org/10.1016/j.watres.2017.01.005

Khan I, Saeed K, Khan I (2019) Nanoparicles: properties, applications and toxicities. Arab J Chem 12(7):908–931. https://doi.org/10.1016/j.arabjc.2017.05.011

Li YH, Di ZC, Ding J, Wu DH, Luan ZK, Zhu YQ (2005) Adsorption thermodynamic, kinetic and desorption studies of Pb2+ on carbon nanotubes. Water Res 39:605–609

Liu MZ, Wang S, Zong S (2018) SERS detection and removal of mercury(II)/silver(I) using oligonucleotide-functionalized core/shell magnetic silica sphere@Au nanoparticles. ACS Appl Mater Interfaces 6(10):7371–7379

Rizwan M, Singh M, Chanchal KM, Roshan MF (2014) Ecofriendly application of nanomaterials: nanobioremediation. J Nanopart Article ID 431787, 7 pages. https://doi.org/10.1155/2014/431787

Singh T, Shukla S, Kumar P, Wahla V, Bajpai VK, Rather IA (2017) Application of nanotechnology in food science: perception and overview. Front Microbiol 8. https://doi.org/10.3389/fmicb.2017.01501

Suolale O, Okoh A (2017) Human enteric bacteria and viruses in five wastewater treatment plants in the Eastern Cape, South Africa. J Infect Pub Health 10(5):541–547. https://doi.org/10.1016/j.jiph.2016.11.012

Tanvir S, Merlier F, Pulvin S (2012) Biosensing of reactive intermediates produced by the photocatalytic activities of titanium dioxide nanoparticles. J Photoch Photobio B 110:22–27

Wang M, Shen W, Yan L, Wang XH, Xu H (2017) Stepwise impact of urban wastewater treatment on the bacterial community structure, antibiotic contents and prevalence of antimicrobial resistance. Environ Pollut 231:1578–1585. https://doi.org/10.1016/j.envpol.2017.09.055

Wang Y, Xiong DB, Zhang W, Su H, Liu Q, Gu J, Zhu S, Zhang D (2016) Surface plasmon resonance of gold nanocrystals coupled with slow-photon-effect of biomorphic TiO_2 photonic crystals for enhanced photocatalysis under visible light. Catal 274:15–21

Xiao HY, Ai ZH, Zhang LZ (2009) Nonaqueous sol-gel synthesized hierarchical CeO_2 nanocrystal microspheres as novel adsorbents for wastewater treatment. J Phys Chem C 113:16625–16630

Chapter 10
Nano-Engineered Gold Particles for Pesticide Mineralization in Contaminated Water

Puja Goel and Manju Arora

Abstract The remediation of pesticides from water is a very challenging problem to render sustainable agriculture and ecosystem. Extensive work has been reported on the usage of biocompatible engineered nanoparticles (NPs) for the mineralization/degradation of pesticides in polluted water. This chapter highlights the present status and future of the biocompatible nano-engineered Gold particles (Au NPs) for pesticides mineralization in contaminated water. Synthesis of Au NPs by various physical and chemical routes, characterization of Au NPs by different analytical techniques and degradation/mineralization mechanism in terms of the photocatalytic activity of Au NPs at ambient conditions have also been described. As a case study, degradation pathway of organochlorine pesticide endosulfan (ES) using Au NPs is discussed. The adsorption of ES on Au NPs brings visible change in colloidal solution coloration from deep purplish pink to blue owing to the Au NPs aggregates formation. The conformation of the included guest molecule, the nature of the host–guest and guest–guest interactions (e.g., hydrogen bonding/hydrophobic groups) and the role of reorganization of Au NPs are observed through FTIR spectroscopic analysis and a mechanism of the mineralization of ES molecules into non-toxic reaction products is also proposed.

10.1 Introduction

Water covers over 70% of the Earth's surface and acts as a very important life sustaining resource for all inhabitants and the environment. Water pollution affects drinking water, rivers, lakes and oceans all over the world which consequently harms human health as well as natural environment. India possesses 2.45% of land, 4% of

P. Goel (✉)
Physics Department, College of Basic Sciences and Humanities, G.B. Pant University of Agriculture and Technology, Pantnagar, India
e-mail: puja.goel@gbpuat-cbsh.ac.in

M. Arora
CSIR- National Physical Laboratory, Dr. K.S. Krishnan Marg, Delhi, New Delhi, India

J. P. N. Rai and S. Saraswat (eds.), *Nano-biotechnology for Waste Water Treatment*, Water Science and Technology Library 111,
https://doi.org/10.1007/978-3-031-00812-2_10

water resources with ~17.7% of world population as per United Nations 2020 data survey. 2010 Central Water Commission (CWC), India projected a demand of 688 billion cubic meter (BCM) of water in agriculture sector and 56 BCM, 12 BCM, 5 BCM and 52 BCM for different sectors like: drinking water, industry, energy and other fields, respectively. Owing to the huge population, urbanization and industrial growth, this demand may further rise to (910, 1072) BCM, (73, 102) BCM, (23, 63) BCM, (15, 130) BCM, (72, 80) BCM by (2025, 2050) (CWC 2010). Contrary to this, availability of fresh water all over India has been drastically reduced due to irregular and less rainfall during monsoon season which causes drying of natural water resources like ponds, lakes, rivers, etc. In addition to this, availability of fresh water further fell off owing to severe water pollution caused by the dissolution of harmful chemicals discharged from industries and agriculture fields during irrigation, washing or field processing directly into river, lake, ocean and natural or man-made reservoirs. The consumption of such contaminated water is the root cause of many deadly diseases not only in human being but in land-dwelling animals and aquatic life forms as well. Considering the gravity of situation, Government of India has also initiated a water mission aiming to provide fresh potable drinking water to all by focusing on various aspects such as treatment of waste water before disposal to nearby water resources, water management, in addition to formulating the laws and regulations to prevent groundwater contamination. Although the contamination stems from both point and nonpoint sources, the present status revealed that most of the water sources nearby cities and urban areas are polluted by garbage and direct dumping of chemicals devoid of any pre-treatment, either legally or illegally. The introduction of nanotechnology may represent a significant advancement in catering efficient and cost-effective water remediation technologies, and therefore the development of suitable nano-engineered bio materials may be considered as a logical next step across the water industry for potential remedial applications.

10.2 Nanomaterials Used for Pesticide Degradation

The nanomaterials exhibit unique mechanical, electrical, optical and magnetic properties as compared to their bulk analogues depending upon their size and morphology. The high surface area to volume ratio provides more active sites on surface. Their catalysis/photocatalysis, adsorption and high reactivity properties are exploited in chemical synthesis and degradation reactions (Parmon 2008; Liang et al. 2012; Kusior et al. 2013; Bujoli et al. 2006). The nanomaterials are successfully used for the removal of heavy metals (Khin et al. 2012), organic pollutants (dyes and pesticides) (Yan et al. 2015) and bacteria/microbes (Kalhapure et al. 2015; Borrego et al. 2016) from contaminated water. The zero-valent metal nanoparticles (Ag, Fe, Al, Ni, Zn and Au), metal oxides nanoparticles (TiO_2, ZnO, Fe_3O_4, γ-Fe_2O_4, α-Fe_2O_3), carbon-based materials like activated carbon (AC), carbon fibers (CF), carbon granules, carbon dots, single wall and multi-wall carbon nanotubes (SWCNTs and MWCNTs), graphene(graphene oxide [GO] and reduced graphene oxide [RGO]), nano porous carbon and nanocomposites consisting of either carbon based materials with zero valent nanometals/metal oxides or with nanoparticles dispersed in metal oxide matrix/organic polymeric membrane, etc., have been used for pesticides detection and degradation.

The small size, large surface area and high extinction coefficient properties of pure and surface-modified Ag and Au NPs-based colorimetric sensors have been used for the detection of residual pesticides (Thakur et al. 2018; Al-Zubaidi et al. 2021; Unser et al. 2015; Singh 2016; Li et al. 2014) present in fruits, food and water. W.Wei et. al. investigated Ag NPs/Nanowires (NWs) embedded into polydimethylsiloxane for on-site detection of thiram pesticide residues on plant leaf surface by Surface-Enhanced Raman Scattering (SERS) technique (Wei et al. 2018). While, D. Xiong and H. Li (2008) reported the Colorimetric detection of iprodione, pyrimethanil, thiabendazole, optunal, parathion-methyl, methomyl and acetamiprid pesticides based on calixarene modified silver nanoparticles in water.

Fe, Al and Ni NPs (Rivero-Huguet and Marshall 2009; Fu et al. 2014; Wang et al. 2013) are also used in contaminated water treatment owing to their reducing capability. Al NPs are very unstable in the aqueous medium and they immediately form oxides/hydroxides layer on the surface which obstruct electron transfer from metal surface to contaminants for degradation (Rivero-Huguet and Marshall 2009). Although, Fe and Zn both are better reducing agents for many hazardous contaminants, Fe^0 has disadvantages of aggregation, oxidation and separation difficulty from the degraded solution (Fu et al. 2014; Wang et al. 2013; Stefaniuk et al. 2016; Lv et al. 2014; Bokare et al. 2013; Tratnyek et al. 2010) whereas Zn NPs are strong reductant and degrades contaminants faster than Fe^0. These are used in dehalogenation reaction of CCl_4 and octachlorodibenzo-p-dioxin (OCDD) in water. Zn NPs are sensitive to solution chemistry rather than on particle size or surface morphology (Tratnyek et al. 2010).

10.2.1 Gold Nanoparticles (Au NPs)

The biocompatible, stable Au NPs colloidal solutions can be easily obtained by capping their surface with organic and biomolecular surfactants. Federal Drug Administration (FDA) has approved its application for diagnostic and clinical purposes. Au NPs may acquire different shapes like spherical, rod, wire, triangular, octahedral, tetrahedral, cubes, pyramids, flower, branches and cage morphological shapes (Xiao et al. 2019; Chen et al. 2008a, b; Grzelczak et al. 2008; Sau and Murphy 2004). The color of spherical-shaped Au NPs aqueous suspension varies from brown to orange to red to purple with increase in particle size from 3 nm to 100 nm. Another very important size dependent optical property of Au NPs is correlated to surface plasmon resonance (SPR) absorption peak which appears in 500–550 nm region. SPR transition occurs when oscillating electric field of electromagnetic radiation interact with a large number of easily polarizable conduction electrons of Au NPs for a collective electron charge oscillation which is in resonance with the frequency of incident radiation. These transitions are obtained in Au NPs with size more than 2 nm and sensitive to shape, solvent, surfactant, core/surface charge, ambient conditions as well as their concentration in dispersed medium. The agglomerated/clustered nanoparticles exhibit red shift in absorption frequency and broadening of SPR signal.

Au NPs play a very vital role in photocatalytic degradation/mineralization of pesticides. This process undergoes in normal sunlight: Au NPs produce hot electrons for the degradation/mineralization of pesticide molecules adsorbed at their surface (Mills and Le Hunte 1997; Castello 2009) into non-toxic products through chemical redox reactions initiated by the reactive oxygen species (ROS). Recently, photocatalysis is found to be one of the best alternative processes for water decontamination as compared to other physical, chemical, biological and sludge methods (Castello 2009).

10.3 Synthesis of Au NPs

Au NPs are synthesized by both "top-down" through physical methods viz. ball milling, sputtering, micro-patterning (Chen et al. 2009), lithography (Yun et al. 2006), pyrolysis (Li et al. 2004), thermolysis (Leff et al. 1995; Tao et al. 2008), laser ablation methods (Sanabria-cala et al. 2018; Wang et al. 2008), ion implantation (Abedini et al. 2013; Misra et al. 2012), microwave radiation (Gangapuram et al. 2018) and "bottom-up" by chemical methods as per application demand. The "bottom-up" techniques involve chemical methods like Turkevich (Turkevich et al. 1951; Frens 1973; Chow and Zukoski 1994; Ojea–Jiménez et al. 2011; Zhao et al. 2013), Brust-Schiffrin (Brust et al. 1994; Hostetler et al. 1998; Goulet and Lennox 2010; Yee et al. 1999; Sardar and Shumaker-Parry 2009), electrochemical (Song et al. 2013; Huang et al. 2006), seed growth (Ziegler and Eychm ller 2011; Nikoobakht and El-Sayed 2003; Ye et al. 2013), polymers and biomolecules (Jeon et al. 2011; Sakai and Alexandridis 2005; Richardson et al. 2006; Sardar et al. 2007), ultrasound assisted (Okitsu et al. 2001; Jiang et al. 2004), green chemistry (Iravani 2011; Muralikrishna et al. 2014; Rajeshkumar et al. 2013; Sujitha and Kannan 2013) and place exchange method for Au NPs monolayer formation (Templeton et al. 1998; Hostetler et al. 1999). In the case study discussed in upcoming section, Au NPs used for degradation of endosulfan pesticide were synthesized by Turkevich method.

10.4 Characterization of Au NPs by Different Techniques

Au NPs are characterized by X-ray diffraction (XRD) (Vasanthi Bathrinarayanan et al. 2013), small-angle X-ray scattering (SAXS) (Allec et al. 2015), Fourier transform infrared spectroscopy (FT-IR) (Lin Vien et al. 1991; Silverstein et al. 2014; Goel and Arora 2018, 2020a, b, 2021), electron microscopic techniques (Allec et al. 2015; Goel and Arora 2018, 2020a, b, 2021; Khanna et al. 2019; Lapresta–Fernández et al. 2014; Rao et al. 2007; Darwich et al. 2011) like scanning electron microscopy (SEM) with EDX accessory (Khanna et al. 2019; Lapresta–Fernández et al. 2014), transmission electron microscopy (TEM) (Allec et al. 2015; Goel and Arora 2018, 2020a, b, 2021; Khanna et al. 2019; Lapresta–Fernández et al. 2014; Rao et al. 2007; Darwich et al. 2011), high-resolution transmission electron microscopy (HRTEM) (Khanna et al. 2019), and atomic force microscopy (AFM) (Rao et al. 2007; Darwich et al. 2011), dynamic light scattering (DLS) (Jans et al. 2009), visual colorimetry (Ramalingam 2019), UV-Visible spectroscopy (UV-Vis) technique in 500–900 nm region (Goel and Arora 2018, 2020a, b, 2021; Roduner 2006; Ray et al. 2015; Shukla and Iravani 2017; Amendola et al. 2017; Zuber et al. 2016), Mass spectroscopy (MS) (Harkness et al. 2010), Inductively coupled plasma-mass spectrometry (ICP -MS) (Trapiella-Alfonso et al. 2014), Nuclear magnetic resonance (NMR) (Guo and Yarger 2018) techniques to confirm their formation, particle size, shape, morphology and degradation/mineralization reaction products.

10.5 Pesticides

Pesticides are basically hazardous organic compounds or mixtures of compounds which are used by farmers on agriculture fields, forestry and horticulture to protect crop from pests, weeds, insects and diseases. They are categorized as insecticides, fungicides, herbicides, rodenticides, algaecides and plant growth regulators as per their specific application used for the control/remediation of insects, fungus, weed, rodents, algae and reproduction of insects in plants, respectively (World Health Organization 1990; Alewu and Nosiri 2011; NSW EPA 2013). Organochloro (OC), organophosphates (OC), carbamates (CB), pyrethroids (PT), triazines (TZ) and neonicotinoids (NN) pesticides are generally used in agriculture fields all over the world by farmers. The excessive application of pesticides raised alarming health and environmental issues (World Health Organization 1990; Alewu and Nosiri 2011; NSW EPA 2013; Hayes et al. 2006; Goulson 2014). Their residues are found in many daily use food products and beverages like instant cooked meals, water, wine, fruit juices, refreshments and domestic animals fodder/livestock, etc. (Cabras and Angioni 2000; Burnett and Welford 2007; Nag and Raikwar 2011). Pesticides residues are also diagnosed in human breast milk samples which affects children health also (Chourasiya et al. 2015; Lu et al. 2015; Eskenazi et al. 1999; Jaeger et al. 1999). Human beings are also exposed to pesticides through skin contact, ingestion or inhalatsion. The type of pesticide, direct or indirect exposure, daily, regularly or irregularly, its duration and the individual body immunity are some factors which decide the effect of pesticides on the human or animal health. They cause dermal, gastrointestinal, carcinogenic, nervous, cardiac, respiratory, reproductive and endocrine systems related problems in human beings (World Health Organization 1990; Alewu and Nosiri 2011; Bretveld et al. 2006; Bassil et al. 2007; Hoppin et al. 2008; Mostafalou and Abdollahi 2013; Gunnell et al. 2007) and excessive exposure may be fatal also (World Health Organization 1990; Gunnell et al. 2007). Hence considering the severity of problem, the environment protection agencies have recommended the ban of many toxic pesticides in agriculture for better life to existing and forthcoming generations on earth as well as in aquatic and marine water.

10.6 Degradation/Mineralization of Endosulfan Pesticide (ESP) by Au NPs

10.6.1 Literature Review

In this section, the work done by the authors for the mineralization of different concentrations ESP solution with Au NPs is presented. Earlier work carried out on the degradation/mineralization of ESP by different workers is briefly reviewed. Two mechanisms are proposed for ESP degradation/mineralization: (i) oxidation and (ii) hydrolysis reactions initiated by biological or chemical entities. ESP is degraded to toxic endosulfan sulfate (very long life) in the soil and less toxic endosulfan diol through oxidation and hydrolysis reactions by *Phanerochaete chrysosporium* (Kim et al. 2001) fungi microorganisms. While, *Mucor thermohyalospora* MTCC 1384 (Shetty et al. 2000), and *Trichoderma harzianum* (Katayama and Matsumura 1993) form intermediate Endosulfan diol which then converts into endosulfan ether to endosulfan hydroxyl ether to endosulfan dialdehyde and finally to less toxic endosulfan lactone (Martens 1977; Guerin and Kennedy 1992; Guerin 1999). Natural adsorbents such as sediments are also used to adsorb pesticides on their surface from contaminated water during treatment process. In chemical process, different nano and bulk materials have been used for the hydrolysis, oxidation and reduction of ESP into less toxic or non-toxic reaction products.

Au NPs as photocatalyst initiate oxidation (Chen et al. 2008a, b; Hallett-Tapley et al. 2011; Christopher et al. 2011; Wee et al. 2012; Huang et al. 2014; Xiao et al. 2015), reduction (Zhu et al. 2010; Padilla et al. 2017; Zheng et al. 2014) and degradation (Adleman et al. 2009; Bakhtiari et al. 2009; Fasciani et al. 2011; Mukherjee et al. 2014) chemical reactions. The photocatalytic activity of plasmonic Au NPs causes (a) near field enhancement, in which excited light intensity increases drastically around the particles for improving the linear and nonlinear absorption efficiencies of nearby photo active reagents, (b) the photothermal effect (Adleman et al. 2009; Bakhtiari et al. 2009; Fasciani et al. 2011), i.e., the conversion of light energy into heat energy for thermally induced chemical reactions and (c) the generation of hot electrons (Mukherjee et al. 2014) which are transferred to surrounding adsorbed molecules (in the present case ESP molecules adsorbed at the surface of Au NPs) to facilitate redox reactions for the degradation/mineralization of adsorbed pesticide molecules. These reaction mechanisms have a common feature of localization and feasible only in nanoscale materials to capture maximum optimal photocatalytic efficiencies.

In the colloidal Au NPs, the pesticide molecules adsorb on their surface through functional groups with lone pair of electrons for catalytic reduction. In addition to this, these NPs can be further embraced in fabricating highly selective nano-sensors for exquisite field detection of environmental contaminants. D. Du et al. developed CdTe quantum dots and Au NPs modified chitosan microspheres interface based acetylcholinesterase biosensor with improved detection sensitivity of thiocholine through enhanced electron transfer process catalyzed by electro-oxidation reaction (Du et al. 2008a) and another electrochemical biosensor based on colloidal Au NPs modified sol-gel interface for the detection of monocrotophos, carbaryl and methyl parathion (Du et al. 2008b) pesticides. S. Anandhakumar et al. reported the non-enzymatic detection of methyl parathion pesticide using gold atomic cluster modified electrode by employing square wave voltammetry (Anandhakumar et al. 2014). A. Sreekumaran et al. demonstrated the use of Au NPs and silver NPs in the detection and removal of OC pesticide, endosulfan by spectrophotometry (Sreekumaran Nair et al. 2003; Sreekumaran Nair and Pradeep 2003) on the basis of adsorption of complex halocarbons on the nanoparticle surface and shift in the plasmon excitation wavelength. These studies demonstrated the successful use of noble metal nanoparticles and their nanocomposites in sensing and removal of pesticides from the soil and natural resources. However, limited information is available on the structural and morphological changes taking place during pesticide degradation reaction.

Nowadays, the electrochemical, optical, Infrared, piezoelectric and mechanical signal-based biosensing analytical techniques (Sassolas et al. 2012; Andreescu and Marty 2006; Luque de Castro and Herrera 2003; Jaffrezic-Renault 2001) have been widely used for the pesticides characterization as compared to conventional chromatography techniques (Sancho et al. 2006; van der Hoff and van Zoonen 1999; Balinova 1996). In lieu of this fact, transmission electron microscopy (TEM), non-destructive infrared (IR) transmittance and UV-Vis spectroscopy techniques have been used to investigate the final products obtained after degradation of ESP and the variations in clustered Au NPs into different small and big size-shaped ensembles like rods, petals and floral patterns in the present case.

10.6.2 Synthesis and Measurements

10.6.2.1 Materials Used in Au NPs Synthesis and ESP Solutions Preparation

For the synthesis of gold nanoparticles (Au NPs), analytical grade CDH make tetrachloro auric acid ($HAuCl_4.3H_2O$, 2% w/v), Sodium Citrate Solution ($C_6H_5Na_3O_7.2H_2O$, 3% w/v) were used without further purification. Technical grade endosulfan flakes and 2-Propanol (HPLC and Spectroscopy grade) were mixed in stoichiometric ratio to obtain 20 ml of each 1 ppm, 2 ppm, 5 ppm, 10 ppm, 20 ppm, 50 ppm and 100 ppm endosulfan solutions.

10.6.2.2 Synthesis of Au NPs by Citrate Route

The colloidal solution of Au NPs was synthesized by the reduction of tetra-chloro auric acid (TCAA) through citrate reduction method as reported by Turkevich et al. (1951). For 150 ml of colloidal Au NPs synthesis, 0.5 mM TCAA and 38.8 mM sodium citrate solutions were obtained by gradually diluting 1.275 ml of TCAA and 4.5 ml sodium citrate solutions with double distilled water up to 150 ml and 15 ml, respectively. The magnetically stirred TCAA solution was heated at 180 °C on hot plate, then 15 ml solution of diluted sodium citrate solution was quickly added to it. The resulting solution turned slightly bluish after 5 min of this reaction and at the end of the reduction process (~20 min), Au NPs colloidal solution of wine red color was obtained.

10.6.2.3 Analytical Techniques Used for Characterization

TEM images recorded on JEOL-JEM (1011) microscope have been used to characterize shape, and size of synthesized Au NPs. For case studies, the samples were prepared in three different batches to analyze the reaction of ESP and Au NPs: (a) 5 ml of 1 ppm ES solution reacted with (0.125 ml, 0.25 ml, 0.5 ml) Au NPs solution (b) 5 ml each of (1 ppm to 100 ppm) ES solutions reacted with 0.5 ml Au NPs solution and (c) 5 ml of 100 ppm ES solution reacted with variable volume of Au NPs solution in 0.1 ml to 5 ml range.

IR transmittance spectra of pure ESP, Au NPs and Au NPs treated ES solutions were recorded in 4000–500 cm^{-1} region by ALPHA T (Bruker make) Fourier transform infrared (FTIR) spectrophotometer. UV-Vis absorption spectra of pure and treated samples were recorded in 200 nm–800 nm range on Shimadzu make UV-Vis spectrometer (Model: UV-2401 PV).

10.6.3 Results and Discussion

10.6.3.1 TEM Analysis of Synthesized Au NPs

TEM images of Au NPs were recorded after drying a small amount of colloidal Au NPs solution on a carbon coated copper grid. Figure 10.1 (a) shows randomly distributed, uniform spherical-shaped well separated Au NPs with narrow size variations in the range of ~8.7 ± 0.17 nm. The size distribution was estimated from large number of nanoparticles captured in different TEM images and presented as histogram in Fig. 10.1 (b). This can be well fitted by the following log normal distribution function (Bootharaju and Pradeep 2012)

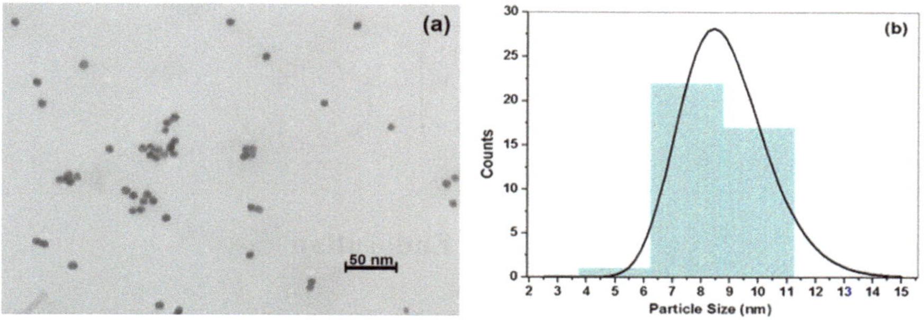

Fig. 10.1 (a) TEM image of GNPs and (b) Particle size distribution histogram (mean particle size 8.7 ± 0.17 nm)

$$y = y0 + \frac{A}{w \times x\sqrt{2\pi}} \exp\left(-\frac{\left(\ln\left({}^{x}/_{xc}\right)\right)^2}{2 \times w^2}\right)$$

where "*w*" represents the standard deviation and "x_c" is the mean particle size with a fitted value of 0.17 nm and 8.7 nm, respectively

10.6.3.2 FTIR Analysis of Au NPs, ES and Au NPs Treated Final Solutions of ES Pesticide

The non-destructive infrared spectroscopy has been used for drawing information about the structural details by the assignment of peaks pertaining to different bonding groups present in the materials (Lin Vien et al. 1991; Silverstein et al. 2014; Goel and Arora 2018, 2020a, b, 2021). In this work, Au NPs, ES pesticide and their coordinated complex salts were characterized from their recorded IR transmittance spectra in 4000–500 cm^{-1} region at ambient temperature. This information helps in understanding the reaction mechanism of ES pesticide mineralization by Au NPs solution. The used powdered ES pesticide sample consists of a mixture of 70% alpha (α) and 30% beta (β) isomers and their chemical structures are given below:

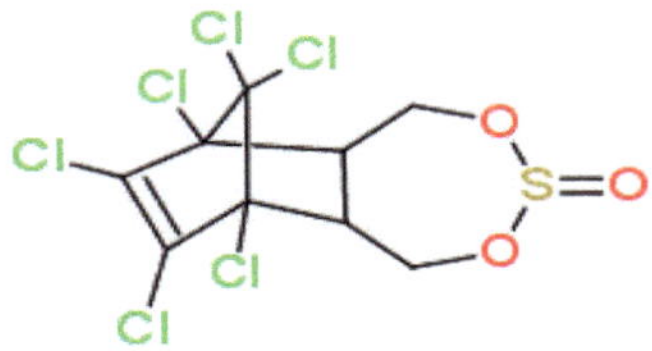

Endosulfan

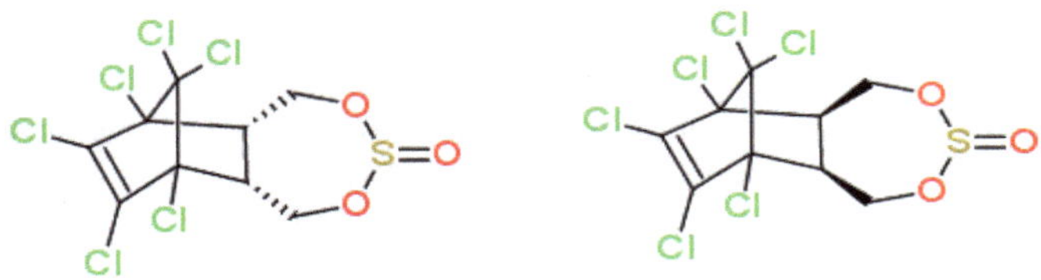

αEndosulfan (70%)βEndosulfan (30%)

Chemical Structure of Endosulfan along with its α and βisomers

IR transmittance spectrum of ES in 4000–500 cm^{-1} region (as shown in Fig. 10.2) exhibits peaks at (3386, 3251), (2935, 2898, 2858), 1605, (1481, 1464, 1452, (1269), 1245), 1193, 1180, 1168, 1145, 1095, 1006, 979, 966, 940, 917, 879, 858, 840, 814, (792, 783), 753, 700, 673, (637, 622, 600), (580 and 538) cm^{-1} pertaining to water antisymmetric and symmetric stretching, methylene stretching, C=C, CH_2

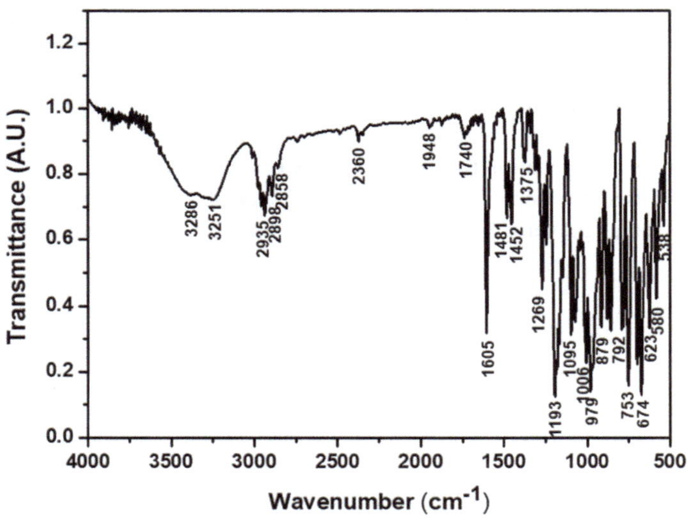

Fig. 10.2 IR transmittance spectrum of ES in 4000–500 cm^{-1} region

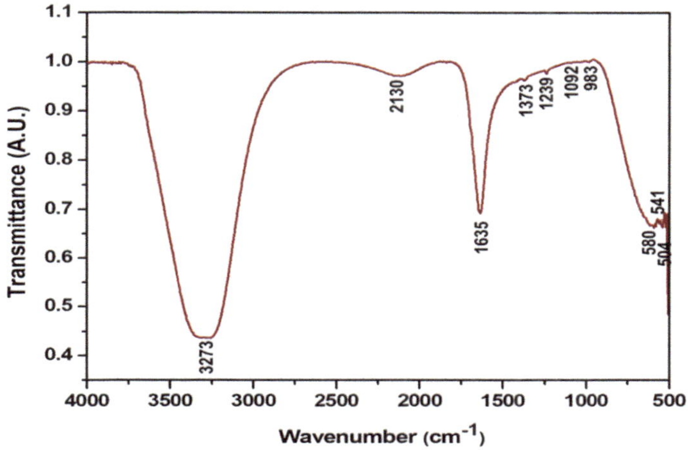

Fig. 10.3 IR transmittance spectrum of Au NPs in 4000–500 cm^{-1} region

scissoring, carboxylate ion stretching, HOH bending/C=C stretching, alcoholic (C–OH), S–O and C–O–S stretching modes of sulfite for two different confirmations, S=O asymmetric stretching of $-SO_3$, C–O rocking, C–Cl stretching, O–S=O, C–Cl, R–SO_3 and SO_3 bending modes (Lin Vien et al. 1991; Silverstein et al. 2014). While IR transmittance spectrum of Au NPs as presented in Fig. 10.3 has strong and broad water stretching, CO_2 stretching, water bending/conjugated –COOH stretching of capped citrate, very very weak peaks of carboxylate anion stretching and Au–O stretching modes at 3273, 2130, 1635, 1373, 1239, 1092, 983, 580 and 541 cm^{-1}, respectively.

Further, IR transmittance spectra in 4000–500 cm^{-1} region were recorded for three sets of ES and Au NPs solutions, e.g., (i) 5 ml of 1 ppm ES with 0.125, 0.25 and 0.5 ml Au NPs (Fig. 10.4), (ii) 0.5 ml Au NPs and 5 ml of 1, 2, 5, 10, 20, 50, 100 ppm ES solution (Fig. 10.5) and (iii) 5 ml of 100 ppm ES with 0.1, 0.2, 0.5, 0.7, 1.0 and 5.0 ml Au NPs. These spectra exhibit the absence of carbonyl, $-SO_3$ and C–Cl functional groups peaks in 1120–1150 cm^{-1}, 850–800 cm^{-1} and 800–550 cm^{-1} regions, respectively. This reveals the mineralization of ES molecules adsorbed on citrate-capped Au NPs.

The characteristic νC–H stretching modes of hydrocarbon and hydrogenated amorphous carbon molecules appear in 3000–2800 cm^{-1} region. In ES + Au NPs (Fig. 10.4), the asymmetric (ν_{as}) and symmetric (ν_s) stretching modes of $-CH_3$ group and $-CH_2$ group ν_{as} stretching mode are obtained at (2971 and 2886) cm^{-1} and 2933 cm^{-1}, respectively. The peak intensity of these modes are sensitive to the conformation of hydrocarbon chain having some of CH_2 groups in *gauche* configuration and confirms the formation of amorphous C or less crystalline phase in ES:Au NPs solution.

CH_2 scissoring mode ($\delta(CH_2)$) provides additional information about the conformation of alkyl chain, side-by-side chain interactions and the arrangement of

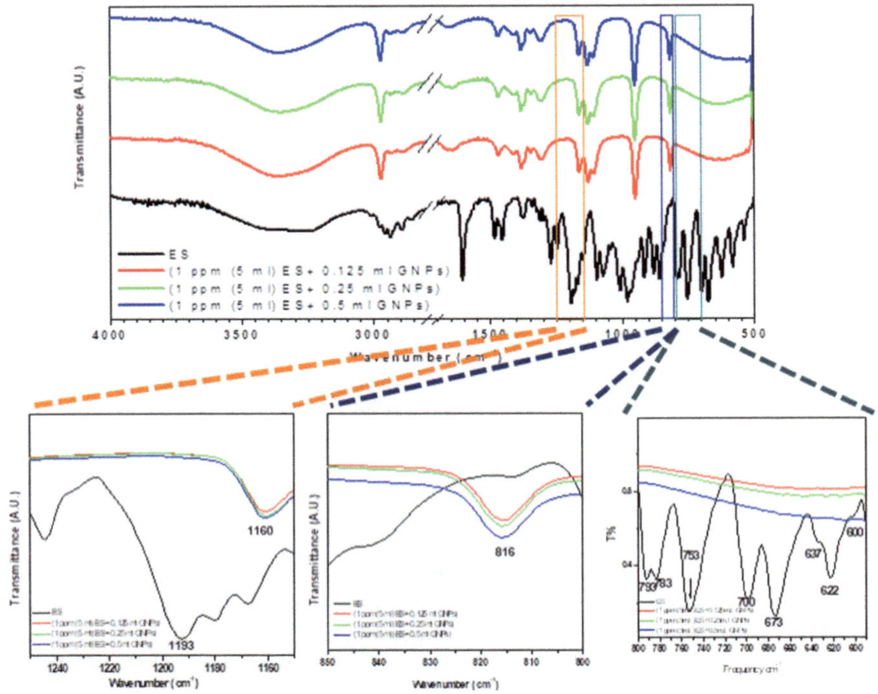

Fig. 10.4 IR transmittance spectrum of 5 ml of 1 ppm ES: x ml Au NPs (GNPs) ($x = 0.125$, 0.25 and 0.5 ml) complex in 4000–500 cm^{-1} region with insets for 1250–1150 cm^{-1}, 850–800 cm^{-1} and 800–590 cm^{-1} regions

the methylene chains. The medium intensity δ (CH_2) peak is observed in 1500–1450 cm^{-1} region with three components with maxima at 1481, 1464 and 1452 cm^{-1} in ES IR transmittance spectrum. The broadness of this band shows the weak side-by-side chains interaction for easy chain movement in ES solution. The presence of 1481 cm^{-1} component arises from *trans* conformation chains of C–C skeletal vibrations while the symmetric bending mode components at 1464 and 1452 cm^{-1} pertains to –CH_3 and partial conformation ordering by the combination of *gauche* defects and the asymmetric deformation of terminal CH_3 group. On adding Au NPs solution, this peak appears at 1468 cm^{-1} with loss in intensity. This further supports the degradation of ES pesticide and clustering of Au NPs as chain and petal or floral shapes in treated solution along with visual change in solution color as seen in TEM images and optical absorption spectra.

CH_2 wagging modes of *gauche* conformation are observed in 1400–1300 cm^{-1} region. In IR transmittance spectrum of ES, this mode appears as medium intensity band at 1373 cm^{-1} which transforms into a doublet peak with maxima at 1379 and 1364 cm^{-1} in Au NPs treated solution along with a gain in intensity of 1379 cm^{-1} component (Fig. 10.4). The occurrence of these components confirms the presence

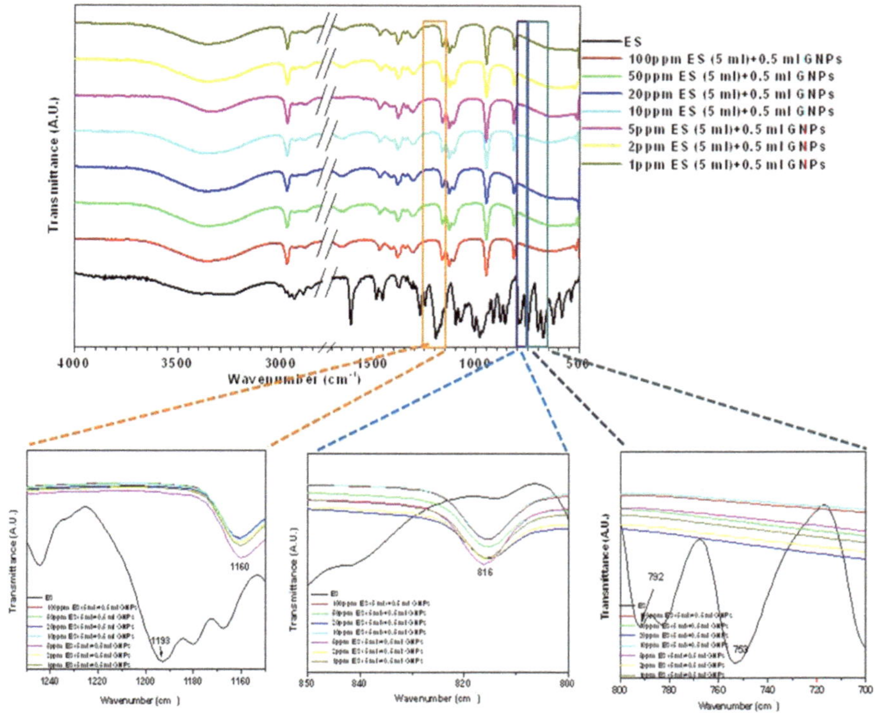

Fig. 10.5 IR transmittance spectrum of x ppm ES ($x = 1, 2, 5, 10, 20, 50, 100$ ppm): 5 ml GNPs complex in 4000–500 cm^{-1} region with insets for 1250–1150 cm^{-1}, 850–800 cm^{-1} and 800–700 cm^{-1} regions

of methyl groups in gem- or iso- dimethyl form in final degraded solution of ES. The bands at 1364 cm^{-1} and 1379 cm^{-1} correspond to the peak of *gauche* conformation having –CH_3 groups in gem- and iso- forms, respectively.

Further, SO_2 asymmetric stretching mode in ES spectrum is observed as a strong double band at 1269 and 1245 cm^{-1} respectively with peaks separation of 24 cm^{-1} owing to conformational alignment along the horizontal and vertical positions. The intensity of 1269 cm^{-1} band is higher than 1245 cm^{-1}, due to strong in-plane hydrogen bonding between –SO_3 oxygen and propanol –OH, hydrogen as compared to the lateral interactions between ESP molecules. In ES IR transmittance spectrum, terminal –SO_3 functional group υSO_2 asymmetrical, symmetrical stretching and bending modes for two conformational configurations are observed as strong and medium intensity peaks at (1193,1095) cm^{-1}, (1006,979) cm^{-1} and (580, 538) cm^{-1}, respectively (Fig. 10.2). While, in ES:Au NPs treated solutions IR transmittance spectra in Fig. 10.4, these modes are not observed which confirms the mineralization of ESP –SO_3 groups.

The very strong broad transmittance peak in 1225–1100 cm^{-1} region with components at 1193, 1145 cm^{-1} pertaining to C–$H_{bend.}$ and CH_{3rock} modes are observed at 1180 and 1168 cm^{-1} in ESP (Fig. 10.2). On adding ESP solution to variable amount of Au NPs (Fig. 10.4), three peak components are observed at 1162, 1128 and 1106 cm^{-1}, respectively. The shift in band positions on addition of Au NPs shows the structural re-orientational ordering in ESP molecules. C–$H_{wag.}$ mode is obtained as shoulder band with components at 840 and 814 cm^{-1} in ESP spectrum and as medium intensity peak at 816 cm^{-1} in Au NPs:ESP solutions.

νC–$C_{str.}$ and νC–$O_{str.}$ peaks are observed as shoulder components at 966 and 940 cm^{-1}, respectively in ESP IR transmittance spectrum. In ESP:Au NPs treated solution, a very strong peak of νC–$C_{str.}$ at 950 cm^{-1} is observed. νC–O_{str} and νO–C_{str} are obtained as medium intensity peaks at 917, 879, 858 cm^{-1} in ESP. On adding variable amount of Au NPs to 1 ppm ESP solution, these peaks disappeared, which further confirms the mineralization of ESP molecules on treatment with Au NPs.

In ESP molecule, six C–Cl bonds are present in three different environments which exhibit νC–Cl stretching modes at (792, 783), 753, 700, 673, (637, 622, 600) cm^{-1} (Fig. 10.2), for 800–590 cm^{-1} region. The strong peak with components at 793 and 783 cm^{-1} are assigned to νC–Cl asymmetric stretching mode of –C(Cl_2) group. νC–Cl asymmetric stretching band at 753 cm^{-1} arises from Cl–C=C–Cl group and (700, 673) cm^{-1} peaks are asymmetric stretching vibrations of two C–Cl groups. The symmetric stretching modes of C–Cl group are observed as strong peaks with maximum at 622 cm^{-1} (Cl–C=C–Cl group) and shoulders at 637, 600 cm^{-1} (C–Cl_2 group) and 582 cm^{-1} (C–Cl group). However, after adding different concentrations of Au NPs to 1 ppm ESP solution, none of these bands are observed which further confirms the complete dehalogenation of ESP molecule in a very short time as compared to earlier reports.

10.6.3.3 UV-Vis Spectroscopy

The optical spectra of Au NPs colloidal solution (Fig. 10.6(a)), ESP solution (Fig. 10.6(b)) and Au NPs treated ESP samples were recorded at ambient temperature in 350–1100 nm region. These measurements were carried out for the following three sets of samples (i) 5 ml of 1 ppm ESP: (0.125 ml, 0.25 ml, 0.5 ml) (Fig. 10.7), (ii) 5 ml (1 ppm, 2 ppm, 5 ppm, 10 ppm, 20 ppm, 50 ppm, 100 ppm) ESP: 0.5 ml Au NPs, (Fig. 10.8) and (iii) 5 ml of 100 ppm ESP: (0.1 ml, 0.2 ml, 0.5 ml, 0.7 ml, 1.0 ml, 5.0 ml) Au NPs, (Fig. 10.9).

The absorption maxima of ESP molecules is observed at a wavelength of 242 nm with shoulder at 290 nm (Fig. 10.6(b)) (Goel and Arora 2018, 2020a, b, 2021; Ray et al. 2015; Shukla and Iravani 2017; Amendola et al. 2017; Zuber et al. 2016; Bootharaju and Pradeep 2012; Kumar and Philip 2006).

A very intense dark brown red color of Au NPs solution originates from the coherent electron motion, which gives strong Surface Plasmon Resonance (SPR) band at 529 nm in the UV-Visible spectrum as can be seen from Fig. 10.6 (a). On adding different concentration of Au NPs (from 0.1 ml solution to 5.0 ml) in 5 ml of 100 ppm ESP solution, the SPR peak shifts to the longer wavelength from 530 to 535 nm (Fig. 10.7). The color of ES + Au NPs solutions change from dark brown red

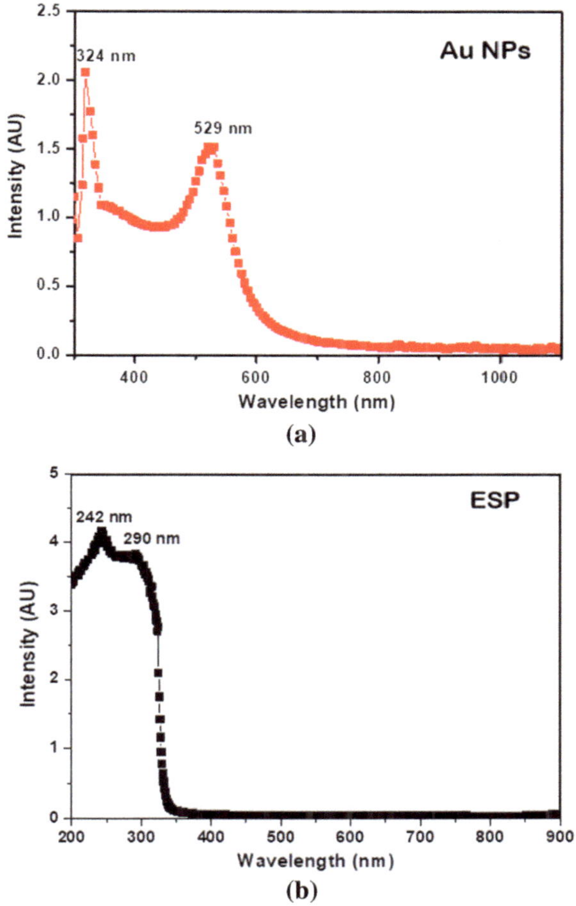

Fig. 10.6 (a) Absorption spectrum of Au NPs colloidal solution in 200–1100 nm range, (b) Absorption spectrum of ESP solution in 200–1100 nm range

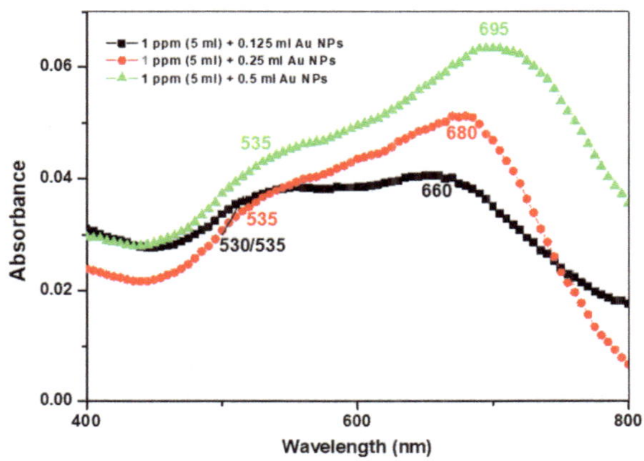

Fig. 10.7 Absorption spectrum of 1 ppm ESP solution with variable concentration of Au NPs colloidal solution in 400–800 nm range

Fig. 10.8 Absorption spectrum of 5 ml (1, 2, 5, 10, 20, 50, 100) ppm Endosulfan pesticide (ES):0.5 ml Au NPs (GNPs) in 300–1100 nm range

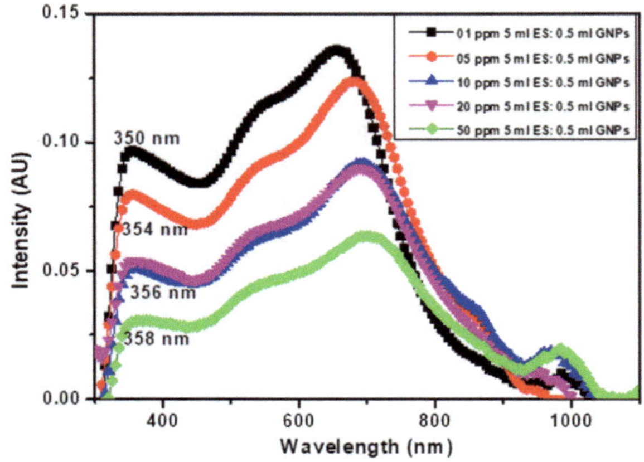

Fig. 10.9 Absorption spectrum of 100 ppm ESP samples with varying Au NPs concentrations in 200–1100 nm range

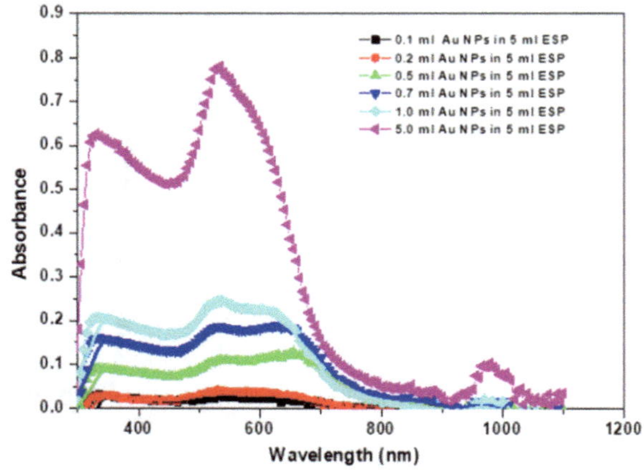

to light blue colored solution. At the same time, a second band due to the formation of Au NPs aggregates is observed in 660–695 nm region and shifts to lower wavelength with decreasing Au NPs concentration.

In 2nd case on increasing Au NPs concentration in 1 ppm ESP solution (Fig. 10.8), we observed two things: (i) the rise in intensity of the SPR signal of Au NPs; and (ii) the presence of a second band centered between 660 and 695 nm, due to the aggregated Au NPs formed during the adsorption/absorption of ESP on the surface of the Au NPs as described in proposed mechanism in forthcoming Sect. 6.4. It can be seen that for higher concentrations of endosulfan, the absorption of the aggregates shift to lower energies \with decrease in the optical absorption intensity which suggests the formation of bigger-sized structures as compared to the individual size of Au NPs.

In 3rd case (Fig. 10.9) on increasing ESP concentration from 1 to 100 ppm in 0.5 ml Au NPs solution, UV-Vis spectra recorded in 300–1000 nm region exhibits two broad overlapped strong absorptions in 300–500 nm, 500–900 nm and third weak absorption in 900–1100 nm region. The UV band in 300–500 nm region with maximum at 357 nm in 1 ppm 5 ml ESP + 0.5 ml Au NPs is attributed to $\pi-\pi^*$ transitions of ESP cyclohexadiene and benzodi-oxathiepin rings. Whereas, there is a very broad strong absorption having shoulder of gold surface plasmon resonance absorption in 532 to 540 nm wavelength region. Another overlapped strong peak maxima varies from 693 to 646 nm and weak peak with maxima at about 980 nm on increasing ESP concentration from 1 to 100 ppm 5 ml ESP in 0.5 ml GNP solution. These absorption arises from the optical absorption of the aggregates formed by the adsorption of ESP on Au NPs. The shift in peak position to low wavelength and change in intensity with increase in ESP concentration have been attributed to increase in aggregates size. This fact is also confirmed by TEM images that exhibit spherical Au NPs of few μm to mm sized agglomerations of different structures.

10.6.4 Proposed Mechanism for Degradation of ESP Using Au NPs

M. Kumar and L. Philip (Anandhakumar et al. 2017; Kumar and Philip 2006) and others (Guerin 1999; Sreekumaran Nair et al. 2003; Beaudette et al. 1998; Kullman and Matsumura 1996; Sutherland et al. 2002; Weir et al. 2006) discussed the mineralization of ES pesticide by microorganisms and possible degradation pathway. Kullman and Matsumura (1996), Sutherland et al. (2002), and Weir et al. (2006) reported the degradation of ES pesticide with white rot fungi and bacterial isolate, respectively. They found that ES degradation was initiated by the oxidation reaction to form toxic endosulfan sulfate, followed by hydrolysis reaction to form ES diol which easily oxidizes to less toxic endosulfan monoaldehyde or it releases H_2 and form endosulfan ether. Then these two products form endosulfan lactone which spontaneously degrades by two ways: either it forms 1, 2, 5, 6-hexachloro-5-chloromethyl-3-carbenium ion ($C_8H_{12}Cl_5$) or via hexachloro cyclohexadiene to diethylcarboxylates ($C_4H_8O_2$) which finally disintegrates into hydrocarbons (Kumar and Philip 2006). However, the complete mineralization of ES is not observed from these culture methods. Another method was proposed by Nair et al. (2003) for the mineralization of halocarbons by silver and gold nanoparticles. ES pesticide molecules first adsorb on the surface of Au NPs through sulfite oxygen followed by an exothermic catalytic reduction reaction.

This is a thermoneutral reaction for noble metals. However, in nano-sized Au NPs, the availability of a very large number of energetic atoms at surface overcome the thermochemical and entropic barriers needed for the mineralization reaction. Au NPs catalytically disintegrate ES pesticide at ambient temperature to form amorphous carbon and metal halides as reaction products in solution.

In the present work, different concentration Au NPs solutions have been used for the mineralization of different concentration (1 ppm to 100 ppm) ES solutions. The removal of ES by Au NPs is initiated by the adsorption of ES on Au NPs followed

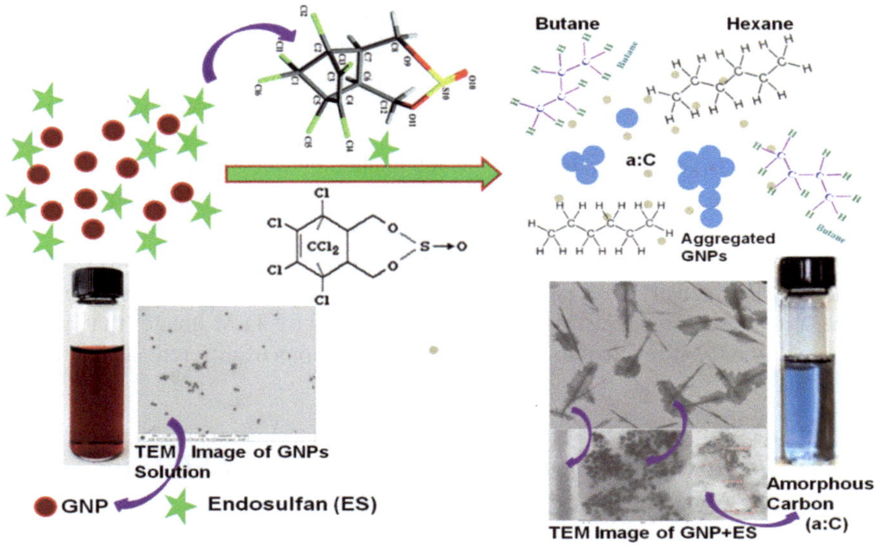

Fig. 10.10 Proposed schematic mechanism for ESP degradation by Au NPs

by photocatalytic decomposition of ESP within few minutes after mixing the two solutions in normal sunlight. Au NPs clusters are present as final reaction product in dispersed/precipitated form in the treated solution. TEM images have been used to explore the variations in shape and morphology of Au NPs by adding different concentration of ESP solution or vice versa. The systematic variations in solution color from purplish-pink to purplish-blue has been observed depending upon the ratio of Au NPs and ESP after 2–4 h of reaction. TEM image of one such sample containing 5 ml of 100 ppm ES and 5 ml Au NPs, which turned into a purplish-blue color after 3 h, is shown in Fig. 10.10.

TEM image shows the formation of petal-like sturcture comprising of a micron length chain-like structure with adsorbed agglomerated particles and ultra fine amorphous carbon surrounding these petals. A magnified view of micron-sized chain, agglomerates and amporhous compound are presented as zoomed-in-region of the TEM image. It is evident that nearly monodispersed spherical Au NPs have grown into large clusters and tend to align in the chain-like structure due to head to head interaction of hydrocarbon chains, as one of the final byproduct of this degradation reaction and Au NPs are trapped in the interstices of these chains and flower petal-shaped strcutures in solution when Au NPs citrate capping leaches out upon reaction with ESP. IR transmittance spectra exhibit the disappearance of vibrational peaks pertaining to carbonyl, $-SO_3$ and C–Cl groups which confirm the mineralization of ESP by Au NPs in 1120–1150 cm^{-1}, 1300–500 cm^{-1} and 800–550 cm^{-1} regions, respectively. These investigations showed that even the lowest concentration of Au NPs is suitable for removal of ESP from the solution. C–H stretching and H–C–H bending vibrations of methyl and methylene groups present in hydrocarbon and amorphous carbon are sensitive to chain conformation which involve partial ordering of chains emerging from the combination of gauche defects and asymmetric disordering of terminal $-CH_3$ groups. This encourages reorganization of Au NPs in solution as chain and petal or floral-shaped arrangement as well as change in solution coloration as described in TEM section and UV-Vis spectroscopy section. The methyl groups may also exist in gem- and iso- forms in hydrocarbon end products.

10.7 Future of Problem

In future, the aim will be to optimize Au NPs synthesis parameters for initiating development of low cost Au NPs impregnated membrane/bed which can be efficiently used as pre-treatment in water purifiers/purification for use in domestic washing, cleaning and agriculture fieldworks. The process can be scaled to pilot level to attract low budget customers/farmers to earn revenue for institute self-sustainment. These business aspects will also be taken care of while pursuing research and technology development for tackling the tedious and complex water purification problem, so that it has direct benefit to society.

The quantitative estimation of ESP present in prepared and final treated solutions if any left or its metabolites along with uncertainty in measurement as per ISO/IEC 17,025 standard guidelines is also very important. This can be achieved by procuring standard reference material of Au NPs from National Metrology Institute with traceability to generate a calibration curve for global acceptance of measured value.

10.8 Conclusion

This book chapter throws light on the usage of nano-engineered Au NPs for pesticide degradation/mineralization. The brief discussions are made on the water pollution, types and sources of water pollution, severity of problem, nanomaterials used in water purification, Au NPs synthesis by different physical and chemical routes, characterization techniques used for exploring their structural, morphological, optical and other properties, quality assurance, pesticides used. This chapter also presented work of Au NPs treated ESP solutions as case study along with proposed mechanism. These investigations provide evidence for the complete mineralization of ESP by Au NPs. TEM and FTIR techniques were used to understand ES degradation process into non-toxic reaction products. The conformation of the included guest molecule, the nature of the host–guest and guest–guest interactions (e.g., hydrogen bonding/hydrophobic groups) and the role of reorganization of Au NPs are observed. The adsorption of ES on Au NPs brings visible change in colloidal solution coloration from deep purplish pink to blue due to the Au NPs aggregates formation. IR transmittance spectra reveal the disintegration of endosulfan, formation of amorphous carbon and hydrocarbons. UV-Vis spectra of these series of samples indicate the change of Au NPs treated ESP solution from wine red to light blue color with shift in SPR and aggregated Au NPs peak position and intensity. These results are also confirmed by the TEM images. The proposed method has reported the easy way to eradicate highly persistent toxic halogenated pesticide from the environment and the application of Au NPs for the swift degradation of ESP molecules in a very short span of time. Then, chapter is finally concluded by considering future aspects of the problem in brief with suggestions.

References

Abedini A, Daud AR, Hamid MAA, Othman NK, Saion E (2013) A review on radiation-induced nucleation and growth of colloidal metallic nanoparticles. Nanoscale Res Lett 8:474

Adleman JR, Boyd DA, Goodwin DG, Psaltis D (2009) Heterogenous catalysis mediated by plasmon heating. Nano Lett 9:4417–4423

Alewu B, Nosiri C (2011) Pesticides and human health. In: Stoytcheva M (ed) Pesticides in the modern world-effects of pesticides exposure. InTech pp 231–250

Allec N, Choi M, Yesupriya N, Szychowski B, White MR, Kann MG et al (2015) Small-angle X-ray scattering method to characterize molecular interactions: proof of concept. Sci Rep 5:12085

Al-Zubaidi LA, Alnuaimi MTH, Aljanabi ZZ, Adel MM (2021) Biodegradation of chlorpyrifos pesticide using silver bionanoparticles Bacillus thuringiensis israelensis extracts, 5ISCESD 2021. IOP Conf. Series: Earth Environ Sci 779(13):012113

Amendola V, Pilot R, Frasconi M, Marago OM, Iati MA (2017) Surface plasmon resonance in gold nanoparticles: a review. J Phys Condens Matter 29:203002

Anandhakumar S, Dhanalakshmi K, Mathiyarasu J (2014) Non-enzymatic organophosphorus pesticide detection using gold atomic cluster modified electrode. Electrochemistry Commun 38:15–18

Andreescu S, Marty JL (2006) Twenty years research in Cholinesterase biosensors: from basic research to practical applications. Biomol Eng 23:1–15

Bakhtiari ABS, Hsiao D, Jin G, Gates BD, Branda NR (2009) An efficient method based on the photothermal effect for the release of molecules from metal nanoparticle surfaces. Angew Chem Int Ed 48:4166–4169

Balinova A (1996) Strategies for chromatographic analysis of pesticide residues in water. J Chromatogr A 754:125–135

Bassil KL, Vakil C, Sanborn M, Cole DC, Kaur JS, Kerr KJ (2007) Cancer health effects of pesticides. Systematic review. Can Fam Physician 53:1704–1711

Beaudette LA, Davies S, Fedorak PM, Ward OP, Pickard MA (1998) Comparison of gas chromatography and mineralization experiments for measuring loss of selected polychlorinated biphenyl congeners in cultures of white rot fungi. Appl Environ Microbiol 64:2020–2025

Bokare V, Jung JL, Chang Y-Y, Chang Y-S (2013) Reductive dechlorination of octachlorodibenzo-p-dioxin by nanosized zero-valent zinc: modeling of rate kinetics and congener profile. J Hazard Mater 250:397–402

Bootharaju MS, Pradeep T (2012) Understanding the degradation pathway of the pesticide, chlorpyrifos by noble metal nanoparticles. Langmuir 28:2671

Borrego B, Lorenzo G, Mota-Morales JD et al (2016) Potential application of silver nanoparticles to control the infectivity of Rift Valley fever virus in vitro and in vivo. Nanomedicine: nanotechnology. Biol Med 12(5):1185–1192

Bretveld RW, Thomas CMG, Scheepers PTJ, Zielhuis GA, Roeleveld N (2006) Pesticide exposure: the hormonal function of the female reproductive system disrupted? Reprod Biol Endocrinol 4:30

Brust M, Walker M, Bethell D, Schiffrin DJ, Whyman R (1994) Synthesis of thiol-derivatized gold nanoparticles in a two -phase liquid–liquid system. J Chem Soc Chem Commun 7:801–802
Bujoli B, Roussiere H, Montavon G et al (2006) Novel phosphate—phosphonate hybrid nanomaterials applied to biology. Prog Solid State Chem 34(2–4):257–266
Burnett M, Welford R (2007) Case study: Coca-cola and water in India: episode 2. Corp Soc Responsib Environ Mgmt 14:298–304
Cabras P, Angioni A (2000) Pesticide residues in grapes, wine, and their processing products. J Agric Food Chem 48:967–973
Castello GK (2009) Handbook of photocatalysts: preparation, structure and applications, Nova Science Pub.
Chen H, Kou X, Yang Z, Ni W, Wang J (2008a) Shape- and size dependent refractive index sensitivity of gold nanoparticles. Langmuir 24:5233–5237
Chen X, Zhu H-Y, Zhao J-C, Zheng Z-F, Gao X-P (2008b) Visible-light-driven oxidation of organic contaminants in air with gold nanoparticle catalysts on oxide supports. Angew Chem Int Ed 47:5353–5356
Chen J, Mela P, Moller P, Lensen MC (2009) Microcontact deprinting: a technique to pattern gold nanoparticles. ACS Nano 3:1451–1456
Chourasiya S, Khillare PS, Jyethi DS (2015) Health risk assessment of organochlorine pesticide exposure through dietary intake of vegetables grown in the peri urban sites of Delhi, India. Environ Sci Pollut Res Int 22:5793–5806
Chow MK, Zukoski CF (1994) Gold sol formation mechanisms: role of colloidal stability. J Colloid Interface Sci 165:97–109
Christopher P, Xin H, Linic S (2011) Visible-light-enhanced catalytic oxidation reactions on plasmonic silver nanostructures. Nat Chem 3:467–472
CWC (2010). Water and related statistics. Water planning and project wing, Central Water Commission, India
Darwich S, Mougin K, Rao A, Gnecco E, Jayaraman S, Haidara H (2011) Manipulation of gold colloidal nanoparticles with atomic force microscopy in dynamic mode: influence of particle – substrate chemistry and morphology, and of operating conditions. Beilstein J Nanotechnol 2:85–98
Du D, Chen S, Cai J, Zhang A (2008a) Electro chemical pesticide sensitivity test using Acetyl cholinesterase biosensor based on colloidal gold nanoparticle modified sol-gel interface. Talanta 74:766–772
Du D, Chen S, Song D, Li H, Chen X (2008b) Develop- ment of Acetylcholinesterase biosensor based on CdTe quantum dots/gold nanoparticles modified Chitosan mi- crospheres interface. Biosens Bioelectron24:475–479
Eskenazi B, Bradman A, Castorina R (1999) Exposures of children to organophosphate pesticides and their potential adverse health effects. Environ Health Perspect 107:409–419
Fasciani C, Bueno-Alejo CJ, Grenier M, Netto-Ferreira JC, Scaiano JC (2011) High-temperature organic reactions at room temperature using plasmon excitation: decomposition of dicumyl peroxide. Org Lett 13:204–207
Frens G (1973) Controlled nucleation for the regulation of the particle size in monodisperse gold suspensions. Nat: Phys Sci 241:20–22
Fu F, Dionysiou DD, Liu H (2014) The use of zero-valent iron for groundwater remediation and wastewater treatment: a review. J Hazard Mater 267:194–205
Gangapuram BR, Bandi R, Alle M, Dadigala R, Kotu GM, Guttena V (2018) Microwave assisted rapid green synthesis of gold nanoparticles using Annona squamosa L peel extract for the efficient catalytic reduction of organic pollutants. J Mol Struct 1167:305–315
Goel P, Arora M (2018) Fabrication of chemical sensor for organochlorine pesticide detection using colloidal gold nanoparticles. MRS Communications 8:1000–1007
Goel P, Arora M (2020a) Mechanism of visible-light-driven photocatalytic degradation of endosulfan pesticide by gold nanoparticles. Environ Res Commun 2(12):075004

Goel P, Arora M (2020b) Remediation of wastewater from chlorpyrifos pesticide by nano-gold photocatalyst. MRS Advances. https://doi.org/10.1557/adv.2020.264

Goel P, Arora M (2021) Photocatalytically driven mineralization of chlorpyrifos pesticide by copper nanoparticles. MRS Advances 6:774–779

Goulet PJG, Lennox RB (2010) New insights into Brust-Schiffrin metal nanoparticle synthesis. J Am Chem Soc 132:9582–9584

Goulson D (2014) Ecology: pesticides linked to bird declines. Nature 511:295–296

Grzelczak M, Pérez-Juste J, Mulvaney P, Liz-Marzán LM (2008) Shape control in gold nanoparticle synthesis. Chem Soc Rev 37:1783–1791

Guerin TF (1999) The anaerobic degradation of endosulfan by indigenous microorganisms from low-oxygen soils and sediments. Environ Pollut 106:13–21

Guerin TF, Kennedy IR (1992) Distribution and dissipation of endosulfan and related cyclodienes in sterile aqueous systems: implications for studies on biodegradation. J Agric Food Chem 40:2315–2323

Gunnell D, Eddleston M, Phillips MR, Konradsen F (2007) The global distribution of fatal pesticide self-poisoning: systematic review. BMC Public Health 7:357

Guo C, Yarger JL (2018) Characterizing gold nanoparticles by NMR spectroscopy. Magn Reson Chem 56(11):1074–1082

Hallett-Tapley GL, Silvero MJ, Gonza´lez-Be´jar M, Grenier M, Netto-Ferreira JC, Scaiano J C, (2011) Plasmon-mediated catalytic oxidation of sec-phenethyl and benzyl alcohols. J Phys Chem C 115:10784–10790

Harkness KM, Cliffel DE, McLean JA (2010) Characterization of thiolate -protected gold nanoparticles by mass spectrometry. Analyst 135(5):868–874

Hayes TB, Case P, Chui S, Chung D, Haeffele C, Haston K et al (2006) Pesticide mixtures, endocrine disruption, and amphibian declines: are we underestimating the impact? Environ Health Perspect 114:40–50. https://doi.org/10.1289/ehp.8051

Hoppin JA, Umbach DM, London SJ, Henneberger PK, Kullman GJ, Alavanja MCR et al (2008) Pesticides and atopic and nonatopic asthma among farm women in the agricultural health study. Am J Respir Crit Care Med 177:11–18. https://doi.org/10.1164/rccm.200706-821OC

Hostetler MJ, Templeton AC, Murray RW (1999) Dynamics of place-exchange reactions on monolayer-protected gold cluster molecules. Langmuir 15:3782–3789

Hostetler MJ, Wingate JE, Zhong CJ, Harris JE, Vachet RW, Clark MR, Londono JD, Green SJ, Stokes JJ, Wignall GD, Glish GL, Porter MD, Evans ND, Murray RW (1998) Alkanethiolate gold cluster molecules with core diameters from 1.5 to 5.2 nm: core and monolayer properties as a function of core size. Langmuir 14:17–30

Huang CJ, Chiu PH, Wang YH, Chen KL, Linn JJ, Yang CF (2006) Electrochemically controlling the size of gold nanoparticles. J Electrochem Soc 153:D193–D198

Huang Y-F, Zhang M, Zhao L-B, Feng J-M, Wu D-Y, Ren B, Tian Z-Q (2014) Activation of oxygen on gold and silver nanoparticles assisted by surface plasmon resonances. Angew Chem Int Ed 53:2353–2357

Iravani S (2011) Green synthesis of metal nanoparticles using plants. Green Chem 13:2638–2650

Jaeger JW, Carlson IH, Porter WP (1999) Endocrine, immune, and behavioral effects of aldicarb (carbamate), atrazine (triazine) and nitrate (fertilizer) mixtures at groundwater concentrations. Toxicol Ind Health 15:133–151

Jaffrezic-Renault N (2001) New trends in biosensors for organophosphorus pesticides. Sensors 1:60–74

Jans H, Liu X, Austin L, Maes G, Huo Q (2009) Dynamic light scattering as a powerful tool for gold nanoparticle bioconjugation and biomolecular binding studies. Anal Chem 81(22):9425–9432

Jeon S-H, Xu P, Zhang B, Mack NH, Tsai H, Chiang LY et al (2011) Polymer -assisted preparation of metal nanoparticles with controlled size and morphology. J Mater Chem 21:2550–2554

Jiang LP, Xu S, Zhu JM, Zhang JR, Zhu JJ, Chen HY (2004) Ultrasonic-assisted synthesis of monodisperse single-crystalline silver nanoplates and gold nanorings. Inorg Chem 43:5877–5883

Kalhapure RS, Sonawane SJ, Sikwal DR et al (2015) Solid lipid nanoparticles of clotrimazole silver complex: an efficient nano antibacterial against Staphylococcus aureus and MRSA. Colloids Surf B: Biointerfaces 136:651–658

Katayama A, Matsumura F (1993) Degradation of organochlorine pesticides, particularly endosulfan, by *Trichoderma harzianum*. Environ Toxicol Chem 12:1059–1065

Khanna P, Kaur A, Goyal D (2019) Algae-based metallic nanoparticles: synthesis, characterization and applications. J Microbiol Methods 163:105656

Khin MM, Nair AS, Babu VJ, Murugan R, Ramakrishna S (2012) A review on nanomaterials for environmental remediation. Energy Environ Sci 5(8):8075–8109

Kim Y-K, Kim S-H, Choi S-C (2001) Kinetics of endosulfan degradation by *Phanerochaete chrysosporium*. Biotechnol Lett 23:163–166

Kullman SW, Matsumura F (1996) Metabolic pathways utilized by Phanerochaete chrysosporium for degradation of the cyclodiene pesticide Endosulfan. Appl Environ Microbiol 62:593–600

Kumar M, Philip L (2006) Endosulfan mineralization by bacterial isolates and possible degradation pathway identification. Bioremediat J 10:179

Kusior A, Klich-Kafel J, Trenczek-Zajac A, Swierczek K, Radecka M, Zakrzewska K (2013) TiO_2–SnO_2 nanomaterials for gas sensing and photocatalysis. J Eur Ceram Soc 33(12):2285–2290

Lapresta -Fernández A, Salinas -Castillo A, Anderson De La Llana S, Costa -Fernández JM, Domínguez-Meister S, Cecchini R, et al (2014) A general perspective of the characterization and quantification of nanoparticles: imaging, spectroscopic, and separation techniques. Crit Rev Solid State 39(6):423–458

Leff DV, Ohara PC, Heath JR, Gelbart WM (1995) J Phys Chem 99:7036–7041

Li C, Hsieh JH, Hung M, Huang BQ, Song YL, Denayer J et al (2004) Ultrasonic spray pyrolysis for nanoparticles synthesis. J Mater Sci 9:3647–3657

Li X, Zhang S, Yu Z, Yang T (2014) Surface-enhanced Raman spectroscopic analysis of phorate and fenthion pesticide in apple skin using silver nanoparticles. Appl Spectroscopy 68:483–487

Liang X-J, Kumar A, Shi D, Cui D (2012) Nanostructures for medicine and pharmaceuticals. J Nanomater Article ID 921897, 2 pages

Lin Vien D, Colthup NB, Fateley WG, Grassellie JG (1991) Infrared and Raman characteristic frequencies of organic molecules. Academic Press, London

Lu D, Wang D, Ni R, Lin Y, Feng C, Xu Q et al (2015) Organochlorine pesticides and their metabolites in human breast milk from Shanghai, China. Environ Sci Pollut Res Int 22:9293–9306

Luque de Castro MD, Herrera MC (2003) Enzyme_ inhibition-based bio-sensors and biosensing systems: questionable analytical de-vices. Biosen Bioelectron 18:279–294

Lv XS, Xue XQ, Jiang GM et al (2014) Nanoscale zero-valent iron (nZVI) assembled on magnetic Fe_3O_4/graphene for chromium (VI) removal from aqueous solution. J Colloid Interface Sci 417:51–59

Martens R (1977) Degradation of [8,9–^{14}C] endosulfan by soil microorganisms. Appl En Viron Microbiol 31:853–858

Mills A, Le Hunte S (1997) An overview of semiconductor photocatalysis. J Photochem Photobiol A 108:1–35

Misra N, Biswal J, Gupta A, Sainis JK, Sabharwal S (2012) Gamma radiation induced synthesis of gold nanoparticles in aqueous polyvinyl pyrrolidone solution and its application for hydrogen peroxide estimation. Radiat Phys Chem 81:195–200

Mostafalou S, Abdollahi M (2013) Pesticides and human chronic diseases: evidences, mechanisms, and perspectives. Toxicol Appl Pharmacol 268:157–177

Mukherjee S, Zhou L, Goodman AM, Large N, Ayala-Orozco C, Zhang Y, Nordlander P, Halas NJ (2014) Hot-electron-induced dissociation of H_2 on gold nanoparticles supported on SiO_2. J Am Chem Soc 136:64–67

Muralikrishna T, Pattanayak M, Nayak P (2014) Green synthesis of gold nanoparticles using (Aloe vera) aqueous extract. World J Nano Sci Technol 3:45–51

Nag SK (2011) Raikwar MK (2011) Persistent organochlorine pesticides residues in animal feed. Environ Monit Assess 174:327–335

Nikoobakht B, El-Sayed MA (2003) Preparation and growth mechanism of gold nanorods (NRs) using seed -mediated growth method. Chem Mater 15:1957–1962

NSW EPA (2013) What are pesticides and how do they work? http://www.epa.nsw.gov.au/pesticides/pestwhatrhow.htm

Ojea-Jiménez I, Bastús NG, Puntes V (2011) Influence of the sequence of the reagents addition in the citrate -mediated synthesis of gold nanoparticles. J Phys Chem C 115:15752–15757

Okitsu K, Yue A, Tanabe S, Matsumoto H, Yobiko Y (2001) Formation of colloidal gold nanoparticles in an ultrasonic field: control of rate of gold (III) reduction and size of formed gold particles. Langmuir 17:7717–7720

Padilla M, Peccati F, Bourdelande JL, Monfort XS, Guirado G, Sodupe M, Hernando J (2017) Enhanced photocatalytic activity of gold nanoparticles driven by supramolecular host–guest chemistry. Chem Commun 53:2126–2129

Parmon V (2008) Nanomaterials in catalysis. Mater Res Innovations 12(2):60–61

Rajeshkumar S, Malarkodi C, Gnanajobitha G, Paulkumar K, Vanaja M, Kannan C et al (2013) Seaweed-mediated synthesis of gold nanoparticles using Turbinaria conoides and its characterization. J Nanostructure Chem 3(44):1–7

Ramalingam V (2019) Multifunctionality of gold nanoparticles: plausible and convincing properties. Adv Colloid Int Sci 271:101989

Rao A, Schoenenberger M, Gnecco E, Glatzel T, Meyer E, Brändlin D et al (2007) Characterization of nanoparticles using atomic force microscopy. J Phys Conf Ser 61:971

Ray TR, Lettiere B, De Rutte J, Pennathur S (2015) Quantitative characterization of the colloidal stability of metallic nanoparticles using UV-vis absorbance spectroscopy. Langmuir 31:3577–3586

Richardson MJ, Johnston JH, Borrmann T (2006) Monomeric and polymeric amines as dual reductants/stabilisers for the synthesis of gold nanocrystals: a mechanistic study. Eur J Inorg Chem 13:2618–2623

Rivero-Huguet M, Marshall WD (2009) Reduction of hexavalent chromium mediated by micron- and nano-scale zerovalent metallic particles. J Environ Monit 11(5):1072–1079

Roduner E (2006) Size matters: why nanomaterials are different. Chem Soc Rev 35:583–592

Sakai T, Alexandridis P (2005) Size -and shape -controlled synthesis of colloidal gold through auto reduction of the auric cation by poly (ethylene oxide)–poly (propylene oxide) block copolymers in aqueous solutions at ambient conditions. Nanotechnology 16(7):S344–353

Sanabria-cala JA, Conde-rodríguez GR, Gauthier GH, Ladeira LO, Laverde-Cataño SA, Pena Ballesteros DY, Merchan-Arenas D (2018) Gold nanoparticles formation mechanism by photochemical synthesis. Chem Eng Trans 64:403–408

Sancho JV, Pozo OJ, Ibanez M, Hernandez F (2006) Potential of liquid chromatography/time-of-flight mass spectrometry for the determination of pesticides and transformation products in water. Anal Bioanal Chem 386:987–997

Sardar R, Park J-W, Shumaker -Parry JS, (2007) Polymer -induced synthesis of stable gold and silver nanoparticles and subsequent ligand exchange in water. Langmuir 23:11883–11889

Sardar R, Shumaker-Parry JS (2009) 9-BBN induced synthesis of nearly monodisperse ω-functionalized alkythiol stabilized gold nanoparticles. Chem Mater 21:1167–1169

Sassolas A, Prieto-Simón B, Marty J-L (2012) Biosensors for pesticide detection: new trends. Am J Anal Chem 3:210–232

Sau TK, Murphy CJ (2004) Room temperature, high-yield synthesis of multiple shapes of gold nanoparticles in aqueous solution. J Am Chem Soc 126:8648–8649

Shetty PK, Mitra J, Murthy NBK, Namitha KK, Savita KN, Raghu K (2000) Biodegradation of cyclodiene insecticide endosulfan by Mucor thermo-hyalospora MTCC 1384. Curr Sci 79:1381–1383

Shukla AK, Iravani S (2017) Metallic nanoparticles: green synthesis and spectroscopic characterization. Environ Chem Lett 15:223–231

Silverstein RM, Webster FX, Kiemle DJ, Bryce DL (2014) Spectrometric identification of organic compounds, 8th edn. Wiley, Hoboken, NJ

Singh P (2016) SPR biosensors: historical perspectives and current challenges. Sens Actuators B Chem 229:110

Song YZ, Li X, Song Y, Cheng ZP, Zhong H, Xu JM et al (2013) Electrochemical synthesis of gold nanoparticles on the surface of multi-walled carbon nanotubes with glassy carbon electrode and their application. Russ J Phys Chem A 87:74–79

Sreekumaran Nair A, Pradeep T (2003) Halocarbon mineralization and catalytic destruction by metal nanoparticles. Curr Sci 84:1560–1564

Sreekumaran Nair A, Tom RT, Pradeep T (2003) Detection and extraction of endosulfan by metal nanoparticles. J Environ Monit 5:363–365

Stefaniuk M, Oleszczuk P, Ok YS (2016) Review on nano zerovalent iron (nZVI): from synthesis to environmental applications. Chem Eng J 287:618–632

Sujitha MV, Kannan S (2013) Green synthesis of gold nanoparticles using Citrus fruits (Citrus limon, Citrus reticulata and Citrus sinensis) aqueous extract and its characterization. Spectrochim Acta A Mol Biomol Spectrosc 102:15–23

Sutherland T, Horne I, Lacey M, Harcourt R, Russell R, Oakeshott J (2002) Isolation and characterization of a Mycobacterium strain that metabolises the insecticide endosulfan. J Appl Microbiol 93:380–389

Tao AR, Habas S, Yang PD (2008) Shape control of colloidal metal nanocrystals. Small 4:310–325

Templeton AC, Hostetler MJ, Kraft CT, Murray RW (1998) Reactivity of monolayer-protected gold cluster molecules: steric effects. J Am Chem Soc 120:1906–1911

Thakur P, Sharma P, Mattei J-L, Queffelec P, Trukhanov AV, Trukhanov SV, Thakur PLV, A, (2018) Influence of cobalt substitution on structural, optical, electrical and magnetic properties of nanosized lithium ferrite. J Mater Sci 29(19):16507

Trapiella -Alfonso L, Costa -Fernández JM, Encinar JR, Pereiro R, Sanz -Medel A (2014) Chapter 8—Mass spectrometry for the characterization of gold nanoparticles. In: Valcárcel M, López -Lorente ÁI (eds) Comprehensive analytical chemistry, vol 66. Elsevier, pp 329–356

Tratnyek PG, Salter AJ, Nurmi JT, Sarathy V (2010) Environmental applications of zerovalent metals: iron vs. zinc In Nanoscale materials in chemistry: environmental applications, vol 1045 of ACS Symposium Series, chapter 9, pp 165–178

Turkevich J, Stevenson PC, Hillier J (1951) A study of the nucleation and growth processes in the synthesis of colloidal gold. Discuss Faraday Soc 11:55–75

Unser S, Bruzas I, He J, Sagle L (2015) Localized surface plasmon resonance biosensing: current challenges and approaches. Sensors 15:15684

van der Hoff GR, van Zoonen P (1999) Trace analysis of pesticides by gas chromatography. J Chromatogr A 843:301–322

Vasanthi Bathrinarayanan P, Thangavelu D, Muthukumarasamy VK, Munusamy C, Gurunathan B (2013) Biological synthesis and characterization of intracellular gold nanoparticles using biomass of Aspergillus fumigatus. Bull Mater Sci 36(7):1201–1205

Wang L, Wei G, Guo C, Sun L, Sun Y, Song Y, Yang T, Li Z (2008) Photochemical synthesis and self-assembly of gold nanoparticles. Colloids Surf A 312:148–153

Wang XY, Zhu MP, Liu HL, Ma J, Li F (2013) Modification of Pd–Fe nanoparticles for catalytic dechlorination of 2,4- dichlorophenol. Sci Total Environ 449:157–167

Wee T-L, Schmidt LC, Scaiano JC (2012) Photooxidation of 9-anthraldehyde catalyzed by gold nanoparticles: solution and single nanoparticle studies using fluorescence lifetime imaging. J Phys Chem C 116:24373–24379

Wei W, Du Y, Zhang L, Yang Y, Gao Y (2018) Improving SERS hot spots for on-site pesticide detection by combining silver nanoparticles with nanowires. J Mater Chem C 6:8793–8803

Weir KM, Sutherland TD, Horne I, Russel RJ, Oakeshott JG (2006) A single Monooxygenase, Ese, is involved in the metabolism of the Organochlorides Endosulfan and Endosulfate in an *Arthrobacter* sp. Appl Environ Microbiol 72:3524–3530

World Health Organization (1990) Public health impact of pesticides used in agriculture. World Health Organization

Xiao T, Huang J, Wang D, Meng T, Yang X (2019) Au and Au based nanomaterials: synthesis and recent progress in electrochemical sensor applications. Talanta 206:120210

Xiao Q, Liu Z, Bo A, Zavahir S, Sarina S, Bottle S, Riches JD, Zhu H (2015) Catalytic transformation of aliphatic alcohols to corresponding esters in O_2 under neutral conditions using visible-light irradiation. J Am Chem Soc 137:1956–1966

Xiong D, Li H (2008) Colorimetric detection of pesticides based on calixarene modified silver nanoparticles in water. Nanotechnology 19(46):465502

Yan J, Han L, Gao W, Xue S, Chen M (2015) Biochar supported nanoscale zerovalent iron composite used as persulfate activator for removing trichloroethylene. Bioresour Technol 175:269–274

Ye X, Zheng C, Chen J, Gao Y, Murray CB (2013) Using binary surfactant mixtures to simultaneously improve the dimensional tunability and monodispersity in the seeded growth of gold nanorods. Nano Lett 13:765–771

Yee CK, Jordan R, Ulman A, White H, King A, Rafailovich M, Sokolov J (1999) Novel one-phase synthesis of thiol-functionalized gold, palladium, and iridium nanoparticles using superhydride. Langmuir 15:3486–3491

Yun SH, Sohn BH, Jumg JC, Zin WC, Ree M, Park JW (2006) Micropatterning of a single layer of nanoparticles by lithographical methods with di-block copolymer micelles. Nanotechnology 17:450–454

Zhao P, Li N, Astruc D (2013) State of the art in gold nanoparticle synthesis. Coord Chem Rev 257:638–665

Zheng Z, Tachikawa T, Majima T (2014) Single-particle study of pt-modified Au Nanorods for plasmon-enhanced hydrogen generation in visible to near-Infrared region. J Am Chem Soc 136:6870–6873

Zhu H, Ke X, Yang X, Sarina S, Liu H (2010) Reduction of nitroaromatic compounds on supported gold nanoparticles by visible and ultraviolet light. Angew Chem Int Ed 49:9657–9661

Ziegler C, Eychm ller A, (2011) Seeded growth synthesis of uniform gold nanoparticles with diameters of 15–300 nm. J Phys Chem C 115:4502–4506

Zuber A, Purdey M, Schartner E, Forbes C, van der Hoek B, Giles D et al (2016) Detection of gold nanoparticles with different sizes using absorption and fluorescence based method. Sens Actuators B Chem 227:117–127

Chapter 11
Nanomaterials for Remediation of Pharmaceutical Containing Aqueous Medium and Wastewaters

Rozi Sharma, Pradeep Kumar Sharma, Anoop Singh, Dheeraj Rathore, and Piyush Malaviya

Abstract Disposal of untreated pharmaceutical waste into water bodies is a growing threat to the aquatic ecosystem and humans as well. Among various physicochemical, advanced oxidation, and bioremediation processes adopted, bioremediation emerges as the most environmentally friendly and economically viable alternative to manage this serious environmental concern. The main focus of the chapter is on nanobiotechnology that utilizes nanoscale materials for removing toxic compounds from wastewater. The nanomaterials have a size even smaller than one billionth of a meter and possess unique properties including high absorption potential, huge surface area, eco-friendly fabrication, and strong affinity toward organic and inorganic compounds. The use of novel nanomaterials for the sustainable treatment of pharmaceutical wastewater is a favorite option for environmental engineers these days. Different nanomaterials used for the treatment of pharmaceutical containing aqueous medium and wastewaters like electrospun lignin nanofibers, polyaniline/ZrO_2 nanocomposite, Cu-TiO_2 single-walled carbon nanotubes, curcumin-loaded nanostructured lipid carrier, Fe_3O_4/Au nanocomposite, Fe_3O_4-chitosan nano-adsorbent, TiO_2/ONLH nanocomposites, ZnO nanoparticles, ZnO/sepiolite heterostructured nanomaterials, Ni-Al layered double hydroxide nanoparticles, tungsten/carbon nanotube heterojunction nanocomposite, organic-inorganic hybrid TiO_2 doped with molybdenum, MnOx/ nano-graphite/2-ethylanthraquinone/activated carbon cathode, graphene-based nanomaterials, and multi-wall carbon nanotubes/titanium oxide

R. Sharma · P. Malaviya (✉)
Department of Environmental Sciences, University of Jammu, Jammu, J&K, India
e-mail: piyushmalaviya@jammuuniversity.ac.in; piyushmalaviya@rediffmail.com

P. K. Sharma
Department of Environmental Science, Graphic Era Deemed to be University, Dehradun, Uttarakhand, India

A. Singh
Department of Scientific and Industrial Research (DSIR), Ministry of Science and Technology, Government of India, New Delhi, India

D. Rathore
School of Environment and Sustainable Development, Central University of Gujarat, Gandhinagar, Gujarat, India

J. P. N. Rai and S. Saraswat (eds.), *Nano-biotechnology for Waste Water Treatment*, Water Science and Technology Library 111,
https://doi.org/10.1007/978-3-031-00812-2_11

(MWCNT/TiO_2) nanocomposite have been discussed in the literature. Classification of the nanomaterials has been also highlighted. The chapter attempts to provide all the useful information on novel nanotechnology employed for pharmaceutical wastewater management.

11.1 Introduction

The industrialization has long been accepted as a characteristic of civilization, as an industry contributes appreciably to the national attempt for self-reliance. However, the rapid development in the industrial sector leads to drastic changes in the environmental equilibrium and drastically affects precious natural resources (Malaviya et al. 2012; Parihar and Malaviya 2013; Muthusaravanan et al. 2018). Pollution load from industry depends on the methods used during the manufacturing of products and the size of the industry. The existing pattern of industrial activity changes the natural flow of materials and introduces deadly chemicals into the environment (Sharma and Malaviya 2022; Sharma et al. 2022a). Water pollution is increasing day by day as industries are releasing a huge amount of potentially poisonous chemical compounds into the hydrosphere, and this is the biggest curse to the human race (Sharma and Malaviya 2013; Gowri et al. 2014). Waterborne infections, resulted because of rapid industrialization, claim the lives of 3.4 million people worldwide each year (Rana et al. 2017).

Pharmaceutical industries produce several drugs that are important for improving life expectancy and quality over a period of time. A variety of pharmaceutical drugs are used for treating many diseases in humans as well as animals (Sofowora et al. 2013). However, the wastewater generated during drug and formulation development processes produces menacing consequences on ecological and human health if disposed of untreated into the environment. Pharmaceutical wastewater has severe color, pungent smell, high chemical oxygen demand (COD), and low biochemical oxygen demand (BOD) (Farhadi et al. 2012). The wastewater contains substances like buffers, soluble organics, suspended solids, chelators, antibiotics, heavy metals (lead, mercury, fluoride, arsenic, cadmium, chromium). 15–20% of the medicinal waste of the hospitals (secondary source of pharmaceutical waste) is bio-accumulative and non-degradable, and due to their improper use and disposal practices, they become persistent in the environment (Wilkinson et al. 2017). The hazardous pharmaceutical contaminants alter the properties of water where they are discharged and pose a serious threat to human and ecosystem health. Endocrine disruption, microflora imbalance, chronic diseases, allergies, metabolic perturbations, tumors are some common health ailments (Rasheed et al. 2020; Cincinelli et al. 2015). Damaged fish spawning, reduced fertility is among disorders of aquatic fauna (Yaqoob et al. 2020). The elimination of pharmaceutical toxicants is, therefore, important to conserve aquatic resources, minimize adverse effects on human health, and prevent environmental damage.

Different physicochemical and biological methods can be used to treat pharmaceutical wastewater. Ion exchange, precipitation, solvent extraction, adsorption, electrochemical processes, and liquid membrane are traditional methods of wastewater treatment (Zaied et al. 2020). Though the traditional physicochemical treatment methods are highly efficient, their high operation and installation costs, and incapability to degrade complex compounds divert the attention of environmentalists toward the development of newer, more powerful, sustainable, and cost-effective technologies for wastewater treatment and conservation of freshwater resources (Sharma and Malaviya 2021; Singh et al. 2016). Biological methods are found better for pharmaceutical industry wastewater treatment over conventional techniques despite being less effective in eliminating stubborn carbon-based pollutants (Farhadi et al. 2012). However, neither physicochemical nor biological treatment methods are suggested as the most suitable procedures for high-strength and complicated pharmaceutical contaminants (Crouse et al. 2012).

Nanotechnology has emerged as a recent alternative for wastewater treatment that employs new devices, structures, and materials that possess at least one nanoscale dimension ranging between 1–100 nm, large surface area, high reactivity and porosity, a large number of active sites, high catalytic and redox potential, strong solution mobility and adsorption ability, and photocatalytic properties. All these features of nanomaterials make nanotechnology an astounding option for wastewater treatment with higher efficiency, lesser cost, and in a more ecologically acceptable manner (Karri et al. 2019). In recent times, researchers mostly use this technology owing to its efficient and multifunctional process, insignificant and inexpensive infrastructures, and more satisfying results during pharmaceutical wastewater treatment compared to other existing physicochemical and biological technologies (Qu et al. 2012).

In this chapter, we have discussed the composition of pharmaceutical wastewater, its generation, and resulting environmental hazards. Small discussion is made on different treatment techniques adopted for pharmaceutical wastewater treatment including physicochemical methods, advanced oxidation processes, bioremediation. However, the chapter is mainly oriented around nanotechnology that utilizes nanoscale materials for removing toxic compounds from wastewater. The most recently studied nanomaterials for pharmaceutical wastewater treatment have been coalesced to serve the readers with the most attractive and the most efficient tools for wastewater management.

11.2 Pharmaceutical Wastewater as an Environmental Concern: Composition, Generation and Hazards

An increase in fatal diseases and advances in medical technology have resulted in the production of a variety of pharmaceuticals and the generation of wastewater during different stages of drug formulation. Wastewater generated and released

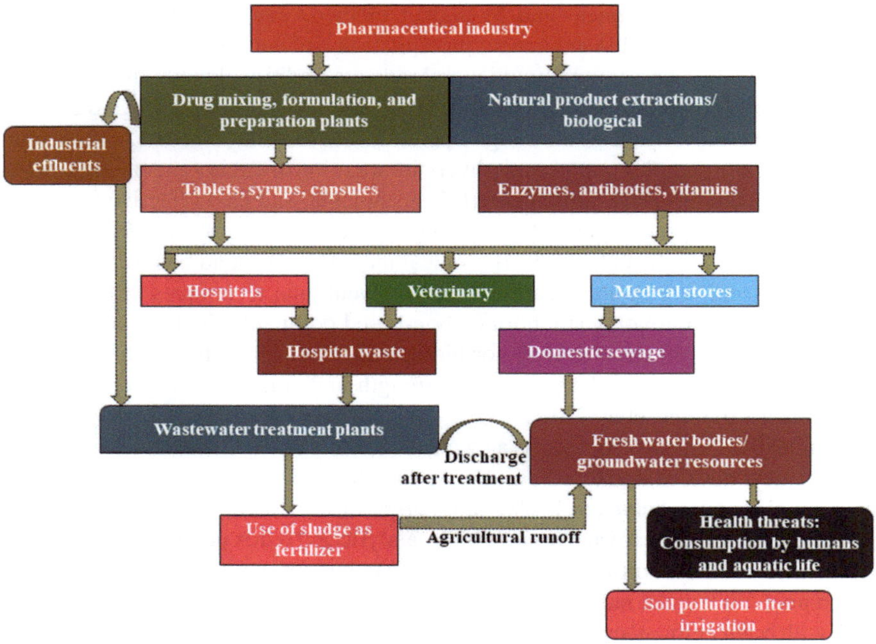

Fig. 11.1 Overview of the pharmaceutical industry, wastewater generated, and ultimate hazards on mankind and the environment

from pharmaceutical industries is associated with the presence of different persistent compounds, heavy metals, personal care products, etc. The source of pharmaceuticals may either be industries or sewage and agricultural runoffs. Addition of the untreated wastewater into freshwater resources has a considerable impact on aquatic-ecosystem as well as human health. Figure 11.1 represents different processes involved in the pharmaceutical industry, wastewater generated, and resulting hazards on the environment.

11.2.1 Composition

Pharmaceutical wastewater includes a diverse group of pharmaceutical chemicals and their metabolites together with transformation products. Antibiotics (penicillins, tetracyclines, lincomycins, quinolones, sulfonamides), anti-inflammatories (aspirin, ibuprofen, diclofenac, mefenamic acid, naproxen), antidepressants (bupropion, duloxetine, fluoxetine, paroxetine), hormones (estradiol, fulvestrant, mestranol, insulin, follitropin, oxytocin, liothyronine, prednisone), anticonvulsants (phenytoin, carbamazepine, ethosuximide, perampanel, vigabatrin), beta-blockers (propranolol, atenolol, betaxolol, nadolol, nebivolol, esmolol hydrochloride), lipid regulators (colestipol, fenofibric acid, niacin, rosuvastatin, probucol, ezetimibe, cerivastatin), and X-ray contrast are the most prevalent pollutants in pharmaceutical wastewater (Rasheed et al. 2020).

11.2.2 Wastewater Generation

The compounds in pharmaceutical industries are formed in the batch process, therefore, a large volume of different toxic products are added to wastewaters during different operations. Moreover, a large amount of water is being used for washing of solid cake, or extraction, or cleaning of equipment (Dixit and Parmar 2013). There are two main sources of pharmaceutical compounds in potable water: production processes of the pharmaceutical compounds, and usage of these compounds that make them available in urban and farm wastewaters. The main source of pharmaceutical wastewater is during the synthesis and formulation of the drug. Pharmaceutical wastewaters are determined by the raw materials, type of equipment used, manufacturing process, preparation season, and also the time of generation (Benotti and Brownawell 2009). Depending upon their source, pharmaceutical wastewater can be classified into the following types: waste generated by the pharmaceutical firms, and secondary wastes produced from treatment and recycling plants, spent liquors from the fermentation process, medical waste from hospitals, floor, and laboratory washing waste, and the municipal wastewater which contaminates the sewage systems significantly (Pal 2018).

11.2.3 Hazards of Pharmaceutical Wastewater

Pharmaceutical wastewater contains not only pharmaceutical components but also intermediates, solvents, catalysts, and many other raw materials used during the manufacturing of medications. These pharmaceutical compounds are unfavorable for catchment water bodies where they are eventually disposed in (Zaman et al. 2014). Since pharmaceutical compounds are lipophilic and persist in the environment for a longer duration even at low concentrations, therefore, they produce chronic effects (Wilkinson et al. 2016). Pharmaceutical pollutants' toxicity varies based on the species exposed, the duration of exposure, the compound concentration, and the developmental window during which organisms are exposed (Gerbersdorf et al. 2015). Growth, mortality, and reproduction had been the main endpoints for toxicity testing. However, attention has also been given in recent times to potential effects at levels of development, molecular biology, histology, and behavior (Godoy and Kummrow 2017). Pharmaceutical contaminants have been investigated in fish tissues such as gills, muscles, blood plasma, liver, and brain. Antibiotics have been observed in algae, bivalves, and aquatic invertebrates. Antibiotic resistance is a serious hazard to human health, and it has received special consideration in recent decades. At a concentration of 5 μg/L, diclofenac produced gill abnormalities and kidney lesions in rainbow trout (Schwaiger et al. 2004), and delay in Zebrafish hatching at concentration 1000–2000 μg/L (Hallare et al. 2004). Oxidative DNA damage and increased gill histopathological index on exposure to erythromycin in rainbow trout and gilthead seabream has been observed by Rodrigues et al. (2016, 2019a, b). When *Daphnia magna* was exposed to propranolol, its heart rate, reproductive tissues, and hepatic tissues were all significantly altered (Jeong et al. 2018). Environmental concentrations (0.2–123 μg/l) of cyclophosphamide and 5-fluorouracil inhibited the growth in *Pseudokirchneriella subcapitata* and *Synechococcus leopoliensis*, influenced the eyesight, and produced mutagenic consequences such as the formation of melanocytes in the gastrointestinal tract of *Lithobates catesbeianus* (Brezovsek et al. 2014). Ibuprofen has been reported to produce a reproductive disorder called compensated hypogonadism in human males (Kristensen et al. 2018). Further investigation is required to determine the effects on human health when they are exposed to such pharmaceutical toxins. After increased awareness about their potential hazards to aquatic organisms and humans, pharmaceutical contaminants have become an important environmental concern. Therefore, this industry wastewater has gotten extensive attention for efficient treatment to reduce potential health hazards, environmental risks, and increase wastewater reuse for healthy aquatic ecosystems (Kanakaraju et al. 2018).

11.3 Treatment Methods for Pharmaceutical Wastewater

Accounting for their health and ecological threats, pharmaceutical wastewater must be treated before its discharge into the environment. Adequate treatment not only minimizes the environmental issues but the treated wastewater can also be applied for irrigational purposes. However, long-term studies are required before large-scale utilization of the treated wastewater in agricultural fields (Malaviya and Sharma 2011). Commonly used treatment methods for pharmaceutical wastewater include physicochemical processes, biological processes (aerobic and anaerobic processes), bioremediation; bioremediation appears to be the most sustainable and economically viable option, among others. Aerobic or anaerobic biological treatment methods greatly inhibit the activity of the microbes due to the presence of stubborn components in the wastewater (Gebhardt and Schroder 2007).

11.3.1 Physicochemical Methods

The physicochemical methods are utilized as primary treatment processes to reduce the waste stream's organic load and followed by biological processes that are used as secondary treatment processes. Commonly used physicochemical methods are electrocoagulation, ozone treatment, activated carbon adsorption. Advanced oxidation processes generate in situ reactive oxygen species like hydroxyl species, hydrogen peroxide, ozone, superoxide radical species that successfully mineralize the substances into CO_2, H_2O, and inorganic ions or acids (Kanakaraju et al. 2018). Fenton oxidation process can also be used for the treatment of pharmaceutical wastewater (Aljuboury et al. 2017), and remove up to 90% of COD from wastewater within just the first ten minutes. Fenton, Photo-Fenton, and Solar-Photo-Fenton, sonolysis techniques have been found effective in the removal of antibiotics (like ciprofloxacin, nalidixic, ofloxacin acid). Antibiotics (especially sulphonamides, certain analgesics, and antiarrhythmic agents) can also be treated with UV-peroxide methods and radiation techniques which are found highly efficient in the elimination of carbamazepine and fluoxetine from pharmaceutical wastewater (Garcia-Espinoza et al. 2018). Though physicochemical methods contribute toward pollutant removal from pharmaceutical waste in most cases, the effective clearance of pharmaceutical medications was determined to be insufficient with these methods. Another limitation of these methods is the high cost of operation and maintenance, and the generation of harmful by-products left at the end of the chemical processes (Amin et al. 2013; Sharma et al. 2021). Hence, for better treatment results, biological or bioremediation strategies have been sought.

11.3.2 Bioremediation

Bioremediation is the removal of contaminants from influent wastewater with the help of living entities and is dependent on the factors including stereochemistry of the drug present, its toxicity, its concentration; the efficiency of the microbe or plant utilized; the conditions present during the degradation process; retention duration; the presence of additional compounds and their concentrations (Misal et al. 2011; Singh and Malaviya 2019; Singh et al. 2021; Sharma et al. 2021, 2022b). Commonly used bioremediation techniques are bacterial bioremediation, phytoremediation, mycoremediation, cyanobacterial bioremediation, phycoremediation, and aerobic and anaerobic methods of bioremediation. Phytoremediation is an emerging environment-friendly and cost-effective technology used for remediating polluted ecosystems and has long-term applicability in the transformation or stabilization of a variety of contaminants from different environmental matrices (Saha et al. 2017; Malaviya et al. 2020; Sharma et al. 2021). Much research has been done on the remediation of pharmaceutical compounds (like diclofenac, ibuprofen, acetaminophen, paracetamol, clofibric acid) from the medium by using aquatic plants like *Armoracia rusticana*, *Linum usitatissimum*, *Typha* species, and *Phragmities australis* (Kummerova et al. 2016).

11.3.3 Nanotechnology for Pharmaceutical Wastewater Treatment

Nanotechnology deals with the imaging, modeling, designing, characterization, and production of materials by controlled manipulation of size and shape at the nanometer scale, and use of these novel nanostructures with superior characteristics in different spheres; including medicine, commercial, engineering, environment (Bawa et al. 2005). The technology is an innovative method for treating wastewater contaminated with metal ions, pharmaceutical medications, organic and inorganic solutes, microbes, and other contaminants with help of nanoparticles that attach to the toxicants and cause their inactivation and destruction. Nanoparticles are materials with sizes even smaller than one billionth of a meter. These nanoscale materials are categorized by their physical, chemical, and biological properties. Nanomaterials are sorbents with unique qualities such as high absorption, a greater surface area, environmentally benign production, and a high affinity for organic and inorganic chemicals. Metal oxides, polymers, zeolites, bio-based, and carbon derivatives have a higher surface area, and other better electrical, mechanical, magnetic, and optical properties at the nanoscale level as compared to the macroscopic scale level (Lingamdinne et al. 2018). This feature may be attributed to a large number of molecules tightly packed together leading to a decrease in the size of the particles. Figure 11.2 represents important characteristics of nanomaterials, their broad classification, applications of nanotechnology, and their advantages and disadvantages. The technology has been gaining researchers' attention worldwide to treat pharmaceutical pollutants like antibiotics, pesticides, hormones, anti-viral drugs, toxic dyes in a more effective, useful, and sustainable manner. Nanotechnology is a simple, effective, and simple-to-use tool for removing toxins from pharmaceutical wastewater. It can identify and screen new drugs quickly and easily, helps in improved understanding of chemicals at the molecular scale, and improves visualization of drugs interactions. It is used in the filtration, remediation, and purification of water. Water filtration using semi-permeable nanoporous membranes (with a thickness of 0.001 to 0.01 μm) was carried out for the removal of undesirable organic molecules and micropollutants (Cincinelli et al. 2015). In water remediation, nanomaterials eliminate chemical pollutants by the phenomenon of adsorption (Aguilar-Perez et al. 2020). Water purification removes pathogens using nanomaterials with antibacterial activity (like graphene, metallic and bio-based composites) (Mohmood et al. 2013). In the following sections, different kinds of nanomaterials, and important nanomaterials used so far for pharmaceutical wastewater treatment have been discussed.

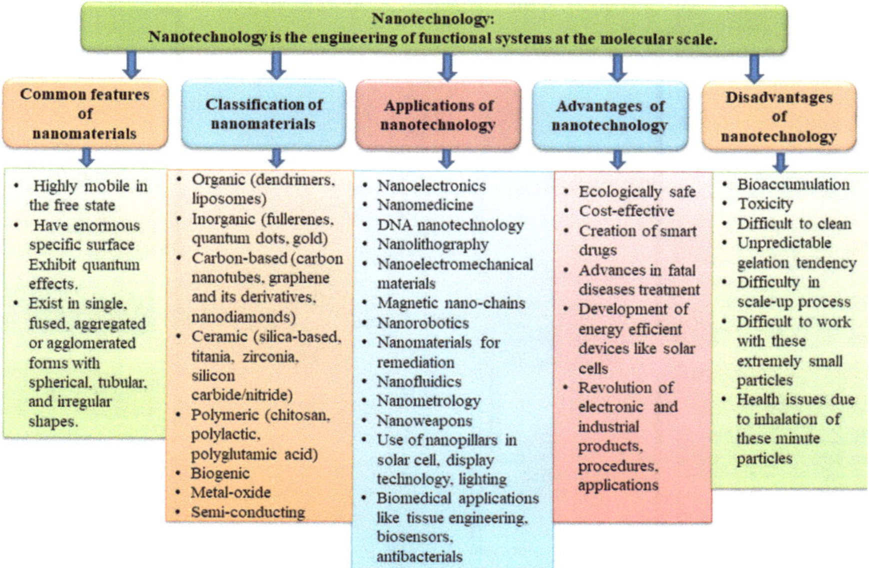

Fig. 11.2 Representation of important applications, advantages and disadvantages of nanotechnology, characteristics of nanomaterials, and their broad classification

11.4 Nanomaterials for Treatment of Pharmaceutical Effluents

Nanomaterials are small-scale particles with a dimension measuring 1–100 nm. They have unique physical, optical, mechanical, thermal, and electronic properties. Owing to their characteristic features, they can be used for various purposes including wastewater treatment, healthcare, industrial sector. They can occur naturally or can be created by different physical, chemical, and biological methods. In Fig. 11.3, methods by which nanomaterials are developed, and mechanisms adopted by them to eliminate toxicants from polluted matrices are represented. Nanomaterials are broadly classified into organic, inorganic, carbon, ceramic, metal oxide, and biogenic-based kinds. The use of nanotechnology for wastewater management has gained a lot of attention and popularity in recent times. In the following subsection, important categories of nanomaterials, and innovative nanomaterials explored so far for their potential to treat pharmaceutical wastewater are discussed.

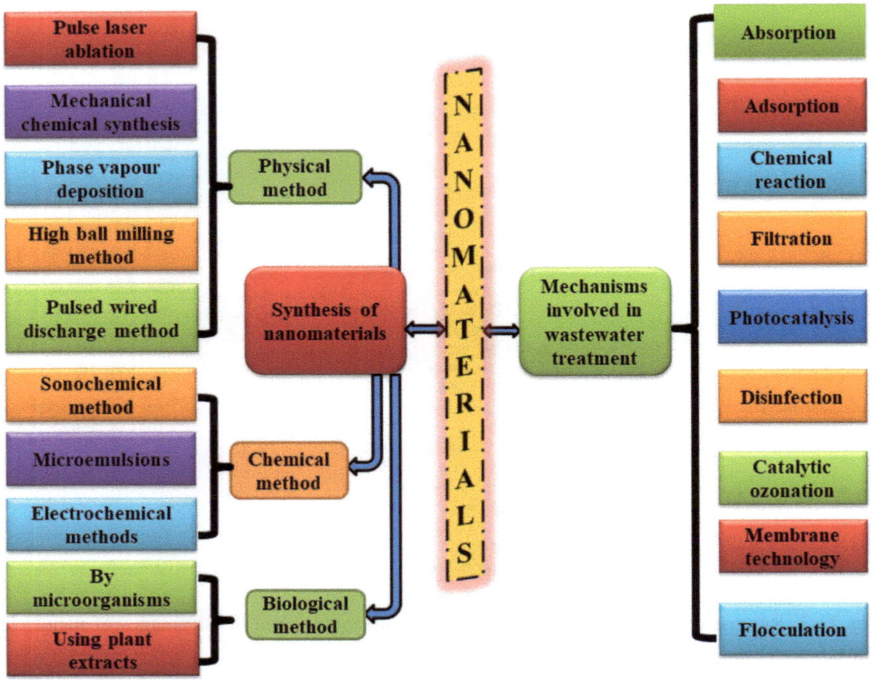

Fig. 11.3 Methods of synthesis of nanomaterials and important mechanisms involved during treatment

11.4.1 Classification of Nanomaterials for Wastewater Treatment

11.4.1.1 Carbon-Based Nanosorbents

Carbon-derived nanomaterials that are made up of carbon atoms have high mechanical strength, large surface area and volume, strong electron affinity, flexibility during functionalization, and many useful properties that remove pollutants from wastewater in a more efficient way (Ali et al. 2019a, b; Adeleye et al. 2016). However, they are not cost-efficient, have exceptional specificity and environmental toxicity. Single-walled and double-walled carbon nanotubes, carbon nanofibers, graphene, and carbon composites are commonly used carbon-based nanomaterials. Carbon nanotubes are cylindrical-shaped (rolled graphene sheets) carbon allotrope that is composed of hexagonal lattice in sp^2 hybridization (Mohmood et al. 2013). These rolled nanotubes may contain single or multiple layers of carbon atoms. Graphitic nanomaterials consist of sp^2 hybridized carbon atoms placed in a honeycomb lattice in a two-dimensional single layer (Thines et al. 2017). Graphene-based nanomaterials possess a larger surface area with more adsorption potential and lesser production costs and remove the pollutants with higher efficiency (Santhosh et al. 2016).

11.4.1.2 Ceramic-Based Nano-Adsorbents

This category mainly includes silica-based and chitosan-derived nanomaterials. Silica-based nano adsorbents are in the size range of 50 nm–1 μm and are made with silicon alkoxides under the catalytic action of ammonia. Nano-sized silica is highly stable, biocompatible, non-toxic, highly selective, and possesses high surface area and adsorption potential. Silica particles with the size range of 2–50 nm are mesoporous silica-based nanomaterials, and particles with sizes lesser than this range are termed nanoporous nanoparticles (Thirunavukkarasu et al. 2020). These nanoparticles have been found highly effective in removing pollutants from the waste stream. Chitosan, a polymer of de-acetylated-b-glucosamine prepared by de-acetylation of chitin is a good adsorbent and removes the colloidal pollutants by coagulation or flocculation method. This chitosan-based nanomaterial is biocompatible and easy to develop, thus a promising option for pollutant removal (Thirunavukkarasu et al. 2020).

11.4.1.3 Metal Oxide-Based Nano-Adsorbents

Metal-based nanoparticles have a higher surface area, high adsorption potential (though poor in continuous column studies), charged surface, are easy and cheap to synthesize, are capable of mediating electrostatic interactions and solute transfer process, and generate little secondary pollutants (Aguilar-Perez et al. 2020). All these distinguishing features make them suitable for use in organic pollutants removal from wastewater, and antibiotics (amoxicillin, doxorubicin, rhodamine) from pharmaceutical effluents (Kadu et al. 2017; Dodoo-Arhin et al. 2020; Karimnezhad et al. 2020). Physical approaches like electrolysis, pyrolysis, evaporation- condensation, sputter deposition, laser ablation; chemical methods including microemulsion, reduction, electrochemical and thermal decomposition, co-precipitation can be used to make metallic nanosorbents (Vijayaraghavan and Ashokkumar 2017).

11.4.1.4 Biogenic Nanosorbent Materials

These nanomaterials are prepared from biological sources like plant extracts, bacteria, fungi in combination with metallic salt solution, in presence of bioactive substances like enzymes, vitamins, alkaloids, phenolics, peptides, and under different experimental conditions (Ali et al. 2017). Though these nanomaterials are environment-friendly and can be recovered from waste also, the limitation is that they require skilled research endeavors and a controlled environment for their fabrication and utilization (Jafari and Aghamiri 2011). Biogenic nanomaterials have been found efficient for the treatment of wastewater from the pharmaceutical sector. Aguilar-perez et al. (2020) studied that during the growth of *Pseudomonas aeruginosa* (bioactive source), secondary metabolites were formed which helped in the formation of zirconium nanoparticles by contributing capping and stabilizing agents. These nanomaterials were found with an adsorption capacity of 526.32 mg/g for tetracycline.

11.4.2 Commonly Used Nanomaterials for Pharmaceutical Wastewater Treatment

11.4.2.1 Electrospun Lignin Nanofibers

Lignin is distorted through electrospinning into anionic nanofibers under the influence of electric current, to eliminate pollutants from wastewater through adsorption. Nanofibers synthesized by electrospinning of lignin are green, inexpensive, non-toxic, and reusable, and hence are gaining popularity in this era where sustainable development is desired. Camire et al. (2020) used nonwoven electrospun nanofibrous lignin (156 nm in diameter) and polyvinyl alcohol to extract fluoxetine (pharmaceutical compound) from wastewater, and they found that 70% of the compound was adsorbed by the nanomaterial.

11.4.2.2 Zirconium Oxide with Polyaniline Nanocatalyst

The semiconducting metal oxide materials like TiO_2, ZrO_2, and ZnO can decompose organic pollutants (Shah et al. 2018). The low-cost zirconium oxides are heterogenous photocatalysts that have high redox potential, excellent chemical and optical properties, which make them suitable for photo-decomposition of pollutants. Polyaniline is also a promising material in the advanced oxidation process for the degradation of a broad range of organic pollutants. However, ZrO_2 nanoparticles modified with polyaniline can decompose the contaminants photocatalytically in presence of photons belonging to the UV-light region with higher efficiencies (Shirmardi et al. 2018). Vijayalakshmi et al. (2020) used nanocomposite of polyaniline and zirconium oxide for decomposition of ciprofloxacin and found that the pharmaceutical antibiotic was completely removed from wastewater in just 2 h. They prepared zirconium oxide nanoparticles by heating a white homogenous solution of zirconium oxychloride at 180 °C for 12 h and calcinating the obtained catalysts at 600 °C for 4 h. Zirconium oxide/polyaniline nanocomposite was then prepared by in situ oxidative polymerization of aniline in acidic medium (2M HCl), in presence of zirconium oxide nanoparticle, and ammonium disulfate as an oxidant. Stirring is necessary till the solution separates green-colored precipitate. The precipitates obtained are of zirconium oxide/polyaniline nanocomposite that is then filtered and collected.

11.4.2.3 Graphene-Based Nanomaterials

Nanocomposites made from graphene-derived materials (for example, graphene oxide, reduced graphene oxide, pure graphene) can efficiently remove pharmaceutical compounds like aspirin, caffeine, acetaminophen through adsorption. Al-Khateeb et al. (2014) found adsorption capacities of 18.1, 13.0, and 19.7 mg/g for aspirin, acetaminophen, and caffeine, respectively. In another study, caffeine, carbamazepine, diclofenac, and ibuprofen were removed by 98.2, 97, 97, and 95.5%, respectively with combined graphene-based nanomaterials (Rizzo et al. 2015). Graphene oxide can remove tetracycline with 313 mg/g adsorption capacity (Gao et al. 2012). Lin et al. (2013) used graphene oxide with magnetic particles for pharmaceutical wastewater treatment and reported 39.1 mg/g adsorption capacity for tetracyclines. Li et al. (2017) also used graphene oxide—Fe_3O_4 nanocomposite for adsorption of chlorpheniramine. It was found that a trace amount of magnetic nanoparticles onto graphene oxide improved the adsorption capacity, but their high concentration significantly reduced the adsorption capacity of graphene or graphene oxide for pharmaceuticals.

11.4.2.4 Curcumin-Loaded Nanostructured Lipid Carrier

Nanoscale lipid carriers loaded with curcumin can decrease the growth of bacterial strains in autoclaved hospital wastewater (Shajari et al. 2020) at 25 °C. These curcumin-based lipid nanomaterials can be prepared by the method given by Ali et al. (2010), i.e., hot high-speed homogenization method. In this method, cetyl palmitate and black seed oil (solid and liquid forms of lipid, respectively), and curcumin are first dissolved in ethanol to make an oily phase. Hot aqueous phase [tween 80 (0.25%)] is then added to oily phase at 80 °C, left for half an hour, homogenized for 10 min at 14,000 rpm, and finally cooled to room temperature to obtain curcumin loaded lipid carrier nanoparticles.

11.4.2.5 Titanium Oxide/Oxygen or Nitrogen Linked Heptazine-Base Polymer (TiO_2/ONLH) Nano-Photocatalyst

Titanium oxide (titania/TiO_2) is an effective semiconductor with photocatalytic properties, a high level of heat and chemical resistance, cost-effectiveness, and non-toxic and eco-friendly nature (Choi et al. 2018). However, there are a few limitations of using it including that titania performs only in UV region, quantum yield is low, structural tunability is poor, and pollutants degrade slowly during the photocatalytic process. Therefore, to overcome these disadvantages, scientists proposed several new approaches including doped titania, organic semiconductor nano-photocatalysts, polymeric photocatalysts (Xiang et al. 2018). Owing to their low price and controllable structures, polymeric nano-photocatalysts are exploited for wastewater treatment nowadays.

ONLH is a heptazine-base polymer with oxygen or nitrogen links that is integrated into titanium oxide nanoparticles to generate a TiO_2/ONLH nanocomposite which functions well in natural sunlight (Zhang et al. 2018). In a study by Zhang et al. (2021), TiO_2/ONLH nanocomposite was used to degrade pharmaceuticals under the effect of sunlight. Particularly, the fluoroquinolone containing compounds like N-piperazinyl, and certain cardiovascular drugs containing long aromatic side chains were easily degraded by the nanocomposite. Propranolol degradation was aided by the presence of -OH- and $-O_2$, owing to the hydroxyl radical addition, ipso substitution, and ring-opening reactions. TiO_2/ONLH was also found to reduce the toxic transformation products formation. The nanocomposite was prepared from semicarbazide hydrochloride by heating at a heating rate of 5 °C/min for 5 h in a muffle furnace at a constant temperature of 550 °C.

11.4.2.6 Multi-Wall Carbon Nanotubes/Titanium Oxide (MWCNT/TiO_2) Nanocomposite

MWCNT are non-metallic multi-wall carbon nanotubes. The MWCNT/TiO_2 nanocomposite is synthesized using the method described by Gao et al. (2009). Purified MWCNT are first sonicated to get dispersed in 1 M acid solution [H_2SO_4:HNO_3 (3:1)], then heated to 140 °C, cooled to room temperature, sonicated again, washed, filtered, and the dark precipitates of MWCNTs are collected. The acid-treated MWCNT and TiO_2 were then added to an isopropanol solution, sonicated for 1 h, vigorously mixed with a solution of 7 ml titanium isopropoxide in 40 ml isopropanol, left at room temperature for 2 h, precipitates were filtered, washed with a mixture of ethanol/water (1:1 v/v), dried at 100 °C for 2 h, and produced MWCNTs/TiO_2 nanocomposite was stored for further use. Tetracycline degradation utilizing a MWCNT/TiO_2 nanocomposite and ultraviolet-C radiation was investigated by Ahmadi et al. (2017). Complete removal of tetracycline (up to 10 mg/L) at 3:2 ratio of MWCNT:TiO_2 after 100 min, 3% mineralization after 300 min. In the same operational settings, the COD concentration of 2267 mg/L in genuine pharmaceutical wastewater dropped to 342 mg/L after 240 min.

11.4.2.7 Biogenic Fe_3O_4/Au Nanocomposite

Fe_3O_4/Au nanocomposite can efficiently degrade imatinib and imipenem drugs under UV and visible radiations. Mirsadeghi et al. (2020) reported 92 and 96% degradation of imatinib and imipenem after exposure of 1200 s to UV light, and 82 and 84% after exposure of 3600 s to visible light. Firstly, 5 mM $HAuCl_4.3H_2O$ was prepared in distilled water; Fe_3O_4 nanoparticles were dispersed in 100 mL of *Carum carvi* L. seeds extract; and then for preparation of Fe_3O_4/Au nanocomposite, 20 mL solution of $HAuCl_4.3H_2O$ was added to Fe_3O_4 nanoparticles. The solution was continued to mix for up to an hour, then left for one day so that biosynthesized Fe_3O_4/Au nanocomposite is separated.

11.4.2.8 Manganese Oxide/Nano-Graphite /2-Ethylanthraquinone/Activated Carbon (MnOx/nano-G/2-EAQ/AC) Cathode

Electrochemical advanced oxidation processes employed for wastewater treatment use cathode made from highly conductive and chemically resistant materials like graphite, porous carbon, activated or fibrous carbon (Liu et al. 2013; Le et al. 2015). However, oxidation efficiency of cathodes can be improved by impregnating electrodes with catalysts including anthraquinone (Tammevesk et al. 2001; Zhang et al. 2007), 2-ethylanthraquinone (2-EAQ) (Wang et al. 2011), tert-butyl-anthraquinone (TBAQ) (Valim et al. 2013). The presence of catalysts improves the generation of OH^- radicals and enhances oxygen reduction. MnOx is also a suitable semiconductor for use in electrochemical advanced oxidation processes, but it has poor electrical conductivity. However, MnOx can be used in the processes after doping on carbon material with high conductivity (nano-graphite). Doped MnOx decorated with catalysts is found more efficient for oxygen reduction. Recently, novel manganese oxide/nano-graphite/2-ethylanthraquinone/activated carbon (carbon material with high conductivity and better oxidation efficiency) cathode has been used by researchers to treat wastewater containing pharmaceuticals. The prepared cathode's activity was evaluated by measuring H_2O_2 and -OH yields, as well as phenol degradation (Wang et al. 2016). The authors found that after 6 h of electrolysis, the biodegradability of antibiotics in pharmaceutical wastewater was increased from 0.01 to 0.31. The following steps were taken to make the novel nanocomposite cathode: in a water bath, 0.75 gms of 2-EAQ/AC, 0.75 gms of MnOx/nano-G, and 4.5 gms of polytetrafluoroethylene (10%) are mixed (at 65 °C). This combination is subsequently pressed over stainless steel mesh, yielding MnOx/nano-G/2- EAQ/AC cathode. The wet electrode is then dried, and used in wastewater treatment studies.

11.4.2.9 ZnO/Sepiolite Heterostructured Nanomaterials

ZnO nanoparticles are incorporated on the surface of a fibrous clay material (sepiolite), and the surface is modified using cetyltrimethylammonium bromide, tetramethoxyortosilicate, or magnetite (Fe_3O_4) nanoparticles to generate ZnO/sepiolite heterostructured nanomaterials that can be used in photocatalytic degradation of pharmaceutical drugs like ibuprofen, acetaminophen, antipyrine (Akkari et al. 2018). The ZnO/sepiolite nanocomposite is prepared in two steps: first, sepiolite is modified in an aqueous medium using cetyltrimethylammonium bromide, and subsequently, ZnO nanoparticles are precipitated from Zn-acetate in methanol with potassium hydroxide (Aranda et al. 2008; Akkari et al. 2016). ZnO nanoparticles are then dispersed in 2-propanol and added to sepiolite modified with cetyltrimethylammonium bromide also dispersed in 2-propanol (5% w/w) to obtain ZnO: organoclay (0.5:1 weight ratio). To generate the ZnO/sepiolite heterostructure, the mixture is ultrasonicated, the recovered solid is dried overnight at 60 °C, and calcined at 500 °C for 1 h under nitrogen and 4 h under air flux. This nanomaterial is of superparamagnetic nature.

11.4.2.10 Tungsten Oxide and Carbon Nanotube (WO_3/CNT) Heterojunction Nanocomposite

WO_3/CNT nanocomposite can be prepared by sol–gel/hydrothermal method given by Yu et al. (2017). For the synthesis of the nanomaterial, 1 ml lactic acid is added to a 60 ml aqueous solution of sodium-tungsten-dehydrate. The colorless solution turns to a yellow gel-type suspension after some time, which is then autoclaved and washed with ethanol to eliminate residual impurities. Centrifugation is used to collect the final products, which are then dried in an oven at 85 °C for 30 h. WO_3/CNT nanocomposite together with visible and ultrasound radiations can be applied for the degradation of tetracycline in pharmaceutical wastewater (Isari et al. 2020a). Results revealed that 60 mg/L of tetracycline can be degraded in one hour at pH: 9, US power: 250 W/m^2, and light intensity: 120 W/m^2. Moreover, 90.6% of COD and 83.7% of TOC can also be eliminated after 220 min.

11.4.2.11 Ni-Al Layered Double Hydroxide Nanoparticles

LDHs (layered double hydroxides) are a type of synthetic ionic inorganic substance that has a clay-like structure and contains divalent or trivalent metal ions such as Zn, Mg, Cu, Co, Ni, Al, Fe, Cr (Evans and Slade 2006). Nickel-aluminum layered double hydroxide (Ni-Al LDH) nanomaterial consists of two layers Ni and Al with nitrate anions in the interlayer. Ion exchange reactions or surface adsorption can exchange interlayer anions with various inorganic or organic anions. In this way, this nanomaterial helps in the removal of pollutants from wastewater. The co-precipitation method is used to make Ni-Al LDH. Under vigorous stirring at room temperature, 0.581 g $Ni(NO_3)_2.6H_2O$ and 0.375 g $Al(NO_3)_3.9H_2O$ are dissolved in deionized water, and the reaction is maintained for 30 min under nitrogen protection. Then, the reaction mixture is subjected to hydrothermal treatment at 90 °C for one day, then centrifuged for 10 min at 4000 rpm, washed, dried at 60 °C, and then stored for future use (Abdolmohammad-Zadeh et al. 2013). Ni-Al layered double hydroxide nanomaterials have been used by Abdolmohammad-Zadeh et al. (2014) for the degradation of mefenamic acid in human serum and real pharmaceutical wastewater.

11.4.2.12 ZnO Nanoparticles

Zinc oxide nanoparticles are widely used by environmental engineers for the removal of pollutants from environmental matrices (Kulkarni et al. 2016; Khezamia et al. 2017). Zinc oxide can be synthesized from zinc nitrate [$Zn(NO_3)_2$] by a wet chemical method. 0.5 M aqueous solution of $Zn(NO_3)_2$ is continuously stirred for 1 h at 70 °C. When this solution is poured dropwise into a 1M solution of sodium hydroxide, a milky white solution is obtained which is then left overnight after stirring for 2 h. Prepared nanoparticles of ZnO (in precipitate forms) are separated, washed with ethanol, oven-dried at 70 °C, and preserved for use (Sridara et al. 2018). Sridara et al. (2018) reported a maximum of 90.1% removal of phenol from pharmaceutical wastewater at pH 6. The percentage removal of phenol increased with an increase in pH and temperature.

11.4.2.13 Nanoparticles Prepared from N, Cu Co-doped TiO_2 Decorated on Single-Walled Carbon Nanotube

Titania has a low adsorption capacity but a high capacity for agglomeration and photo-induced charge mobility which limit its application in industrial uses due to lower oxidative potential. However, co-doping titania with certain metals or non-metals can enhance the light absorption potential and photocatalytic performance of titania in the visible light window, and effectively decrease the band gap value (Eslami et al. 2016; Adyani and Ghorbani 2018). Furthermore, the adsorption capacity of titania can be increased with adsorbent support of carbon nanotubes. Carbon nanotubes possess admirable electron mobility, larger surface area, mechanical strength, and thermal stability (Payan et al. 2017). The large surface area of carbon nanotubes creates more sorption sites for target pollutants, and they increase the lifetime of electrons by enhancing the charge separation potential of titania (Dahl et al. 2014). Therefore, the technique of co-doping of metals/non-metals on carbon nanotubes impregnated with titania nanoparticles is a promising approach for the removal of pharmaceutical drugs from wastewater. Isari et al. (2020a) used N, Cu co-doped TiO_2 nanoparticles decorated on single-walled carbon nanotube for pharmaceutical wastewater treatment, and reported 93% and 89% reduction in COD and TOC, respectively in 3 h. This novel hybrid nanomaterial was synthesized by sol-gel/hydrothermal method. 0.15 gms of carbon nanotube were added to the solution containing 20 ml distilled Mili-Q water and 40 ml absolute ethanol, mixed well for 1 h. After 1 h, 2 gms fine powder of NCuT were added to the above mixture, mixed for 2 h by the ultrasound probe, autoclaved, oven-heated at 120 °C for 3 h, cooled, and dried to get N, Cu-TiO_2-carbon nanotube nanocomposite.

11.4.2.14 Molybdenum-Doped Organic-Inorganic Hybrid TiO_2 Nanoparticles

TiO_2 has been recognized as the most effective nano-photocatalyst. Furthermore, doping TiO_2 with metals can improve the treatment potential of TiO_2-based nanoparticles (Sayilkan et al. 2007; Zhang et al. 2008). Doping with molybdenum has been also explored in recent times and applied for pharmaceutical wastewater. Pure TiO_2 was employed as the starting material for the molybdenum-doped TiO_2 hybrid nanoparticles, and then the dopant (MoO_3) was added. About 10 mL 1 N HCl was added as a mineralizer. n-butylamine was added to the aforesaid mixture in various concentrations (0.8, 1.0, 1.2, and 1.4 M) and vigorously swirled for a few minutes. The final mixture was then autoclaved, cooled to room temperature, washed, allowed to settle, washed again to remove any remaining surfactant and solvent. The collected product was dried for a few hours in a hot air oven at 40–50 °C, and the dried particles were exposed to photocatalytic testing. Using molybdenum-doped TiO_2 organic-inorganic hybrid nanoparticles (modified with n-butylamine at 150 °C at autogeneous pressure for 18 h), 93% degradation of pharmaceutical effluents was reported by Shahmoradi et al. (2010).

11.4.2.15 Cu-Doped TiO_2 Decorated on Single-Walled Carbon Nanotube Ternary Porous Nanocomposite

Cu-doped TiO_2 impregnated on single-walled carbon nanotube nanocomposites have been currently investigated for pharmaceutical effluents treatment. The nanocomposites were synthesized through the hydrothermal method given by Yang et al. (2013). Single-walled carbon nanotubes were added to the solution of 20 ml distilled water and 40 ml absolute ethanol, exposed to ultrasonic treatment for 1 h, uniformly mixed with Cu-doped TiO_2, stirred for 2 h. The homogenous milky suspension was obtained which was then autoclaved, oven-heated at 120 °C for 3 h, cooled, dried, and finally calcinated to get finely powdered nanocomposite. Payan et al. (2019) used Cu-doped TiO_2-carbon nanotube nanocomposite with 4% Cu content and 10% carbon nanotubes for sulfamethazine removal at pH 7, and an initial drug concentration of 30 mg/l. They found complete degradation of sulfamethazine after 135 min, and 59.3% of TOC, and 74.6% of COD after 390 min exposure to solar radiations.

11.5 Conclusion

Water pollution due to pharmaceutical drugs is a serious environmental threat. Conventional technologies are less capable of removing pollutants from pharmaceutical wastewater whereas nanotechnology has been perceived as the most interesting and effective solution in the area of wastewater management. The technology has been gaining spotlight ascribed to the ecologically safe, energy-effective, low cost, small size, greater surface area, high porosity, high reactivity, and better catalytic and adsorption properties of the nanomaterials. The composition and source/process stage of drug formulation decides the level of toxicity of wastewater generated and resultant adverse hazards on environmental and human health. Different types of nanomaterials are used for treatment purposes including carbon-based, chitosan-based, metallic-oxide-based, and biogenic-based. Researchers have developed many pristine, doped, and hybrid nanomaterials for the enhanced treatment of pharmaceutical wastewater. Scientists have also fabricated innovative biological nanomaterials with high capability, selectivity, affinity, intensity, and capacity for the treatment of wastewater. With the use of different nanomaterials possessing unique physicochemical and surface properties, nanotechnology has shown efficient degradation of almost all the pollutants present in wastewater.

Acknowledgements The corresponding author (PM) gratefully acknowledges the financial support received from JK Science Technology & Innovation Council, Department of Science & Technology, India (Grant no. JKST&IC/SRE/945-47).

References

Abdolmohammad-Zadeh H, Jouyban A, Amini R, Sadeghi G (2013) Nickel-aluminum layered double hydroxide as a nano-sorbent for the solid phase extraction of selenium, and its determination by continuous flow HG-AAS. Microchim Acta 180:619–626

Abdolmohammad-Zadeh H, Morshedzadeh F, Rahimpour E (2014) Trace analysis of mefenamic acid in human serum and pharmaceutical wastewater samples after pre-concentration with Ni-Al layered double hydroxide nano-particles. J Pharm Anal 4(5):331–338

Adeleye AS, Conway JR, Garner K, Huang Y, Su Y, Keller AA (2016) Engineered nanomaterials for water treatment and remediation: costs, benefits, and applicability. Chem Eng 286:640–662

Adyani SM, Ghorbani M (2018) A comparative study of physicochemical and photocatalytic properties of visible light responsive Fe, Gd and P single and tri-doped TiO_2 nanomaterials. J Rare Earths 36:72–85

Aguilar-Perez KM, Aviles-Castrillo JI, Ruiz-Pulido G (2020) Nano-sorbent materials for pharmaceutical-based wastewater effluents—an overview. Case Stud Chem Environ Eng 2:100028

Ahmadi M, Motlagh HR, Jaafarzadeh N, Mostoufi A, Saeedi R, Barzegar G, Jorfi S (2017) Enhanced photocatalytic degradation of tetracycline and real pharmaceutical wastewater using MWCNT/TiO2 nanocomposite. J Environ Manage 186:55–63

Akkari M, Aranda P, Belver C, Bedia J, Amara ABA, Ruiz-Hitzky E (2018) ZnO/sepiolite heterostructured materials for solar photocatalytic degradation of pharmaceuticals in wastewater. Appl Clay Sci 156:104–109

Akkari M, Aranda P, Ben Rhaiem H, Amara ABH, Ruiz-Hitzky E (2016) ZnO/ clay nanoarchitectures: synthesis, characterization and evaluation as photocatalysts. Appl Clay Sci 131:131–139

Ali AM, Ronning HT, Alarif W, Kallenborn R, Al-Lihaibi SS (2017) Occurrence of pharmaceuticals and personal care products in effluent-dominated Saudi Arabian coastal waters of the Red Sea. Chemosphere 175:505–513

Ali I, Peng C, Khan ZM, Naz I, Sultan M, Ali M, Abbasi IA, Islam T, Ye T (2019a) Overview of microbes based fabricated biogenic nanoparticles for water and wastewater treatment. J Environ Manage 230:128–150

Ali S, Rehman SAU, Luan HY, Farid MU, Huang H (2019b) Challenges and opportunities in functional carbon nanotubes for membrane-based water treatment and desalination. Sci Total Environ 646:1126–1139

Ali H, Shirode AB, Sylvester PW, Nazzal S (2010) Preparation, characterization, and anticancer effects of simvastatin–tocotrienol lipid nanoparticles. Int J Pharm 389:223–231

Al-Khateeb LA, Almotiry S, Salam MA (2014) Adsorption of pharmaceutical pollutants onto graphene nanoplatelets. Chem Eng 248(Suppl. C):191–199

Aljuboury DA, Palaniandy P, Aziz HBA, Feroz S (2017) Evaluation of the photocatalyst of TiO2/Fenton/ZnO to treat the petroleum wastewater. AIP Conference Proceedings 1892:040024

Amin A, Naik ATR, Azhar M, Nayak H (2013) Bioremediation of different waste waters—a review. Continent. J Fish Aquat Sci 7(2):7–17

Aranda P, Kun R, Martin-Luengo MA, Letaief S, Dekany I, Ruiz-Hitzky E (2008) Titania–sepiolite nanocomposites prepared by a surfactant templating colloidal route. Chem Mater 20:84–91

Bawa R, Bawa SR, Maebius SB, Flynn T, Chiming Wei C (2005) Protecting new ideas and inventions in nanomedicine with patents. Nanomed Nanotechnol Biol Med 1(2):150–158

Benotti MJ, Brownawell B (2009) Microbial degradation of pharmaceuticals in estuarine & coastal seawater. Environ Pollut 157(3):994–1002

Brezovsek P, Elersek T, Filipic M (2014) Toxicities of four anti-neoplastic drugs and their binary mixtures tested on the green alga *Pseudokirchneriella subcapitata* and the cyanobacterium *Synechococcus leopoliensis*. Water Res 52:168–177

Camire A, Espinasse J, Chabot B, Lajeunesse A (2020) Development of electrospun lignin nanofibers for the adsorption of pharmaceutical contaminants in wastewater. Environ Sci Pollut Res 27:3560–3573

Choi S, Johnston M, Wang G-S, Huang CP (2018) A seasonal observation on the distribution of engineered nanoparticles in municipal wastewater treatment systems exemplified by TiO_2 and ZnO. Sci Total Environ 625:1321–1329

Cincinelli A, Martellini T, Coppini E, Fibbi D, Katsoyiannis A (2015) Nanotechnologies for removal of pharmaceuticals and personal care products from water and wastewater: a review. J Nanosci Nanotechnol 15(5):3333–3347

Crouse BA, Ghoshdastidar AJ, Tong AZ (2012) The presence of acidic and neutral drugs in treated sewage effluents and receiving waters in the Cornwallis and Annapolis River watersheds and the mill CoveSewage treatment plant in Nova Scotia Canada. Environ Res 112:92–99

Dahl M, Liu Y, Yin Y (2014) Composite titanium dioxide nanomaterials. Chem Rev 114:9853–9889

Dixit D, Parmar N (2013) Treatment of pharmaceutical waste water by electro-coagulation & natural coagulation process: review. VSRD Int J Nanotechnol 4(5):79–88

Dodoo-Arhin D, Asiedu T, Agyei-Tuffour B, Nyankson E, Obada D, Mwabora JM (2020) Photocatalytic degradation of Rhodamine dyes using zinc oxide nanoparticles. Mater Today Proc 38(2):809–815

Eslami A, Amini MM, Yazdanbakhsh AR, Mohseni-Bandpei A, Safari AA, Asadi A (2016) N, S co-doped TiO_2 nanoparticles and nanosheets in simulated solar light for photocatalytic degradation of non-steroidal anti-inflammatory drugs in water: a comparative study. J Chem Technol Biotechnol 91:2693–2704

Evans DG, Slade RCT (2006) Structural aspects of layered double hydroxides. In: Duan X, Evans DG (eds) Layered double hydroxides. Springer, Berlin, Germany, pp 1–5

Farhadi S, Aminzadeh B, Torabian A, Khatibikamal V, Alizadeh Fard M (2012) Comparison of COD removal from pharmaceutical wastewater by electrocoagulation, photoelectrocoagulation, peroxi-electrocoagulation and peroxiphotoelectrocoagulation processes. J Hazard Mater 219–220:35–42

Gao B, Chen GZ, Puma GL (2009) Carbon nanotubes/titanium dioxide (CNTs/TiO2) nanocomposites prepared by conventional and novel surfactant wrapping sol–gel methods exhibiting enhanced photocatalytic activity. Appl Catal B 89(3–4):503–509

Gao Y, Li Y, Zhang L, Huang H, Hu J, Shah SM, Su X (2012) Adsorption and removal of tetracycline antibiotics from aqueous solution by graphene oxide. J Colloid Interface Sci 368(1):540–546

Garcia-Espinoza JD, Mijaylova-Nacheva P, Aviles-Flores M (2018) Electrochemical carbamazepine degradation: effect of the generated active chlorine, transformation pathways and toxicity. Chemosphere 192:142–151

Gebhardt W, Schroder HF (2007) Liquid chromatography–(tandem) mass spectrometry for the follow-up of the elimination of persistent pharmaceuticals during wastewater treatment applying biological wastewater treatment and advanced oxidation. J Chromatogr 1160:34–43

Gerbersdorf SU, Cimatoribus C, Class H, Engesser K-H, Helbich S, Hollert H, Lange C, Kranert M, Metzger J, Nowak W (2015) Anthropogenic trace compounds (ATCs) in aquatic habitats—research needs on sources, fate, detection and toxicity to ensure timely elimination strategies and risk management. Environ Int 79:85–105

Godoy AA, Kummrow F (2017) What do we know about the ecotoxicology of pharmaceutical and personal care product mixtures? A critical review. Crit Rev Environ Sci Technol 47:1453–1496

Gowri RS, Vijayaraghavan R, Meenambigai P (2014) Microbial degradation of reactive dyes—a review. Int J Curr Microbiol App Sci 3(3):421–436

Hallare A, Kohler HR, Triebskorn R (2004) Developmental toxicity and stress protein responses in zebrafish embryos after exposure to diclofenac and its solvent DMSO. Chemosphere 56:659–666

Isari AA, Hayati F, Kakavandi B, Rostami M, Motevassel M, Dehghanifar E (2020a) N, Cu co-doped TiO_2@functionalized SWCNT photocatalyst coupled with ultrasound and visible-light: an effective sono-photocatalysis process for pharmaceutical wastewaters treatment. Chem Eng J 392:123685

Isari AA, Mehregan M, Mehregan S, Hayati F, Kalantary RR, Kakavandi B (2020b) Sono-photocatalytic degradation of tetracycline and pharmaceutical wastewater using WO_3/CNT heterojunction nanocomposite under US and visible light irradiations: a novel hybrid system. J Hazard Mater 390:122050

Jafari M, Aghamiri SF (2011) Evaluation of carbon nanotubes as solid-phase extraction sorbent for the removal of cephalexin from aqueous solution. Desalin Water Treat 28(1–3):55–58

Jeong TY, Yoon D, Kim S, Kim HY, Kim SD (2018) Mode of action characterization for adverse effect of propranolol in Daphnia magna based on behavior and physiology monitoring and metabolite profiling. Environ Pollut 233:99–108

Kadu BS, Wani KD, Kaul-Ghanekar R, Chikate RC (2017) Degradation of doxorubicin to non-toxic metabolites using Fe-Ni bimetallic nanoparticles. Chem Eng 325:715–724

Kanakaraju D, Glass BD, Oelgemoller M (2018) Advanced oxidation process mediated removal of pharmaceuticals from water: a review. J Environ Manage 219:189–207

Karimnezhad H, Navarchian AH, Gheinani TT, Zinadini S (2020) Amoxicillin removal by Fe-based nanoparticles immobilized on polyacrylonitrile membrane: individual nanofiltration or Fenton reaction, vs. engineered combined process. Chem Eng Res Des 153:187–200

Karri RR, Shams S, Sahu JN (2019) Overview of potential applications of nano-biotechnology in wastewater and effluent treatment. In: Ahsan A, Ismail AF (eds) Micro and nano technologies, nanotechnology in water and wastewater treatment. Elsevier, pp 87–100

Khezamia L, Taha K, Amami E, Ghiloufi I, Lassaad EMM, El Z (2017) Removal of cadmium (II) from aqueous solution by zinc oxide nanoparticles: kinetic and thermodynamic studies. Desalin Water Treat 62:346–354

Kristensen DM, Desdoits-Lethimonier C, Mackey AL, Dalgaard MD, De Masi F, Munkbol CH, Styrishave B, Antignac JP, Le Bizec B, Platel C (2018) Ibuprofen alters human testicular physiology to produce a state of compensated hypogonadism. Proc Natl Acad Sci USA 115:715–724

Kulkarni AV, Chavhan A, Bappakhane A, Chimmankar J (2016) ZnO nanoparticles as adsorbent for removal of methylene blue dye. Res J Chem Environ Sci 4:158–163

Kummerova M, Zezulka S, Babula P, Triska J (2016) Possible ecological risk of two pharmaceuticals diclofenac and paracetamol demonstrated on a model plant Lemna minor. J Hazard Mater 302:351–361

Le TXH, Charmette C, Bechelany M (2015) Facile preparation of porous carbon cathode to eliminate paracetamol in aqueous medium using electro-Fenton system. Electrochim Acta 188:378

Li CM, Chen CH, Chen WH (2017) Different influences of nanopore dimension and pH between chlorpheniramine adsorptions on graphene oxide-iron oxide suspension and particle. Chem Eng 307:447–455

Lin Y, Xu S, Li J (2013) Fast and highly efficient tetracyclines removal from environmental waters by Graphene oxide functionalized magnetic particles. Chem Eng 225(Suppl. C):679–685

Lingamdinne LP, Koduru JR, Chang YY, Karri RR (2018) Process optimization and adsorption modeling of Pb(II) on nickel ferritereduced graphene oxide nano-composite. J Mol Liq 250:202–211

Liu S, Zhao XR, Sun HY (2013) The degradation of tetracycline in a photo-electro-Fenton system. Chem Eng 231:441

Malaviya P, Hali R, Sharma N (2012) Impact of dyeing industry effluent on germination and growth of Pea (Pisum sativum). J Environ Biol 33:1075–1078

Malaviya P, Sharma A (2011) Effect of distillery effluent on yield attributes of *Brassica napus* L. J Environ Biol 32(3):385–389

Malaviya P, Singh A, Anderson TA (2020) Aquatic phytoremediation strategies for chromium removal. Rev Environ Sci Bio/technol 19:897–944

Mirsadeghi S, Zandavar H, Yousefi M, Rajabi HR, Pourmortazavi SM (2020) Green-photodegradation of model pharmaceutical contaminations over biogenic Fe3O4/Au nanocomposite and antimicrobial activity. J Environ Manage 270:110831

Misal SA, Lingojwar DP, Shinde RM, Gawai KR (2011) Purification and characterization of azoreductase from alkaliphilic strain Bacillus badius. Process Biochem 46(6):1264–1269

Mohmood I, Lopes CB, Lopes I, Ahmad I, Duarte AC, Pereira E (2013) Nanoscale materials and their use in water contaminants removal—a review. Environ Sci Pollut Res 20:1239–1260

Muthusaravanan S, Sivarajasekar N, Vivek JS, Paramasivan T, Naushad M, Prakashmaran J, Al-Duaij OK (2018) Phytoremediation of heavy metals: mechanisms, methods & enhancements. Environ Chem Lett 16(4):1339–1359

Pal P (2018) Treatment and disposal of pharmaceutical wastewater: towards sustainable strategy. Sep Purif Rev 47(3):179–198

Parihar A, Malaviya P (2013) Textile wastewater treatment using sawdust as adsorbent. Int J of Environ Sci 2:110–113

Payan A, Fattahi M, Jorfi S, Roozbehani B, Payan S (2017) Synthesis and characterization of titanate nanotube/single-walled carbon nanotube (TNT/SWCNT) porous nanocomposite and its photocatalytic activity on 4-Chlorophenol degradation under UV and solar irradiation. Appl Surf Sci 434:336–350

Payan A, Isari AA, Gholizade N (2019) Catalytic decomposition of sulfamethazine antibiotic and pharmaceutical wastewater using Cu-TiO_2@functionalized SWCNT ternary porous nanocomposite: influential factors, mechanism, and pathway studies. Chem Eng 361:1121–1141

Qu X, Brame J, Li Q, Alvarez PJ (2012) Nanotechnology for a safe and sustainable water supply: enabling integrated water treatment and reuse. Acc Chem Res 46(3):834–843

Rana RS, Singh P, Kandari V, Singh R, Dobhal R, Gupta S (2017) A review on characterization & bioremediation of pharmaceutical industries' wastewater: an Indian perspective. Appl Water Sci 7(1):1–12

Rasheed T, Bilal M, Hassan AA, Nabeel F, Bhargava RN, Ferreira LFR, Tran HN, Iqbal HMN (2020) Environmental threatening concern and efficient removal of pharmaceutically active compounds using metal-organic frameworks as adsorbents. Environ Res 185:109436

Rizzo L, Fiorentino A, Grassi M, Attanasio D, Guida M (2015) Advanced treatment of urban wastewater by sand filtration and graphene adsorption for wastewater reuse: effect on a mixture of pharmaceuticals and toxicity. J Environ Chem Eng 3(1):122–128

Rodrigues S, Antunes S, Correia A, Nunes B (2016) Acute and chronic effects of erythromycin exposure on oxidative stress and genotoxicity parameters of *Oncorhynchus mykiss*. Sci Total Environ 545:591–600

Rodrigues S, Antunes SC, Correia AT, Golovko O, Zl abek V, Nunes B (2019a) Assessment of toxic effects of the antibiotic erythromycin on the marine fish gilthead seabream (*Sparus aurata* L.) by a multi-biomarker approach. Chemosphere 216:234–247

Rodrigues S, Antunes SC, Correia AT, Nunes B (2019b) Toxicity of erythromycin to *Oncorhynchus mykiss* at different biochemical levels: detoxification metabolism, energetic balance, and neurological impairment. Environ Sci Pollut Res 26:227–239

Saha K, Maheshwari RU, Sikder J, Chakraborty S, da Silva SS, dos Santos JC (2017) Membranes as a tool to support biorefineries: applications in enzymatic hydrolysis, fermentation and dehydration for bioethanol production. Renew Sustain Energy Rev 74:873–890

Santhosh C, Velmurugan V, Jacob G, Jeong SK, Grace AN, Bhatnagar A (2016) Role of nanomaterials in water treatment applications: a review. Chem Eng 306:1116–1137

Sayilkan F, Asilturk M, Sener S, Erdemoglu S, Erdemoglu M, Syilkan H (2007) Hydrothermal synthesis, characterization and photocatalytic activity of nanosized TiO_2 based catalysts for Rhodamine B degradation. Turk J Chem 31:211–221

Schwaiger J, Ferling H, Mallow U, Wintermayr H, Negele R (2004) Toxic effects of the non-steroidal anti-inflammatory drug diclofenac: part I: histopathological alterations and bioaccumulation in rainbow trout. Aquat Toxicol 68:141–150

Shah AUHA, Akhlaq S, Sayed M, Bilal S, Ali N (2018) Synthesis and characterization of polyaniline-zirconium dioxide and polyaniline–cerium dioxide composites with enhanced photocatalytic degradation of rhodamine B dye. Chem Pap 72:2523–2538

Shahmoradi B, Ibrahim IA, Sakamoto N, Ananda S, Row TNG, Soga K, Byrappa K, Parsons S, Shimizu Y (2010) In situ surface modification of molybdenum-doped organic-inorganic hybrid TiO_2 nanoparticles under hydrothermal conditions and treatment of pharmaceutical effluent. Environ Technol 31(11):1213–1220

Shajari M, Rostamizadeh K, Shapouri R, Taghavi L (2020) Eco-friendly curcumin-loaded nanostructured lipid carrier as an efficient antibacterial for hospital wastewater treatment. Environ Technol Innov 18:100703

Sharma R, Kumari R, Pant D, Malaviya P (2022a) Bioelectricity generation from human urine and simultaneous nutrient recovery: role of microbial fuel cells. Chemosphere 292:133437. https://doi.org/10.1016/j.chemosphere.2021.133437

Sharma R, Mishra A, Pant D, Malaviya P (2022b) Recent advances in microalgae-based remediation of industrial and non-industrial wastewaters with simultaneous recovery of value-added products. Bioresour Technol 344:126129. https://doi.org/10.1016/j.biortech.2021.126129

Sharma S, Malaviya P (2013) Bioremediation of tannery wastewater by Aspergillus niger SPFSL2-a isolated from tannery sludge. Int J Basic Appl Sci 2:88–93

Sharma R, Malaviya P (2021) Management of stormwater pollution using green infrastructure: the role of rain gardens. Wires Water 8(2):1507

Sharma R, Malaviya P (2022) Constructed wetlands for textile wastewater remediation: a review on concept pollutant removal mechanisms and integrated technologies for efficiency enhancement. Chemosphere 290:133358. https://doi.org/10.1016/j.chemosphere.2021.133358

Sharma R, Mishra A, Pant D, Malaviya P (2021) Recent advances in microalgae-based remediation of industrial and non-industrial wastewaters with simultaneous recovery of value-added products. Bioresour Technol 344:126129

Sharma R, Sharma R, Parveen K, Pant D, Malaviya P (2021) Comprehensive and critical appraisal of plant-based defluoridation from environmental matrices. Chemosphere 281:130892. https://doi.org/10.1016/j.chemosphere.2021.130892

Sharma R, Vymazal J, Malaviya P (2021) Application of floating treatment wetlands for stormwater runoff: a critical review of the recent developments with emphasis on heavy metals and nutrient removal. Sci Total Environ 777:146044. https://doi.org/10.1016/j.scitotenv.2021.146044

Shirmardi A, Teridi MAM, Azimi HR, Basirun WJ, Sheini FJ, Yousefi R (2018) Enhanced photocatalytic performance of ZnSe/ PANI nanocomposites for degradation of organic and inorganic pollutants. Appl Surf Sci 462:730–738

Singh A, Malaviya P (2019) Chromium phytoaccumulation and its impact on growth and photosynthetic pigments of Spirodela polyrrhiza (L.) Schleid.on exposure to tannery effluent. Environ Sustain 2(2):157–166. https://doi.org/10.1007/s42398-019-00062-4

Singh A, Sharma R, Pant D, Malaviya P (2021) Engineered algal biochar for contaminant remediation and electrochemical applications. Sci Total Environ 774:145676. https://doi.org/10.1016/j.scitotenv.2021.145676

Singh A, Vyas D, Malaviya P (2016) Two-stage phyto-microremediation of tannery effluent by *Spirodela polyrrhiza* (L.) Schleid. and chromium resistant bacteria. Biores Technol 216:883–893

Sofowora A, Ogunbodede E, Onayade A (2013) The role and place of medicinal plants in the strategies for disease prevention. Afr J Tradit Complement Altern Med 10:210–229

Sridara R, Ramanane UU, Rajasimman M (2018) ZnO nanoparticles—synthesis, characterization and its application for phenol removal from synthetic and pharmaceutical industry wastewater. Environ Nanotechnol Monit Manage 10:388–393

Tammevesk K, Kontturi K, Nichols RJ (2001) Surface redox catalysis for O2 reduction on quinine modified glassy carbon electrodes. J Electroanal Chem 515:101

Thines RK, Mubarak NM, Nizamuddin S, Sahu JN, Abdullah EC, Ganesan P (2017) Application potential of carbon nanomaterials in water and wastewater treatment: a review. J Taiwan Inst Chem Eng 72:116–133
Thirunavukkarasu, A, Nithya R, Sivashankar R (2020) A review on the role of nanomaterials in the removal of organic pollutants from wastewater. Rev Environ Sci Biotechnol 19:751–778
Valim RB, Reis RM, Castro PS, Lima AS, Rocha RS, Bertotti M, Lanza MR (2013) Electrogeneration of hydrogen peroxide in gas diffusion electrodes modified with tert-butyl-anthraquinone on carbon black support. Carbon 61:236
Vijayalakshmi S, Kumar E, Venkatesh PS, Raja A (2020) Preparation of zirconium oxide with polyaniline nanocatalyst for the decomposition of pharmaceutical industrial wastewater. Ionics 26(10):1–7
Vijayaraghavan K, Ashokkumar T (2017) Plant-mediated biosynthesis of metallic nanoparticles: A review of literature, factors affecting synthesis, characterization techniques and applications. J Environ Chem Eng 5(5):4866–4883
Wang ZX, Li G, Yang F, Chen YL, Gao P (2011) Electro-Fenton degradation of cellulose using graphite/PTFE electrodes modified by 2-ethylanthraquinone. Carbohyd Polym 86:1807
Wang W, Yu J, Zou J, Yu X (2016) Mechanism for enhancing biodegradability of antibiotic pharmacy wastewater by in-situ generation of H_2O_2 and radicals over MnOx/ nano-G/2-EAQ/AC cathode. Electrochim Acta 191:426–434
Wilkinson JL, Hooda PS, Barker J, Barton S, Swinden J (2016) Ecotoxic pharmaceuticals, personal care products, and other emerging contaminants: a review of environmental, receptor-mediated, developmental, and epigenetic toxicity with discussion of proposed toxicity to humans. Crit Rev Environ Sci Technol 46:336–381
Wilkinson J, Hooda PS, Barker J, Barton S, Swinden J (2017) Occurrence, fate and transformation of emerging contaminants in water: an overarching review of the field. Environ Pollut 231:954–970
Xiang Y, Wang X, Zhang X, Hou H, Dai K, Huang Q, Chen H (2018) Enhanced visible light photocatalytic activity of TiO_2 assisted by organic semiconductors: a structure optimization strategy of conjugated polymers. J Mater Chem 6:153–159
Yang N, Liu Y, Wen H, Tang Z, Zhao H, Li Y, Wang D (2013) Photocatalytic properties of graphdiyne and graphene modified TiO_2: from theory to experiment. ACS Nano 7:1504–1512
Yaqoob AA, Parveen T, Umar K, Ibrahim MNM (2020) Role of nanomaterials in the treatment of wastewater: a review. Water 12(2):495
Yu W, Chen J, Shang T, Chen L, Gu L, Peng T (2017) Direct Z-scheme g-C_3N_4/WO_3 photocatalyst with atomically defined junction for H_2 production. Appl Catal B 219:693–704
Zaied BK, Rashid M, Nasrullah M, Zularisam AW, Pant D, Singh L (2020) A comprehensive review on contaminants removal from pharmaceutical wastewater by electrocoagulation process. Sci Total Environ 726:138095
Zaman MF, Akter MS, Muhit IB (2014) Pharmaceutical wastewater management: a review. In: 2nd International Conference on Advances in Civil Engineering. CUET, Chittagong, Bangladesh, pp 201–206
Zhang Q, Chen P, Tan C, Chen T, Zhuo M, Xie Z, Wang F, Liu H, Cai Z, Liu G (2018) A photocatalytic degradation strategy of PPCPs by a heptazine-based CN organic polymer (OCN) under visible light. Environ Sci Nano 5:2325–2336
Zhang S, Chen Y, Yu Y, Wu H, Wang S, Zhu B, Huang W, Wu S (2008) Synthesis, characterization of Cr-doped TiO_2 nanotubes with high photocatalytic activity. J Nanopart Res 10:871–875
Zhang Q, Du R, Tan C, Chen P, Yu G, Shubo Deng S (2021) Efficient degradation of typical pharmaceuticals in water using a novel TiO2/ONLH nano-photocatalyst under natural sunlight. J Hazard Mater 403:123582
Zhang G, Yang W, Yang F (2007) Electrochemical behavior and electrocatalytic activity of anthraquinone disulphonate in solution phase and as doping species at polypyrrole modified glassy carbon electrode. J Electroanal Chem 602:163

Chapter 12
Photocatalytic Decolourization of Dyes Using Nanoparticles-Assisted Composite Catalysts

Sameena Mehtab, Shubham Sharma, Minakshi Pandey, and M. G. H. Zaidi

Abstract Over the years, demand of dyes is tremendously increased due to the continuous growth of textile, paper and leather market size. Effluents from these industries are challenging sources of residual dye pollutants into the environmental water. Coloured water deteriorate the water quality, decline dissolved oxygen levels, damage photosynthesis, enter the food chain and may lead to severe health hazards. Treatment and recycling of dyes-polluted water can help in conservation of water and preventing water pollution footprint in environment. There exist a remarkable method known as photocatalysts which may be utilized for achieving various goals such as antifogging, antifouling, production of hydrogen, antibacterial activity, degradation of different kinds of pollutants in wastewater, sterilization, self-cleaning, deodorization, conservation and storage of energy and purification of air. This process is gaining more focus towards wastewater treatment for complete mineralization of the pollutants under mild temperature and pressure requirements. A photocatalyst must have good light UV–Vis light absorption capability for better photocatalytic performance. Recently, a process of dye decolourization with the help of nanocomposites has been turning into a prominent technology in the direction of environmental remediation. Due to the ability to stabilize the excited electron in the conducting band via reduction in rate of hole/electron recombination and decreasing the semiconductor's band energy, the nanocomposites have gained significant high photocatalytic efficiency over the nanoparticles. Enhanced photocatalytic efficiency can be achieved by synthesizing specific nanocomposites through conducting polymers with metal/metal oxides nanoparticles.

S. Mehtab (✉) · S. Sharma · M. Pandey · M. G. H. Zaidi
Department of Chemistry, College of Basic Sciences and Humanities, G. B. Pant University of Agriculture and Technology, Pantnagar, Uttarakhand 263145, India
e-mail: smiitr@gmail.com

J. P. N. Rai and S. Saraswat (eds.), *Nano-biotechnology for Waste Water Treatment*, Water Science and Technology Library 111,
https://doi.org/10.1007/978-3-031-00812-2_12

Graphical Abstract

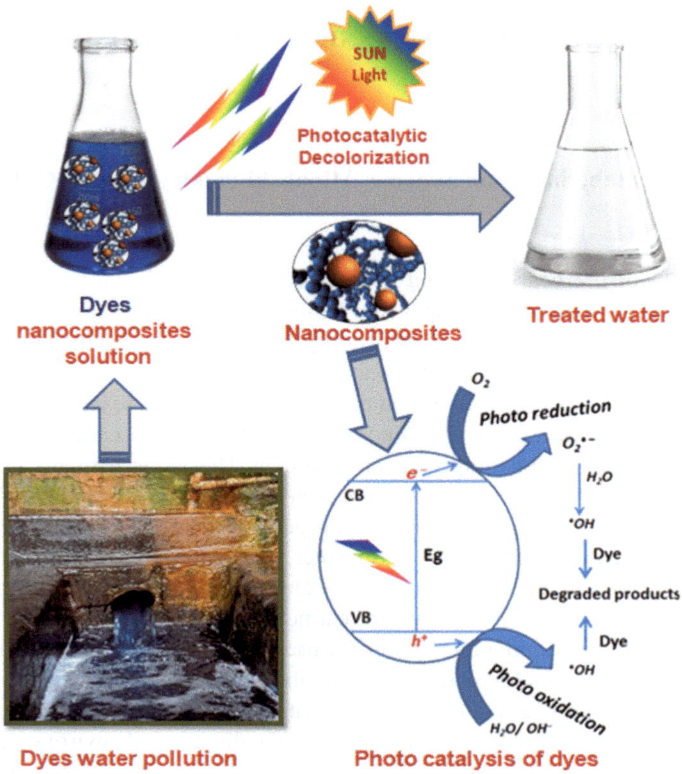

Keywords Photocatalyst · Classification · Decolourization · Nanocomposites · Dyes · Photoexcitation

12.1 Dyes and their Classifications

Dyes are natural or synthetic chemicals used to impart different colours in any substance. Chemical structure of dyes possesses conjugated system, electron resonance and one or more colour-bearing group (chromophore) that can undergo π–π^* and n–π^* transitions, that absorb light in visible region of spectrum (400–700 nm) and impart colours in dye molecules. There are many ways to classify dyes; a very common classification of dyes is based on the origin by which it is made, accordingly classification is given as below (Allen 2013; Hunger 2007).

12.1.1 Natural Dyes

Dyes are derived from plants, fungi, lichens and minerals sources without any chemical processing. Since very beginning man started dyeing their clothes with dyes of natural origins. The majority of natural dyes are derived from plant sources like root, seeds, leaves, flowers, barriers, bark, wood and fruits. Natural dyes do not require any special care, they are wonderful and rich in tones, act as health cure especially as food dyes have which fortunately come with no problems related to disposal, and also with no effects such as carcinogenic effect with the quality of being an easily biodegradable. The requirements for Natural dyes are as simple as a dye house to use for applying on matrix, and along with it, a mild reaction condition has to be there for the purpose of their extraction and application. However, Natural dyes also come with some limitations such as poor colour yield, poor fastness properties, limited colour patterns and difficult process of blending dyes (Merdan et al. 2017).

12.1.2 Synthetic Dyes

They are primarily used worldwide for their application in printing textiles which attach themselves via chemical bonding that takes place between the molecules of dyes and the fibres. These are basically a class of highly coloured organic substances. However, there are various manners in which the classification of dyes can be done such as, according to the application of the materials, chemical structure and constitution, etc. (Chiu et al. 2019). A lot of organic dyes fall under the category of synthetic dyes that are used in the textile industries. Apart from organic dyes, inorganic dyes like sodium dichromate are also utilized for the extraction of certain specific colours. The classification of organic dyes is generally based on their chemical constitution. But when it comes to textile processing, dyes classification takes place on the basis of their usage and some important well-known types of dyes are briefly described here.

12.1.2.1 Reactive Dyes

If we talk about their proper names for the reactive dyes, then they are referred to with the name of "fibre reactive dyes", which actually are the water-soluble anionic compounds. When it comes to obtaining the desired bright shades, the dyes belonging to this class are the largest ones which are extensively used on the cellulosic (cotton) for the same purpose. Reactive dyes are also very well-suited for nylon, wool and silk too. The auxiliaries that are needed include sodium chloride, sodium carbonate, urea and buffers (Benkhaya et al. 2020).

12.1.2.2 Disperse Dyes

Disperse dyes holds the property of being water insoluble, therefore, they are mainly used for dyeing polyester. Dyeing synthetic fibres such as cellulose acetate and polyamides is one of the minor use cases of them. Among all the dyes, the title of the smallest dye molecule is held by the disperse dye molecules. The composition of dispersed dye molecules is azobenzene or anthraquinone molecule based with attached nitro, amine, hydroxyl, etc. (Harfi and Harfi 2017).

12.1.2.3 Acid Dyes

When it comes to dyeing wool and other animal fibres, the acid dyes are the exclusive pick for it. They are widely used on nylon in the scenario where high wash fastness is a must. Acid dyes are water-soluble anionic salts of sodium and sulphuric acid or organic acids. The other requirements for the perfect dyeing via acid dyes are: sulphuric acid, formic acid, ammonium phosphate, ammonium sulphate, sodium sulphate, acetic acid, ammonium acetate and levelling agents.

12.1.2.4 Sulphur Dyes

These dyes are water-insoluble in nature and composed of either sulphur or sodium sulphide. The application of sulphur dyes is usually carried out at high temperature of 600–1000 °C in alkaline solution medium, mainly on cotton for economical dark shades. Reduced sodium sulphide solution of sulphur dyes is applied, afterwards it gets reoxidized to insoluble form on the fibre. Sulphur dyes are mainly applied to cotton for economical dark shades. When it comes to resistance, they are excellent for washing but perform poorly in sunlight. They usually provide very poor fastness to chlorine. The other requirements for the perfect dyeing via acid dyes are: sodium sulphate, acetic acid, sodium sulphide, sodium chloride, sodium carbonate, hydrogen peroxide and sodium dichromate.

12.1.2.5 Basic Dyes

Basic dyes also referred to as "cationic dyes" are water-soluble in nature. The dyes belonging to this class are known for their bright colours. It also has good fastness and is majorly applied on polysters, silk, wool and acrylic fibres. Most often these dyes fall under the category of hydrochlorides or salts of organic bases. Their application follows along with organic acids such as sodium sulphate, formic acid, acetic acid, oxalic acid, tannic acid and sodium acetate.

12.1.2.6 Direct Dyes

They are also referred to as "substantive dyes". These are mainly azo compounds with sodium salts of sulphonic acids. They are anionic compounds which are soluble in water. Their primary use case is for dyeing cotton. This class of dyes is used by most of the small dyeing houses because of the low cost they offer, and how easy they are to apply. And they even require less auxiliary chemicals which include sodium chloride, sodium sulphate, sodium nitrite, sequestering agents, hydrochloric acid and aromatic amines. Direct dyes offer colours that are stable for washing but the colour fastness they offer is not up to the mark.

12.1.2.7 Mordant Dyes

This category comprises both types of dyes—natural dyes and synthetic dyes. Because of the fact that they show no affinity for textile fibres, the fibres are mordant with a metallic oxide before being applied. Chromium oxide is the most commonly used mordant, therefore giving these dyes the other name of chrome dyes. Other chemicals needed are sulphuric acid or formic acid, sodium (or potassium) dichromate, acetic acid, sodium sulphate, ammonium sulphate and penetrating agents.

12.1.2.8 Vat Dyes

If we talk about resistance to both washing and sunlight, the Vat dyes are considered to be the most resistant towards both. They are made from indigo, anthraquinone and carbazole and are water-soluble in nature. After this process, they are applied to fibre and then reoxidized to the original insoluble form. Other chemicals which are needed for dyeing with vat dyes include caustic soda, sodium hydrosulphite, dispersing agents, hydrogen peroxide and perborate.

12.1.2.9 Azoic Dyes

Azo dyes are composed of azo component (–N = N–) and used for serving the purpose of cotton fabrics dyeing. They are widely used dyes and largest component of textile dyes that comprise 6–70% of all textile dyestuff in practice. In the dyeing process, colours are formed inside fibres, by in situ reaction of coupler followed by application of azo dyes.

12.1.2.10 Pre-metallized Dyes

These are mordant dyes that carry metallic oxide (usually chromic) in their structure. This vanishes any need for the adding of dichromates in the entire process. These dyes are applied on nylon, silk and wool.

12.2 Dyes-Based Water Pollution

In the growing world, requirements of synthetic dyes are continuously increasing due to the wide range of applications in the textile, paper, leather, paints, cosmetics, pharmaceuticals and food industries. Worldwide, up to 10,000 different dyes are used and their total annual production is around 7×10^5 metric tons. Among all industries, the textile industry contributes towards the most elevated release of dye effluents to the environment (Lellis et al. 2019). Wastewater discharge from dyes processing units is one of the major environmental concerns. Dyes impart different colours in water bodies that hindered the photosynthesis by decreasing the light penetration in water and in turn affecting the food source of aquatic life form (Hassan and Carr 2018). Dyes water pollution also disturbs the aquatic biota by decreasing the dissolved oxygen levels in water bodies (Imran et al. 2015). The presence of even a small measure of dye in water causes deterioration in water bodies likewise lessens the solubility of dissolved oxygen subsequently expanding the natural oxygen demand of water, upgrades the turbidity of water and annihilates the food web of that specific region. In addition, the effluent of dyes components from industries deteriorate the aquatic flora and fauna by generating a thin deposition over water bodies.

Dyes are being designed with a combination of complex chemicals and when these dyes get released in water they undergo oxidation and reduction, which produce hazardous by-products which further increase need for their removal from wastewater. Dyes and their by-products when leach out in water do not break down and stay for a very long period of time and disturb the water ecosystem (Chequer et al. 2013). Dyes are serious environmental pollutants and can cause severe toxic, allergic, mutagenic and carcinogenic effects if entered in food chains (Sandhya 2010). Dyeing industries released huge amount of dyes containing coloured effluents into environmental bodies. Dyes have adverse effects on all forms of life and it is important to treat dyeing water effluent before their final disposal. Dyes are complex, long-lasting and toxic pollutants in the climate, there is a high scope for investigating and creating different strategies for their degradation.

12.3 Existing Treatment Options

In the course of recent years, substantial investigations were carried out for the decolourization of industrial effluents from dyeing effluents. Both industrial and municipal wastewater required diverse treatment methods to eliminate dyes, prior to their discharge into aquatic environment. Depending on the complexity of polluted water, so far different methods have been used for dye degradation in the environmental samples. Some important ways for dye degradation reported are sono-catalytic degradation (Salavati et al. 2012; Li et al. 2014; Taghizadeh and Seifi-Aghjekohal 2015) ozonation (Hu et al. 2016), chemical oxidation (Esteves et al. 2016; Al Angari et al. 2019), adsorption (Hashem and El-Shishtawy 2001; El-Shishtawy and Melegy 2001; El-Shishtawy and Soltan 2002; Sathian et al. 2014; El-Zahhar and Awwad 2016), photocatalytic degradation (Cai et al. 2015; Chandra et al. 2016; Elango and Roopan 2016; Kumar et al. 2016a, 2016b), coagulation (Wei et al. 2015; Huang et al. 2015), microbiological decomposition (Kumar et al. 2016a, 2016b) and adsorption processes (Hao et al. 2000). However, in some of these methods, dye's pollutants are only transferred from one phase to another and leaving the problem mostly unsolved. Conventional methods of treatments for example, adsorption on activated carbon, compound precipitation, and separation, coagulation are non-damaging strategies and just exchange dyes starting with one phase to the next, causing secondary pollution that requires further treatment. There is a critical requirement for an exceptionally productive framework to treat dyeing water for the total degradation of harmful particles into simpler non-toxic forms. Numerous industries and research centres treat chemical dyeing water utilizing different methodologies before releasing them into environment. Regularly rehearsed approaches incorporate physical methods—sedimentation, floatation and filtration. Microbiological methods can be used for dye degradation, but these methods are quite slow and cannot degrade complicated dyes. The chemical methods include coagulation or flocculation, filtration, ozonization, illumination, adsorption, photolysis and particle pair extraction. These compound strategies are helpful yet uneconomical as they require high energy and enormous usage of the chemical reagents for the degradation of dyes from the water bodies, so they are not in much practice. These techniques are extravagant and demand high set-up expenses and are incapable to meet restrict effluent colour standards. Therefore, considerable consideration has been centred around the complete oxidation of organic compounds to innocuous items, for example, CO_2 and H_2O by the advance oxidation–reduction measures.

Recently, photocatalysis emerged as the latest green technology relying on advanced redox processes and is quite proficient in producing varied reactive species with the likes of oxygen and nitrogen species, oxygen and hydroxyl radicals. Photocatalysis has gained massive attention because of its potential application in degradation of different types of organic dye molecules from wastewater (Cavassin et al. 2015; Ramazanzadeh et al. 2015; Waiskopf et al. 2016; Nosaka and Nosaka 2017; Podporska-Carroll et al. 2017).

12.4 Photocatalysis Approaches: Basics and Types

Photocatalysis has been raised as a capable technology for considering the global environmental issues related to the water pollution caused by the dyes. In the word "photocatalysis", prefix 'photo' means something that is related to photon or light and 'catalysis' refers to the process of accelerating the rate of chemical reaction without altering the end product. The catalyst is a substance, which reduces the activation energy to increase the rate of reaction. Photocatalysis is a process including reactions that take place through absorbing light by photocatalyst for accelerating reactions rate. Generally photocatalysts are semiconductors in nature that generate electron–hole pair on exposure to the light. On the basis of physical state of the reactants, photocatalytic reactions can be categorized in two categories as homogeneous and heterogeneous photocatalysis.

12.4.1 Homogeneous Photocatalysis

Homogeneous photocatalysis involves the reactions in which both photocatalyst and medium are same in same phase. They contain an optimized single phase solution including soluble catalysts as photosensitizer or light absorbing system and act as redox reactions catalytic sites for target substrate (dyes). The presence of a light absorber is essential for this process as it is responsible for chemical modification into desired products. After activation in excited state photon-absorbing species accelerates the target chemical reaction by interacting with light and substrate. Transition-metal complexes show suitable stability and electronic band gap that make them a good choice to use as homogeneous photocatalysis. Most common application of homogeneous catalysis is the formation of $OH^{\bullet}$ radicals using ozone in photo-Fenton system to generate accessible and clean fuel in the form of molecular hydrogen from solar energy (Tahir et al. 2020).

12.4.2 Heterogeneous Photocatalysis

Photocatalytic reactions in which the phases for both catalyst and reactant are not same are known as heterogeneous photocatalysis (Gaya and Abdullah 2008). Heterogeneous photocatalysis has showcased themselves as an efficient tool for various important reactions like, water splitting, photoreduction, hydrogen transfer, organic synthesis, and metal deposition, water and gaseous detoxification, etc. (Carp and Huisman 2004). Semiconductor materials like ZnO, TiO_2, Fe_2O_3, CdS, MgO, WO_3, etc., have been proven as good choice to be used as photocatalysis (Rajeshwar et al. 2008). These materials have photostability, inertness to chemical environment and wide range of band gap, due to which there is acceleration in creating the chemical equilibrium between reagent and products in the heterogeneous system. In this process the photocatalytic efficiency depends on the interaction of the catalyst with the photon to produce electron/hole pairs that generate free radicals for onward reaction. There is a simultaneous transfer of the excited electrons to the reducible specimen which is accepted by catalyst from the holes-occupied oxidizable specimen. In both directions, the net flow of the electrons turns out to be null with the catalyst being unaltered.

Energy difference between lowest unoccupied molecular orbital (conduction band) and highest occupied molecular orbital (valence band) is defined as band gap. Materials with band gap energy (Eg) in the range 1.5–3.0 eV behave as semiconductors and exhibit the ability to conduct electricity even at the room temperature. All photocatalysts are mainly semiconductor in nature and on expose to light filled valence band electron absorb required energy and get excited to empty conduction band to produce an electron–hole (e^-h^+) pair or charge carriers.

$$\text{Catalyst} + hv \rightarrow e^- + H^+$$
$$H_2O + h^+_{vb} \rightarrow OH + H^+$$
$$e^-_{cb} + O_2 \rightarrow O_2^-$$
$$O^{2-} + H^+ \rightarrow HO_2$$
$$2HO_2 \rightarrow H_2O_2 + O_2$$
$$H_2O_2 \rightarrow 2OH$$

Excited electrons work for reducing an acceptor and holes are used for oxidation of donor molecules. The ability to provide an environment for both the oxidation and reduction makes the photocatalyst important for the photocatalysis. The relative position of the semiconductor's valence bands and conduction's together with the substrate's redox levels are deciding factors for destiny of the excited electron and hole. There are four ways for the interaction to take place between the semiconductor and substrate that happens based on the relative positions of conduction and valence bands and also the redox levels. These four different ways are:

- For the reduction of substrate to happen, the redox level of substrate has to be lower compared to the semiconductor conduction band.
- Substrate get oxidized, when substrate redox level is higher than semiconductor valence band.
- When substrate redox level is higher than the conduction band and lower than the valence of the semiconductor, there is no possibility of oxidation as well as reduction.
- When the redox level of the substrate is lower than the conduction band and higher than the valence band, both the reduction and the oxidation of the substrate can happen.

Cu_2O is a p-type photocatalyst with band gap 2.0–2.2 eV in the visible range, which has to deal with the photo-wavering and photo-corrosion. We can utilize the developed Cu_2O nanocomposite with n-type semiconductor like Fe_2O_3 to get rid of these drawbacks.

12.5 Nanocomposites for Photocatalysis Degradation of Dyes

Decolourization and degradation of dyes effluents has been a challenging topic for the researchers all over the world. Nanocomposites based photocatalyst have enhanced light absorption for harvesting more photonic energy and found to be promising materials for degradation of dyes. Photocatalytic nanocomposites are specially designed coupled of two or more semiconductors that display higher redox behaviour than photogenerated charge carriers. Photocatalytic nanocomposites have recombination of photocharge carriers with high surface area that exposes more active sites for catalytic activity (Loddo et al. 2018; Khan and Lee 2021; Xu et al. 2019; Loghambal et al. 2018). The core of this technology lies when nanocomposites are irradiated by light energy (higher than band gap of nanocomposite), there is transfer of an electron to the empty conduction band of the nanocomposites from the fully filled valence band (Kumar et al. 2020).

Nanocomposites are effective materials, when excited with visible or UV light generates a reactive hydroxyl radical and superoxide, which rapidly degrade the dye molecules or organic pollutants (Rajagopalan 2016). So there is a need to develop synergistically coupled nanocomposites that produce reactive oxygen radicals for efficient dye removal from water.

12.5.1 Nanoparticles Zeolite Composite Based Photocatalysts

Catalytic activity of nanoparticle is improved by employing porous supports materials of large area and high ion exchange capacity as matrices or stabilizers in composite. Recently e.g., Zeolites supported $ZnO{-}Fe_2O_3{-}MnO_2$ nanocomposite have been reported greater effects on photocatalytic degradation of dyes to escalate the wavelength of excited light that improves light catalytic activity and utilization efficiency. Investigations suggested that by encapsulating the non-metal ion such as sulphur, phosphorus, nitrogen, carbon or oxygen into oxide fine particles through mechanochemical methods, the band gap of ZnO can be narrowed. These modified ZnO composite have stronger absorption in the visible region which improves the degradation of organic dyes pollutants. The conventional liquid impregnation process synthesizes zeolite immobilized TiO_2 nanocomposite, which were used to efficient degradation of methylene blue dye (Saqib et al. 2019).

12.5.2 Nanoparticles Carbon Material Composite-Based Photocatalysts

Carbonaceous materials composite are excellent choice for synthesizing the modified photocatalysts. Activated carbon, carbon quantum dots, carbon nanotubes (SWCNT or MWCNT)/nanofibres, graphene, fullerene and carbon aerogels have been investigated as carbonaceous photocatalysts. Carbon component increases hetero-conjunctions and surface area of carbon phase in the composite of carbon semiconductors that can trap photogenerated electrons to suppress the recombination of electron and hole (Ge et al. 2019). These modified carbonaceous photocatalyst have capability for enhancing the capacity of visible light absorption, reaction sites, the electron transportation, adsorption capacity of reactants and utilization stability.

Pd co-doped TiO_2 on double-walled carbon nanotube-based nanocomposites was prepared by a modified sol–gel method that results in enhanced photocatalytic activity towards degradation of eosin yellow. Carbon nanotubes change the morphology and increase optical properties of nanocomposites and work as excellent electron transfer channels to provide high electrical conductivity and electron storage capacity.

Manganese dioxide nanoparticles/activated carbon (MnO_2/AC) composites were prepared through chemical reduction method and utilized for photodegradation of Congo red dye. Morphological study also revealed that MnO_2 NPs were well-dispersed on AC and their degradation efficiency to degrade CR dye is more than MnO_2 NPs (Khan et al. 2019).

Graphitic carbon nitride ($g-C_3N_4$)/TiO_2 nanocomposite shows 97% greater efficiency towards the degradation of rhodamine-B dye with modified surface area and higher absorption of light (Monga and Basu 2019).

12.5.3 Nanoparticles Conducting Polymers Composite-Based Photocatalysts

The conducting polymers have π-conjugated system and have high photogenerated electron–hole pairs charge separation to work as excellent nanocomposites. Conducting polymer nanocomposites provides better surface for catalysis under light illumination and has electronic characteristics like σDC and σAC electrical conductivity, low energy optical transition, high electron affinity and photoelectric properties. Metal nanoparticle decorated conducting polymer nanocomposites have gained much attention towards photocatalytic materials for dye degradation. Conducting polymers like polypyrrole (PPy), polyaniline (PANI), polythiophene (PTh) and polyacetylene (PAc) act as photosensitizer for conducting and semiconducting material because of their extended π-conjugated electronic system with high charge carriers.

The composite of polyaniline reinforced with BiOCl have shown increased absorption intensity under visible region by chemisorption method owing to synergistic effect between PANI and BiOCl, primary used for photodegradation of methyl orange dye (Wang et al. 2013).

The in situ oxidative polymerization method was used for synthesizing a nanocomposite of polyaniline with TiO_2 and graphene (PANI@TiO_2/GN) (Kumar et al. 2016c). Characterization evaluates to well-distributed peaks of TiO_2 within PANI and GN nanosheets network, which points towards the successful formation of PANI@TiO_2/GN nanocomposites. Higher visible light absorption of PANI@TiO_2/GN over PANI@TiO_2 and good photocatalytic activity for methylene blue degradation are presented via optical analysis results. The sensitizing effect of PANI and electron scavenging property of GN are the key factors behind the high photocatalytic activity of PANI@TiO_2/GN.

12.5.4 Nanoparticles Biopolymer Composite-Based Photocatalysts

Recently, the photocatalytic capacities have been found to increase using nanoparticles embedded in biopolymeric composites with enhanced surface area and active sites. Due to biodegradable and non-toxic nature of biopolymers, their use has been attracted in the development of nanocomposites. The modification of nanoparticles using biopolymer accelerated photocurrent transport by increasing the number of photogenerated electrons and holes (Sarkar et al. 2020).

Nanocomposite was synthesized by barium-assisted cross-linking of carboxymethyl cellulose (CMC) with sodium alginate and ZnO nanoparticles. 60 mg nanocomposite of developed alginate-CMC:ZnO (in ratio 1:0.8) gives 90.12% photocatalytic degradation activity for congo red dye solution under sunlight irradiation at pH 3 with reaction time of 110 min (Ramadhani and Helmiyati 2020). Cellulose has also been considered as a supporting material to developed nanocomposites with ZnO nanoparticles that reflect improved photocatalytic efficiency for methylene blue photodegradation (Melinte et al. 2019).

Biocomposite-based photocatalyst comprised of nanochitosan, CMC and TiO_2 was reported for crystal violet dye degradation. Chitosan metal nanocomposites have been reported for photocatalytic degradation of several dyes as, chitosan/Pd for congo red (Sargin et al. 2020), chitosan/Fe for basic dye (Kasiri 2019) and chitosan/La for Azo dyes (Sirajudheen and Meenakshi 2019), etc.

The co-precipitation of sodium tungstate and calcium chloride with commercial biochar synthesized the calcium tungstate ($CaWO_4$)-doped biochar nanocomposites, which were used for degrading and decolourizing methyl orange and rhodamine dyes (Zhang et al. 2019) (Fig. 12.1and Table 12.1).

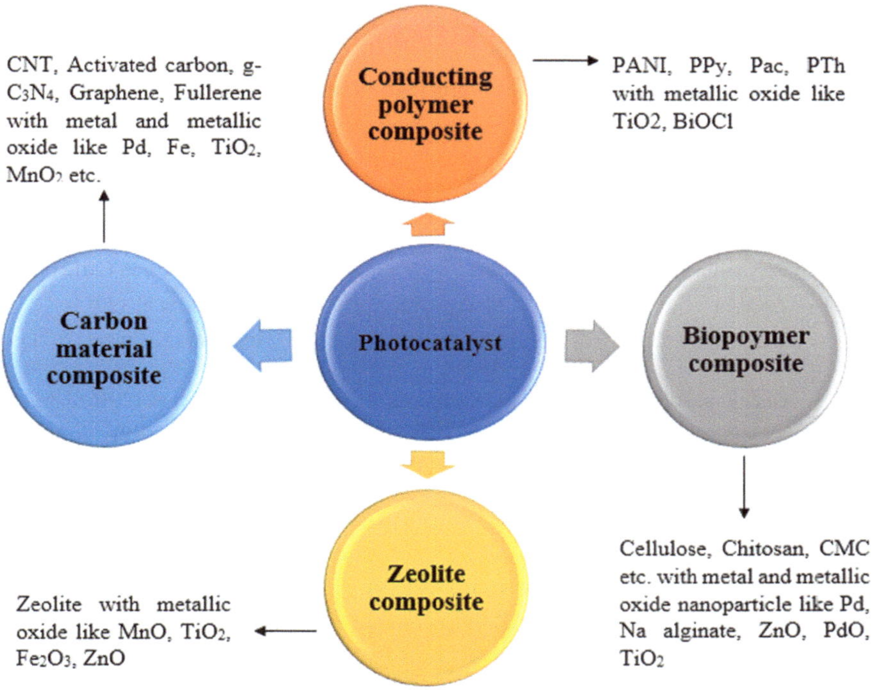

Fig. 12.1 Different Nanocomposites-based Photocatalysts

12.6 Mechanism for Photocatalyzed Dye Degradation

Photocatalysis is a photon-induced catalytic reaction in which oxygen acts as an electron acceptor and creates reactive oxygen species which breaks the dye molecules (Ajmal et al. 2014; Yang et al. 2005). In this eco-friendly and inexpensive method, the polymeric nanomaterials are used for degrading the dyes from wastewater.

There are two types of mechanisms for dye degradation as discussed follows.

- Indirect dye degradation mechanism
- Direct dye degradation mechanism

12.6.1 Indirect Dye Degradation Mechanism

In this mechanism, the radical species (e^- and h^+) are generated by the photoexcitation of the semiconductor material that is responsible for the degradation of dyes. Generally here semiconducting materials act as a photocatalyst, which may irradiate the light energy for stimulating redox process.

Fundamental steps for indirect dye degradation are as follows.

Table12.1 List of nanocomposites utilized for efficient dyes removal

Sr. No	Nanocomposites	Dye employed	References
1	Carbon quantum dots -modified ZnS	Alizarin red S (ARS)	Kaur et al. (2016)
2	Urea Derived $g-C_3N_4/ZnO$	Methylene Blue (MB) Rose Bengal	Alharthi et al. (2020)
3	Graphitic carbon nitride ($g-C_3N_4$) /TiO_2	Rhodamine B (RhB)	Monga and Basu (2019)
4	Polyaniline/ZnO	MB	Olad and Nosrati (2012)
5	ZnO/Ag	Methyl Orange (MO)	Kuriakose et al. (2014)
6	TiO_2/MgO	MO ARS	Arikal and Kallingal (2021)
7	Guar gum/Al_2O_3	Malachite Green (MG)	Pathania et al. (2016)
8	SnSe/SnO_2	MB RhB	Kaur et al. (2019)
9	ZnS/Fe_2O_3/Ag	MG Eosin blue (EB)	Sunkaraa and Botsab (2020)
10	ZnO/CeO_2	MB	Saravanan et al. (2012)
11	GO/TiO_2	Orange ME2RL	Kumaran et al. (2020)
12	Cu_2O/Cu	MB	Li et al. (2018)
13	Al_2O_3/ZrO_2	Reactive Blue 222 Reactive Yellow 145	Yaghoubi et al. (2020)
14	TiO_2/reduced graphene oxide (RGO)	RhB Acid orange II (AO-II)	Wang et al. (2012)
15	TiO_2/diatomite composites	RhB MO MB	Zhang et al. (2016)
16	TiO_2/CuO	MO	Koohestani and Sadrnezhaad (2016)
17	ZnO/MWCNT	RhB	Yan et al. (2009)
18	Polyaniline/bismuth selenide(Bi_2Se_3)	MO RhB MG	Chatterjee et al. (2019)
19	CNT/TiO_2 nanotube	MO	Jiang et al. (2013)
20	Zirconium oxide (ZrO2)/graphene(GR)	MB RhB	Rani et al. (2016)
21	Metal-fullerene/TiO_2	MB	Meng et al. (2012)
22	ZrP/AgBr	RhB	Pica et al. (2019)
23	Zirconium/titanium/chitosan	Orange II	Demircivi and Simsek (2018)

(continued)

Table12.1 (continued)

Sr. No	Nanocomposites	Dye employed	References
24	Iron-modified biochar	MB Acid orange 7	Qiu et al. (2020)
25	Ag/TiO_2/biochar composite	MO	Shan et al. (2020)
26	Nano-zerovalent manganese/biochar composite	Congo red	Iqbal (2021)

12.6.1.1 Photoexcitation

Photoexcitation takes place when there is sufficient energy gap between filled valence and empty conduction band. Due to electron excitation process, a hole generates in the valence band (h_{VB}^+) and an electron in conduction band (e_{CB}^-).

The Eq. (12.1) indicates net result of generated electron and hole pair.

$$\text{Semiconductor material} + h\upsilon \rightarrow e^-(CB) + h^+(VB) \quad (12.1)$$

where, h: Planck's constant 6.626×10^{-34} m^2kg/s and v: frequency of incident light

12.6.1.2 Ionization of Water

The ionization of water molecule takes place by the hole generated in the filled valence band as a result of visible beam irradiation and forms the reactive hydroxyl radicals.

$$h^+ + H_2O \rightarrow H^+ + OH^- \quad (12.2)$$

$$h^+ + OH^- \rightarrow OH^\bullet \quad (12.3)$$

The hydroxyl radicals formed on the surface of irradiated semiconductor are strong oxidizing agent, which can easily oxidize organic dyes contaminants.

12.6.2 Oxygen Ionosorption

The electron formed in conduction band reduces the oxygen molecule by forming superoxide ion electron (O_2^-), which prevents the electron–hole recombination and induces the redox reaction.

$$e^- + O_2 \rightarrow O_2^{-\bullet} \quad (12.4)$$

12.6.2.1 Protonation of Superoxide

The hydroperoxyl radical ($HO_2^{\bullet}$) in conduction band can be formed by protonating superoxide ($O2^{-\bullet}$) ion electron and afterwards H_2O_2. The peroxide is highly unstable hence dissociates into hydroxyl radical.

$$O_2^{-\bullet} + H^+ \rightarrow HOO^{\bullet} \quad (12.5)$$

$$2HOO^{\bullet} \rightarrow H_2O_2 + O_2 \quad (12.6)$$

$$H_2O_2 \rightarrow 2OH^- \quad (12.7)$$

$$\text{Dye} + OH^- \rightarrow CO_2 + H_2O \quad (12.8)$$

$$\text{Dye} + h^+ \rightarrow \text{oxidation products} \quad (12.9)$$

$$\text{Dye} + e^- \rightarrow \text{reduction products} \quad (12.10)$$

On the surface of photoexcited photocatalyst semiconductor, both oxidation and reduction activities are prevalent. The entire procedure is depicted in Fig. 12.2.

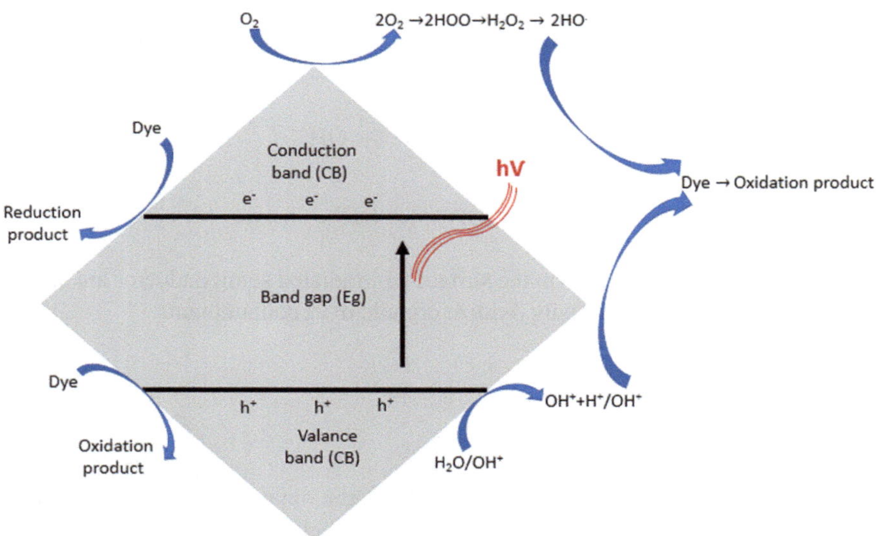

Figure12.2 Indirect dye degradation mechanism

The absorbed surface water of photocatalyst oxidized by h_{VB}^{+}, due to the shifting of electrons from the filled valence band, forms strong oxidative decomposer hydroxyl radicals ($OH^{\bullet}$). The reactive hydroxyl radical react with dyes organic matter and ultimately decomposes it into carbon dioxide and water (Eq. 12.9).

12.6.3 Direct Mechanism for Dye Degradation

In the direct dye degradation mechanism, the excitation of dyes occurs from ground state to excited state after the irradiation under visible light photon (1400–800 nm). This process is followed by injecting an electron in conduction band of semiconductor to convert dye species into a semi-oxidized radical dye cation ($dye^{+\bullet}$). Now the reaction between trapped electrons and dissolved oxygen takes place, which lead to superoxide radical anions ($O_2^{\bullet}$). The produced superoxide ($O_2^{-\bullet}$) gets protonated and forms hydroperoxyl radical ($HOO^{\bullet}$) and afterwards hydrogen peroxide (H_2O_2). The peroxide formed in above step is further dissociated into reactive hydroxyl radicals ($OH^{\bullet}$), which is responsible for the oxidation of organic compounds.

$$\text{Dye} + \text{h} \rightarrow \text{dye}^{*} \tag{12.11}$$

$$\text{Dye}^{*} + \text{TiO}_2 \rightarrow \text{dye}^{+} + \text{TiO}_2 \tag{12.12}$$

12.7 Factors Affecting Photocatalytic Dye Degradation

The reaction that occurs on the surface of photocatalyst depends on various parameters including semiconductor properties, particle size, absorption capacity, reaction temperature, reaction time, pH and concentration of dyes. Surface properties of catalyst also have an influence on the operating efficiency of the entire photodegradation. Some important properties are discussed below.

12.7.1 Surface and Structural Properties of the Semiconductor

Photocatalyst surfaces with suitable redox potentials, good adsorbent properties, high exposed facets with thermal stability and efficient towards solar energy absorption shows better catalytic performances (Ahmed et al. 2010).

12.7.2 Effect of Particle Size

There are various factors on which the efficiency of the photocatalysts depends such as crystal structure, size of the particle, shape, surface area, porosity, composition of crystals, electrical properties, surface functionalities and band gap as well (Salah et al. 2004). The nanomaterial with larger surface area but smaller size results in efficient photocatalytic performance over materials in bulk. There is a huge accumulation of the atoms on the surface of a catalyst when its size is smaller. This is the reason for its increased surface to volume ratio. This property is helpful to improve transfer rates and the number of active sites of interfacial charge carriers, therefore leading to the achievement of higher catalytic activities (Cernuto et al. 2011). The spherical ZnO molecule demonstrates much greater efficiency, when compared to rod-shaped and spindled ZnO samples which is ultimately because of its large surface area (Saravanan et al. 2013a, 2013b).

12.7.3 Effect of pH

The pH of degradation reaction has influenced the ability of reaction rates and surface charge of the photocatalyst. The highest degree of photo decolourization of methyl orange was observed at pH 2, due to electrostatic attraction between the catalyst surface and methyl orange (Zhu et al. 2012). However, the results of dye treatment fetched on the basis of the dye adsorption strength can't be explained due to the presence of various other parameters which operate simultaneously. For example, it has been found that on increasing the pH from 1 to 9 leads to decolourizing of the reactive orange 4 from 25.27% to whooping 90.49% just after 40 min and degradation amount changes from 15.16% to 87.24% just within 80 min, respectively (Muruganandham and Swaminathan 2004).

The dye degradation mechanism generates positive holes as major species for the oxidation reaction at lower pH and hydroxyl radical at higher pH range (Tunesi and Anderson 1991; Goncalves et al. 1999).

It is reported that decolourization and degradation rate of dyes are increasing with increase in pH (Mahvi et al. 2009) and concentration of H^+ ions exceed at lower pH range, which is responsible for reducing the azo group electron densities by interacting with the azo linkage. This ultimately leads to decrease in the reactivity of hydroxyl radicals ($OH^{\bullet}$) by electrophilic mechanism. It's also reported that, when dealing with photocatalysts, a cationic dye shows higher rate of adsorption than anionic dyes (Baran et al. 2003; Tang and Chen 1996; Neppolian et al. 2002). TiO_2 is a strong oxidizing agent at lower pH, but when H^+ is in excess that too at a very low pH, it results in a decreased reaction rate due to agglomeration of TiO_2 particles at lower pH. It may affect the surface area of the catalyst by reducing it for maximum adsorption of the dye as well as photon (Shukla et al. 2010).

In summary, there may be dissimilarity in photocatalytic activities in dye degradation reactions exhibited by different organic dyes because of the differences in their chemical nature. Some dyes can be decomposed quicker in the acidic solutions, while others decompose faster in alkaline solutions. As a result, before calculating the ideal pH for maximum photodegradation rate, it is vital to investigate the nature of the organic dyes to be destroyed.

12.7.4 Effect of Photocatalyst Load

Various studies are reported to analyse the effect of catalyst loading on the photocatalytic reaction particularly in the aqueous suspension form. It is also mandatory to calculate the optimum amount of catalyst for degrading the dyes and this amount of photocatalyst is dependent upon the intensity of visible light, reactor configuration, types of dyes and morphology of catalyst (Reza et al. 2017; Riga et al. 2007). The absorption of dyes on catalytic surface is increased with increasing the loading of photocatalyst (Kaur and Singh 2007). The loading of TiO_2 nanoparticle up to 1000 mgL^{-1} improves the rate of photo-decolourization with an overall 60% total organic carbon removal rate. The reported optimized condition for photocatalytic degradation of reactive red 198 is 0.3 gL^{-1} and consumption of TiO_2 is 0.5 gL^{-1}, which explains a direct relationship of the amount of catalyst with optimal level of light absorption and the amount of decolourization at a steady level (Mahvi et al. 2009).

12.7.5 Effect of Initial Dye Concentration

The concentration of dye is inversely proportional to the rate constant of decolourization which may affect the degradation and decolourization rate. At higher concentration of dyes, the more active sites of catalyst surface is filled with ions of dye which can lead to the decrease in the production of hydroxyl radicals. For reactive yellow dye, the variation in initial concentration of dye also affects its photocatalytic degradation and discolouration in the concentration range (0.025–0.1 gL^{-1}) (Kamat et al. 2000; Ilinoiu et al. 2013). The degradation efficiency of dyes can also be increased by increasing the concentration of initial dye, such as, the increment of the concentration of methyl orange reduces the path length of entering visible light and the lower concentration of methyl orange reverses this mechanism (Zhu et al. 2012).

12.7.6 Effect of Reaction Temperature

The higher temperature of semiconducting material during photocatalytic reaction reduces the degradation rate by recombining electron hole pairs and eluting the active reactant species. For example, the temperature of TiO_2 material is elevated above 80 °C during the photocatalytic reaction decreases the efficiency of photocatalytic activity. The lower value of activation energy (5000–20,000 J mol^{-1}) of photocatalytic reaction reflects the temperature dependence of the degradation rate. Photocatalytic devices do not require heat and can work at ambient temperature because of photonic activation. The optimal reaction temperature of TiO_2 material is observed to be between 20 and 80 °C for photocatalytic activity. The apparent activation energy rises when the temperature is lowered to 0 °C (Pardeshi and Patil 2008). This optimal range is determined mostly by the material's activation energy in the photocatalytic reaction.

12.7.7 Effect of Pollutant Adsorption Strength

There is an important role of the competitive adsorption which takes place between the water molecules and the dye to be generated in the heterogeneous photocatalysis. There is a limit of not more than one nanometre far for photogenerated reactive oxidizing species from where it generates. This causes to reduce the rate of dye degradation with the increasing distance from the surface of the catalyst particles (Chatterjee and Dasgupta 2005). However, in various studies, there hasn't been any report related to the direct relationship between photodegradation and adsorption. The adsorbed molecule or intermediate act as poison on the catalyst surface. The presence of coulombic interactions between the active group of photocatalyst dye and the molecule are necessary for the proper adsorption of the dye (Bizani et al. 2006). The complexity in the layers of dye molecules throughout the surface of the catalyst, resulting in limited interaction between the catalyst and the excited dye molecule in scenario, where there is a direct degradation reaction between the catalyst and the incoming light during the indirect mechanism (Wiszniowski et al. 2002). In both the above situations, there is an expectation of reduction in the photooxidation process and it may decrease the initial rate of dye in the presence of acidic solution, which in the case of alkaline solutions, shows lowering in the degradation rate reflecting the trouble faced by anionic dye molecules in approaching the catalytic surface (Vautier et al. 2001).

12.7.8 Effect of Inorganic Salts

Mineral ions are commonly associated with dye concentrations wastewater. The higher amount of mineral ions such as iron, copper or phosphate, etc., in dye contents of wastewater have the ability to reduce the photodegradation efficiency. The possible and major reasons for this can be the competition of these substances with the dyes for the sites which are active on the surface of the photocatalyst and thus deactivating the photocatalyst, which ultimately results in decreased degradation of targeted dyes. While the higher oxidation state of some cations like zinc, calcium and magnesium does not trigger inhibitory effect on the rate of photodegradation. A significant amount of the inorganic anions like carbonate, chloride, sulphate and nitrate can be found in the wastewater released by the dye industry (Hussein et al. 2014). The presence of salts in wastewater dyes has an adverse effect on the colloidal stability, which leads to rapid transfer of mass and decreases the surface contact of the photocatalyst and dye molecule.

References

Ahmed S, Rasul MG, Martens WN et al (2010) Heterogeneous photocatalytic degradation of phenols in wastewater: a review on current status and developments. Desalination 261:3–18

Ajmal A, Majeed I, Malik RN et al (2014) Principles and mechanisms of photocatalytic dye degradation on TiO_2 based photocatalysts: a comparative overview. RSC Adv 4:37003–37026

Al Angari YM, Ewais HA, El-Shishtawy RM (2019) Kinetics and mechanism of the oxidative decolorization of direct violet 31 in the presence of peroxodisulfate-silver (I) as a redox system. Transit Met Chem 44:57–64

Alharthi FA, Ali Alghamdi A, Alanazi HS (2020) Photocatalytic Degradation of the Light Sensitive Organic Dyes: Methylene Blue and Rose Bengal by Using Urea Derived g-C_3N_4/ZnO Nanocomposites. Catalysts 10:1457

Allen, Reginald L. (2013) Colour chemistry. Springer Science & Business Media

Arikal D, Kallingal A (2021) Photocatalytic degradation of azo and anthraquinone dye using TiO_2/MgO nanocomposite immobilized chitosan hydrogels. Environ Technol 42:2278–2291

Baran W, Makowski A, Wardas W (2003) The influence of $FeCl_3$ on the photocatalytic degradation of dissolved azo dyes in aqueous TiO_2 suspensions. Chemosphere 53:87–95

Benkhaya S, M'rabet S, El Harfi A (2020) A review on classifications, recent synthesis and applications of textile dyes. Inorg Chem Commun 115:107891

Bizani E, Fytianos K, Poulios I et al (2006) Photocatalytic decolorization and degradation of dye solutions and wastewaters in the presence of titanium dioxide. J Hazard Mater 136:85–94

Cai L, Xu T, Shen J et al (2015) Highly efficient photocatalytic treatment of mixed dyes wastewater via visible-light-driven AgI–Ag_3PO_4/MWCNTs. Mater SciSemicond Process 37:19–28

Carp O, Huisman CL et al (2004) Photoinduced reactivity of titanium dioxide. Prog Solid State Ch 32:33–177

Cavassin ED, de Figueiredo LFP, Otoch JP et al (2015) Comparison of methods to detect the in vitro activity of silver nanoparticles (AgNP) against multidrug resistant bacteria. J Nanobiotechnology 13:1–16

Cernuto G, Masciocchi N, Cervellino A et al (2011) Size and shape dependence of the photocatalytic activity of TiO_2 nanocrystals: a total scattering Debye function study. J Am ChemSoc133: 3114–3119

Chandra R, Mukhopadhyay S, Nath M (2016) TiO_2@ZIF-8: A novel approach of modifying micro-environment for enhanced photo-catalytic dye degradation and high usability of TiO_2 nanoparticles. Mater Lett 164:571–574
Chatterjee D, Dasgupta S (2005) Visible light induced photocatalytic degradation of organic pollutants. J Photochem Photobiol 6:186–205
Chatterjee MJ, Ahamed ST, Mitra M et al (2019) Visible-light influenced photocatalytic activity of polyaniline-bismuth selenide composites for the degradation of methyl orange, rhodamine B and malachite green dyes. Appl Surf Sci 470:472–483
Chequer FD, De Oliveira GR, Ferraz EA et al (2013) Textile dyes: dyeing process and environmental impact. Eco-friendly textile dyeing and finishing6:151–176
Chiu YH, Chang TF, Chen CY, Sone M, Hsu YJ (2019) Mechanistic insights into photodegradation of organic dyes using heterostructure photocatalysts. Catalysts 9:430
Chong MN, Jin B, Chow CW (2010) Recent developments in photocatalytic water treatment technology: a review. Water Res 44:2997–3027
Demircivi P, Simsek EB (2018) Fabrication of Zr-doped TiO_2/chitosan composite catalysts with enhanced visible-light-mediated photoactivity for the degradation of Orange II dye. Water Sci Technol 78:487–495
El Harfi S, El Harfi A (2017) Classifications, properties and applications of textile dyes: A review. Applied Journal of Environmental Engineering Science 3:3
Elango G, Roopan SM (2016) Efficacy of SnO_2 nanoparticles toward photocatalytic degradation of methylene blue dye. J PhotochemPhotobiol B Biol 155:34–38
El-Shishtawy RM, Melegy AA (2001) Geochemistry and utilization of montmorillonitic soil for cationic dye removal. AdsorpSciTechnol 19:609–620
El-Shishtawy RM, Soltan AM (2002) Bypass kiln dust as adsorbent for anionic dye and heavy metal ions removal from aqueous solution. Toxicol Environ Chem 82:1–10
El-Zahhar AA, Awwad NS (2016) Removal of malachite green dye from aqueous solutions using organically modified hydroxyapatite. J Environ Chem Eng 4:633–638
Esteves BM, Rodrigues CS, Boaventura RA et al (2016) Coupling of acrylic dyeing wastewater treatment by heterogeneous Fenton oxidation in a continuous stirred tank reactor with biological degradation in a sequential batch reactor. J Environ Manage 166:193–203
Gaya UI, Abdullah AH (2008) Heterogeneous photocatalytic degradation of organic contaminants over titanium dioxide: a review of fundamentals, progress and problems. J PhotochemPhotobiol c: Photochem 9:1–12
Ge J, Zhang Y, Park SJ et al (2019) Recent advances in carbonaceous photocatalysts with enhanced photocatalytic performances: a mini review. Materials 12:1916
Goncalves MS, Oliveira-Campos AM, Pinto EM et al (1999) Photochemical treatment of solutions of azo dyes containing TiO_2. Chemosphere 39:781–786
Hao OJ, Kim H, Chiang PC (2000) Decolorization of wastewater. Crit Rev Environ SciTechnol 30:449–505
Hashem A, El-Shishtawy RM (2001) Preparation and characterization of cationized cellulose for the removal of anionic dyes. AdsorpSciTechnol 19:197–210
Hassan MM, Carr CM (2018) A critical review on recent advancements of the removal of reactive dyes from dye house effluent by ion-exchange adsorbents. Chemosphere 209:201–219
Hu E, Wu X, Shang S et al (2016) Catalytic ozonation of simulated textile dyeing wastewater using mesoporous carbon aerogel supported copper oxide catalyst. J Clean Prod 112:4710–4718
Huang X, Gao B, Yue Q et al (2015) Compound bioflocculant used as a coagulation aid in synthetic dye wastewater treatment: the effect of solution pH. Sep PurifTechnol 154:108–114
Hunger, K. (2007) Industrial dyes: chemistry, properties, applications
Hussein BS, Waheeb AS, Umran AN et al (2014) Investigation of Degradation of Toluidine Blue and Rose Bengal Dyes from Aqueous Solution Using TiO_2Uv. J *Kufa* Phys 9:25
Ilinoiu EC, Pode R, Manea F et al (2013) Photocatalytic activity of a nitrogen-doped TiO_2 modified zeolite in the degradation of Reactive Yellow 125 azo dye. J Taiwan InstChemEng 44:270–278

Imran M, Crowley DE, Khalid A et al (2015) Microbial biotechnology for decolorization of textile wastewaters. Rev Environ Sci Biotechnol 14:73–92
Iqbal J, Shah NS, Sayed M (2021) Nano-zerovalent manganese/biochar composite for the adsorptive and oxidative removal of Congo-red dye from aqueous solutions. J Hazard Mater 403:123854
Jiang T, Zhang L, Ji M et al (2013) Carbon nanotubes/TiO_2 nanotubes composite photocatalysts for efficient degradation of methyl orange dye. Particuology 11:737–742
Kamat PV, Murakoshi K, Wada Y et al (2000) Semiconductor nanoparticles. In Handbook of Nanostructured Materials and Nanotechnology 291–344
Kasiri MB (2019) Application of chitosan derivatives as promising adsorbents for treatment of textile wastewater. The Impact and Prospects of Green Chemistry for Textile Technology 417–469
Kaur S, Singh V (2007) TiO_2 mediated photocatalytic degradation studies of Reactive Red 198 by UV irradiation. J Hazard Mater 141:230–236
Kaur D, Bagga V, Behera N (2019) SnSe/SnO_2 nanocomposites: novel material for photocatalytic degradation of industrial waste dyes. Adv Compos Mater 2:763–776
Kaur S, Sharma S, Kansal SK (2016) Synthesis of ZnS/CQDs nanocomposite and its application as a photocatalyst for the degradation of an anionic dye ARS. SuperlatticesMicrostruct98 :86–95
Khan I, Sadiq M, Khan I et al (2019) Manganese dioxide nanoparticles/activated carbon composite as efficient UV and visible-light photocatalyst. ESPR 26:5140–5154
Khan SB, Lee SL (2021) Nanomaterials significance; contaminants degradation for environmental applications Nano Express2:022002
Koohestani H, Sadrnezhaad SK (2016) Photocatalytic degradation of methyl orange and cyanide by using TiO_2/CuO composite. Desalin Water Treat 57:22029–22038
Kumar R, El-Shishtawy RM, Barakat MA (2016a) Synthesis and characterization of Ag-Ag_2O/TiO_2@polypyrrole heterojunction for enhanced photocatalytic degradation of methylene blue. Catalysts 6:76
Kumar SS, Shantkriti S, Muruganandham T et al (2016b) Bioinformatics aided microbial approach for bioremediation of wastewater containing textile dyes. Ecol Info 31:112–121
Kumar R, Ansari MO, Parveen N et al (2016c) Facile route to a conducting ternary polyaniline@ TiO_2/GN nanocomposite for environmentally benign applications: photocatalytic degradation of pollutants and biological activity. RSC advances6: 111308–111317
Kumar R, Travas-Sejdic J, Padhye LP (2020) Conducting polymers-based photocatalysis for treatment of organic contaminants in water. CEJ Advances 100047
Kumaran V, Sudhagar P, Konga AK (2020) Photocatalytic Degradation of Synthetic Organic Reactive Dye Wastewater Using GO-TiO_2 Nanocomposite Pol J Environ. Stud 29:1683–1690
Kuriakose S, Choudhary V, Satpati B (2014) Facile synthesis of Ag–ZnO hybrid nanospindles for highly efficient photocatalytic degradation of methyl orange. PCCP 16:17560–17568
Lellis B, Fávaro-Polonio CZ, Pamphile JA et al (2019) Effects of textile dyes on health and the environment and bioremediation potential of living organisms. Biotechnol Res Innov 3:275–290
Li S, Wei C, Wang J et al (2014) Sonocatalytic activity of Yb, B, Ga-codoped Er^{3+}: $Y_3Al_5O_{12}$/TiO_2 in degradation of organic dyes. Mater SciSemicond Process 26:438–447
Li S, Lin Q, Liu X (2018) Fast photocatalytic degradation of dyes using low-power laser-fabricated Cu_2O–Cu nanocomposites. RSC Adv 8:20277–20286
Loddo V, Bellardita M, Camera-Roda G et al (2018) Heterogeneous photocatalysis: A promising advanced oxidation process. In Current trends and future developments on (bio-) membranes. Elsevier 1–43
Loghambal S, Catherine AA, Subash SV (2018) Analysis of langmuir-hinshelwood kinetics model for photocatalytic degradation of aqueous direct blue 71 through analytical expression 55:7
Mahvi AH, Ghanbarian M, Nasseri S et al (2009) Mineralization and discoloration of textile wastewater by TiO_2 nanoparticles. Desalination 239:309–316
Melinte V, Stroea L, Chibac-Scutaru AL (2019) Polymer Nanocomposites for Photocatalytic Applications. Catalysts 9:986
Meng ZD, Zhang FJ, Zhu L et al (2012) Synthesis and characterization of M-fullerene/TiO_2 photocatalysts designed for degradation azo dye. Mater Sci Eng C 32:2175–2182

Merdan N, Eyupoglu S, Duman MN (2017) Ecological and sustainable natural dyes In Textiles and Clothing Sustainability. Springer, Singapore, pp 1–41

Monga D, Basu S (2019) Enhanced photocatalytic degradation of industrial dye by g-C_3N_4/TiO_2 nanocomposite: Role of shape of TiO_2. Adv Powder Technol 30:1089–1098

Muruganandham M, Swaminathan M (2004) Solar photocatalytic degradation of a reactive azo dye in TiO2-suspension. Sol Energy Mater Sol Cells 81:439–457

Neppolian B, Choi HC, Sakthivel S et al (2002) Solar light induced and TiO_2 assisted degradation of textile dye reactive blue 4. Chemosphere 46:1173–1181

Nosaka Y, Nosaka AY (2017) Generation and detection of reactive oxygen species in photocatalysis. Chem Rev 117:11302–11336

Olad A, Nosrati R (2012) Preparation, characterization, and photocatalytic activity of polyaniline/ZnO nanocomposite. Res Chem Intermed 38:323–336

Pardeshi SK, Patil AB (2008) A simple route for photocatalytic degradation of phenol in aqueous zinc oxide suspension using solar energy. Sol Energy 82:700–705

Pathania D, Katwal R, Sharma G (2016) Novel guar gum/Al_2O_3 nanocomposite as an effective photocatalyst for the degradation of malachite green dye. Int J Biol Macromol 87:366–374

Pica M, Calzuola S, Donnadio A et al (2019) De-ethylation and cleavage of rhodamine B by a zirconium phosphate/silver bromide composite photocatalyst. Catalysts 9:3

Podporska-Carroll J, Myles A, Quilty B et al (2017) Antibacterial properties of F-doped ZnO visible light photocatalyst. J Hazard Mater 324:39–47

Qiu Y, Xu X, Xu Z et al (2020) Contribution of different iron species in the iron-biochar composites to sorption and degradation of two dyes with varying properties. Chem Eng J 389:124471

Rajagopalan V (2016) A new synergetic nanocomposite for dye degradation in dark and light. Sci Rep 6:1–10

Rajeshwar K, Osugi ME, Chanmanee W et al (2008) Heterogeneous photocatalytic treatment of organic dyes in air and aqueous media. J PhotochemPhotobiol c: Photochem 9:171–192

Ramadhani S, Helmiyati H (2020) Alginate/CMC/ZnO nanocomposite for photocatalytic degradation of Congo red dye. In AIP Conference Proceedings (Vol. 2242, No. 1, p 040026). AIP Publishing LLC.

Ramazanzadeh B, Jahanbin A, Yaghoubi M et al (2015) Comparison of antibacterial effects of ZnO and CuO nanoparticles coated brackets against Streptococcus mutans. J Dent 16:200

Rani S, Aggarwal M, Kumar M et al (2016) Removal of methylene blue and rhodamine B from water by zirconium oxide/graphene. Water Sci 30:51–60

Reza KM, Kurny ASW, Gulshan F (2017) Parameters affecting the photocatalytic degradation of dyes using TiO_2: a review. Appl Water Sci 7:1569–1578

Riga A, Soutsas K, Ntampegliotis K et al (2007) Effect of system parameters and of inorganic salts on the decolorization and degradation of Procion H-exl dyes. Comparison of H_2O_2/UV, Fenton, UV/Fenton, TiO_2/UV and TiO_2/UV/H_2O_2 processes. Desalination 211:72–86

Salah NH, Bouhelassaa M, Bekkouche S et al (2004) Study of photocatalytic degradation of phenol. Desalination 166:347–354

Salavati H, Tavakkoli N, Hosseinpoor M (2012) Preparation and characterization of polyphosphotungstate/ZrO_2 nanocomposite and their sonocatalytic and photocatalytic activity under UV light illumination. UltrasonSonochem 19:546–553

Sandhya S (2010) Biodegradation of azo dyes under anaerobic condition: role of azoreductase. Biodegradation of azo dyes. 39–57

Saqib NU, Adnan R, Shah I (2019) Zeolite supported TiO_2 with enhanced degradation efficiency for organic dye under household compact fluorescent light. Mater Res Express 6:095506

Saravanan R, Karthikeyan N, Govindan S (2012) Photocatalytic degradation of organic dyes using ZnO/CeO_2 nanocomposite material under visible light. Adv Mat Res 584:381–385

Saravanan R, Gupta VK, Narayanan V et al (2013a) Comparative study on photocatalytic activity of ZnO prepared by different methods. J Mol Liq 181:133–141

Saravanan R, Karthikeyan S, Gupta VK et al (2013b) Enhanced photocatalytic activity of ZnO/CuO nanocomposite for the degradation of textile dye on visible light illumination. Mater Sci Eng C 33:91–98
Sargin I, Baran T, Arslan G (2020) Environmental remediation by chitosan-carbon nanotube supported palladium nanoparticles: Conversion of toxic nitroarenes into aromatic amines, degradation of dye pollutants and green synthesis of biaryls. Sep Purif Technol 247:116987
Sarkar S, Ponce NT, Banerjee A et al (2020) Green polymeric nanomaterials for the photocatalytic degradation of dyes: a review. Environ ChemLett 1–12
Sathian S, Rajasimman M, Rathnasabapathy CS (2014) Performance evaluation of SBR for the treatment of dyeing wastewater by simultaneous biological and adsorption processes. J. Water Process. Eng 4:82–90
Shan R, Lu L, Gu J (2020) Photocatalytic degradation of methyl orange by Ag/TiO_2/biochar composite catalysts in aqueous solutions. Mater SciSemicond Process 114:105088
Shukla PR, Wang S, Ang HM et al (2010) Photocatalytic oxidation of phenolic compounds using zinc oxide and sulphate radicals under artificial solar light. Sep Purif Technol 70:338–344
Sirajudheen P, Meenakshi S (2019) Facile synthesis of chitosan-La^{3+}-graphite composite and its influence in photocatalytic degradation of methylene blue. Int J Biol Macromol 133:253–261
Sunkaraa JR, Botsab SM (2020) Visible Light Driven Catalyst ZnS/Fe_2O_3/Ag Ternary Nanocomposite for the Degradation of Dyes. Russ J PhysChem B 94.
Taghizadeh MT, Seifi-Aghjekohal P (2015) Sonocatalytic degradation of 2-hydroxyethyl cellulose in the presence of some nanoparticles. UltrasonSonochem 26:265–272
Muhammad Bilal Tahir, Tahir Iqbal, Muhammad Rafique et al (2020) Nanomaterials for photocatalysis. In Micro and Nano Technologies,Nanotechnology and Photocatalysis for Environmental Applications Elsevier 65–76
Tang WZ, Chen RZ (1996) Decolorization kinetics and mechanisms of commercial dyes by H_2O_2/iron powder system. Chemosphere 32:947–958
Tunesi S, Anderson M (1991) Influence of chemisorption on the photodecomposition of salicylic acid and related compounds using suspended titania ceramic membranes. J PhysChemA 95:3399–3405
Vautier M, Guillard C, Herrmann JM (2001) Photocatalytic degradation of dyes in water: case study of indigo and of indigo carmine. J Catal 201:46–59
Waiskopf N, Ben-Shahar Y, Galchenko M (2016) Photocatalytic reactive oxygen species formation by semiconductor–metal hybrid nanoparticles. Toward light-induced modulation of biological processes. Nano Lett 16:4266–4273
Wang D, Li X, Chen J et al (2012) Enhanced photoelectrocatalytic activity of reduced graphene oxide/TiO_2 composite films for dye degradation. Chem Eng Sci 198:547–554
Wang Q, Hui J, Li J et al (2013) Photodegradation of methyl orange with PANI-modified BiOCl photocatalyst under visible light irradiation. Appl Surf Sci 283:577–583
Wei Y, Ding A, Dong L et al (2015) Characterisation and coagulation performance of an inorganic coagulant-poly-magnesium-silicate-chloride in treatment of simulated dyeing wastewater. Colloids Surf APhysicochem Eng 470:137–141
Wiszniowski J, Robert D, Surmacz-Gorska J et al (2002) Photocatalytic decomposition of humic acids on TiO_2: Part I: Discussion of adsorption and mechanism. J Photochem Photobiol A 152:267–273
Xu C, Anusuyadevi PR, Luque AC et al (2019) Nanostructured materials for photocatalysis. ChemSoc Rev 48:3868–3902
Yaghoubi A, Ramazani A, TaghaviFardoodS, (2020) Synthesis of Al_2O_3/ZrO_2 Nanocomposite and the Study of Its effects on Photocatalytic Degradation of Reactive Blue 222 and Reactive Yellow 145 Dyes. Chemistry Select 5:9966–9973
Yan Y, Chang T, Wei P et al (2009) Photocatalytic activity of nanocomposites of ZnO and multi-walled carbon nanotubes for dye degradation. J DispersSciTechnol 30:198–203

Yang J, Chen C, Ji H et al (2005) Mechanism of TiO_2-assisted photocatalytic degradation of dyes under visible irradiation: photoelectrocatalytic study by TiO_2-film electrodes. J Phys Chem 109:21900–21907

Zhang G, Wang B, Sun Z et al (2016) A comparative study of different diatomite-supported TiO_2 composites and their photocatalytic performance for dye degradation. Desalin Water Treat 57:17512–17522

Zhang Y, Fan R, Zhang Q et al (2019) Synthesis of $CaWO_4$-biochar nanocomposites for organic dye removal. Mater Res Bull 110:169–173

Zhu H, Jiang R, Fu Y et al (2012) Effective photocatalytic decolorization of methyl orange utilizing TiO_2/ZnO/chitosan nanocomposite films under simulated solar irradiation. Desalination 286:41–48

Chapter 13
Zinc Sulphide Nanoparticles as a Bacteriostatic and Invigorated Catalytic Tool for Multiple Dye Degradation: An Approach Towards Environment Remediation

Aayasha Negi, Reena Gangwar, and Devendra Singh Negi

Abstract The objective of this research is to assess a green strategy to synthesize multifunctional Zinc sulphide nanoparticles (ZnS-NPs), a material that has attracted significant attention in recent years. An aqueous extract of *Girardinia diversifolia* (*G. diversifolia*) leaves was utilized to form ZnS nanoparticles and their properties were evaluated using various analytical techniques such as UV-Vis, Tauc plot, XRD, FTIR, EDX, and TEM. The extract's phytoconstituents serve as a reducing and capping agent, as determined by FTIR. UV spectrum exhibited a peak near 262.31 nm, which is corresponding to characteristic peak of ZnS nanoparticles. ZnS-NPs are cubic crystalline in nature as shown by XRD investigation. Moreover, morphological examinations using TEM and EDX revealed that the integrated spherical particles (11.28 nm particle size) are largely constituted of Zinc (52.38%) and Sulphur (31.23%). The as-prepared ZnS-NPs show potential antibacterial abilities against 5 bacterial strains in this investigation and the maximum zone of inhibition were determined using agar well diffusion method. ZnS-NPs acted as an efficient catalyst for the degradation of hazardous dyes (4.124 eV band gap energy). The dye degradation of Methylene blue (MB), Acridine orange (AO), and Rose bengal (RB) was examined utilizing nanoparticles. As a result, the ZnS-NPs produced in this study are appropriate for industrial wastewater treatment. Increased reaction rate as well as % degradation efficiency were attained by increasing the concentration of dye.

Keywords *Girardinia diversifolia* · ZnS-NPs · Toxic dye deprivation · Wastewater treatment · Antibacterial assay

A. Negi · D. S. Negi (✉)
Department of Chemistry, H.N.B. Garhwal University (A Central University), Srinagar Garhwal, Uttarakhand 246174, India
e-mail: devendra_negi@yahoo.com

R. Gangwar
Department of Microbiology, H.N.B. Garhwal University (A Central University), Srinagar Garhwal, Uttarakhand 246174, India

J. P. N. Rai and S. Saraswat (eds.), *Nano-biotechnology for Waste Water Treatment*, Water Science and Technology Library 111,
https://doi.org/10.1007/978-3-031-00812-2_13

13.1 Introduction

Enormous population explosion in the twenty-first century has prompted a series of obstacles, including metabolic stress, medical hurdle, and pollution. Chemical pollutants emitted by the textile sector have become a major source of water contamination, causing terrible diseases transmission throughout the world. Scientists are working over a range of physical and chemical techniques to assist in resolving these concerns but all the efforts to combat the treatment of effluents from textile factories are proving ineffective. Nanomaterials due to high surface area to volume ratio show enhanced properties than their bulk parts. The high reactivity of nanomaterials is critical for eliminating contaminant from polluted water through photocatalytic degradation (Jabeen et al. 2017). Numerous novel photocatalytic chemicals and materials have been developed over the last few decades. Synthesis of inorganic semiconductor nanomaterials has gained researcher's interest due to their features, which include high reactivity, low melting point, electrical, optical, and magnetic competences. Inorganic semiconductor nanomaterials, band gap engineering, and particle size control are regarded as critical tools for tailoring properties for various photocatalysts and other light emitting materials across a variety of sectors. Due to the efficiency and possible applicability of semiconductor photocatalysis, it has the potential to completely eradicate harmful compounds. In environmental remedy, metal oxide is frequently employed as a photocatalyst (Hoffmann et al. 1995). Recently, transition metal sulphides, particularly ZnS, have been extensively studied and researched due to their superior catalytic properties than TiO_2 (Yanagida et al. 1986). Formalized paraphrase Zinc sulphide was among the first semiconductors discovered, having a wide band gap of >3.72 eV found in cubic or wurtzite nature (Fujiwara et al. 1998). Potential applications in ZnS nanostructures have been reported, comprising nanorods, nanowires, and nanobelts (Biswas and Kar 2008; Kar et al. 2005). These findings demonstrate that ZnS nanoparticles (NPs) are effective photocatalysts due to the rapid formation of electron–hole pairs by photoexcitation and the highly negative reduction potentials of excited electrons. Photocatalytic properties are demonstrated not only in the photoreductive production of H_2 from water and the photoreduction of CO_2, but also in the phototransformation of various organic substrates, including the oxidative formation of carbon–carbon bonds from organic electron donors, cis-trans photoisomerization of alkenes, and photoreduction of aldehydes and their derivatives. ZnS nanoparticles have previously been synthesized by a wide range of chemical methods, including solvothermal/hydrothermal methods (Gou et al. 2004; Yanagida et al. 1986). As a result, ZnS has been found in various nanoform using different techniques to synthesize technologically essential nanostructures in a simple and environmentally friendly manner. Furthermore, contamination by microbiological species is a big concern. According to the study, the most well-known microbes (*Staphylococcus aureus*, *Escherichia coli*, *Kelbsiella species*, and *Pseudomonas aeruginosa*) are responsible for 14.8% to 60% of post-injury contaminations. There was a perceived need for the creation of creative solutions to these difficulties over the time (Dessalegn et al. 2014; Elumalai and Velmurugan 2015). With the passage of time, there has been a dire need for the development of techniques that pose minimal risk and are unlikely to introduce harmful materials into the environment. The "Green Method," which synthesize NPs by utilizing plant resources, has discovered a comprehensive solution to this unresolved difficulty. Green Method focuses on producing the desired product without the use of hazardous intermediates since green techniques are cost-effective, ecologically friendly, and chemical-free (Zheng et al. 2015).

In this study, ZnS nanoparticles were synthesized utilizing *G. diversifolia* leaves. *G. diversifolia*, also known as Bichoo Ghas, Nilgiri nettle, and Himalayan huge nettle, is a plant native to India. Girardinia genus, Urticaceae family, grows abundantly in moist and damp hill forests. *G. diversifolia* has traditionally been used to treat gastrointestinal illnesses, chest discomfort, rheumatism, tuberculosis, headaches, joint pain, diabetes, gonorrhoea, and delivery issues. It has also been used in ancient days to heal bone fractures, internal traumas, and for blood purification (Liu et al. 2012; Shrestha and Dhillion 2003; Rana et al. 2021).

13.2 Materials and Methodology

13.2.1 Plants Collection Site

Disease-free *G. diversifolia* were collected from Lansdowne (29.8377° N, 78.6871° E) Pauri District, Uttarakhand, India as shown in Fig. 13.1. By thoroughly washing the leaves twice with distilled water, concomitant dirt were removed, which are then air-dried and crushed using hand grinder.

13.2.2 Required Chemicals

Methylene Blue ($C_{16}H_{18}ClN_3S$), Rose Bengal ($C_{20}H_2Cl_4I_4Na_2O_5$), Acridine Orange ($C_{17}H_{19}N_3$), and Sodium sulphide ($Na_2S.5H_2O$) were acquired from Fischer Scientific while highly pure chemical Zinc nitrate ($Zn(NO_3)_2$) was obtained from Sigma Aldrich.

13.2.3 Procured Bacterial Strains

V.C.S.G. Gov. Institute of Medical Science and Research, Srinagar Garhwal, Uttarakhand provided clinical isolate *E. coli*. *S. aureus* (MTCC-1144), *Streptococcus pneumoniae* (MTCC-655), *Klebsiella pneumoniae* (MTCC-4030), and *P. aeruginosa* (MTCC-2474).

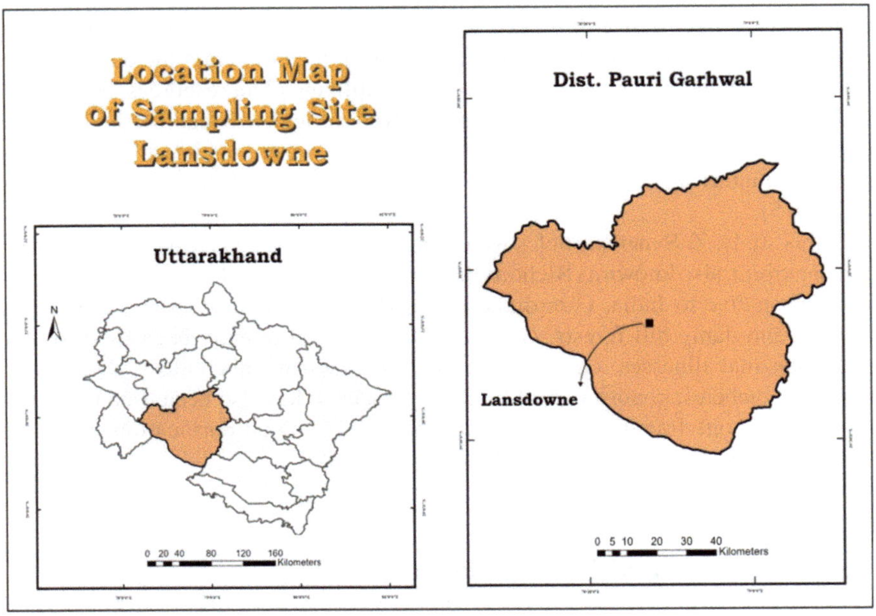

Fig. 13.1 Location map

13.3 Experimental Details

13.3.1 Green Synthesis of Zinc Sulphide Nanoparticles

To remove any remaining dust or moisture, fresh *G. diversifolia* leaves were washed many times using distilled water and air-dried. For 40 min, 20 g of powdered leaves were mixed in distilled water and heated at 45 °C. The extract was chilled and centrifuged at 4000 rpm to eliminate undesirable residues. It was then adjusted to 100 ml by combining it with double distilled water before filtration. The concentrate was maintained at 4 °C for further research.

Following plant extraction, 200 mL of 0.1 M Zinc nitrate solution was added in controlled calibration to 100 mL of 0.1 M $Na_2S.5H_2O$ solution while continuously stirring for an hour at 35 °C. After that, the white-coloured solution was swirled for 2 h at 25 °C. Subsequently, 20 mL of the resulting aq. extract was carefully added to the white-coloured solution, which was then incubated at 50 °C for 3 h. The resulting mixture was centrifuged at 2000 rpm for 15 min. The resultant was then dried at a temperature of 50 °C and crushed to get fine particle. Following that, the properties of the resultant ZnS NPs have been investigated utilizing several characterization techniques (Alijani et al. 2019) (Fig. 13.2).

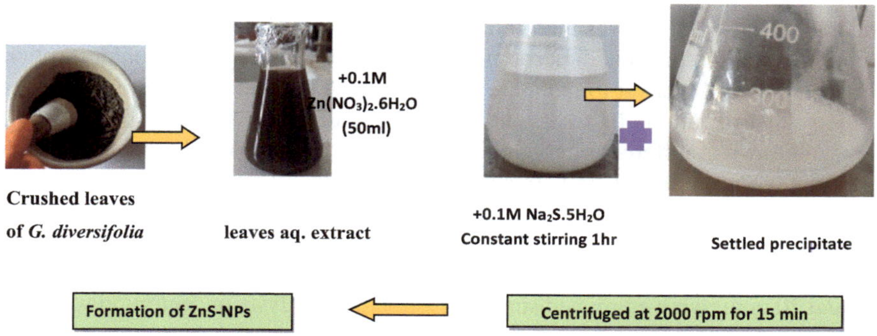

Fig. 13.2 Schematic presentation

13.3.2 Characterization

The nanoparticles were air-dried before being analysed. Our previously published paper (Negi et al. 2021) mentions the equipment used to characterize the created ZnS-NPs. To study the functional groups in the leaf extract and NPs, air-dried nanoparticles were grounded with KBr and scanned in the wavelength range of 4000–500 cm^{-1} using a FTIR spectrometer (Perkin Elmer Spectrum). An FEI Tecnai (HR-TEM) instrument has been used to assess the dimensions of the generated NPs working at 40–120 kV voltage and equipped with Olympus soft imaging software, Germany.

13.3.3 Bactericidal Activity

The antimicrobial potential of the NPs depends on two basic characteristics which are particle size and the material employed to synthesize NPs. According to the literature review, the antimicrobial potential of ZnS nanoparticles can be described as follows: (1) bacterial outer membrane denaturation; (2) creation of holes in the bacterial cell membrane, leading to cell membrane breakdown; and (3) interactions among both NPs and enzyme disulphide or sulphhydryl groups interrupt metabolic processes, causing cell death.

13.3.4 Estimation of Photocatalytic Potential

Various impacting factors such as catalyst concentration, pH, dye concentration, and solution temperature were investigated to assess catalytic activity. To generate a colloidal suspension of ZnS-NPs, 1 mg of each dye (Methylene blue (MB), Rose bengal (RB), Acridine orange (AO)) was continuously mixed in 100 ml of distilled water. The dye solution was therefore mixed with 5 mg ZnS nanoparticles and the solution was then stirred in dark room for 20 min to make the absorption/desorption between dye and catalyst to maintain equilibrium. After that the solution was kept at the solar irradiation. To assess the rate of degradation, a known volume (5 ml) was withdrawn at different intervals (0, 30, 60, 120, 150 min) and recorded using a spectrophotometer at 200–650 nm and the degradation rate was calculated in terms of change in intensity at max, i.e. 663 nm (MB), 450–465 nm (AO), and 450–650 nm (RB).The usual equation was used to compute the per cent degradation. A variable catalytic amount of samples (100, 150, and 200 mg) and dye concentrations (10, 15, and 20 ppm) were employed similarly, while the catalyst concentration was kept constant.

Many organic pollutants can be destroyed by irradiation using semiconductor nanoparticles on exposing to solar irradiation. When NPs activate by light, electron–hole pairs are generated, which act as potent oxidizing and reducing agents.

$$\mathrm{ZnS} + \mathrm{Energy} \rightarrow \mathrm{h}^{+} + \mathrm{e}^{-}$$

The redox processes are denoted as follows:

$$\mathrm{OH}^{-} + \mathrm{h}^{+} \rightarrow \bullet\mathrm{OH},\ \mathrm{O}_2 + \mathrm{e}^{-} \rightarrow \mathrm{O}_2^{-}$$

The hydroxyl radical formed by the oxidation of adsorbed water or adsorbed OH is the primary oxidant in dye degradation, and the presence of oxygen can prevent the recombination of electron-hole pairs. CO_2 and H_2O are among the final products of a complete reaction.

$\mathrm{Dyes(MB)} + \bullet\mathrm{OH} \rightarrow CO_2 + H_2O + NH_4^{+} + NO_3^{-} + SO_4^{2-} + Cl^{-}$ (Soltani et al. 2012) (Fig. 13.3).

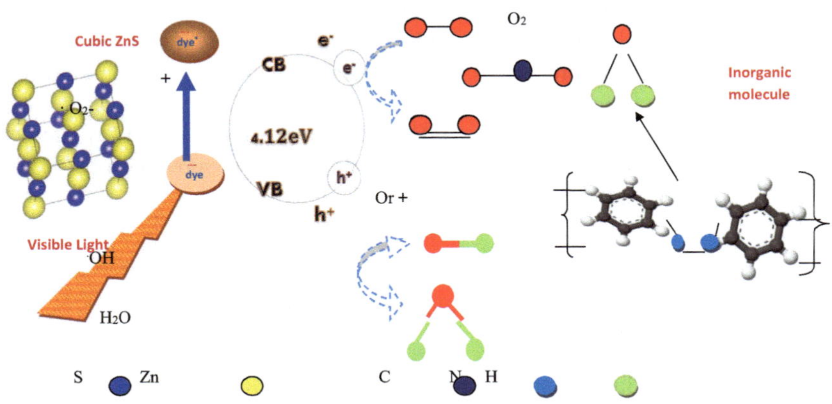

Fig. 13.3 Mechanism of photodegradation

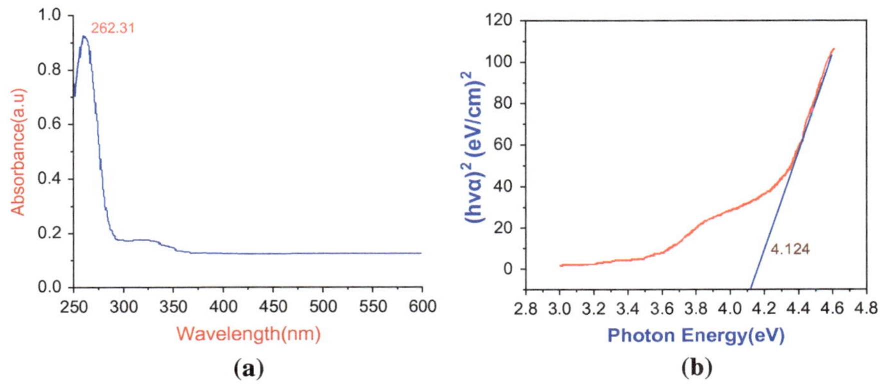

Fig. 13.4 (**a**) UV-Vis spectra. (**b**) Tauc-Plot

13.4 Results and Discussion

13.4.1 Band Gap Evaluation from Optical Observations

The UV-Visible absorption spectral band at 262.31 nm has been projected which is quite similar to the characteristic band of ZnS accounted earlier (Moussaoui et al. 2010). The band gap energy (E_{bg}) of these NPs has been estimated using the equation $E_{bg} = hc/\lambda$, where h refers to plank constant and c, λ are velocity of light and wavelength, respectively (Fig. 13.4).

13.4.2 Phase/Structure Analysis

Powder XRD analysis was performed to investigate crystallographic phases and NPs XRD patterns. Peaks in the XRD graph at 2θ of 13.96°, 40°, and 53.37° correspond to lattice planes (111), (220), and (311), respectively, and reveal a crystalline and cubic structure with a crystalline size of 29.86 nm, which is similar to the characteristic structure of ZnS. The equation is used to calculate the lattice parameter of the ZnS as $1/d^2_{h,k,l} = (h^2 + K^2 + l^2)/a^2$. The average crystalline size of the synthesized nanoparticles was determined using Scherrer's formula (Negi et al. 2021). To investigate the mechanism of ZnS nanoparticles formation as well as provide information on the chemical changes that occur in the functional groups involved in the reduction, an FTIR spectrum was examined, which revealed a very strong peak at 3424.4 cm^{-1}, which was most likely assigned to the plant extract's -OH stretching groups. Due to the changes in the chemical environment and a decrease in the concentration of covalent bonds in the solution, the intensity of this band declined substantially in the broad peak at 2934.10 cm^{-1}. At 1593.02 cm^{-1}, the existence of (N-H groups or

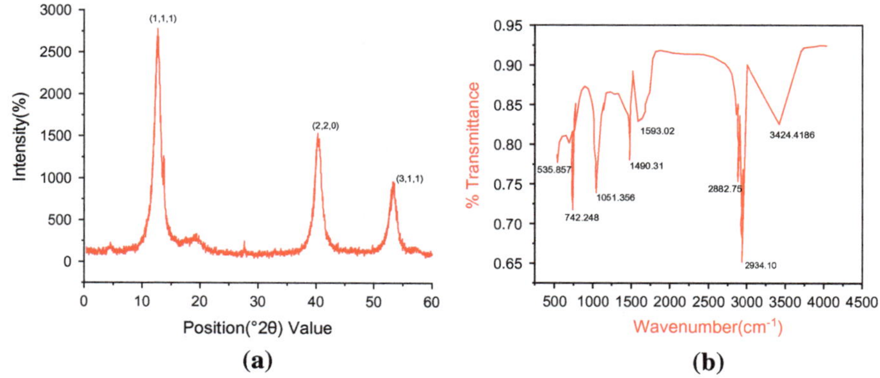

Fig. 13.5 (**a**) XRD analysis. (**b**) FTIR spectrum

aromatic rings) linked to ZnS nanoparticles has also been detected. The observed peak at 1490.31 cm^{-1} was assigned to the C = S stretching band, and ZnS vibrations at 1051.356, 742.248, and 535.857 cm^{-1} resulted in a reduction in band intensity, indicating a reduction in binding dosage or reducing agent adsorption on the surface of metal NPs (Fig. 13.5).

13.4.3 Morphological Analysis

EDS spectral analysed the elemental compositions of synthesized nanoparticles as depicted in Fig. 13.6(a). EDX analysis demonstrated the composition of ZnS-NPs, revealing a Zinc content of 52.38%, Sulphur content of 31.23%, as well as carbon

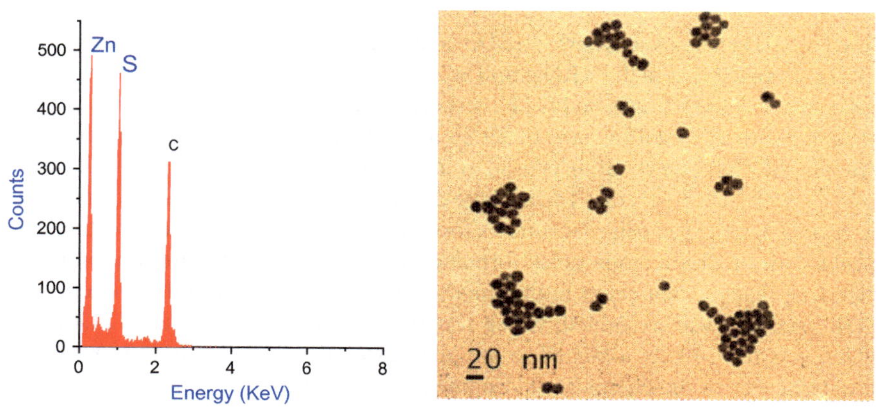

Fig. 13.6 (**a**) EDX Spectrum. (**b**) HR-TEM micrograph

content of 16.39% as in Fig. 13.6(a). Furthermore, TEM scans were widely used to explain shapes, surface morphology, and size of synthesized NPs. These scannings were used to analyse size using digimizer software, which reported that the mean size of these spherical particles is around 11.28 nm as shown in Fig. 13.6(b).

13.4.4 Bactericidal Potential

The antibacterial activity was determined using the agar well diffusion method earlier employed in previously published paper (Negi et al. 2021) to evaluate Ag-NPs against two gram-positive pathogens, *Staphylococcus aureus* and *Streptococcus pneumonia* as well as three gram-negative pathogens, *K. pneumoniae*, *E. coli*, and *P. aeruginosa*, which is briefly as, each well was loaded with 100 L of the working suspension/solution of different plant extracts, and the wells were filled with the same amount of extraction solvent for control using a micropipette. Plates were left for a short period of time until the extracts diffused in the medium before being incubated at 37 °C for 24 h. The tests were repeated three times, and the inhibition zones for each extract were measured and the findings were recorded (Fig. 13.7 and Table 13.1).

13.4.5 Catalytic Potential

To degrade the dyes, distinct interaction boundaries such as starting dye dose, temperature, pH were fluctuated. At pH 7.0, the absorbance intensity was found to decrease steadily with increasing time intervals, with the greatest degradation of MB 53.8% in Fig. 13.8(a), AO 63% in Fig. 13.8(b) and RB 76% in Fig. 13.8(c). The normalized change in concentration and degradation efficiency was used to determine the photo degradation of dye. After 30 min, the degradation effectiveness of a pure AO was 13%, and after 5 h, it was 23%. After 15 min of mixing ZnS-NPs, the efficiency was stabilized at 35% which increased to 63% after 25 min. of consistent presentation of sunlight. Degradation of pure RB was evaluated at 13%, which increased to 23% after 2 h. After mixing ZnS-NPs for 10 min, degradation was 34% and increased to 76% after 15 min. After 20 min of solar light irradiation, the photocatalyst's MB degradation potential was about 53.8% which was 11% in the early 5 min. It could be because the catalyst utilized is a semiconductor that generates electron-hole pairs when exposed to sunlight. Moreover, it was observed that photodegradation is dye concentration-dependent. In the case of Methylene blue, the kinetic rate was $3.15 \times 10^2\ \text{min}^{-1}$. For RB and Methylene orange, the rate constant was estimated at $6.3 \times 10^5\ \text{min}^{-1}$ and $4.39 \times 10^2\ \text{min}^{-1}$, respectively.

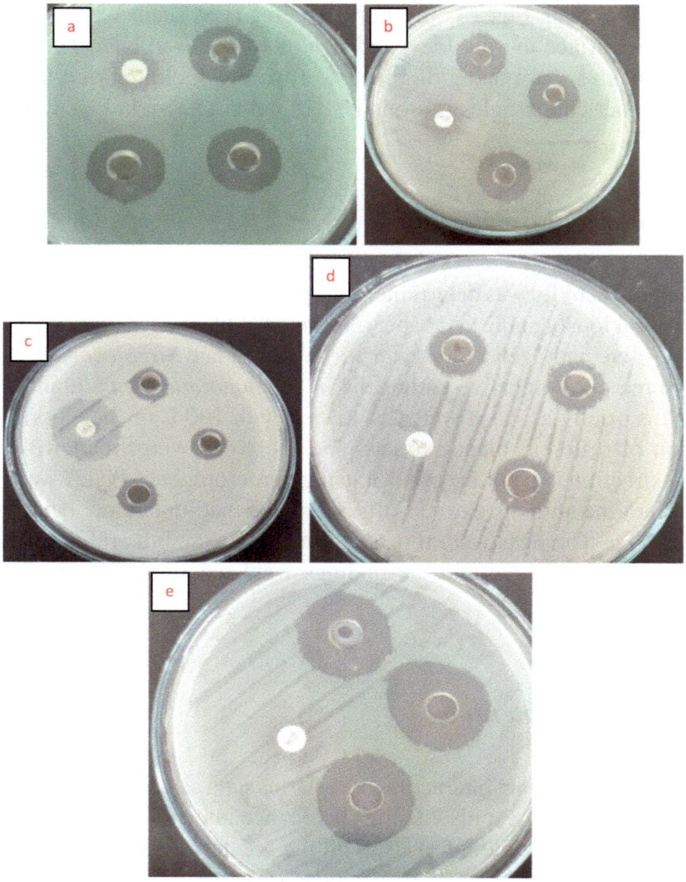

Fig. 13.7 Bacterial inhibition zone

Table 13.1 Inhibition zone against bacterial strain at various concentration

S.No	Pathogen	CONTROL	(Erythromycin)	50 μL	100 μL	150 μL
1	*k. pneumonia*	–	9 mm	15 mm	16 mm	18 mm
2	*S. pneumonia*	–	12 mm	13 mm	13 mm	15 mm
3	*P. aeruginosa*	–	10 mm	9 mm	8 mm	9 mm
4	*E. coli*	–	–	13 mm	17 mm	19 mm
5	*S. aureus*	–	9.5 mm	13 mm	19 mm	20.1 mm

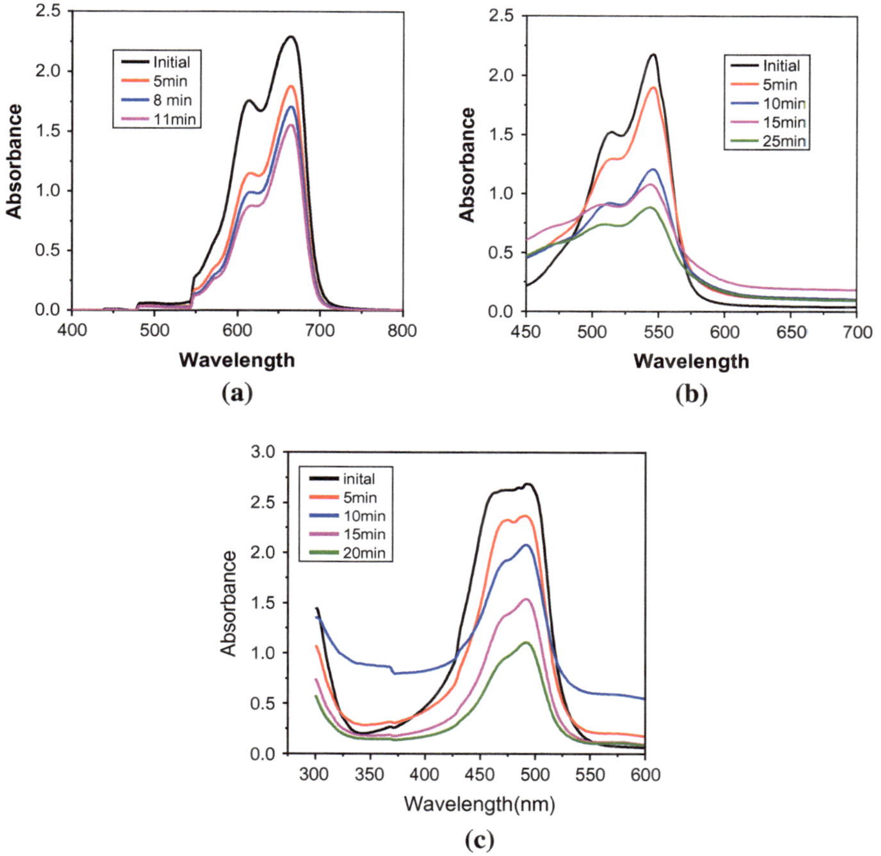

Fig. 13.8 (**a**) MB degradation by ZnS NPs. (**b**) Spectral variation of AO by ZnS NP at various time. (**c**) RB degradation utilizing ZnS-NPs at multiple intervals

13.5 Conclusion

The production of Zinc sulphide by *G. diversifolia* is simple, quick, and environmentally friendly. The plant is easy to obtain, and because it is abundant and has no financial application, it could be used to generate efficient visible light photocatalytic material with high purity ZnS. The goal of the synthesis was to determine antibacterial medication resistance as well as the growing threat of environmental toxicity by utilizing plants with well-known preferences. XRD, SEM, and TEM have been used to characterize the cubic crystalline phase of ZnS-NPs synthesized at 60 °C from *G. diversifolia* with a size restriction of approximately 11.28 nm and spherical shape. The photocatalyst NPs are being used to manage contamination of water and keep the environment clean. Recent research found that biosynthesized Ag-NPs could suppress a bacterial disease. As a result of these realities and findings, it is frequently asserted that the current framework encompasses a broader range of potential uses.

References

Alijani HQ, Pourseyedi S, Mahani MT, Khatami M (2019) Green synthesis of zinc sulfide (ZnS) nanoparticles using Stevia rebaudiana Bertoni and evaluation of its cytotoxic properties. J Mol Struct 1175:214–218

Biswas S, Kar S (2008) Fabrication of ZnS nanoparticles and nanorods with cubic and hexagonal crystal structures: a simple solvothermal approach. Nanotechnology 19:045710

Dessalegn L, Shimelis T, Tadesse E, Gebre-selassie S (2014) Aerobic bacterial isolates from post-surgical wound and their antimicrobial susceptibility pattern: a hospital based cross-sectional study. J Med Res 3:18–23

Elumalai K, Velmurugan S (2015) Applied surface science green synthesis, characterization and antimicrobial activities of zink oxide nanoparticles from the leaf extract of Azadirachta indica (L.). Appl Surf Sci 345:329–336

Fujiwara H, Murakoshi K, Wada Y, Yanagida S (1998) Observation of adsorbed N, N-dimethylformamide molecules on colloidal ZnS nanocrystallites Effect of coexistent counteranion on surface structure. Langmuir 14:4070–4073

Gou Y, Su Z, Xue Z (2004) A study on preparation and photocatalytic characterization of conjugated polymer/ZnS complex. Mater Res Bull 39:2203–2208

Hoffmann MR, Martin ST, Choi W, Bahnemann DW (1995) Environmental applications of semiconductor photocatalysis. Chem Rev 95:69–96

Jabeen K, Su L, Li J, Yang D, Tong C, Mu J, Shi H (2017) Microplastics and mesoplastics in fish from coastal and fresh waters of China. Environ Pollut 221:141–149

Kar S, Biswas S, Chaudhuri S (2005) Catalytic growth and photoluminescence properties of ZnS nanowires. Nanotechnology 16:737

Liu WW, Shen XC, Tao L, Gao YQ (2012) Chemical constituents and quality standard of Girardinia diversifolia (Link) Friis. Chin Pharm J 47(23):1943–1946

Moussaoui M, Saoudi R, Tishchenko AV, Chassagneux F (2010) Tunable optical properties of ZnS nanoparticles. Electromagn Light Scattering XII:178

Negi A, Vishwakarma RK, Gangwar R, Negi DS (2021) Synthesis of Zinc Oxide nanoparticles utilizing leaves of taraxacum officinale radix: an adaptable tool against bacterial, fungal strains and a methodology towards ecological remediation as a photocatalyst. J Adv Sci Res 12(3).

Rana D, Bhatt A, Lal B, Parkash O, Kumar A, Uniyal SK (2021) Use of medicinal plants for treating different ailments by the indigenous people of Churah subdivision of district Chamba Himachal Pradesh, India. Environ Dev Sustain 23:1162–1241

Shrestha PM, Dhillion SS (2003) Medicinal plant diversity and use in the highlands of Dolakha district Nepal. J Ethnopharmacol 86:81–96

Soltani N, Saion E, Hussein MZ, Erfani M, Abedini A, Bahmanrokh G, Navasery M, Vaziri P (2012) Visible light-induced degradation of methylene blue in the presence of photocatalytic ZnS and CdS nanoparticles. Int J Mol Sci 13:12242–12258

Yanagida S, Mizumoto K, Pac C (1986) Semiconductor photocatalysis. Part 6. Cis-trans photoisomerization of simple alkenes induced by trapped holes at surface states. J Am Chem Soc 108:647–654

Zheng Y, Fu L, Han F, Wang A, Cai W, Yu J, Yang J, Peng F (2015) Green biosynthesis and characterization of zinc oxide nanoparticles using Corymbia citriodora leaf extract and their photocatalytic activity. Green Chem Lett Rev 8:59–63

Chapter 14
The Current Scenario in Chitosan Nanocomposite Application in Wastewater Treatment

Shivani Uniyal and Jai Prakash Narain Rai

Abstract An intensified value of different pollution, in water resources associated with their various consecutive environmental concerns, has engendered researchers to seek suitable remediation strategies for wastewater treatment. The traditional technologies used in wastewater treatment encompass sorption, ion-exchange, precipitation, membrane filtration, reverse osmosis, and solvent extraction. Among these approaches, sorption has emanated as simple, robust, and efficient means for the removal of a vast range of pollutants from wastewater. Over the past decades, nanocomposites have been applied successfully as sorbents in various wastewater treatment studies. Chitosan is an ample cationic biopolymer, with unique structure, nontoxicity, biocompatibility, pluri-dimensional characteristic, with diverse applications. This chapter reviews recent developments in the application of chitosan nanocomposites in the removal of a broad range of organic and inorganic contaminants from wastewater which comprehend synthesis and modification of chitosan-based hydrogels, associated advantages, and, analytical performance. Furthermore, innovative approaches prerequisite for surface modification and selectivity improvement of chitosan nanocomposites to alleviate the pollutant removal efficiency are addressed. Finally, the challenges and prospects for escalating the performance of chitosan nanocomposite-based wastewater treatment methods are also discussed.

Keywords Biopolymer · Chitosan · Nanocomposite · Sorption · Wastewater treatment

S. Uniyal (✉)
Department of Botany, Banaras Hindu University, Varanasi 221005, India
e-mail: shivi273@gmail.com

J. P. N. Rai
Department of Environmental Science, College of Basic Sciences, G.B. Pant University of Agriculture and Technology, Pantnagar 263145, India

J. P. N. Rai and S. Saraswat (eds.), *Nano-biotechnology for Waste Water Treatment*, Water Science and Technology Library 111,
https://doi.org/10.1007/978-3-031-00812-2_14

14.1 Introduction

An extensive discharge of wastewater containing toxic elements resulted from rapid industrialization and agricultural practices have escalated the demand for clean water (Zhu et al. 2013; Huang et al. 2017). The World Health Organization and United Nations are currently focussing on the development of sustainable wastewater treatment solutions (Morsi et al. 2017). Conventionally, numerous methods are used for wastewater treatment, such as coagulation/flocculation (Li et al. 2013; Wang et al. 2015), membrane-based methods (Huang et al. 2009; Gogoi et al. 2016), etc. However, adsorption has been the preferred method owing to its economic feasibility, simplicity, and the potential applicability of using efficient and inexpensive adsorbents as sludge ash, clay, zeolites, and resins (Inbaraj et al. 2016; Ahmad et al. 2019). Nanomaterials, materials having at least one dimension less than approximately 100 nm, are being used in various fields such as biomedical, sensor development, catalysis, etc. (Mostafa et al. 2020). Owing to the large surface-to-volume ratio, several kinds of nanomaterials are being employed for the adsorption of heavy metal ions such as CeO_2, Fe_3O_4, TiO_2, and ZnO (Singh et al. 2010; Recillas et al. 2011).

Chitosan (CHI) is a poly-cationic polymer that attracts ubiquitous interest due to its biocompatibility, low cost, nontoxicity, biodegradability, excellent film-forming ability, great adsorption potential, and abundance (Ravi Kumar 2000; Ahmed and Mirza 2018). The chemical structure of this well-known polysaccharide consists of β(1 → 4)-2- amino-2-deoxy-β-d-glucan units (Abdou et al. 2008) and is derived from the deacetylation of chitin, a constituent of the exoskeletons of some crustacean shells (Komi and Hamblin 2016). The application of chitosan as a potential adsorbent in wastewater treatment studies can be ascribed to the presence of amino and hydroxyl groups and the resultant adsorption between chitosan and contaminant (Ahamad et al. 2020). Despite the promising results achieved by the utilization of chitosan as an adsorbent in the wastewater treatment strategies, there are still some formidable bottlenecks in the practical application of alone chitosan such as weak bonding potential and mechanical strength, poor reusable performance, low chemical stability, and potential production of secondary pollutants (Pan et al. 2011; Reddy and Lee 2013; Wu et al. 2017). Moreover, one of the prominent shortcomings cognate with the use of raw chitosan that limits its use in an industrial scale is pH sensitivity owing to the presence of $-NH_2$ and -OH groups. To conquer these impediments, magnetic materials are being applied to produce chitosan-based nanocomposite with more stability and reusability (Sowmya and Meenakshi 2014; Chagas et al. 2019).

14.2 Recent Methods Used in the Synthesis of Nanocomposite

Coupled with technological development several physical, chemical, and biological approaches have been developed for nanocomposite synthesis.

14.2.1 *Solvothermal Process*

Solvothermal synthesis encompasses a chemical reaction performed in a solvent at temperatures and pressure above the boiling point and 1 bar, respectively (Nunes et al. 2019). The methods ensure high activity of the chemical reactant and synthesis of products in a specific phase (Rane et al. 2018). A magnetic/graphene/chitosan-based nanocomposite was developed through a facile solvothermal process (Rebekah et al. 2020). Successively, this adsorbent was employed in batch mode adsorption studies and showed good results for the adsorption of 2-naphthol from aqueous solution, exhibiting almost 99.8% adsorption.

14.2.2 *Biogenic Synthesis*

Biogenic synthesis of nanocomposites is another strategy to develop nanocomposite. Sathiyavimal et al. used *Psidium guajava* aqueous leaf extract for the biogenic synthesis of chitosan (CS)-coated copper oxide NPs (CS-CuO nanocomposite) (Sathiyavimal et al. 2020). The developed nanocomposite was evaluated for congo red and methylene blue degradation and the results showed 90.8% and 91.3% degradation for congo red and methylene blue, respectively. The study advocated the use of biosynthesized chitosan as an eco-friendly approach for industrial wastewater treatment without affecting the aquatic systems. In a similar work, a simple one-pot green route was utilized to synthesize chitosan-iron oxide (CS-Fe_2O_3) nanocomposite (Ahmad and Mirza 2018). The developed nanocomposites were investigated and applied for Pb(II) and Cd(II) removal from industrial and hospital wastewater. In this work, novel nanocomposites showed 72–83% and 75 to 86% removal efficiency for Pb(II) and Cd(II), respectively.

14.2.3 *Microwave Heating*

The microwave heating technique has also been explored in the nanocomposite preparation owing to less time required as compared to conventional heating methods (Mostafa et al. 2020). The technique led to the formation of nanocomposites via the complexation reaction between functional groups of zinc oxide nanoparticles and chitosan.

14.3 Removal of Pollutants via Nanocomposites

Table 14.1 summarizes the nanocomposites currently designed to remediate a vast range of organic and inorganic contaminants in wastewater.

14.3.1 Organic Contaminants

Substantial research has been dedicated to the development and successful application of novel nanocomposites for the treatment of organic contaminants in wastewater.

14.3.1.1 Drugs

In recent years, the development of novel nanocomposites and their applications in drug removal from wastewater have drawn more attention. For instance, Li et al. synthesized a novel chitosan-glyoxal/polyvinylpyrrolidone/MoS_2 (CSG/PVP/MoS2) nanocomposites and evaluated their adsorption and photocatalytic potential for the drug from water (Li et al. 2019). The synthesized nanocomposite demonstrated remarkable diclofenac degradation (~94.5%) indicating potential and effective industrial applications. Another similar application described by Ahamad et al. was to fabricate chitosan, thiobarbituric acid, malondialdehyde, and Fe_3O_4 (CTM@Fe3O4)-based nanocomposite for the adsorption of tetracycline (TC) (Ahamad et al. 2020). By using the batch technique, an optimized method for excellent adsorption of TC was established, which resulted in ~215.31 mg g^{-1} TC adsorption. Li et al. fabricated a nanocomposite film by using $NiFe_2O_4$-COF-chitosan-terephthalaldehyde and applied it to treat antibiotic contaminated wastewater (Fig. 14.1) (Li et al. 2020).

14.3.1.2 Proteins

Polyaniline–metal hybrid has been evident to be a more efficient functional material, which offers a promising platform for nanocomposite fabrication and application in protein-rich wastewater treatment. For example, Mostafa and Darwish, reported tuned chitosan polyaniline–metal (Cu and Ag) hybrid nanocomposites for bovine

Table 14.1 Chitosan-based nanocomposites recently designed to remediate a vast range of organic and inorganic contaminants in wastewater

Class of contaminant	Type of contaminant	Contaminant	Chitosan-based nanocomposite used for the treatment	Treatment efficiency	Reference
Organic	Drugs	Diclofenac	Chitosan-glyoxal/Polyvinylpyrrolidone/MoS_2 (CSG/PVP/MoS_2)	94.5%	Li et al. (2019)
		Tetracycline	Chitosan, thiobarbituric acid, malondialdehyde and Fe_3O_4 nanoparticles (CTM@Fe3O4)	215.31 mg g^{-1}	Ahamad et al. (2020)
		Tetracycline	$NiFe_2O_4$-COF-chitosan-terephthalaldehyde	388.52 mg g^{-1}	Li et al. (2020)
		Cefotaxime		309.26 mg g^{-1}	
	Proteins	Bovine serum albumin (BSA)	Chitosan polyaniline–metal (Cu and Ag) hybrid nanocomposites	100%	Mostafa and Darwish, (2014)
	Dyes	Disperse Orange 25	Chitosan cenospheres	97.30%	Markandeya et al. (2017)
		Disperse Blue 79:1		94.22%	
		Rose Bengal dye	Molecularly imprinted polymer (MIP) chitosan-TiO_2 nanocomposite	98.89%	Ahmed et al. (2018)
		Malachite green	Chitosan-wrapped multiwalled carbon nanotube as filler within PEBA thin film nanocomposite (CWNT-PEBA/PES TFN)	98.7%	Mousavi, et al. (2020)
	Polyaromatic hydrocarbon (PAHs)	Anthracene	Iron-oxide-based chitosan-nanocomposites	95%	Rani et al. 2020
		Phenanthrene		92%	

(continued)

Table 14.1 (continued)

Class of contaminant	Type of contaminant	Contaminant	Chitosan-based nanocomposite used for the treatment	Treatment efficiency	Reference
Inorganic	Heavy metals	Cu(II), Cd(II), and Pb(II)	ZnO@Chitosan core/organically shell nanocomposite (ZOCS)	117.6 – 476.1 mg g^{-1}	Saad et al. 2018
		Cu(II)	Epichlorohydrin cross-linked chitosan Schiff base-Fe_3O_4 nanocomposite (ECSBNC)	90.90 mg g^{-1}	Anush and Vishalakshi 2019
		Cr(VI)		83.33 mg g^{-1}	
		Cr(VI)	Triethylenetetramine-modified hollow Fe_3O_4/SiO_2/chitosan magnetic nanocomposites (Fe_3O_4/SiO_2/CS-TETA)	254.6 mg g^{-1}	Wang et al. (2020)

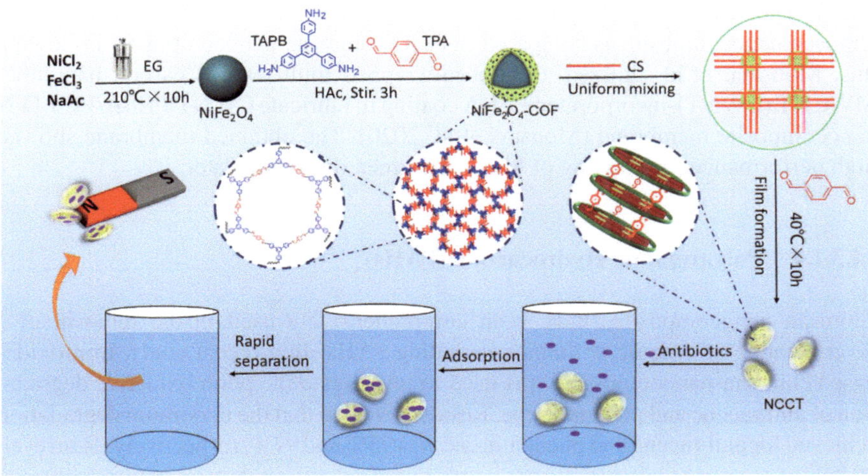

Fig. 14.1 Schematic illustration of preparation of $NiFe_2O_4$/COF/CS/TPA nanocomposites film (NCCT). Reprinted with permission from Li et al. (2020)

serum albumin (BSA) degradation (Mostafa and Darwish 2014). The developed nanocomposites demonstrated excellent oxidative catalytic degradation with 100% tryptophane loss with simultaneous bactericidal activities against *Bacillus subtilis* and *Escherichia coli*.

14.3.1.3 Dyes

To perpetuate the implementation of nanocomposites, copious attempts have been devoted to developing a selective nanoparticle layer for the treatment of dye industry wastewater. For example, Markandeya et al. developed an economical solution for dye wastewater by the synergistic use of cenospheres, a waste product along with chitosan (Markandeya et al. 2017). The rational combination of chitosan, cenospheres (10:3), and glutaraldehyde, a cross-linking agent, leads to the formation of the novel nanocomposite. The synthesized nanocomposite demonstrated about 97.30% and 94.22% removal for Disperse Orange 25 and Disperse Blue 79:1 dyes, respectively. Unlike the economical use of a waste product, Ahmed et al. used a novel molecularly imprinted polymer (MIP) chitosan-TiO_2 nanocomposite for the selective adsorption of Rose Bengal dye from wastewater (Ahmed et al. 2018). The obtained nanocomposite also showed superior adsorption toward the analyte of interest with 98.89% adsorption compared to pure chitosan with 53.73% removal efficiency. Besides the MIP decoration to boost the pollutant adsorption efficiency, doping with rare earth elements have been used to increase the catalytic activity of nanocomposites. Alimard synthesized a novel Nd-Ce doped Fe_3O_4-chitosan nanocomposite and evaluated the Fenton degradation of Direct Red 81 dye using it as a catalyst (Alimard

2019). The results revealed increased catalytic activity of Nd-Ce doped Fe_3O_4-chitosan nanocomposite as compared to Fe_3O_4-chitosan nanocomposite. Following this, Mousavi, et al. utilized a chitosan-wrapped multi-walled carbon nanotube-MWCNT (CWNT)-incorporated PEBA coating to fabricate CWNT-PEBA/PES TFN nanocomposite membrane (Mousavi et al. 2020). The obtained membrane showed high performance for effacing of Malachite green dye from water.

14.3.1.4 Polyaromatic Hydrocarbon (PAHs)

Chitosan nanocomposites have been comprehensively used in the removal of a diverse range of organic pollutants including PAHs. In a recent study, iron-oxide-based chitosan-nanocomposite was used to investigate the photooxidative degradation of anthracene and phenanthrene. Results revealed that the maximum degradation achieved for anthracene and phenanthrene was 95% and 92%, respectively (Rani et al. 2020).

14.3.2 Inorganic Contaminants

Owing to its interaction with metal ions and concomitant nanoparticles, considerable interest lies in the employment of chitosan-based nanocomposites in heavy metal adsorption studies (Anush and Vishalakshi 2019). Several demonstrations of improved adsorption performance of inorganic contaminants through the synthesis of novel nanocomposites paved the way for further research in the applications of these materials in heavy metal removal from wastewater. For example, the role of ZnO@Chitosan core/organically shell nanocomposite (ZOCS) in the treatment of heavy metal contaminated waste has been demonstrated through the determination of adsorption capacity by the use of the Langmuir isotherm model (Saad et al. 2018). In this work, adsorption-desorption cyclic results suggested its reusability multiple times without loss in the efficiency. Furthermore, the metal ions adsorption capacity obtained using the fabricated nanocomposite was 117.6–476.1 mg g^{-1}. Alsabagh et al. designed multifunctional chitosan-based bi-nanocomposites and ter-nanocomposites through structural modifications (Alsabagh et al. 2015). In the subsequent work, the synthesized nanocomposites were investigated for the removal of Cu, Cd, and Pb from the wastewater and the results indicated efficient removal of heavy metals, thereby advocating their potential use in wastewater treatment systems (Morsi et al. 2017). Moussout et al. reported a nanocomposite synthesized with chitosan (CS) and bentonite/chitosan nano biocomposite (nano 5%Bt/CS) (Moussout et al. 2018). In this work, obtained nanocomposite was investigated for the chromium (VI) ions adsorption capacity in tanning industry wastewater. Based on the adsorption data, nanocomposite showed significant removal of chromium (VI) ions via physical adsorption.

14.4 Advancements in Chitosan Nanocomposites

Significant attention has been given to chitosan as an excellent dispersant due to its ability to evenly disperse nanoparticles throughout its structure, during chelation (Singh et al. 2011). The adsorption capacity of chitosan-based nanocomposites can be further upgraded by the incorporation of new functional groups through various modifications (Anush and Vishalakshi 2019). These modifications include chemical modification, the use of different filters, and hydrogel-based modification (Mostafa et al. 2020).

14.4.1 Chemical Modification

14.4.1.1 Incorporation of Organic Compounds and Surfactants

An innovative approach to develop modified chitosan was realized by Vakili et al., who modified chitosan beads with hexadecylamine (HDA) and 3-aminopropyl triethoxysilane (APTES) surfactants (Vakili et al. 2017). The chemically modified chitosan beads showed a clear change in the RB4 adsorption behavior with an improved adsorption capacity of 1.48 times. To further boost the adsorption performance of chitosan-based nanocomposites, the triethylenetetramine (TETA) can be deposited on these materials. For example, Wang et al. synthesized triethylenetetramine-modified hollow Fe_3O_4/SiO_2/chitosan magnetic nanocomposites (Fe_3O_4/SiO_2/CS-TETA) (Fig. 14.2) (Wang et al. 2020). Owing to hollow and porous nano skeleton, as well as TETA-based modification, an excellent adsorption capacity of Cr(VI) ions, i.e., 254.6 mg g 1^{-1} was achieved by the fabricated nanocomposites. The results further confirmed rapid adsorption with good magnetic recovery and reusability, making them suitable candidates for practical application.

14.4.1.2 Cross-Linking

In the cross-linking procedure, participating counterions led to the genesis of intermolecular or intramolecular linkages by diffusing into the polymer followed by subsequent reaction (Hasirci et al. 2017). An ideal cross-linking agent used in the process of nanocomposite synthesis must keep the active functional groups as in pristine chitosan (Kucuka and Urucu 2019). Shankar et al. modified chitosan beads by the 2-hydroxy-1-naphthaldehyde cross-linking agent (Shankar et al. 2020). With this modification, good results were obtained in terms of adsorption efficiency of pentachlorophenol (PCP) pesticide from aqueous solution. In a similar study, chitosan was modified by reaction with 3-methyl-1-phenyl-5-(piperidin-1-yl)-1H-pyrazole-4-carbaldehyde (Anush and Vishalakshi 2019). The obtained Schiff base was then

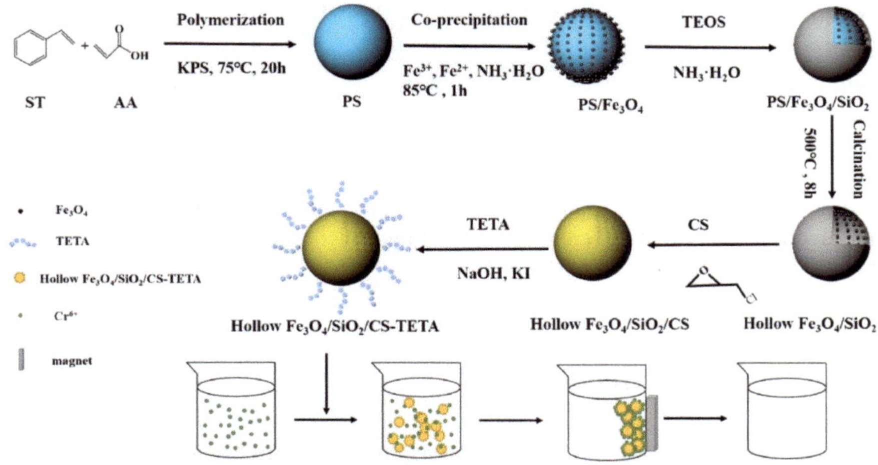

Fig. 14.2 Schematic illustration of the synthesis of hollow Fe_3O_4/SiO_2/CS-TETA nanocomposites and their application in recycle removal of Cr(VI) ions. Reprinted with permission from Wang et al. (2020)

used to synthesize a cross-linked gel by cross-linking with epichlorohydrin. Modified chitosan gel was further decorated with Fe_3O_4 nanoparticles to prepare magnetic nanocomposite, which showed improved adsorption for Cu(II) and Cr(VI) ions with values 90.90 and 83.33 mg g^{-1}, respectively. In another study, the cross-linker (TEOS; Tetraethyl orthosilicate) was used in a different ratio to synthesize three different matrices (PSCS-(1,2,3)) of polyhedral oligomeric silsesquioxane (POSS)-modified chitosan nanocomposites. The prepared matrices were investigated for the removal of Cd (II) from wastewater. Owing to the perfect balance between available active functional sites and the stability of synthesized matrices, PSCS-2 showed the most superior performance among all the nanocomposite matrices (Kucuka and Urucu 2019).

14.4.2 Nanofilters

The remarkable characteristics of nanoparticles can be transferred to the composite only after attaining a strong interfacial bonding (Hu et al. 2014), which could be achieved by employing a vast range of nanofilters as reinforcement phase such as, silica-based modification and metal-based modification (Moura et al. 2016).

14.4.2.1 Silica-Based Modification

Incorporation of silica materials in chitosan is reported to improve its thermal and structural stability (Chew et al. 2011; Chen et al. 2012). Silica material can be used in the form of layered silicates (Moura et al. 2016). The chemical structure of layered silicates composed of tetrahedrally bonded silicon (Si) atoms and octahedrally bonded aluminum (Al) or magnesium (Mg) atoms surrounded by four and eight oxygen atoms, respectively (Reddy 2011). Copello et al. produced a layer-by-layer silicate–chitosan composite and evaluated its stability and potential to adsorb heavy metals viz. Cd(II), Cr(III), and Cr(VI) (Copello et al. 2008). The results displayed that the non-covalent immobilization method used in the investigation improved the stability and adsorption capacity of the synthesized composites. Montmorillonite (MMT) is an aluminosilicate, being used in the modification of chitosan-based nanocomposites (Park et al. 2012; Ngwabebhoh et al. 2016). In subsequent work, montmorillonite (MMT) was used for the synthesis of chitosan composite hydrogels and the developed nanocomposite showed improved removal capacity for both Cu^{2+} and nitrazine yellow (NY) dye (Ngwabebhoh et al. 2016).

14.4.2.2 Metal-Based Modification

The mechanical properties and adsorption capacity of a chitosan matrix can significantly be enhanced by the insertion of inorganic particles (Caridade et al. 2013). For instance, in a study, chitosan beads were modified with zirconium and the modified chitosan beads displayed efficient phosphate adsorption (Liu and Zhang 2015).

14.4.3 Hydrogel-Based Modification

Hydrogels are a unique class of polymers, which can imbibe an enormous amount of water, with simultaneous conservation of their shape (Jeong et al. 2012). The application of hydrogels has been a very common strategy for the improved adsorption performance. Chitosan-based hydrogels are known to adsorb pollutants through the process of complexation, electrostatic, and hydrogen bonding (Pakdel and Peighambardoust 2018). There are numerous reports on the utilization of chitosan-based hydrogels in the field of wastewater treatment (Kim et al. 2012; Mondal et al. 2015). For instance, Yu et al. developed a hydrogel-based method to synthesize magnetic chitosan iron(III) (MCh-Fe) hydrogel nanocomposites (Yu et al. 2013). The synthesized nanocomposite demonstrated prompt and complete Cr^{6+} removal. In a study, zirconium-chitosan hydrogel beads were fabricated and evaluated for the removal of boron(III) (Kluczka et al. 2018). The results indicated significant boron(III) adsorption from wastewater. In subsequent work, a chitosan/AgCl/ZnO (CS/AgCl/ZnO) nanocomposite hydrogel was designed for the degradation of methylene blue (MB) (Taghizadeh et al. 2020). The results of the degradation study further demonstrated effective MB degradation by hydrogel beads system.

14.5 Outlook and Conclusion

Chitosan-based nanocomposites have been immensely utilized in the adsorption studies of organic and inorganic contaminants from wastewater. To alleviate the poor stability and increase the sorption efficiency, numerous amendments have been done to construct modified nanocomposites. A variety of methods including chemical modification, application of nanofilters, and hydrogel-based methods are developed for preparing modified chitosan-based nanocomposites with improved performance. The modified nanocomposites could be produced by the solvothermal process, biogenic methods, or by employing microwave heating. Despite the efficient pollutant adsorption achieved by the chitosan-based nanocomposites, there are still some bottlenecks associated with its use in practical application. For instance, efforts are required to determine the long-term stability and robustness of the chitosan-based nanocomposites for real-time use (Ahmad and Mirza 2018). In addition, regeneration potential of chitosan-based hydrogels should be thoroughly studied to determine the economic feasibility (Pakdel and Peighambardoust 2018). Moreover, process optimization for chitosan-based nanocomposites for each system must be accomplished to modulate the eventual quality and parameters of the fabricated nanocomposite (Yousefi et al. 2020). Since the functional groups in the chitosan could be engaged in the cross-linking reaction, the adsorption efficiency can be affected in a cross-linking procedure (Chen et al. 2012). Therefore the research studies should be directed toward developing innovative techniques to increase the sorption efficiency.

We believe that chitosan-based nanocomposites with their excellent pollutant adsorption characteristics will provide a new vision to the researchers and lead to practical application with further modifications.

References

Abdou ES, Nagy KSA, Elsabee MZ (2008) Extraction and characterization of chitin and chitosan from local sources. Bioresour Technol 99(5):1359–1367. https://doi.org/10.1016/j.biortech.2007.01.051. Epub 2007 Mar 26

Ahamad T, Naushada M, Al-Al-Shahrani T, Al-hokbanya N, Alshehri SM (2020) Preparation of chitosan based magnetic nanocomposite for tetracycline adsorption: kinetic and thermodynamic studies. Int J Biol Macromol 147:258–267

Ahmad R, Mirza A (2018) Facile one pot green synthesis of chitosan-iron oxide (CS-Fe_2O_3) nanocomposite: removal of Pb(II) and Cd(II) from synthetic and industrial wastewater. J Clean Prod 186:342–352

Ahmad ZU, Yao L, Wang J, Gang DD, Zappi ME (2019) Neodymium embedded ordered mesoporous carbon (OMC) for enhanced adsorption of sunset yellow: characterizations, adsorption study and adsorption mechanism. Chem Eng J 359:814–826

Ahmed MA, Abdelbar NM, Mohamed AA (2018) Molecular imprinted chitosan-TiO_2 nanocomposite for the selective removal of Rose Bengal from wastewater. Int J Biol Macromol 107:1046–1053

Alimard P (2019) Fabrication and kinetic study of Nd-Ce doped Fe_3O_4-chitosan nanocomposite as catalyst in Fenton dye degradation. Polyhedron 171:98–107

Alsabagh AM, Fathy M, Morsi RE (2015) Preparation and characterization of chitosan/silver nanoparticle/copper nanoparticle/carbon nanotube multifunctional nano-composite for water treatment: heavy metals removal; kinetics, isotherms and competitive studies. RSC Adv 5:55774–55783

Anush SM, Vishalakshi B (2019) Modifed chitosan gel incorporated with magnetic nanoparticle for removal of Cu(II) and Cr(VI) from aqueous solution. Int J Biol Macromol 133:1051–1062

Caridade SG, Merino EG, Alves NM et al (2013) Chitosan membranes containing micro or nano-size bioactive glass particles: evolution of biomineralization followed by in situ dynamic mechanical analysis. J Mech Behav Biomed Mater 20:173–183

Chagas PMB, Caetano AA, Rossi MA, Goncalves MA, Ramalho TDC, Correa AD, Guimaraes IDR (2019) Chitosan-iron oxide hybrid composite: mechanism of hexavalent chromium removal by central composite design and theoretical calculations. Environ Sci Pollut Res Int 26:15973–15988

Chen AW, Zeng GM, Chen GQ, Hu XJ, Yan M, Guan S, Shang C, Lu LH, Zou ZJ, Xie GX (2012) Novel thiourea-modified magnetic ion-imprinted chitosan/TiO2 composite for simultaneous removal of cadmium and 2,4-dichlorophenol. Chem Eng J 191:85–94. https://doi.org/10.1016/j.cej.2012.02.071

Chew SL, Wang K, Chai SP, Goh KL (2011) Elasticity, thermal stability and bioactivity of polyhedral oligomeric silsesquioxanes reinforced chitosan-based microfibres. J Mater Sci Mater 22(6):1365–1374

Copello GJ, Varela F, Vivot RM, Diaz LE (2008) Immobilized chitosan as biosorbent for the removal of Cd(II), Cr(III) and Cr(VI) from aqueous solutions. Bioresour Technol 99(14):6538–6544

Gogoi P, Thakur AJ, Devi RR, Das B, Maji TK (2016) A comparative study on sorption of arsenate ions from water by crosslinked chitosan and crosslinked chitosan/MMT nanocomposite. J Environ Chem Eng 4:42484257. https://doi.org/10.1016/j.jece.2016.09.027

Hasirci V, Huri PY, Tanir TE, Eke G, Hasirci N (2017) 1.22 Polymer fundamentals: polymer synthesis. In: Ducheyne P, Healy K, Hutmacher DW, Grainger DW, Kirkpatrick CJ (eds) Comprehensive biomaterials II. Elsevier, Oxford, pp 478–506

Hu KS, Kulkarni DD, Choi I et al (2014) Graphene-polymer nanocomposites for structural and functional applications. Prog Polym Sci 39:1934–1972. https://doi.org/10.1016/j.progpolymsci.2014.03.001

Huang C, Zeng G, Huang D, Lai C, Xu P, Zhang C, Cheng M, Wan J, Hu L, Zhang Y (2017) Effect of Phanerochaete chrysospoium inoculation on bacterial community and metal stabilization in lead-contaminated agricultural waste composting. Biores Technol 243:294–303

Huang G, Zhang H, Shi JX, Langrish TAG (2009) Adsorption of chromium(VI) from aqueous solutions using cross-linked magnetic chitosan beads. Ind Eng Chem Res 48:2646–2651. https://doi.org/10.1021/ie800814h

Inbaraj BS, Chen BH (2016) Nanomaterial-based sensors for detection of foodborne bacterial pathogens and toxins as well as pork adulteration in meat products. J Food Drug Anal 24:15–28

Jeong B, Kim SW, Bae YH (2012) Thermosensitive sol-gel reversible hydrogels. Adv Drug Deliv Rev 54(1):37–51

Kim TY, Park SS, Cho SY (2012) Adsorption characteristics of Reactive Black 5 onto chitosan beads crosslinked with epichlorohydrin. J Ind Eng Chem 18:1458–1464

Kluczka J, Gnus M, Kazek-Kęsik A, Dudek G (2018) Zirconium-chitosan hydrogel beads for removal of boron from aqueous solutions. Polymer 150:109–118

Komi DEA, Hamblin MR (2016) Chitin and chitosan: production and application of versatile biomedical nanomaterials. Int J Adv Res 4:411–427

Kucuka AC, Urucu OA (2019) Silsesquioxane-modified chitosan nanocomposite as a nanoadsorbent for the wastewater treatment. React Function Polym 140:22–30

Liu X, Zhang L (2015) Removal of phosphate anions using the modified chitosan beads: Adsorption kinetic, isotherm and mechanism studies. Powder Technol 277:112–119

Li L, Fan L, Sun M, Qiu H, Li X, Duan H, Luo C (2013) Adsorbent for chromium removal based on graphene oxide functionalized with magnetic cyclodextrin-chitosan. Colloids Surf B 107(2013):76–83. https://doi.org/10.1016/j.colsurfb.2013.01.074

Li Z, Liu Y, Zou S, Lu C, Bai H, Mu H, Duan J (2020) Removal and adsorption mechanism of tetracycline and cefotaxime contaminants in water by $NiFe_2O_4$-COF-chitosan-terephthalaldehyde nanocomposites film. Chem Eng J 382:123008

Li X, Zhang Z, Fakhri A, Gupta VK, Agarwal S (2019) Adsorption and photocatalysis assisted optimization for drug removal by chitosan-glyoxal/Polyvinylpyrrolidone/MoS_2 nanocomposites. Int J Biol Macromol 136:469–475

Markandeya, Dhiman N, Shukla SP, Kisku GC (2017) Statistical optimization of process parameters for removal of dyes from wastewater on chitosan cenospheres nanocomposite using response surface methodology. J Clean Prod 149:597–606

Mondal S, Li C, Wang K (2015) Bovine serum albumin adsorption on gluteraldehyde crosslinked chitosan hydrogels. J Chem Eng Data 60(8):2356–2362

Morsi RE, Alsabagh AM, Nasr SA, Zaki MM (2017) Multifunctional nanocomposites of chitosan, silver nanoparticles, copper nanoparticles and carbon nanotubes for water treatment: antimicrobial characteristics. Int J Biol Macromol 97:264–269

Mostafa TB, Darwish AS (2014) An approach toward construction of tuned chitosan/polyaniline/metal hybrid nanocomposites for treatment of meat industry wastewater. Chem Eng J 243:326–339

Mostafa MH, Elsawy MA, Darwish MSA, Hussein LI, Abdaleem AH (2020) Microwave-Assisted preparation of Chitosan/ZnO nanocomposite and its application in dye removal. Mater Chem Phys 248:122914

Moura D, Mano JF, Paiva MC, Alves NM (2016) Chitosan nanocomposites based on distinct inorganic fillers for biomedical applications. Sci Technol Adv Mater 17(1):626–643. https://doi.org/10.1080/14686996.2016.1229104

Mousavi SR, Asgharia M, Mahmoodi NM (2020) Chitosan-wrapped multiwalled carbon nanotube as filler within PEBA thin film nanocomposite (TFN) membrane to improve dye removal. Carbohyd Polym 237:116128

Moussout H, Ahlafi H, Aazza M, Amechrouq A (2018) Bentonite/chitosan nanocomposite: preparation, characterization and kinetic study of its thermal degradation. Thermochimica Acta 659:191–202

Ngwabebhoh FA, Erdem A, Yildiz U (2016) Synergistic removal of Cu(II) and nitrazine yellow dye using an eco-friendly chitosan-montmorillonite hydrogel: optimization by response surface methodology. J Appl Polym Sci 133(29):43664–43678

Nunes D, Pimentel A, Santos L, Barquinha P, Pereira L, Fortunato E, Martins R (2019) Synthesis, design, and morphology of metal oxide nanostructures. In: Nunes D, Pimentel A, Santos L, Barquinha P, Pereira L, Fortunato E, Martins R (eds) Metal Oxide Nanostructures. Elsevier, pp 21–57

Pakdel PM, Peighambardoust SJ (2018) Review on recent progress in chitosan-based hydrogels for wastewater treatment application. Carbohyd Polym 201:264–279

Pan Y, Wu T, Bao H, Li L (2011) Green fabrication of chitosan films reinforced with parallel aligned graphene oxide. Carbohydr Polym 83:1908–1915

Park SH, Lee HS, Choi JH et al (2012) Improvements in barrier properties of poly(lactic acid) flms coated with chitosan or chitosan/clay nanocomposite. J Appl Polym Sci 125:E675–E680

Rane AV, Kanny K, Abitha VK, Thomas S (2018) Methods for synthesis of nanoparticles and fabrication of nanocomposites. In: Bhagyaraj SM, Oluwafemi OS, Kalarikkal N, Thomas S (eds)

Micro and Nano technologies, synthesis of inorganic nanomaterials. Woodhead Publishing, pp 121–139

Rani M, Rachna, Shanker U (2020) Metal oxide-chitosan based nanocomposites for efficient degradation of carcinogenic PAHs. J Environ Chem Eng 8:103810

Ravi Kumar MNV (2000) A review of chitin and chitosan applications. React Funct Polym 46:1–27. https://doi.org/10.1016/S1381-5148(00)00038-9

Rebekah A, Bharath G, Naushad M, Viswanathan C, Ponpandian N (2020) Magnetic graphene/chitosan nanocomposite: a promising nanoadsorbent for the removal of 2-naphthol fromaqueous solution and their kinetic studies. Int J Biol Macromol 159:530–538

Recilla S, García A, González E, Casals E, Puntes V, Sánchez A, Font X (2011) Use of CeO_2, TiO_2 and Fe_3O_4 nanoparticles for the removal of lead from water: toxicity of nanoparticles and derived compounds. Desalination 277(1–3):213–220

Reddy B (2011) Advances in diverse industrial applications of nanocomposites. Polymer/Clay Nanocomposites Intech, pp 139–176

Reddy DHK, Lee S (2013) Application of magnetic chitosan composites for the removal of toxic metal and dyes from aqueous solutions. Adv Colloid Interface 201–202:68–93

Saad AHA, Azzam AM, El-Wakeel ST, Mostafa BB, El-latif MBA (2018) Removal of toxic metal ions from wastewater using ZnO@Chitosan coreshell nanocomposite. Environ Nanotechnol Monit Manag 9:67–75

Sathiyavimal A, Vasantharaj S, Kaliannanc T, Pugazhendhi A (2020) Eco-biocompatibility of chitosan coated biosynthesized copper oxide nanocomposite for enhanced industrial (Azo) dye removal from aqueous solution and antibacterial properties. Carbohyd Polym 241:116243

Shankar A, Kongot M, Saini VK, Kumar A (2020) Removal of pentachlorophenol pesticide from aqueous solutions using modified chitosan. Arab J Chem 13(1):1821–1830

Singh J, Ima J, Whitten JE, Soares JW, Steeves DM (2010) Chemisorption of a thiol-functionalized ruthenium dye on zinc oxide nanoparticles: implications for dye-sensitized solar cells. Chem Phys Lett 497(4–6):196–199

Singh J, Srivastava M, Dutta J, Dutta PK (2011) Preparation and properties of hybrid monodispersed magnetic-Fe2O3 based chitosan nanocomposite film for industrial and biomedical applications. Int J Biol Macromol 48:170–176

Sowmya A, Meenakshi S (2014) Zr (IV) loaded cross-linked chitosan beads with enhanced surface area for the removal of nitrate and phosphate. Int J Biol Macromol 69:336–343

Taghizadeh MT, Siyahi V, Ashassi-Sorkhabi H, Zarrini G (2020) ZnO, AgCl and AgCl/ZnO nanocomposites incorporated chitosan in the form of hydrogel beads for photocatalytic degradation of MB, *E. coli* and *S. aureus*. Int J Biol Macromol 147:1018–1028

Vakili M, Rafatullah M, Ibrahim MH, Abdullah AZ, Gholami Z, Salamatinia B (2017) Enhancing reactive blue 4 adsorption through chemical modification of chitosan with hexadecylamine and 3-aminopropyl triethoxysilane. J Water Process Eng 15:49–54

Wang YM, Duan L, Sun Y, Hu N, Gao JY, Wang H, Xie XM (2015) Adsorptive removal of Cr(VI) from aqueous solutions with an effective adsorbent:crosslinked chitosan/montmorillonite nanocomposites in the presence of hydroxyl aluminium oligomeric cations. Desalin Water Treat 57:1–9

Wang X, Liu X, Xiao C, Zhao H, Zhang M, Zheng N, Kong W, Zhang L, Yuan H, Zhang L, Lu J (2020) Triethylenetetramine-modified hollow Fe_3O_4/SiO_2/chitosan magnetic nanocomposites for removal of Cr(VI) ions with high adsorption capacity and rapid rate. Micropor Mesopor Mat 297:110041

Wu S, Kan J, Dai X, Shen X, Zhang K, Zhu M (2017) Ternary carboxymethyl chitosanhemicellulose-nanosized TiO_2 composite as effective adsorbent for removal of heavy metal contaminants from water. Fibers Polym 18:22–32

Yousefi V, Mohebbi-Kalhori D, Samimi A (2020) Start-up investigation of the self-assembled chitosan/montmorillonite nanocomposite over the ceramic support as a low-cost membrane for microbial fuel cell application. Int J Hydrog Energy 45:4804–4820

Yu Z, Zhang X, Huang Y (2013) Magnetic chitosan–iron(III) hydrogel as a fast and reusable adsorbent for chromium(VI) removal. Ind Eng Chem Res 52(34):11956–11966
Zhu J, Wei S, Chen M, Gu H, Rapole SB, Pallavkar S, Ho TC, Hopper J, Guo Z (2013) Magnetic nanocomposites for environmental remediation. Adv Powder Technol 24:459–467

Chapter 15
Biochar Impregnated Nanomaterials for Environmental Cleanup

Prachi Dixit, Kajal S. Rathore, and Shivom Singh

Abstract Biochar-based nanomaterials have shown unbelievable prospective in managing various environmental contaminations and display immense enhancement in functional groups, surface active sites, pore size properties, catalytic degradation properties, etc. Their applications gaining increase interest due to the simplicity of preparation methods and their enhanced physicochemical properties. Therefore, the development of such nanomaterials has targeted in various research objectives including water/wastewater treatment, soil remediation, agriculture, and pollution. Sorption, catalysis, redox reaction and other structural and functional properties of biochar showed the alteration or removal of contaminations from the environment and are considered as incredible efficient ways in handling various challenging environmental issues. Considering the quick expansion of biochar-based nanomaterials. This chapter will serve the helpful information to summarize the various applications of this magical material in environmental cleanup, along with wastewater treatment, soil remediation, and agriculture advancement. In addition, synthesizing such efficient nanoparticles will provide a novel approach against various environmental concern and promise vast applications of nanotechnology.

Keywords Biochar · Nanomaterial · Wastewater treatment · Soil remediation · Agriculture advancement

P. Dixit
Department of Biotechnology, ITM University, Gwalior, India

K. S. Rathore
Department of Biotechnology, Government Kamla Raja Post Graduate College, Gwalior, India

S. Singh (✉)
Department of Environmental Science, ITM University, Gwalior, India
e-mail: shivomsingh@itmuniversity.ac.in

J. P. N. Rai and S. Saraswat (eds.), *Nano-biotechnology for Waste Water Treatment*, Water Science and Technology Library 111,
https://doi.org/10.1007/978-3-031-00812-2_15

15.1 Introduction

Biochar is a carbon containing solid material produced by pyrolysis of bio-organic supplies at <700°C under anoxic conditions (Lehmann et al. 2006). The crude supplies of biochar are primarily biomass squander (straw, defecation, or muck), with advantages squander asset use as well as adequately ease natural weakening (Chen and Chen 2009). The strategies to make biochar, including pyrolysis, aqueous carbonization, gasification, etc., distinctive bio-natural materials, pyrolysis temperature, change techniques, and different components will impact the exhibition of biochar. The most examinations predominantly center on these territories, for example, soil execution guideline (Lehmann et al. 2011), worldwide environmental change (Woolf et al. 2010), and sustainable biofuels (Abdullah and Wu 2009). Lately, biochar in the field of wastewater treatments has become an interesting research (Uchimiya et al. 2010; Chen et al. 2011; Fang et al. 2014).

Groundwater and surface water sullying with poisonous minor components is an overall ecological issue on account of harmful, non-degradable and cancer-causing nature of these components (Adrees et al. 2015; Rizwan et al. 2016; Antoniadis et al. 2017). Arsenic (As) is a metal found bounteously in the earth hull (Khalid et al. 2017; Shakoor et al. 2019). It is the 12th bountiful component in human body and 20th in the subsurface land arrangements (Shakoor et al. 2016). In watery conditions, As is basically found in two inorganic structures: arsenite III and arsenate V (Niazi et al. 2015; LeMonte et al. 2017; Awad et al. 2019). It is delivered from enduring of As containing minerals, normal cycles, anthropogenic exercises, and As containing pesticides (Niazi et al. 2018; Shahid et al. 2018). It is assessed that > 70 nations have As fixation from under 0.5 to more than 5000 μg/L which is disturbing for future (Shahid et al. 2020). As tainting causes numerous well-being impacts in people, plants, and creatures (Khalid et al. 2017; Waqas et al. 2017). It is unavoidable to remediate As tainted water with economical and condition amicable procedures attributable to intense and interminable As harming, significant cost-effective, eco-accommodating, and supportable methodologies have been investigated by the specialists for the appropriation of weighty metals from sullied fluid frameworks (Shakoor et al. 2016; Niazi et al. 2018; Hina et al. 2019).

Heavy metals are released from various industries including mining, metal completing the process of, electroplating, glass, materials, earthenware production, and capacity batteries. As of late, there have been developing worries over water contamination by substantial metals delivered from modern effluents because of the poisonous impacts of their particles on creatures and their collection in biota. Among these heavy metals, copper (Cu) and mercury (Hg) have gotten extensive consideration because of their harmfulness. The presence of these metal particles in water jeopardizes people and amphibian lives since they collect in the natural ways of life (Wang et al. 2012). In this way, there is the need to decontaminate and reuse wastewater defiled by substantial metal particles to make sure about elective wellsprings of water (Ali 2010) and to ensure our evolved ways of life (Wang et al. 2012). A few treatment innovations have been recommended to eliminate these heavy metal particles from polluted wastewater including adsorption, particle trade, synthetic precipitation, layer advances, switch assimilation, and electrochemical treatment (Demirbas 2008). Among these advances, adsorption is a widespread, generally advantageous, and quick strategy for treating heavy metals sullied arrangements economically and effectively (Dupont and Guillon 2003). The significant points of interest of this treatment innovation are its low buildup age and the possibility to recuperate and reuse the pre-owned adsorbent (Yu et al. 2013). Moreover, bio sorbents are ecologically cordial as well as promptly accessible in adequate amounts, and a very notable bio sorbent is biochar. The biochar is a fine-grained, permeable, and carbon-rich material delivered from the warm debasement of natural materials under oxygen-restricted conditions. Biochar has likewise been demonstrated to be a powerful sorbent for the sorption of a wide assortment of inorganic and natural toxins from fluid arrangements (Xue et al. 2012). The advancement in biochar creation (Cha et al. 2016; Liu et al. 2017) and its application in compound combination and sorption science (Qian et al. 2015; Inyang and Dickenson 2015; Inyang et al. 2016) have been assessed.

15.2 Preparation of Nanobiochar

Before understanding the formation process of nanobiochar, it is necessary to understand the procedure of biochar formation. This is as follows:

Biochar arrangement techniques are for the most part pyrolysis, aqueous carbonization, gasification, and different strategies. Pyrolysis is a decay response under high temperature and anoxic condition (Fig. 15.1). In light of pyrolysis time, temperature, and warming rate, it tends to be partitioned into moderate pyrolysis, quick pyrolysis, and "blaze" pyrolysis (Onay and Kockar 2003). Studies have demonstrated that lower pyrolysis temperature and more slow warming rates add to the development of strong items, in the moderate warm splitting, the substance of strong item up to 35%, and in this way, slow pyrolysis is considered as a fundamental planning of biochar (Kambo and Dutta 2015). Hydrothermal carbonization (HTC) alludes to the response of biomass in a submerged stale framework for 5 minutes to 16 hours at a weight of 2–6 MPa and < 350°C temperature (Kambo and Dutta 2015; Tan et al. 2015; Berge et al. 2015). Hydrothermal carbonization measure utilizes H_2O as the response medium under high pressure and warming conditions, it is difficult to create destructive substances. Subsequently, the biochar arranged by this strategy is more appropriate for the adsorption of water poisons (Regmi et al. 2012; Zhang et al. 2013). In any case, this technique is restricted by the arrangement conditions, the requirement for high weight and high temperature of the costly reactor. In light of its high planning cost (Regmi et al. 2012), the functional application is hard to promote. Different techniques, for example, drying, gasification, fast pyrolysis, and "flash" pyrolysis, are fundamentally used to create bio-oil or vaporous materials (Tan et al. 2015) as a result of the moderately little yield of strong items acquired, for example,

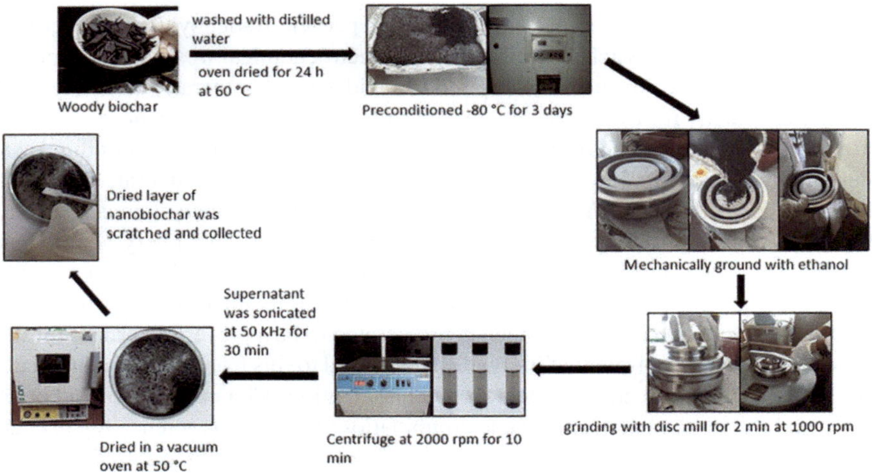

Fig. 15.1 Production process of nanobiochar (*Source:* Ramanayaka et al. [2019])

gas-item substance of gasification about 90%, in which the higher the temperature the higher the substance of vaporous items.

15.2.1 Factors Affecting the Pyrolysis Procedure

The crude materials for the preparation of biochar are bountiful. Fundamentally, any type of natural material can be pyrolyzed (Laird et al. 2009). Because of the huge yield of biomass strong waste assets, biomass is a typical crude material for biochar, predominantly including wheat and corn straw, wood chips, nut shell, rice husk, animals and poultry fertilizer, kitchen squander, muck, organic product skin, and so forth (Tan et al. 2017).

Biochar arranged from various materials contains various extents of cellulose, hemicellulose, and lignin, so its yield, component piece, and debris content are unique (Crombie et al. 2013). Enders et al. (2012), found that the debris substance of straw biochar is higher than that of other biochar, which is basically brought about by the high Si substance of straw. Yuan et al. (2011), looked at the physicochemical properties of biochar arranged from various feed stocks. The debris substance of biochar from corn straw arranged at 700°C was the most elevated, contrasted with canola, soybean, and nut straw biochar.

15.2.2 Other New Methods

Notwithstanding the pyrolysis, hydrothermal carbonization, and microwave carbonization examined above, streak carbonization and torrefaction (Chen et al. 2015) are different techniques for biomass change. During the flash carbonization measure, the flash fire is lighted at a high weight (1–2 Mpa) on the biomass pressed bed to change over the biomass into the gas and strong stage items (Cha et al. 2016). It is accounted for that about 40% of biomass is changed over to strong stage items at 1 Mpa (Mochidzuki et al. 2003). Plasma pyrolysis innovation is predominantly applied in the readiness of syngas and coke. Contrasted and the customary splitting innovation, it can enormously expand the syngas and diminish the yield of bio-oil (Tang and Huang 2005). Be that as it may, it is hard to advocate the new pyrolysis innovation because of its significant expense and vitality utilization.

The physical and synthetic properties of biochar rely principally upon the kinds of feedstock and pyrolysis conditions, i.e., temperature, habitation time, reactor type, and warming rate. Ordinary carbonization, quick pyrolysis, flash carbonization, gasification, and microwave helped pyrolysis are the fundamental thermo-compound cycles that are normally used to create biochar (Manyà 2012). It has been exhibited that the biochar delivered at 600–700°C has less H and O functional groups because of drying out and deoxygenation of the biomass yet shows profoundly fragrant nature with efficient carbon layers and has lower particle trade abilities (Uchimiya et al. 2011; Ahmad et al. 2014). Then again, the biochar was created at lower temperature at around 300–400°C by Glaser et al. 2002. It is unpredictable and heterogenous physical and substance organization of biochar that gives its amazing properties of contaminant evacuation by means of sorption (Vithanage et al. 2015).

A blend of biochar and deionized water (1:50, w/w) was sonicated for 40 min. at a recurrence of 37 kHz. In the wake of agreeing to 36 hours, the suspension was taken out and a similar measure of refined water was added to the encourage to be sonicated for the subsequent time. This technique was rehashed multiple times until Tyndall wonder not evident in the suspension under static infrared shaft light, and the suspension from every treatment was gathered in a similar compartment. The suspension was separated utilizing a polyethersulfone layer channel with a width of 0.22 mm and the filtrate was gathered. So as to dodge the misfortune by direct centrifugation and to gather little measure particles, a between molecule charge guideline strategy was utilized to gather the bio-nanoparticles (BNPs). The pH esteem directs the charge properties of the molecule surface. Therefore, a technique for pH changing was utilized to alter molecule surface electron. Particles effortlessly grouped could be gathered subsequently to diminish their cooperation power. Gathered particles were dried before yield figuring, and after that scattered by pH alteration, where these dispersive nanoparticles were utilized for additional investigations.

15.3 Adsorption Mechanism of Nanobiochar

Adsorption is a significant natural process, which impacts the biogeochemistry of contaminations. Biochar-based nanomaterials can adsorb numerous sorts of toxins, similar to natural mixes, weighty metals, and microorganisms (Lu et al. 2020). The adsorption system is relying upon the nature of the contaminants and the compound properties of the adsorbent surface (Rosales et al. 2017). The significant courses of adsorption could be generally isolated into physical course where the adsorbate chooses the outside of the adsorbent, the precipitation course where the adsorbent structure layers on the adsorbent surface, and the pore-filling course portrayed by the buildup of the adsorbate into the pores of the adsorbent (Pignatello 2011). When all is said in done, biochar with bigger surface zone has more sorption destinations, encouraging the sorption limit. The biochar-based composites additionally have a bigger surface zone by impregnating biochar with explicit materials. For this situation, the biochar principally assumes a part as a platform with high surface region on which different materials are kept (Sizmur et al. 2017).

The electrostatic fascination capacity on the outside of biochar assumes a significant part in the adsorption of poisons. By and large, the surface power of biochar is negative, so it has a decent adsorption execution for positive particles, for example, smelling salts, weighty metals. In the event that the biochar is adjusted, so the surface power is decidedly charged, anions, for example, phosphate can be adsorbed (Deng et al. 2017).

As far as component arrangement, C, H, O, N are the significant components that structure the network of biochar, and different components like Si, P, S show diverse mass rates in explicit biochars, which play a unique or even significant part in sorption of explicit poisons by shaping sorption destinations with various structures that have solid fondness toward toxins (Lu et al. 2020). For instance, the sorption of Pb and Al on biochar was ascribed to the co-precipitation with P and Si in the biochar as $Pb_5(PO_4)_3(OH)$ and $KAlSi_3O_8$, respectively (Cao et al. 2009, 2011; Qian and Chen 2013). In contrast to some other carbonaceous fuel, nanobiochar is commonly utilized as a dirt alteration and in water treatment. The physicochemical properties of nanobiochar, for example, surface zone, charged surface, and utilitarian gathering differ obviously dependent on the biomass source and pyrolysis condition, which influence the adsorptive limit of nanobiochar to substantial metals and natural mixes (Nartey and Zhao 2014). There has been accounted for that the sorption conduct between natural mixes and carbon-based material for the most part follows systems, for example, pore-filling, electrostatic connection, hydrophobic communication, hydrogen holding association, and π-EDA cooperations, and some of the time the synchronous event of a few sorption instruments (Ji et al. 2011).

By and large, the adsorption of natural contaminations by nanobiochar-based material is basically through the mix of pore immobilization and electrostatic fascination of natural practical gatherings. The adsorptions of hefty metals are fundamentally through electrostatic fascination, particle trade, and complexation response of surface utilitarian gatherings (Fig. 15.2) (Deng et al. 2017).

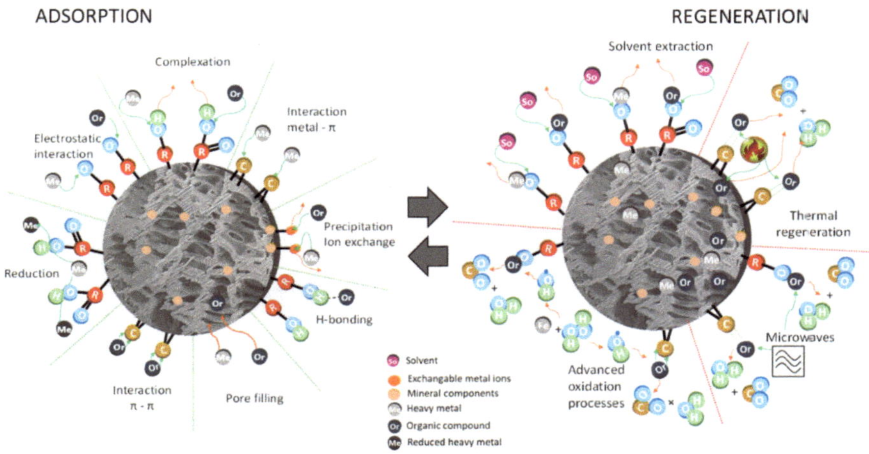

Fig. 15.2. Adsorption and regeneration mechanism in nanobiochar. (*Source* Sanromán et al. [2020])

15.4 Applications of Nanobiochar in Environmental Cleanup

Among the increased number of publications, nanobiochar materials can be used for environmental cleanup such as wastewater/water treatment, pollution, soil remediation, agriculture advancement, and constructed wetlands, etc. (Fig. 15.3). In this section, some applications of nanobiochar are discussed below:

15.4.1 Nanobiochar in Wastewater Treatment

The evacuation of substantial metals has been one of the principle research interests in use of biochar-based nanocomposites for cleaning of wastewater.

For the most part, biochar has a higher surface zone, high pH, and a negative surface charge. This encourages explicit adsorption of metal particles by means of oxygenated practical gatherings, electrostatic appreciation for fragrant gatherings, and precipitation on the mineral debris segments of the biochar. And yet biochar is normally a helpless adsorbent for oxy-anions contaminants like NO_3^-, PO_4^{3-}, and AsO_4^{3-} (Lima and Marshall 2005).

The impacts of various nanomaterials on the adsorption of substantial metals are primarily shown in the accompanying viewpoints:

1. Influence on a superficial level utilitarian gatherings of biochar.
2. Improvement or decay on the pore property of biochar.
3. Serving as dynamic destinations themselves.

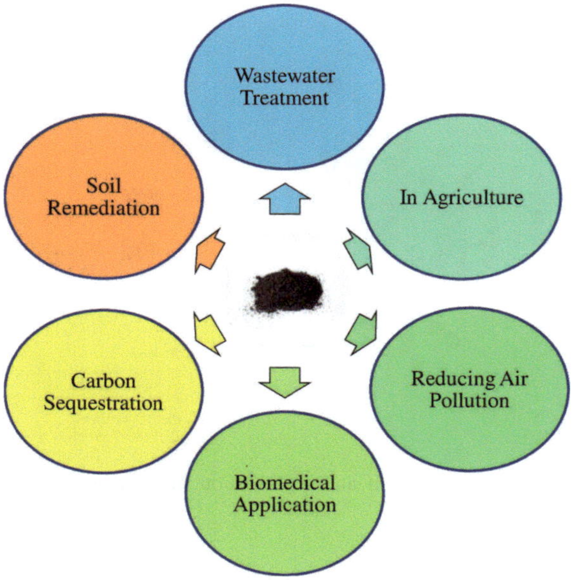

Fig. 15.3 Varoius applications of nanobiochar

Two principle impacts of nanomaterials on biochar are existed during the time spent inorganic contaminants expulsion: upgrading the surface zone of biochar and impregnating nanocomposites on the biochar surface, which may drastically expand the adsorption of inorganic contaminants from water (Zhang and Gao 2013; Jung and Ahn 2016). Biochar got from various biomass have been seriously and successfully utilized as sorbents for natural contaminants in water and wastewater. The best worry of natural contaminants in watery arrangements has been centered on phenols and colors. For example, pyrolysis biochar got from macroalgae have been adequately used to eliminate material colors (malachite green, gem violet, and Congo red), which are not really debased because of their security to light and oxidizing operators, and protection from vigorous absorption (Chen et al. 2018). Biochar has additionally been accounted for as a viable adsorbent for phenols in wastewater (Thang et al., 2019; Mohammed et al. 2018). Inorganic contamination in wastewater incorporates Cr, Cu, Pb, Cd, Hg, Fe, Zn, and As particles and mixes like nitrate and nitrite of nitrogen, NH_4, P, and sulfides of hydrogen that cause critical danger to well-being (Cao et al. 2009).

15.4.2 Nanobiochar in Soil Remediation

Organic pollutants, for example, polycyclic aromatic hydrocarbons (PAHs) and ionizable natural contaminants and substantial metals in soil can be immobilized through nanobiochar application. The toxin immobilization contrasts in reliance on contamination type, ecological conditions, and biochar properties. Likewise, the type of biochar and application rate is subject to the contaminant blend and soil explicit properties. The alteration of soils for their remediation targets diminishing at that point danger of poison move to waters or receptor creatures in vicinity. The natural material, for example, biochar may fill in as a mainstream decision for this reason since its source is organic and it might be legitimately applied to soils with little pre-treatment (Beesleyet al. 2011). There are two viewpoints which make biochar alteration better than other natural materials: the first is the high soundness against rot, with the goal that it can stay in soil for longer occasions giving long haul advantages to soil and the second is having greater capacity to hold the supplements. Biochar correction improves soil quality by expanding soil pH, moisture holding capacity, cation-trade limit, and microbial flora (Mensah and Frimpong 2018).

Nanobiochar can settle heavy metals in the defiled soil, improve the nature of the sullied soil (Ippolito et al. 2012), and has a noteworthy decrease in crop take-up of substantial metals. If there should be an occurrence of acidic sullied soils, contingent upon the sort of biochars and replaceable cations (Na, Mg, K, and Ca) present in it could hold the key for the arrival of a portion of these cations during sorption measure with the weighty metal, and in this manner may advance the adjustment cycle.

Numerous investigations have shown that nanobiochar-revised soil can help retain an assortment of natural contaminants (Table 15.1), consequently diminishing their take-up by plants. Use of a limited quantity of biochar to soil can essentially decrease the aggregation of pesticides and other natural contaminations in plants (Hilber et al. 2009; Kookana 2010). Biochar is applied to soil for the molding and treatment purposes; application can likewise be positive in decrease of harmful parts. Ongoing examinations have demonstrated that biochar is additionally equipped for adsorbing substantial metals, for example, lead, cadmium, nickel, and some striking natural contaminants that polluted soils which can hurt human, plants, and animals (Ameloot et al. 2013). Subsequently, utilization of nanobiochar can conceivably give another answer for remediation of the soil contamination.

Table 15.1 Impacts of nanobiochar in application on bioavailability of toxins in soils

Contaminants	Effect	Reference
Diuron chlorpyrifos and carbofuran	Reductions of chlorpyrifos and carbofuranin total plant residues, respectively	Yu et al. (2009)
PAHs	Pore water concentrations of PAHs were reduced by biochar, with greater than 50% decrease of the heavier, more toxicologically relevant PAHs	Beesley et al. (2010)
Pentachlorophenol	Biochar reduced PCP bioavailability in soil	Xu et al. (2011)
Chlorobenzenes (CBs)	Biochar amendment significantly reduced the bioavailability of CBs	Song et al. (2012b)
Hexachlorobenzene (HCB)	Biochar amendment of soil resulted in a rapid reduction in the bioavailability of HCB, even at 0.1% biochar application rate	Song et al. (2012a)

15.4.3 Nanobiochar in Agriculture

There are a few reports which show that biochar has the ability to invigorate the dirt microflora, which brings about more prominent amassing of carbon in soil. Other than adsorbing natural substances, supplements, and gases, biochars are probably going to offer a territory for microbes, actinomycetes, and organisms (Thies and Rillig 2009). The utilization of biochar expanded mycorrhizal development in clover bioassay plants by giving the appropriate conditions to colonization of plant roots (Solaiman et al. 2010). Carrots and vegetables developed on steep inclines and in soils with fewer than 5.2 pH demonstrated essentially improved development by the expansion of biochar (Rondon et al., 2004). It was discovered that biochar expanded the natural N_2 fixation of *Phaseolus vulgaris* (Rondon et al. 2007) chiefly because of more prominent accessibility of micronutrients after use of biochar. Lehmann et al. (2003) reported that biochar reduced leaching of NH^{4+} by supporting it in the surface soil where it was available for plant uptake. Biochar can build the estimation of non-gathered agrarian items (Major et al. 2005) and advance the plant development (Lehmann et al. 2003; Oguntunde et al. 2004).

15.4.4 Nanobiochar in Air Remediation

Nanobiochar could reduce air pollution by adsorbing the emissions of gaseous air pollutants. A large number of studies have shown that nanobiochar can inhibit the emissions of NO_2, CO_2, and CH_4 (Liu et al. 2011). It also has the potential to reduce greenhouse gases emissions by reducing soil organic carbon (Lehmann 2007). Besides greenhouse gases, biochar from different feedstock has also been modified by the combination of physical and chemical modification as an alternative method of removing Hg^o from flue gas, H_2S gas adsorption and Volatile Organic Compounds (VOCs) (Shen et al. 2015; Zhang et al. 2017).

15.4.5 Nanobiochar in Carbon Sequestration

Carbon sequestration via nanobiochar production is technically feasible. The production of biochar hacks this carbon cycle by transforming carbon into a stable form that can persist degradation, preventing emissions to the atmosphere (Qambrani et al. 2017). By combining the photosynthetic route with pyrolytic conversion, an efficient carbon removal system can be designed. At scale, biochar production should ultimately influence the atmospheric carbon balance by reducing atmospheric carbon concentrations (Fawzy et al. 2021).

15.5 Conclusion

In the present study, the synthesized biochar-based nanocomposites exhibited excellent capacity for environmental cleanup. It is practical and ecological inviting with improved qualities of nanocomposites for the disinfecting of sullied water and other natural cordial processes. Nanobiochar has been used as likely impetus in ecological remediation with noteworthy progression in the ongoing years. Nanobiochar creation, functionalization, and their utilization in heterogeneous redox reduction of tenacious inheritance and rising contaminants have been assessed. Significant accomplishments and energizing discoveries of biochar-substrate cooperation have been examined. The adsorption component of nanobiochar on contaminations is the aftereffect of synergism of numerous sorts of adsorption measures. In spite of the fact that the advancement of nanobiochar is still in its outset. Nanobiochar is a practical and reusable material that presents a wide scope of opportunities for the further advancement of natural cleanup.

References

Abdullah H, Wu H (2009) Biochar as a fuel: 1. Properties and grindability of biochars produced from the pyrolysis of mallee wood under slow-heating conditions. Energy Fuels 23(8):4174–4181

Adrees M, Ali S, Rizwan M et al (2015) The effect of excess copper on growth and physiology of important food crops: a review. Environ Sci Pollut Res Int 22:8148–8162

Ahmad M, Rajapaksha AU, Lim JE et al (2014) Biochar as a sorbent for contaminant management in soil and water: a review. Chemosphere 99:19–33

Ali I (2010) The quest for active carbon adsorbent substitutes: inexpensive adsorbents for toxic metal ions removal from wastewater. Sep Purif Rev 39(3/4):95–171

Ameloot N, deNeve S, Jegajeevagan K et al (2013) Short-term CO_2 and N_2O emissions and microbial properties of biochar amended sandy loam soils. Soil Biol Biochem 57:401–410

Antoniadis V, Shaheen SM, Boersch J et al (2017) Bioavailability and risk assessment of potentially toxic elements in garden edible vegetables and soils around a highly contaminated former mining area in Germany. J Environ Manage 186:192–200

Awad YM, Vithanage M, Niazi NK et al (2019) Potential toxicity of trace elements and nanomaterials to Chinese cabbage in arsenic- and lead-contaminated soil amended with biochars. Environ Geochem Health 41:1777–1791

Beesley L, Moreno-Jiménez E, Gomez-Eyles JL (2010) Effects of biochar and greenwaste compost amendments on mobility, bioavailability and toxicity of inorganic and organic contaminants in a multi-element polluted soil. Environ Pollut 158:2282–2287

Beesley L, Moreno-Jimenez E, Gomez-Eyles JL et al (2011) A review of biochars' potential role in the remediation, revegetation and restoration of contaminated soils. Environ Pollut 159:3269–3282

Berge ND, Li L, Flora JRV et al (2015) Assessing the environmental impact of energy production from hydrochar generated via hydrothermal carbonization of food wastes. Waste Manage 43:203–217

Cao XD, Ma LN, Gao B et al (2009) Dairy-manure derived biochar effectively sorbs lead and atrazine. Environ Sci Technol 43(9):3285–3291

Cao XD, Ma LN, Liang Y et al (2011) Simultaneous immobilization of lead and atrazine in contaminated soils using dairy-manure biochar. Environ Sci Technol 45(11):4884–4889

Cha JS, Park SH, Jung SC et al (2016) Production and utilization of biochar: A review. J Ind Eng Chem 40:1–15
Chen B, Chen Z (2009) Sorption of naphthalene and 1-naphthol by biochars of orange peels with different pyrolytic temperatures. Chemosphere 76(1):127–133
Chen B, Chen Z, Lv S (2011) A novel magnetic biochar efficiently sorbs organic pollutants and phosphate. Bioresour Technol 102(2):716–723
Chen D, Zheng Z, Fu K et al (2015) Torrefaction of biomass stalk and its effect on the yield and quality of pyrolysis products. Fuel 159:27–32
Chen Y, Lin YC, Ho SH et al (2018) Highly efficient adsorption of dyes by biochar derived from pigments-extracted macroalgae pyrolyzed at different temperature. Bioresour Technol 259:104–110
Crombie K, Mašek O, Sohi SP et al (2013) The effect of pyrolysis conditions on biochar stability as determined by three methods. Gcb Bioenergy 5:122–131
Demirbas A (2008) Heavy metal adsorption onto agro-based waste materials: a review. J Hazard Mater 157(2/3):220–229
Deng Y, Zhang T, Wang Q (2017) Biochar Adsorption Treatment for Typical Pollutants Removal in Livestock Wastewater: A Review. Engineering Applications of Biochar. 71–82.
Dupont L, Guillon E (2003) Removal of hexavalent chromium with a lignocellulosic substrate extracted from wheat bran. Environ Sci Technol 37(18):4235–4241
Enders A, Hanley K, Whitman T et al (2012) Characterization of biochars to evaluate recalcitrance and agronomic performance. Bioresour Technol 114:644–653
Fang C, Zhang T, Li P et al (2014) Application of magnesium modified corn biochar for phosphorus removal and recovery from swine wastewater. Int J Environ Res Public Health 11(9):9217–9237
Fawzy S, Osman AI, Yang H et al (2021) Industrial biochar systems for atmospheric carbon removal: a review. Environ Chem Lett 19:3023–3055
Glaser B, Lehmann J, Zech W (2002) Ameliorating physical and chemical properties of highly weathered soils in the tropics with charcoal: a review. Biol Fertil Soils 35(4):219–230
Hilber I, Wyss GS, Mäder P et al (2009) Influence of activated charcoal amendment to contaminated soil on dieldrin and nutrient uptake by cucumbers. Environ Pollut 157:224–2230
Hina K, Abbas M, Hussain Q et al (2019) Investigation into arsenic retention in arid contaminated soils with biochar application. Arab J Geosci 12:1–8
Inyang M, Dickenson E (2015) The potential role of biochar in the removal of organic and microbial contaminants from potable and reuse water: a review. Chemosphere 134:232–240
Inyang MI, Gao B, Yao Y et al (2016) A review of biochar as a low-cost adsorbent for aqueous heavy metal removal. Crit Rev Environ Sci Technol 46:406–433
Ippolito JA, Laird DA, Busscher WJ (2012) Environmental benefits of biochar. J Environ Qual 41:967–972
Ji L, Wan Y, Zheng S et al (2011) Adsorption of tetracycline and sulfamethoxazole on crop residue-derived ashes: implication for the relative importance of black carbon to soil sorption. Environ Sci Technol 45:5580–5586
Jung KW, Ahn KH (2016) Fabrication of porosity-enhanced MgO/biochar for removal of phosphate from aqueous solution: application of a novel combined electrochemical modification method. Bioresour Technol 200:1029–1032
Kambo HS, Dutta A (2015) A comparative review of biochar and hydrochar in terms of production, physico-chemical properties and applications. Renew Sustain Energy Rev 45:359–378
Khalid S, Shahid M, Niazi NK et al (2017) Arsenic behaviour in soil-plant system: biogeochemical reactions and chemical speciation influences. Enhancing Cleanup of Environmental Pollutants. Springer, pp 97–140
Kookana RS (2010) The role of biochar in modifying the environmental fate, bioavailability, and efficacy of pesticides in soils: a review. J Aust Soil Res 48:627–637
Laird DA, Brown RC, Amonette JE et al (2009) Review of the pyrolysis platform for coproducing bio-oil and biochar. Biofuels Bioprod Biorefining 3:547–562
Lehmann J (2007) A handful of carbon. Nature 447:143–144

Lehmann J, da Silva JP, Jr Steiner C et al (2003) Nutrient availability and leaching in an archaeological anthrosol and a ferrasol of the Central Amazon basin: Fertilizer, manure, and charcoal amendments. Plant Soil 249:343–357

Lehmann J, Gaunt J, Rondon M (2006) Bio-char sequestration in terrestrial ecosystems-A review. Mitig Adapt Strateg Glob Change 11(2):403–427

Lehmann J, Rillig MC, Thies J et al (2011) Biochar effects on soil biota–a review. Soil Biol Biochem 43(9):1812–1836

LeMonte JJ, Stuckey JW, Sanchez JZ et al (2017) Sea level rise induced arsenic release from historically contaminated coastal soils. Environ Sci Technol 51:5913–5922

Lima IM, Marshall WE (2005) Adsorption of selected environmentally important metals by poultry manure-based granular activated carbons. J Chem Technol Biotechnol: Int Res Process, Environ Clean Technol 80(9):1054–1061

Liu WJ, Li WW, Jiang H et al (2017) Fates of chemical elements in biomass during its pyrolysis. Chem Rev 117:6367–6398

Liu Y, Yang M, Wu Y et al (2011) Reducing CH_4 and CO_2 emissions from waterlogged paddy soil with biochar. J Soil Sediment 11:930–939

Lu L, Yu W, Wang Y et al (2020) Application of biochar-based materials in environmental remediation: from multi-level structures to specific devices. Biochar 2:1–2

Major J, Steiner C, Ditommaso A et al (2005) Weed composition and cover after three years of soil fertility management in the central Brazilian Amazon: Compost, fertilizer, manure and charcoal applications. Weed Biol Manag 5:69–76

Manyà JJ (2012) Pyrolysis for biochar purposes: a review to establish current knowledge gaps and research needs. Environ Sci Technol 46(15):7939–7954

Mensah AK, Frimpong KA (2018) Biochar and/or compost applications improve soil properties, growth, and yield of maize grown in acidic rainforest and coastal savannah soils in Ghana. Int J Agron 1–8.

Mochidzuki K, Soutric F, Tadokoro K et al (2003) Electrical and Physical Properties of Carbonized Charcoals. Ind Eng Chem Res 42:5140–5151

Mohammed NAS, Abu-Zurayk RA, Hamadneh I et al (2018) Phenol adsorption on biochar prepared from the pine fruit shells: Equilibrium, kinetic and thermodynamics studies. J Environ Manag 226:377–385

Nartey OD, Zhao B (2014) Biochar preparation, characterization, and adsorptive capacity and its effect on bioavailability of contaminants: an overview. Adv Mater Sci Eng 1–12.

Niazi NK, Bibi I, Shahid M et al (2018) Arsenic removal by perilla leaf biochar in aqueous solutions and groundwater: an integrated spectroscopic and microscopic examination. Environ Pollut 232:31–41

Niazi NK, Singh B, Minasny B (2015) Mid-infrared spectroscopy and partial least squares regression to estimate soil arsenic at a highly variable arsenic-contaminated site. Int J Environ Sci Technol 12:1965–1974

Oguntunde PG, Fosu M, Ajayi AE et al (2004) Effects of charcoal production on maize yield, chemical properties and texture of soil. Biol Fertil Soils 39:295–299

Onay O, Kockar OM (2003) Slow, fast and flash pyrolysis of rapeseed. Renew Energy 28(15):2417–2433

Pignatello JJ (2011) Interactions of anthropogenic organic chemicals with natural organic matter and black carbon in environmental particles. Biophysico-Chemical Processes of Anthropogenic Organic Compounds in Environmental Systems. Wiley, Hoboken. NJ, USA , pp 1–50

Qambrani NA et al (2017) Biochar properties and eco-friendly applications for climate change mitigation, waste management, and wastewater treatment: a review. Renew Sustain Energy Rev 79:255–273

Qian K, Kumar A, Zhang H et al (2015) Recent advances in utilization of biochar. Renew Sustain Energy Rev 42:1055–1064

Qian L, Chen B (2013) Dual role of biochars as adsorbents for aluminum: the effects of oxygen-containing organic components and the scattering of silicate particles. Environ Sci Technol 47(15):8759–8768
Ramanayaka S, Tsang DCW, Hou D et al (2019) Green synthesis of graphitic nanobiochar for the removal of emerging contaminants in aqueous media. Sci Total Environ 706:135725
Regmi P, Moscoso JLG, Kumar S et al (2012) Removal of copper and cadmium from aqueous solution using switchgrass biochar produced via hydrothermal carbonization process. J Environ Manag 109(17):61–69
Rizwan M, Ali S, Adrees M et al (2016) Cadmium stress in rice: toxic effects, tolerance mechanisms, and management: a critical review. Environ Sci Pollut Res Int 23:17859–17879
Rondon M, Lehmann J, Ramírez J et al (2007) Biological nitrogen fixation by common beans (Phaseolus vulgaris L.) increases with bio-char additions. Biol Fertil Soils 43:699–708
Rondon M, Ramirez A, Hurtado M (2004) Charcoal Additions to High Fertility Ditches Enhance Yields and Qvuality of Cash Crops in Andean Hillsides of Columbia (CIAT Annual Report 2004). Cali, Colombia
Rosales E, Meijide J, Pazos M et al (2017) Challenges and recent advances in biochar as low cost biosorbent: From batch assays to continuous-flow systems. Bioresour Technol 246:176–192
Sanromán AF, Pazos M, Rosales E et al (2020) Unravelling the Environmental Application of Biochar as Low-Cost Biosorbent: A Review. Appl Sci 10:7810
Shahid M, Imran M, Khalid S et al (2020) Arsenic environmental contamination Status in South Asia. Arsenic in Drinking Water and Food. Springer 13–39.
Shahid M, Khalid M, Dumat C et al (2018) Arsenic level and risk assessment of groundwater in Vehari, Punjab Province, Pakistan. Expo Health 10:229–239
Shakoor MB, Niazi NK, Bibi I et al (2016) Remediation of arsenic-contaminated water using agricultural wastes as biosorbents. Crit Rev Environ Sci Technol 46:467–499
Shakoor MB, Niazi NK, Bibi I et al (2019) Exploring the arsenic removal potential of various biosorbents from water. Environ Int 123:567–579
Shen B, Li G, Wang F et al (2015) Elemental mercury removal by the modified bio-char from medicinal residues. Chem Eng J 272:28–37
Sizmur T, Fresno T, Akgül G et al (2017) Biochar modification to enhance sorption of inorganics from water. Bioresour Technol 246:34–47
Solaiman ZM, Blackwell P, Abbott LK et al (2010) Direct and residual effect of biochar application on mycorrhizal root colonisation, growth and nutrition of wheat. Aust J Soil Res 48:546–554
Song Y, Wang F, Bian YR et al (2012a) Bioavailability assessment of hexachlorobenzene in soil as affected by wheat straw biochar. J Hazard Mater 217–218:391–397
Song Y, Wang F, Yang XL et al (2012b) Influence and assessment of biochar on the bioavailability of chlorobenzenes in soil. Chin J Environ Sci 33:169–174
Tan X, Liu Y, Zeng G et al (2015) Application of biochar for the removal of pollutants from aqueous solutions. Chemosphere 125:70–85
Tan Z, Lin CS, Ji X et al (2017) Returning biochar to fields: A review. Appl Soil Ecol 116:1–11
Tang L, Huang H (2005) Plasma Pyrolysis of Biomass for production of Syngas and Carbon adsorbent. Energy Fuels 9:1174–1178
Thang PQ, Jitae K, Giang NMVBL et al (2019) Potential application of chicken manure biochar towards toxic phenol and 2, 4-dinitrophenol in wastewaters. J Environ Manag 251:109556
Thies JE, Rillig M (2009) Characteristics of biochar: Biological properties. In: Lehmann M, Joseph S (eds) Biochar for Environmental Management Science and Technology. Earthscan, London, pp 85–105
Uchimiya M, Chang S, Klasson KT (2011) Screening biochars for heavy metal retention in soil: role of oxygen functional groups. J Hazard Mater 190(1/2/3): 432–441.
Uchimiya M, Lima IM, Thomas KK et al (2010) Immobilization of heavy metal ions (CuII, CdII, NiII, and PbII) by broiler litter-derived biochars in water and soil. J Agric Food Chem 58(9):5538–5544

Vithanage M, Rajapaksha AU, Ahmad M et al (2015) Mechanisms of antimony adsorption onto soybean stover-derived biochar in aqueous solutions. J Environ Manage 151:443–449
Wang NX, Zhang XY, Wu J et al (2012) Effects of microcystin-LR on the metal bioaccumulation and toxicity in Chlamydomonas reinhardtii. Water Res 46(2):369–377
Waqas H, Shan A, Khan YS et al (2017) Human health risk assessment of arsenic in groundwater aquifers of Lahore, Pakistan. Hum Ecol Risk Assess 23:836–850
Woolf D, Amonette JE, Street-Perrott AF et al (2010) Sustainable biochar to mitigate global climate change. Nat Comm 1(5): 56–64
Xu T, Lou LP, Luo L et al (2011) Effect of bamboo biochar on pentachlorophenol leachability and bioavailability in agricultural soil. Sci Total Environ 414:727–731
Xue YW, Gao B, Yao Y et al (2012) Hydrogen peroxide modification enhances the ability of biochar (hydrochar) produced from hydrothermal carbonization of peanut hull to remove aqueous heavy metals: batch and column tests. Chem Eng Technol 200(201/202):673–680
Yu JX, Wang LY, Chi RA et al (2013) Competitive adsorption of Pb^{2+} and Cd^{2+} on magnetic modified sugarcane bagasse prepared by two simple steps. Appl Surf Sci 268:163–170
Yu XY, Ying GG, Kookana RS (2009) Reduced plant uptake of pesticides with biochar additions to soil. Chemosphere 76:665–671
Yuan JH, Xu RK, Zhang H (2011) The forms of alkalis in the biochar produced from crop residues at different temperatures. Bioresour Technol 102:3488–3497
Zhang M, Gao B (2013) Removal of arsenic, methylene blue, and phosphate by biochar/AlOOH nanocomposite. Chem Eng J 226:286–292
Zhang X, Gao B, Creamer AE et al (2017) Adsorption of VOCs onto engineered carbon materials: A review. J Hazard Mater 338:102–123
Zhang ZB, Cao XH, Liang P et al (2013) Adsorption of uranium from aqueous solution using biochar produced by hydrothermal carbonization. J Radioanal Nucl Chem 295(2):1201–1208

Chapter 16
Application of Biomimetic Membranes for Water Purification

Sneh Gautam, Pushpa Lohani, and Poonam Maan

Abstract The deterioration of water quality is a big global problem at present that is increasing continuously due to the rapid growth of population, uncontrolled industrialization and urbanisation which causing serious water pollution. Various membrane-based technologies like forward osmosis, reverse osmosis and Electrodialysis etc. are being used for water purification but fabrication material and techniques are still the main limitations of these technologies. Recently, the biomimetic membrane is emerged as a potential candidate for membrane-based water purification technology due to its extremely high permeability and selectivity for water molecules. Fundamentally, the biomimetic membrane is mainly composed of three components: aquaporins that are transmembrane protein and selectively works as water channels for passage of water molecules through the membrane; amphiphilic matrix, it works as a housing material for aquaporins in which these proteins are embedded, and the third component is the porous solid substrate that augments the mechanical strength of the membrane. In this chapter, each component has been discussed in detail along with previous examples of biomimetic membrane fabrication techniques. Besides this, the applications of biomimetic membrane for wastewater treatment and challenges associated with this technology have also been focused in this chapter.

Keywords Biomimetic membrane · Aquaporin · Water purification · Reverse osmosis · Forward osmosis · Electrodialysis

S. Gautam (✉) · P. Lohani
Department of Molecular Biology and Genetic Engineering, CBSH, G. B. Pant University of Agriculture & Technology, Pantnagar, India
e-mail: sneh.gautam@gbpuat-cbsh.ac.in

P. Maan
Department of Agriculture Biotechnology, College of Agriculture, Sardar Vallabhbhai Patel University of Agriculture and Technology, Meerut, India

J. P. N. Rai and S. Saraswat (eds.), *Nano-biotechnology for Waste Water Treatment*, Water Science and Technology Library 111,
https://doi.org/10.1007/978-3-031-00812-2_16

Abbreviations

AEM	Anion-exchange membrane
BCPs	Block copolymers
CA	Cellulose acetate
CEM	Cation-exchange membrane
CO_2	Carbon dioxide
CTA	Cellulose triacetate
ED	Electrodialysis
FO	Forward osmosis
IEM	Exchange membrane
kDa	Kilo Dalton
LMH	Litere/m^2/h
MPD	m-phenylene-diamine
NF	Nanofiltration
PA	Polyamide
PAEK	Poly(arylene ether ketone)
PAH	Polyallylamine hydrochloride
PAI	Polyamide-imide
PAN	Polyacrylonitrile
PBI	Polybenzimidazole
PCTE	Polycarbonate rack-etch
PE	Polyethylene
PEG	Polyethylene glycol
PEI	Polyethyleneimine
PES	Polyethersulfone
PMAA	Poly(methacrylic acid)
PP	Polypropylene
PSF	Polysulfone
PVA	Polyvinyl alcohol
PVDF	Polyvinylidene fluoride
RO	Reverse osmosis

16.1 Introduction

Water scarcity, climate change and increase in population are the most serious challenges of the twenty-first century. Fast-growing population, climate change, increasing global warming and industrialisation are enhancing water scarcity day by day. Besides this, continuous complex waste discharge from industries and municipal waste are also causing contamination in freshwater resources like lakes, rivers and oceans. However, the demand for fresh drinking water is increasing with the increase in global population. According to UNICEF and WHO, over 1.1 billion people do not have access to safe drinking water (WHO 2012; Giwa et al. 2017). Hence, water scarcity is a serious threat to our long-term existence that demands prompt action. As a result, one of the most important research areas is creating state of the art for collecting freshwater from urban and industrial wastewaters (Elimelech, 2006). Therefore, innovative separation and purification technologies are essential to purify polluted and contaminated water (Burek et al. 2016; Van Loosdrecht and Brdjanovic, 2014). The most common methods of water purification are thermal processes and membrane purification technology (Giwa et al. 2017). Thermal processes like multi-effect distillation, vapour compression, multi-stage flash and humidification dehumidification are generally based on condensation and evaporation of water (Giwa et al. 2017). However, in membrane purification technologies which include forward osmosis (FO), reverse osmosis (RO), electrodialysis (ED) and nanotechnology-based processes, membranes are used as salt rejection barriers. Membrane technology is more energy-efficient and produces high-quality water than conventional purification technologies (Kesieme et al. 2013). RO is the most commonly used membrane separation technique for water purification, accounting for over 55% of all desalination plants in the world (Miller et al. 2015), still associated with the problem of fouling. In membrane treatment technologies, the membrane performance mainly depends on fabrication techniques and the material used to fabricate the membrane. The performance of the membrane selective layer is limited by the absence of molecular-scale design in existing membrane fabrication technologies (Baker 2012; Gin and Noble 2011). Water purification membranes are typically comprised of polymeric materials. However, with incremental advancements in polymeric membrane materials, there has been little improvement in membrane performance still needed more attention for further improvement in fabrication technology. On the other hand, recent advances in nanotechnology have emerged as alternate techniques to circumvent the limitations of traditional membranes. Nanotechnology advances give the opportunity of biomimicry which imitate natural processes to address arduous scientific and engineering challenges (Conrad et al. 2016; Bar-Cohen, 2011). Biomimetic membranes are a type of membrane that mimics the structure of biological membranes and are the product of these researches. Aquaporin is a foremost transmembrane protein that permits only water to pass through plasma membranes in biological systems (Weig et al. 1997). It works as a water channel that is highly selective and permeable for only water molecules and completely blocks ions and other contaminants from passing through (Giwa et al. 2017). In the last few years, the use of aquaporin for water purification has gained pronounced attention in water treatment industries (13–15). Aquaporins have unique selectivity and permeability for water molecules that are practically incomparable to any man-made membrane or technology. Various techniques like vesicle fusion (Wang et al. 2015), chemical crosslinking (Sun et al. 2013a), vesicle adsorption (Li et al. 2012), magnetic-aided adsorption (Sun et al. 2013b) and interfacial polymerisation (Zhao & Zou, 2012a; Zhao & Qiu, 2012b) have been developed to fabricate of aquaporin-based biomimetic membrane for water purification. Firstly, this chapter summarises the fundamental concepts of conventional technologies, including a review of their benefits and drawbacks to emphasise their current limitations. Further, the concept of biomimetic membrane and its building blocks have been discussed in detail by highlighting the most recent research and development in aquaporin-based biomimetic membranes. Finally, the advantages of aquaporin membrane with their challenges and prospects have been discussed.

16.2 Conventional Membrane for Water Purification

Presently, various membrane technologies are used for water purification which includes RO, FO and ED that can be differentiated according to their driving force which may be high pressure, temperature and electricity. These membrane technologies have some advantages and disadvantages that have been discussed below.

16.2.1 Reverse Osmosis

In RO technology, water flows across a semipermeable membrane against pressure (Wenten, 2016). This system can easily eliminate impurities like salt, solid particles, organic components and biological contaminants etc. (Fane et al. 2015; Fuwad et al. 2019a, b). This system can reject 95–99% impurities. However, RO system is not sufficient to remove the gases like CO_2 due to the lack of ionisation in the solution (Fuwad et al. 2019a, b). RO system needs high pressure to purify the water thus, the efficiency of RO system mainly depends on the pressure-bearing capability of a membrane which is influence by the materials of the membrane. In RO system, to bear high pressure, various polymer-based membranes made up of CA, polyamide, PVA, PES, PMAA, PAEK, PVDF and PSf etc. are being used (Giwa et al. 2019). Among various polymers, polyamide membranes are more popular than others due to their high water permeability, salt rejection efficiency, a wide range of operational pH (4–11) and temperature (Fuwad et al. 2019a, b). Nonetheless, these membranes are highly sensitive towards the oxidising agents which cause fouling of the membrane that is a major drawback of RO system. Apart from polyamide, CA is another option for RO membrane (Fuwad et al. 2019a, b). Initially, cellulose diacetate membrane was used for water purification which rejected 98% salt but showed the deprived water permeation (Lee et al. 2011). Later cellulose diacetate membrane has been replaced with cellulose triacetate which exhibited enhanced mechanical and chemical stability and improved chlorine resistance. Further, the efficiency of the cellulose-based membrane was improved by blending with other polymers which displayed high resistivity against chlorine, oxidising agent and fouling with a narrow operational pH range (4–6).

Apart from numerous advantages of RO system include maximum salt rejection, high permeate flux and ultrapure water filtration, there are some constraints of this system. RO membranes are more susceptible to mechanical damage and fouling as compared to the other existing methods due to high pressure which is continuously

Table 16.1 Various water purification technology with their characteristic features

Purification process	Membrane materials	Driving force	Advantages	Disadvantages	References
Reverse osmosis	CA, PA and thin film composite materials	High hydraulic pressure	Removal of 95–99% dissolve solute, Ultrapure water filtration	Pre-treatment required, high operation cost, stream rejection volume high, membrane fouling	(Fuwad et al. 2019b; Nunes and Peinemann 2010)
Forward osmosis	CA, CTA, PBI, PES and PSf	Differential pressure	highly resistance against fouling, low energy consumption, long membrane life	low water permeation, internal concentration polarisation, high solute rejection	(Xu and Ge 2018)
Electrodialysis	PVDF, PVA, CA, PP and PE	Electric potential differences	Removal of dissolved salt and ions, Antifouling mechanism	Uncharged species remain in the water, scaling, high cost and short life span of the membrane	(Sosa-Fernandez et al. 2018)

applied on the membrane that push the impurities into the membrane. These impurities trap inside the membrane and block it which is not easy or almost impossible to clean. Besides this, RO is not suitable for low salt concentration due to high energy consumption (Bartels et al. 2005). Moreover, a high-energy pumping station and high cost of pre-treatment process and large membrane area dramatically increase the cost of RO system (Wilf and Bartels, 2005; Latorre et al. 2015) (Table 16.1).

16.2.2 Forward Osmosis

FO is popular as an emerging technology for water purification. In this technology, differential pressure is applied across a semipermeable membrane in the direction of water flow to purify the water. FO has some advantages over RO technology. It is highly resistant to fouling and consumes low energy. However, prolonged use may cause surface fouling of membrane which can be easily removed by backwashing without any chemical treatment. The performance of FO membrane mainly depends on the chemical structure of the polymer, pore size and thickness of the membrane (Xu and Ge 2018). The membrane is fabricated by comprising a thin and dense selective layer on a porous support layer which provides mechanical stability to the membrane

(Lu et al. 2014). Various polymers have been explored to fabricate the FO membrane in past few years. CA mainly cellulose triacetate (CTA) and crosslinked polyamide (PA) made by interfacial polymeriszation areis used to fabricate comparatively dense selective layers (Li and Wang 2010; Zhao & Zou 2012a; Lau et al. 2012), however, relatively loose selective layers can be synthesiszed by using PBI and crosslinked polyamide-imide (Wang et al. 2007; Setiawan et al. 2013). On the other hand, the porous support layer is synthesiszed by PSf, PES, PAN, and PVDF (Xu and Ge 2018; Jung et al. 2004). Further, in various previous studies, fouling and low water permeability of FO membrane hasve been overcome by surface modification and blending with different other polymers (Guiver et al. 1990; Duong et al. 2015; Zhong et al. 2013). Still, there are some challenges associated with FO membranes such as draw solution recovery, low water permeation and internal concentration polariszation and high solute rejection (Fig. 16.1).

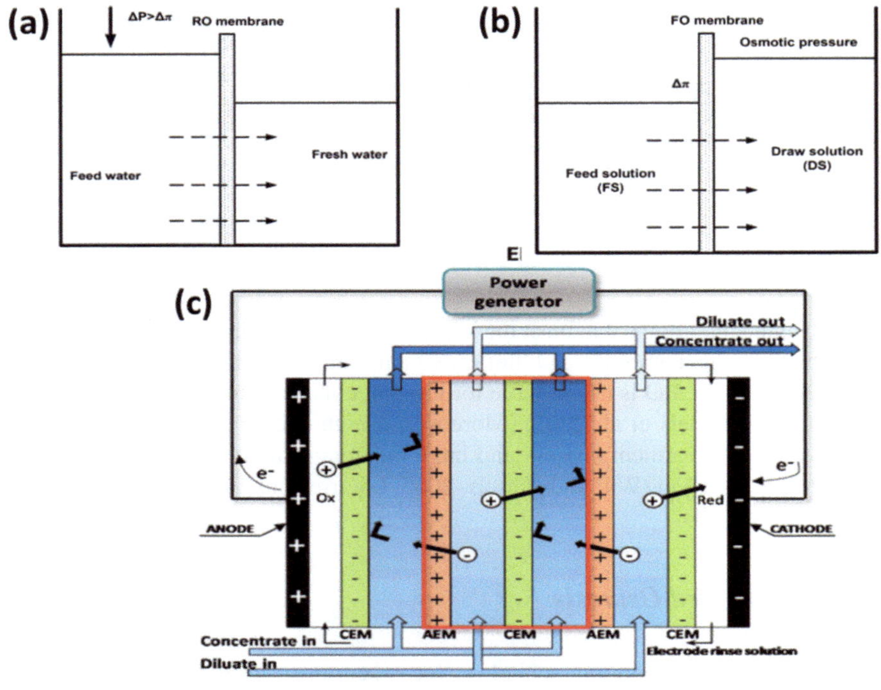

Fig. 16.1 Schematics of (a) Reverse osmosis and (b) Forward osmosis (Mecha, 2018); (c) Electrodialysis techniques (Gurreri et al. 2020). Both References are adopted from Creative Commons Attribution licence

16.2.3 Electrodialysis

ED technology is mainly used for the desalination of seawater or brackish water (Tanaka, 2015). In this technology, dissolve salt and ions are removed from water by passing it through a stack of ion IEM under the influence of electric potential (Jaewon 2021). IEM are two types according to the charge present on the membrane i.e. CEM and AEM. CEM consists of negative charges groups (groups like PO_3H^-, PO_3^{2-}, COO^- and SO_3^-) which only allow passing cations however anions are rejected by these membranes. AFM have positive charges (NH_3^+, NR_3^+, NRH_2^+, PR_3^+, NR_2H^+ and SR_2^+) that only allows the movement of anion and retain the cations (Gurreri et al. 2020). During the process, when the electric potential difference is generated across the electrode, cations migrate towards the cathode by diffusing through CEM whereas anion moves towards the anode through AEM which only allow the passing of anion (Fuwad et al. 2019a, b). IEMs can be differentiated into homogenous and heterogeneous membranes according to their fabrication method. Homogenous membranes are fabricated by blending of polymers or polymerisation of monomers and grafting of polymers with various moieties and charged groups chemically bounded with the membrane (Yaroslavtsev and Nikonenko 2009). Heterogeneous membranes contain physically mixed charge groups that are evenly distributed throughout the membrane and exhibit uniform properties in all regions. For the fabrication of IEM, organic and inorganic both types of materials are used. Organic compounds include various types of polymers like PVDF, PVA, CA, PP and PE etc. (Cheng et al. 2014; Pandey et al. 2014; Hosseini et al. 2012) and block copolymers (Müller et al. 2012) having several charged groups such as acrylonitrile, acrylic and methacrylic (Hosseini et al., 2014; Chakrabarty et al., 2013). Inorganic membranes are synthesised by using zeolites and phosphate salts through these membranes are very costly and possess weak electrochemical properties. Nowadays, for further enhancing the properties of IEMs, the organic and inorganic compounds in an association are being used to fabricate the membranes which have been found high in mechanical and thermal properties and chemically resistance (Khan et al., 2015). Thus, ED is very effective to remove dissolved salts and ions from the water that selectively eliminate many ions, and it is more economic than RO. Besides this, the fouling of these membranes can be easily removed by reversing the electrode polarity. Therefore, the antifouling property is the most significant property of this technology. Still, there are some challenges which include the short life of the membrane, high cost and scaling. Besides these, the membrane only removes the charged particles and there is no effect on uncharged species (Fuwad et al. 2019a, b). Further, the ED consume more energy with an increase the salt concentration which is not feasible.

Thus, various membranes have been used in water purification and desalination. However, It is always a challenge to obtain excellent membrane stability and permeability, high salt rejection and outstanding flux at one time which can reduce energy consumption. Nowadays, aquaporin–based biomimetic membranes have been used as a promising candidate for water purification due to their high stability and water permeability, excellent salt rejection property.

16.3 Biomimetic Membrane

The biomimetic membrane has gained great significance over conventional membrane due to its extraordinary performance in terms of stability, permeability, salt rejection and energy consumption. The biomimetic membrane is mainly influenced by the structure and function of the biological membrane which contains aquaporin transporter protein that allows only water molecules to pass through it. Aquaporin has extraordinary selectivity for water molecules, and it is capable to reject any very small molecules like boric acid, chlorides, dissolved gas, urea as well as proton also (Giwa et al. 2017; Ismail et al. 2015; Tang et al. 2015). Besides this, aquaporin transport protein also showed excellent permeability for water molecules and can transport up to one billion water molecules per second which are 5 to 1000 greater than the conventional membrane (Phongphanphanee et al. 2008; Ma et al. 2012). Due to these properties, aquaporin-based biomimetic membranes are highly efficient for water purification and would be a breakthrough for the desalination world. A commercial biomimetic membrane mainly consists of three components: aquaporin which is a transporter protein for water molecules across the membrane, amphiphilic polymers in which aquaporin molecules are embedded and supportive matric that provide and improve the mechanical stability to the membrane.

16.3.1 Aquaporins

Aquaporins are transmembrane proteins found in all three domains of life, archaea, bacteria and eukaryotes (Zardoya and Villalba, 2001; Lee et al. 2005; Heymann and Engel, 1999). These proteins are necessary for maintaining water homeostasis in cells and allow water molecules to pass through the cell membrane selectively. Aquaporins were identified in plants, animals and microbes after the first finding of aquaporin protein in red blood cells (Denker et al. 1988). Depending on their structure and function, aquaporins are differentiated into three subfamilies. Aquaporins that exclusively allow only water to pass through it is called orthodox aquaporins. Other than water, some aquaporins enable tiny molecules such as carbon dioxide (Endeward et al. 2006) urea, ammonia (Holm et al. 2005) and glycerol (Holm et al. 2005; Litman et al. 2009) to pass through; these proteins are referred to as aquaglyceroporins (Mukhopadhyay et al. 2014). A third subclass called subcellular aquaporins or super-aquaporins was recently identified which has an unusual amino acid sequence from the other two families; however, their function is still unclear due to limited investigations and problems in synthesis and purification (Ishibashi, 2006; Ishibashi et al. 2011; Fuwad et al. 2019a, b). So far, nearly 13 aquaporins have been identified in various organs and tissues of mammals (Gena et al. 2011; Giwa et al. 2017). The kidney contains six of these molecules (Nielsen et al. 2002). Some of the most studied Aquaporins include Aquaporin 1, Aquaporin 2, Aquaporin 3, Aquaporin 4 and Aquaporin Z (Giwa et al. 2017) (Fig. 16.2).

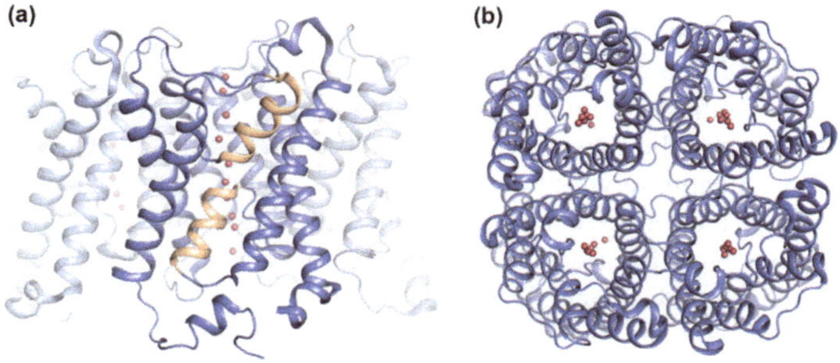

Fig. 16.2 Overall Crystal structure of human aquaporins (AQP5). Representing tetramer viewed of AQP: (a) parallel to the membrane and (b) from the extracellular side (Roche and Törnroth-Horsefield 2017 under Creative Commons Attribution licence)

The sequence of aquaporin-1 exposes its general unit structure in the form of two tandem repeats, each comprising three transmembrane spanning segments (Tang et al. 2015). The six transmembrane segments surround a central pore in an hour-glass shape with pseudo-two-fold symmetry, as revealed by biochemical study and crystal structural analysis. Each repeat denotes a hemi-pore, in which two hemi-pores fold together to produce a water channel (Tang et al. 2015). Each tandem repeat comprises a loop with an asparagine–proline–alanine (NPA) signature motif between TM2 and TM3 (Tang et al. 2015). According to research with mammalian aquaporin-1, it occurs in the plasma membrane as homo-tetramers. Aquaporins have a significant degree of sequence similarity: sequence identity present from 26 to 49% and augmented with 45–60% in the channel region. The three-dimensional structures of all aquaporins are highly similar. The root-mean-square displacements for Cα atoms are almost fixed in the overall chain range which is between 0.17 and 0.23 nm, and 0.07 to 0.13 nm in the membrane-spanning helices which indicating towards the conserved structure of a protein (Hashido et al. 2005). However, different glycosylation patterns is present in different aquaporins. The peptide backbone of aquaporin-1 is around 28 kDa (equivalent to around 112 kDa for the tetramer), while the glycosylated forms range from 40 to 60 kDa (equivalent to up to 240 kDa for the tetramer) (Borgnia et al. 1999; Tang et al. 2015). Most aquaporins have phosphorylation motifs of cytoplasmic protein kinase A, indicating a regulatory function (Han and Patil, 2000).

16.3.2 Amphiphilic Polymers

Aquaporins are transmembrane proteins embedded inside the plasma membranes in living tissues (Von Heijne 2007). Therefore, to develop a biomimetic membrane, they must be integrated into specific amphiphilic membranes made up of lipid bilayers or block copolymers (BCPs) that imitate the structure of the plasma membrane

(Wallin and Heijne, 1998). Lipids (Sessa and Weissmann 1968) and BCPs (Discher and Eisenberg, 2002) are two examples of materials in which aquaporins can be incorporated to synthesise a biomimetic membrane.

Lipids are amphiphilic substances that form bounded vesicular structures termed liposomes in an aqueous environment. By adding various components or modifying their composition, the surface charge and properties of the membrane can be tailored. Many techniques have been developed to reconstitute aquaporin proteins in lipid membranes since their discovery. Detergent-based reconstitution that includes detergent removal, dilution and dialysis is the most popular method (Zeidel et al. 1994; Prasad et al. 1998) to incorporate aquaporin in a lipid membrane. However, the mechanical strength of these lipids is not properly suitable to perform the effective function of the biomimetic membrane. Because the function of transmembrane protein is mainly influenced by membrane mechanical features such as membrane fluidity and stability which is controlled by adding additives like cholesterol (Liu et al. 2000). On the other hand, BCPs are also employed to develop a biomimetic membrane that mimics the plasma membranes by creating polymersomes that provide higher stability and durability due to the presence of polymers. BCPs gives the opportunity to increase the length of a polymer chain, select the right blocks, incorporate functional groups for manipulating the properties like thickness, permeability and durability of the membrane (Fuwad et al. 2019a, b). Stoenescu et al. (2004), His-tagged aquaporin 0 was introduced into a BCP and observed that the protein orientation depends on the symmetry of polymers. In another study, aquaporin was introduced into ABA BCP vesicles which exhibited a twofold rise in water permeability as compared to the commercial membrane (Kumar et al. 2007). Xie et al. (2015) investigated the influence of the length of the BCP hydrophobic chain on the Aquaporin permeability and observed that the hydrophobic mismatch is directly connected to the osmotic permeability of aquaporins. In biomimetic water filtration systems, both lipid-based and synthetic polymeric membranes are currently employed. Recent research has found that BCP membranes surpass lipid membranes in terms of cost, durability, rejection of small molecules like water (Discher et al. 1999), ambient phase transition temperature and lifespan (Jaskiewicz et al. 2012).

16.3.3 Substrate for Membrane Support Used in Various Techniques

Membrane fabricated by using both lipid and synthetic polymeric materials is too delicate to bear hydrostatic pressure since their thickness is only a few nanometers in each case, which necessarily required solid support to improve membrane stability and mechanical strength (Fuwad et al. 2019a, b). Solid supports have a porous structure and are comprised of organic or inorganic materials. Organic substrates provide a lot of versatility in chemical manipulation of surface, high porosity and uniform pore size that makes it easier to evaluate the performance of manufactured membranes

(Fuwad et al. 2019a, b). They also have the ability to retain proteins and lipids, which makes them suitable for use in biomimetic membranes.

Nanofiltration substrates are also popular choices for membrane support because of their higher flux than RO and significantly better retention than ultrafiltration, particularly for low molecular weight molecules including ions, sugars and natural organic materials (Mohammad et al. 2015). Hence, the substrate plays such an essential role in separation, most biomimetic membranes are made on a nanofiltration substrate. Furthermore, nanofiltration substrates have a low surface roughness which favours lipid stability and membrane formation (Freger 2004), and their surface charge may be easily adjusted to meet process requirements (Schaep and Vandecasteele, 2001). Depending on the surface charge and lipid structure, various materials like CA, Psf and PAN are utilised as a substrate to get the best results. According to the substrate structure and surface properties, various methodologies are used to obtain, potent and impeccable membranes. These methods include vesicle fusion (Wang et al. 2015), chemical crosslinking (Sun et al. 2013a), vesicle adsorption (Li et al. 2012), magnetic-aided adsorption (Sun et al. 2013b) and interfacial polymerisation (Zhao & Qiu, 2012b, a). Because of its simplicity and low defect creation, vesicle fusion is the most extensively used approach for stabilising membranes on a substrate. The surface charge and structure of both vesicles and substrate have a role in conjugation methods (Fuwad et al. 2019a, b). By using a vesicle fusion approach, a lipid bilayer membrane was synthesised by Kaufman et al. (2010) on a commercial nanofiltration polymeric membrane. Their research lighted on vesicle fusion on a solid substrate and confirmed lipid bilayer coverage on two membranes, NF-7450 and NF-270. Owing to the presence of sulfonated Psf, NF-7450 surface has strong acidic SO_3^- groups and due to the presence of weakly acidic COO^- groups of NF-270 has a negative charge on its surface. Even though the membranes showed good lipid bilayer coverage (97%) at a low pH (pH 2), still had two issues. At low pH, the stability of aquaporin proteins may affect, and it is also challenging to maintain lipid vesicles with a hydrophilic head on the hydrophobic surface. Wang et al. (2012) used layer-by-layer assembly of polyelectrolytes to synthesise biomimetic membranes. In this study, various alternative PEI and PSS polyelectrolyte layers were used to coat negatively charged H-PAN membrane. To produce a lipid bilayer membrane, on the modified substrate, positively charged liposomes were shattered on PSS layers (Wang et al. 2015). In another study, Sun et al. (2013a, b, c) generate a simple technique to enhance the incorporation of aquaporin vesicles and increase their mechanical strength by applying magnetic force. Magnetic force was used to immobilise magnetic nanoparticles containing proteoliposomes between the polyelectrolytes multilayers on the H-PAN substrate. Further, the stabilisation of liposomes was performed by the PAA-PSS blend and the polyelectrolyte layers of PAH. Further, Li et al. (2014) used the polymer crosslinking method to create an aquaporin-based nanofiltration membrane. In this study, due to adhering capacity of PDA with or without covalent bonding to a surface, it was used to coat proteoliposomes. Subsequently, at high temperature, PAI membrane was used to stabilise PDA-coated proteoliposomes and PEI polyelectrolyte layer was applied for encapsulation (Li et al. 2014). A research was conducted by Zhao and Zou (2012a), and Zhao and Qiu

(2012b) to synthesise aquaporin incorporated biomimetic membrane by interfacial polymerisation. In short, proteoliposomes containing MPD was used to soak Psf substrate and then subjected to TMC to generate a crosslinked polyamide layer encasing proteoliposomes (Zhao & Zou, 2012a; Zhao & Qiu, 2012b). This process was found more effective in terms of chemical cleaning (Li et al. 2017), long-term operations (Xia et al. 2017) and osmotic membrane bioreactors (Luo et al. 2018) as compared to traditional membranes. Besides this, the process was also found suitable for scale-up the production of the membrane (over 200 cm^2) with the least defect. By generating amide bonds between lipid bilayers and a porous Psf substrate, Ding et al. (2015) fabricated an AqpZ incorporated planar biomimetic membrane. The aquaporin bilayer was maintained by an amidation reaction between amino groups of the proteoliposomes and the carboxyl groups of the PDA layer on a Psf substrate that had been modified with a PDA layer. Recently, Fuwad et al. (2019a, b) used PCTE substrate functionalized by PDA to stabilise aquaporin liposomes by electrokinetic immobilisation approach. In this process, the aquaporin liposomes were propelled over the substrate by electrokinetic force which causes liposomes to entrap within or on the pores of the substrate and ensure a uniform coating of liposomes without aggregation or rupture. The fabricated membrane exhibited a high water flux of 7.45 LMH and 97.8% NaCl rejection with an excellent desalination property.

Further, inorganic compounds are also used as a substrate for membrane fabrication. These substrates have more chemical, mechanical, thermal stability than organic substrates and are also more expensive as compared to the organic substrate. Besides this, these materials have more advantages like well-defined pore structure, resistance to solvent and easy to sterilise. Among various inorganic compounds, alumina is frequently used as a substrate to fabricate biomimetic membranes due to its high pore density and uniform pore size. Duong et al. (2012) used alumina as a substrate to fabricate a membrane that was functionalized by gold to deposit a polymer bilayer through covalent interaction between the gold- and disulfide-functionalized ends of polymeric vesicles. However, due to the large pore size (approx. 55 nm), some flaws appeared in the membrane bilayer, despite the fact that the membrane pore size was uniform with high porosity. Wang et al. (2011) anticipated the use of PEG headrest reinforced with alumina to protect aquaporin to denature by contact with the substrate surface.

16.4 Advantages and Challenges of Biomimetic Membrane

Conventional membranes have wide pore size distributions which permit diffusion of species that should be held by the membrane, and inappropriate initial flux blocks the pore of the membrane are the primary problems of these membranes. Further, lack of isoporosity that causes fouling is also a common problem of conventional membranes (Fane et al. 2015; Shi et al. 2014). Additionally, due to the inherent nature of the fabrication material and procedure, all conventional membranes typically have a low pore density (Ulbricht, 2006). Hence, attaining high flux with low pore density is another difficulty associated with the conventional membrane. Owing to intrinsic structural features of aquaporin, aquaporin-based membrane is only the answer to various unresolved problems of conventional membranes However, The aquaporin structure has an ideal pore size, which prevents undesirable species from blocking pores and make membranes resistant to fouling (Li et al. 2015).

For the industrial application of biomimetic membrane, there are still some challenges that need to be overcome. First, aquaporin protein production is critical to the success of these membranes. However, due to their low stability in a specific aqueous environment to control the purity and quantity of these hydrophobic proteins are difficult which require the use of specific detergent. Second, incorporation of intact and functional aquaporin protein in the artificial system is a difficult task. While lipid and BCP-based compounds are the perfect analogues to plasma membrane however these substances are low in mechanical stability which is again a limitation of these membranes. Further, upscaling of fabrication of aquaporin-based biomimetic membrane is also a challenge that needs to resolve in the future research.

16.5 Conclusions and Future Prospects

Various membrane technologies are being used for water purification. Nonetheless, due to several limitations of conventional memnbarne technologies, aquaporin-based biomimetic membrane is emerging as an alternative solution for water purification owing to its advanced properties like high selectivity and permeability for water molecules. However, this membrane is more selective for water molecules but it is still associated with some challenges. The production of aquaporin in intact and pure form, and further incarpoartion in lipid and BCP-based materials are quite difficult. Additionally, upscaling of the biomimetic membrane for industrial use is also challenging. However, several approaches are being utilised to hit the current problem associated with the industrial application of these techniques. Nanolipoprotein particles approach can be used to maintain the purity and stability of aquaporin, and further incorporation efficiency into the membrane without the use of detergents in aqueous environments. Although, 3D bio-printing is also emerging as a potential approach to fabricate defect-free aquaporin membranes.

References

Baker RW (2012) Membrane technology and applications. John Wiley & Sons

Bar-Cohen Y (2011) Biomimetics: nature-based innovation. CRC Press

Bartels C, Franks R, Rybar S, Schierach M, Wilf M (2005) The effect of feed ionic strength on salt passage through reverse osmosis membranes. Desalination 184(1–3):185–195

Borgnia M, Nielsen S, Engel A, Agre P (1999) Cellular and molecular biology of the aquaporin water channels. Annu Rev Biochem 68(1):425–458

Burek, P., Satoh, Y., Fischer, G., Kahil, M. T., Scherzer, A., Tramberend, S., ... & Wiberg, D. (2016). Water futures and solution-fast track initiative.
Chakrabarty T, Prakash S, Shahi VK (2013) End group cross-linked 2-(dimethylamino) ethylmethacrylate based anion exchange membrane for electrodialysis. J Membr Sci 428:86–94
Cheng C, White N, Shi H, Robson M, Bruening ML (2014) Cation separations in electrodialysis through membranes coated with polyelectrolyte multilayers. Polymer 55(6):1397–1403
Conrad, F., Lofthouse, V., & Escobar-Tello, C. (2016). Design based on nature–a literature investigation. In DS 83: Proceedings of the 18th International Conference on Engineering and Product Design Education (E&PDE16), Design Education: Collaboration and Cross-Disciplinarity, Aalborg, Denmark, 8th-9th September 2016 (pp. 696–701).
Denker BM, Smith BL, Kuhajda FP, Agre P (1988) Identification, purification, and partial characterization of a novel Mr 28,000 integral membrane protein from erythrocytes and renal tubules. J Biol Chem 263(30):15634–15642
Ding W, Cai J, Yu Z, Wang Q, Xu Z, Wang Z, Gao C (2015) Fabrication of an aquaporin-based forward osmosis membrane through covalent bonding of a lipid bilayer to a microporous support. Journal of Materials Chemistry A 3(40):20118–20126
Discher BM, Won YY, Ege DS, Lee JC, Bates FS, Discher DE, Hammer DA (1999) Polymersomes: tough vesicles made from diblock copolymers. Science 284(5417):1143–1146
Discher DE, Eisenberg A (2002) Polymer vesicles. Science 297(5583):967–973
Duong PH, Chisca S, Hong PY, Cheng H, Nunes SP, Chung TS (2015) Hydroxyl functionalized polytriazole-co-polyoxadiazole as substrates for forward osmosis membranes. ACS Appl Mater Interfaces 7(7):3960–3973
Duong PH, Chung TS, Jeyaseelan K, Armugam A, Chen Z, Yang J, Hong M (2012) Planar biomimetic aquaporin-incorporated triblock copolymer membranes on porous alumina supports for nanofiltration. J Membr Sci 409:34–43
Elimelech, M. (2006). The global challenge for adequate and safe water. Journal of Water Supply: Research and Technology—AQUA, 55(1), 3–10.
Endeward V, Musa-Aziz R, Cooper GJ, Chen LM, Pelletier MF, Virkki LV, Gros G (2006) Evidence that aquaporin 1 is a major pathway for CO2 transport across the human erythrocyte membrane. FASEB J 20(12):1974–1981
Fane AG, Wang R, Hu MX (2015) Synthetic membranes for water purification: status and future. Angew Chem Int Ed 54(11):3368–3386
Freger V (2004) Swelling and morphology of the skin layer of polyamide composite membranes: an atomic force microscopy study. Environ Sci Technol 38(11):3168–3175
Fuwad A, Ryu H, Lee JH, Kim D, Yoo YE, Kim YR, Jeon TJ (2019a) An electrokinetic approach to fabricating aquaporin biomimetic membranes for water purification. Desalination 452:9–16
Fuwad A, Ryu H, Malmstadt N, Kim SM, Jeon TJ (2019b) Biomimetic membranes as potential tools for water purification: Preceding and future avenues. Desalination 458:97–115
Gena P, Pellegrini-Calace M, Biasco A, Svelto M, Calamita G (2011) Aquaporin membrane channels: biophysics, classification, functions, and possible biotechnological applications. Food Biophys 6(2):241–249
Gin DL, Noble RD (2011) Designing the next generation of chemical separation membranes. Science 332(6030):674–676
Giwa, A., Ahmed, M., & Hasan, S. W. (2019). Polymers for Membrane Filtration in Water Purification. In Polymeric Materials for Clean Water, Springer, Cham. (pp. 167–190).
Giwa A, Hasan SW, Yousuf A, Chakraborty S, Johnson DJ, Hilal N (2017) Biomimetic membranes: A critical review of recent progress. Desalination 420:403–424
Guiver MD, Croteau S, Hazlett JD, Kutowy O (1990) Synthesis and characterization of carboxylated polysulfones. Br Polym J 23(1–2):29–39
Gurreri L, Tamburini A, Cipollina A, Micale G (2020) Electrodialysis applications in wastewater treatment for environmental protection and resources recovery: A systematic review on progress and perspectives. Membranes 10(7):146

Han Z, Patil RV (2000) Protein kinase A-dependent phosphorylation of aquaporin-1. Biochem Biophys Res Commun 273(1):328–332

Hashido M, Ikeguchi M, Kidera A (2005) Comparative simulations of aquaporin family: AQP1, AQPZ, AQP0 and GlpF. FEBS Lett 579(25):5549–5552

Heymann JB, Engel A (1999) Aquaporins: phylogeny, structure, and physiology of water channels. Physiology 14(5):187–193

Holm LM, Jahn TP, Møller AL, Schjoerring JK, Ferri D, Klaerke DA, Zeuthen T (2005) NH 3 and NH 4+ permeability in aquaporin-expressing Xenopus oocytes. Pflugers Arch 450(6):415–428

Hosseini SM, Gholami A, Madaeni SS, Moghadassi AR, Hamidi AR (2012) Fabrication of (polyvinyl chloride/cellulose acetate) electrodialysis heterogeneous cation exchange membrane: Characterization and performance in desalination process. Desalination 306:51–59

Hosseini SM, Rahzani B, Asiani H, Khodabakhshi AR, Hamidi AR, Madaeni SS, Seidypoor A (2014) Surface modification of heterogeneous cation exchange membranes by simultaneous using polymerization of (acrylic acid-co-methyl methacrylate): Membrane characterization in desalination process. Desalination 345:13–20

Ishibashi, K. (2006). Aquaporin superfamily with unusual npa boxes: S-aquaporins (superfamily, sip-like and subcellular-aquaporins). Cellular and molecular biology (Noisy-le-Grand, France), 52(7), 20–27.

Ishibashi K, Kondo S, Hara S, Morishita Y (2011) The evolutionary aspects of aquaporin family. American Journal of Physiology-Regulatory, Integrative and Comparative Physiology 300(3):R566–R576

Ismail AF, Padaki M, Hilal N, Matsuura T, Lau WJ (2015) Thin film composite membrane—Recent development and future potential. Desalination 356:140–148

Jaewon J (2021) Ion Exchange Membrane for Reverse Electrodialysis. Academ J Polym Sci 5(1):00218–0022. https://doi.org/10.19080/AJOP.2021.05.555654

Jaskiewicz K, Makowski M, Kappl M, Landfester K, Kroeger A (2012) Mechanical properties of poly (dimethylsiloxane)-block-poly (2-methyloxazoline) polymersomes probed by atomic force microscopy. Langmuir 28(34):12629–12636

Jung B, Yoon JK, Kim B, Rhee HW (2004) Effect of molecular weight of polymeric additives on formation, permeation properties and hypochlorite treatment of asymmetric polyacrylonitrile membranes. J Membr Sci 243:45–57

Kaufman Y, Berman A, Freger V (2010) Supported lipid bilayer membranes for water purification by reverse osmosis. Langmuir 26(10):7388–7395

Kesieme UK, Milne N, Aral H, Cheng CY, Duke M (2013) Economic analysis of desalination technologies in the context of carbon pricing, and opportunities for membrane distillation. Desalination 323:66–74

Khan MA, Kumar M, Alothman ZA (2015) Preparation and characterization of organic–inorganic hybrid anion-exchange membranes for electrodialysis. J Ind Eng Chem 21:723–730

Kumar M, Grzelakowski M, Zilles J, Clark M, Meier W (2007) Highly permeable polymeric membranes based on the incorporation of the functional water channel protein Aquaporin Z. Proc Natl Acad Sci 104(52):20719–20724

Latorre FJG, Báez SOP, Gotor AG (2015) Energy performance of a reverse osmosis desalination plant operating with variable pressure and flow. Desalination 366:146–153

Lau WJ, Ismail AF, Misdan N, Kassim MA (2012) A recent progress in thin film composite membrane: a review. Desalination 287:190–199

Lee, J. K., Kozono, D., Remis, J., Kitagawa, Y., Agre, P., & Stroud, R. M. (2005). Structural basis for conductance by the archaeal aquaporin AqpM at 1.68 Å. Proceedings of the National Academy of Sciences, 102(52), 18932–18937.

Lee KP, Arnot TC, Mattia D (2011) A review of reverse osmosis membrane materials for desalination—Development to date and future potential. J Membr Sci 370(1–2):1–22

Li D, Wang H (2010) Recent developments in reverse osmosis desalination membranes. J Mater Chem 20(22):4551–4566

Li X, Chou S, Wang R, Shi L, Fang W, Chaitra G, Fane AG (2015) Nature gives the best solution for desalination: Aquaporin-based hollow fiber composite membrane with superior performance. J Membr Sci 494:68–77

Li X, Wang R, Tang C, Vararattanavech A, Zhao Y, Torres J, Fane T (2012) Preparation of supported lipid membranes for aquaporin Z incorporation. Colloids Surf, B 94:333–340

Li X, Wang R, Wicaksana F, Tang C, Torres J, Fane AG (2014) Preparation of high performance nanofiltration (NF) membranes incorporated with aquaporin Z. J Membr Sci 450:181–188

Li Z, Linares RV, Bucs S, Fortunato L, Hélix-Nielsen C, Vrouwenvelder JS, Amy G (2017) Aquaporin based biomimetic membrane in forward osmosis: Chemical cleaning resistance and practical operation. Desalination 420:208–215

Litman, T., Søgaard, R., & Zeuthen, T. (2009). Ammonia and urea permeability of mammalian aquaporins. Aquaporins, 327–358.

Liu DZ, Chen WY, Tasi LM, Yang SP (2000) Microcalorimetric and shear studies on the effects of cholesterol on the physical stability of lipid vesicles. Colloids Surf, A 172(1–3):57–67

Lu X, Boo C, Ma J, Elimelech M (2014) Bidirectional diffusion of ammonium and sodium cations in forward osmosis: role of membrane active layer surface chemistry and charge. Environ Sci Technol 48(24):14369–14376

Luo W, Xie M, Song X, Guo W, Ngo HH, Zhou JL, Nghiem LD (2018) Biomimetic aquaporin membranes for osmotic membrane bioreactors: Membrane performance and contaminant removal. Biores Technol 249:62–68

Ma, H., Burger, C., Hsiao, B. S., & Chu, B. (2012). Highly permeable polymer membranes containing directed channels for water purification ACS Macro Letters 723–726.

McMullen, L. D. (2008). Remediation at the water treatment plant. In Nitrogen in the Environment (pp. 623–629). Academic Press.

Mecha, A. C. (2018). Applications of Reverse and Forward Osmosis Processes in Wastewater Treatment: Evaluation of Membrane Fouling. Osmotically Driven Membrane Processes: Approach, Development and Current Status, 235–253.

Miller S, Shemer H, Semiat R (2015) Energy and environmental issues in desalination. Desalination 366:2–8

Mohammad AW, Teow YH, Ang WL, Chung YT, Oatley-Radcliffe DL, Hilal N (2015) Nanofiltration membranes review: Recent advances and future prospects. Desalination 356:226–254

Mukhopadhyay, R., Bhattacharjee, H., & Rosen, B. P. (2014). Aquaglyceroporins: generalized metalloid channels. Biochimica et Biophysica Acta (BBA)-General Subjects, 1840(5), 1583–1591.

Müller F, Ferreira CA, Franco L, Puiggali J, Aleman C, Armelin E (2012) New sulfonated polystyrene and styrene–ethylene/butylene–styrene block copolymers for applications in electrodialysis. J Phys Chem B 116(38):11767–11779

Nielsen S, Frøkiær J, Marples D, Kwon TH, Agre P, Knepper MA (2002) Aquaporins in the kidney: from molecules to medicine. Physiol Rev 82(1):205–244

Nunes SP, Peinemann KV (2010) 1.06-Advanced Polymeric and Organic-Inorganic Membranes for Pressure-Driven Processes. Comprehensive Membrane Science and Engineering 2017:113–129

Pandey RP, Thakur AK, Shahi VK (2014) Stable and efficient composite anion-exchange membranes based on silica modified poly (ethyleneimine)–poly (vinyl alcohol) for electrodialysis. J Membr Sci 469:478–487

Phongphanphanee S, Yoshida N, Hirata F (2008) On the proton exclusion of aquaporins: a statistical mechanics study. J Am Chem Soc 130(5):1540–1541

Prasad GR, Coury LA, Finn F, Zeidel ML (1998) Reconstituted aquaporin 1 water channels transport CO2 across membranes. J Biol Chem 273(50):33123–33126

Roche JV, Törnroth-Horsefield S (2017) Aquaporin protein-protein interactions. Int J Mol Sci 18(11):2255

Schaep J, Vandecasteele C (2001) Evaluating the charge of nanofiltration membranes. J Membr Sci 188(1):129–136

Sessa G, Weissmann G (1968) Phospholipid spherules (liposomes) as a model for biological membranes. J Lipid Res 9(3):310–318
Setiawan L, Wang R, Tan S, Shi L, Fane AG (2013) Fabrication of poly(amide-imide)-polyethersulfone dual layer hollow fiber membranes applied in forward osmosis by combined polyelectrolyte cross-linking and depositions. Desalination 312:99–106
Shi X, Tal G, Hankins NP, Gitis V (2014) Fouling and cleaning of ultrafiltration membranes: A review. Journal of Water Process Engineering 1:121–138
Sosa-Fernandez PA, Post JW, Bruning H, Leermakers FAM, Rijnaarts HHM (2018) Electrodialysis-based desalination and reuse of sea and brackish polymer-flooding produced water. Desalination 447:120–132
Stoenescu R, Graff A, Meier W (2004) Asymmetric ABC-triblock copolymer membranes induce a directed insertion of membrane proteins. Macromol Biosci 4(10):930–935
Sun G, Chung TS, Chen N, Lu X, Zhao Q (2013a) Highly permeable aquaporin-embedded biomimetic membranes featuring a magnetic-aided approach. RSC Adv 3(24):9178–9184
Sun G, Chung TS, Jeyaseelan K, Armugam A (2013b) Stabilization and immobilization of aquaporin reconstituted lipid vesicles for water purification. Colloids Surf, B 102:466–471
Sun G, Chung TS, Jeyaseelan K, Armugam A (2013c) A layer-by-layer self-assembly approach to developing an aquaporin-embedded mixed matrix membrane. RSC Adv 3(2):473–481
Tanaka Y (2015) Ion exchange membranes: fundamentals and applications. Elsevier
Tang C, Wang Z, Petrinić I, Fane AG, Hélix-Nielsen C (2015) Biomimetic aquaporin membranes coming of age. Desalination 368:89–105
Ulbricht M (2006) Advanced functional polymer membranes. Polymer 47(7):2217–2262
van Loosdrecht MC, Brdjanovic D (2014) Anticipating the next century of wastewater treatment. Science 344(6191):1452–1453
Von Heijne G (2007) The membrane protein universe: what's out there and why bother? J Intern Med 261(6):543–557
Wallin E, Heijne GV (1998) Genome-wide analysis of integral membrane proteins from eubacterial, archaean, and eukaryotic organisms. Protein Sci 7(4):1029–1038
Wang, H., Chung, T. S., Tong, Y. W., Jeyaseelan, K., Armugam, A., Chen, Z., ... & Meier, W. (2012). Highly permeable and selective pore-spanning biomimetic membrane embedded with aquaporin Z. small, 8(8), 1185–1190.
Wang H, Chung TS, Tong YW, Meier W, Chen Z, Hong M, Armugam A (2011) Preparation and characterization of pore-suspending biomimetic membranes embedded with Aquaporin Z on carboxylated polyethylene glycol polymer cushion. Soft Matter 7(16):7274–7280
Wang KY, Chung TS, Qin JJ (2007) Polybenzimidazole (PBI) nanofiltration hollow fiber membranes applied in forward osmosis process. J Membr Sci 300(1–2):6–12
Wang M, Wang Z, Wang X, Wang S, Ding W, Gao C (2015) Layer-by-layer assembly of aquaporin Z-incorporated biomimetic membranes for water purification. Environ Sci Technol 49(6):3761–3768
Weig A, Deswarte C, Chrispeels MJ (1997) The major intrinsic protein family of Arabidopsis has 23 members that form three distinct groups with functional aquaporins in each group. Plant Physiol 114(4):1347–1357
Wenten IG (2016) Reverse osmosis applications: prospect and challenges. Desalination 391:112–125
WHO, UN, Progress on Drinking Water and Sanitation. Joint Monitoring Programme Update, (2012).
Wilf M, Bartels C (2005) Optimization of seawater RO systems design. Desalination 173(1):1–12
Xia L, Andersen MF, Hélix-Nielsen C, McCutcheon JR (2017) Novel commercial aquaporin flat-sheet membrane for forward osmosis. Ind Eng Chem Res 56(41):11919–11925
Xie W, Low JWJ, Armugam A, Jeyaseelan K, Tong YW (2015) Regulation of Aquaporin Z osmotic permeability in ABA tri-block copolymer. AIMS Biophysics 2(3):381–397
Xu W, Ge Q (2018) Synthetic polymer materials for forward osmosis (FO) membranes and FO applications: a review. Rev Chem Eng 35(2):191–209

Yaroslavtsev AB. and Nikonenko VV., 2009. Ion-exchange membrane materials: Properties, modification, and practical application. Nanotechno in Russia, 4(3):137–159.

Zardoya R, Villalba S (2001) A phylogenetic framework for the aquaporin family in eukaryotes. J Mol Evol 52(5):391–404

Zeidel ML, Nielsen S, Smith BL, Ambudkar SV, Maunsbach AB, Agre P (1994) Ultrastructure, pharmacologic inhibition, and transport selectivity of aquaporin CHIP in proteoliposomes. Biochemistry 33(6):1606–1615

Zhao S, Zou L, Tang CY, Mulcahy D (2012a) Recent developments in forward osmosis: opportunities and challenges. J Membr Sci 396:1–21

Zhao Y, Qiu C, Li X, Vararattanavech A, Shen W, Torres J, Tang CY (2012b) Synthesis of robust and high-performance aquaporin-based biomimetic membranes by interfacial polymerization-membrane preparation and RO performance characterization. J Membr Sci 423:422–428

Zhong P, Fu X, Chung TS, Weber M, Maletzko C (2013) Development of thin-film composite forward osmosis hollow fiber membranes using direct sulfonated polyphenylenesulfone (sPPSU) as membrane substrates. Environ Sci Technol 47(13):7430–7436

Chapter 17
Nanobiosensors: Diagnostic Tools for Environmental Contaminants

Rama Pal, I. Rashmi, and Jai Prakash Narain Rai

Abstract Environmental pollution, a growing global concern, adversely affects human health and socio-economic development. Besides the presence of various environmental contaminants like chemical substances, heavy metals, viruses, bacteria, parasitic pathogens and their toxins, emerging environmental contaminants (also called micropollutants) have drawn scientific attention and public concerns. The growing demand for environmental pollution control necessitates the development of rapid analytical tools with greater efficacy and precision for the on-site and real-time monitoring of a broader spectrum of various pollutants without extensive sample preparation. Nanobiosensors thus appear as a powerful alternative to conventional analytical techniques that are associated with the issues such as the requirement of sophisticated and expensive instruments and expert personnel for their operation. Nanobiosensors are the fabrication products of nanoscale hybrid materials, such as complexes consisting of nanoparticles (NPs) and biological molecules which are ideal for the detection of contaminants with ultrahigh sensitivity, selectivity and rapid responses. Thus, the use of nanobiosensors will significantly improve environmental monitoring approaches in the future. This chapter emphasizes on nanobiosensor approaches for environmental pollutants monitoring, challenges and future perspectives.

Keywords Carbon nanotube · Environmental contaminants · Nanowire · Nanobiosensor · Quantum dot

---

R. Pal (✉)
Indian Institute of Soil and Water Conservation, Research Centre, Agra 282006, UP, India
e-mail: rama.env@gmail.com

I. Rashmi
Indian Institute of Soil and Water Conservation, Research Centre, Kota, Rajasthan, India

J. P. N. Rai
Department of Environmental Sciences, College of Basic Sciences, G.B. Pant University of Agriculture and Technology, Pantnagar, India

J. P. N. Rai and S. Saraswat (eds.), *Nano-biotechnology for Waste Water Treatment*, Water Science and Technology Library 111,
https://doi.org/10.1007/978-3-031-00812-2_17

17.1 Introduction

Environmental pollution is a major global problem that adversely affects human health and the sustainable socio-economic development of society. The various environmental samples have been contaminated with heavy metals, organic compounds, pathogens and toxins which is a serious issue to be addressed by environmental scientists and governments. Additionally, the presence of emerging pollutants in the aquatic environment resulting from the point and diffuse pollution has strongly raised scientific interest and public concern (Kvenvolden and Cooper 2003; Duan et al. 2007; Busetti et al. 2005).

The traditional monitoring techniques of heavy metals, pesticides, pathogens and toxic intermediates in soil and water are highly detailed and comprehensive which is time and cost-intensive (Malik et al. 2013; Shahbazi et al. 2018). To overcome the drawbacks associated with traditional techniques, different biosensors (a device comprised of a biologically active element to recognize the analyte of interest) are being developed. Biosensor is an integral unit of a biological recognition element with a signal transducer component to produce an electrical or electronic signal to be detected, amplified and measured by a detector in order to quantify a specific analyte. The accuracy of a biosensor's function depends upon the efficient signal capture of the interaction between the analyte and the biological element by the transducer and converting them into magnetic, optical, gravimetric, fluorescent, electrochemical or electrochemiluminescent signals. Hence, nanomaterial-enabled biosensors provide precise detection, on the nanomolar to sub-picomolar level, of environmental contaminants (Aragay et al. 2012) as nanomaterials are capable of immobilizing bioreceptor elements at higher mass-to-volume ratio and even play the role of transduction element itself. Because of the nanosize (1–100 nm) of nanomaterials, new and better signal transduction technologies have been designed for accurate and rapid analysis of single or multiple components in vivo.

This chapter gives an overview of the different types of nanobiosensors (mechanical, optical, electrochemical, and magnetic) and their potential applications for the detection of various pollutants along with challenges and future perspectives.

17.2 Nanobiosensors

Nanobiosensors are altered version of a biosensor and employ various nanomaterials for immobilizing various biomolecules (recognition component) to detect very small quantities of toxic contaminants in environmental samples. Nanobiosensors can also be a unit comprised of nanosized biologically derived functionalized element linked to a physical/chemical transducer. Thus, a nanobiosensor can be characterized on the basis of three components: a nanomaterial(s), a biological recognition element (bioreceptor) that interacts with an analyte and provides specificity, and a signal

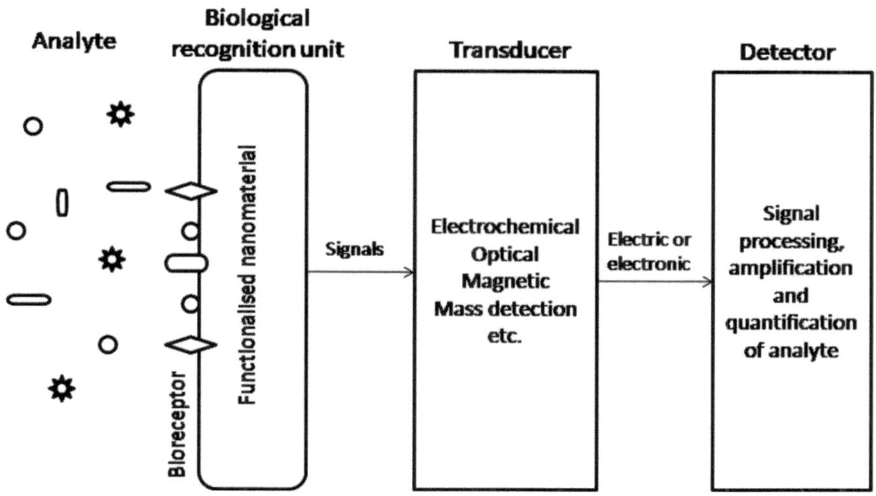

Fig. 17.1 The schematic representation of a nanobiosenser consisting of bioreceptors and transducer

transducer (e.g., electrochemical, optical, mass, etc.) to produce an electronic signal that is proportional to the concentration of a target analyte (Fig. 17.1).

The basic functions of nanomaterials in biosensors can be mainly classified as: (1) immobilization of biomolecules; (2) enhancement of electron transfer or signal production; (3) labeling biomolecules; and (4) acting as reactant. The roles that different nanomaterials have played in biosensor systems are summarized in Table 17.1.

17.3 Nanomaterials for Biosensing

The nanobiosensor technology gained momentum after the exploration, synthesis and possible utilization of nanomaterials. The nanomaterials can be categorized as: (i) nanoparticles, (ii) carbon nanotube and (iii) nanowire.

Nanoparticles provide high surface areas per unit volume due to their extremely small size. They can be synthesized using standard chemical techniques without the requirement of advanced fabrication processes. Most biological molecules can be labeled with metal nanoparticles without compromising their biological activities. The nanoparticles including gold, silver, palladium, platinum, copper and cobalt have been widely used in the manufacturing of nanobiosensors. The gold and silver nanoparticles have unique localized surface plasmon resonance (LSPR) ranging between the visible and infrared spectrum, which is sensitive to the composition, shape, size (aspect ratio), surrounding environment and interparticle spacing. This plasmonic property laid the foundation for the development of colorimetric

Table 17.1 Nanomaterials used in nanobiosensors

No	Nanomaterial type	Advantages	References
1	Carbon nanotubes	Stable at exo- and endohedral functionalization, improved enzyme loading, higher aspect ratios and high electrical conductivity (approximately 100 times greater than copper wires)	Chen et al. (2008), Pedrosa et al. (2010) and Oliveira et al. (2013)
2	Nanoparticles	Better loading of bio-analyte, aid in immobilization and possess good catalytic properties	Li and Wei (2017), Swierczewska et al. (2011), Maruthupandi et al. (2020) and Wang et al. (2016)
3	Quantum dots	Easy surface modification, excellent fluorescence resonance energy transfer donor and quantum confinement of charge carriers	Bala et al. (2018), Jia et al. (2018) and Hu et al. (2019)
4	Nanowires	High electrical conductivity and wide applicability for bio- and chemical sensing	Aluri et al. (2011) and Yunus et al. (2012)
5	Nanorods	Can be made from most elements (metals and nonmetals) and compounds; high excitation of surface plasmons, size tunable energy regulation, can be coupled with MEMS and induce specific field responses	Yu and Irudayaraj (2007), Kuswandi (2018) and Shams et al. (2019)

environmental sensors. The LSPR also intensifies the electromagnetic field near the surface of nanoparticles, which is useful in detection through surface-enhanced Raman spectroscopy. Magnetic nanoparticles made up of iron and its oxides possess para- or superparamagnetic behavior under the influence of a magnetic field at room temperature. The working principle for detection of analyte using magnetic nanoparticles depends upon two relaxation mechanisms, i.e., Brownian and Neel relaxations. The MNPs are functionalized with biocompatible coating to avoid their aggregation in the absence of magnetic field. The use of various magnetic nanoparticles in nanobiosensors is summarized in Table 17.2.

QDs are "0-dimensional" nanosized crystalline semiconductor (ranging from 2 to 10 nm) with unique tunable optical properties, such as a broad absorption spectrum, narrow emission peak and tunable emission wavelength, which is size-dependent. They have replaced organic and fluorescent dyes in biosensing due to their strong chemical stability with extreme resistance to photobleaching. QDs have been widely used in detection of heavy metal ions and organic pollutants. Several kinds of QDs have been synthesized, such as CdSe (Algarra et al. 2012), CdSe/ZnS (Sung and Lo 2012), CdTe (Chao et al. 2013; Gan et al. 2012), CdTe/CdS (Gui et al. 2012), ZnS (Koneswaran and Narayanaswamy 2009) and ZnSe/ZnS (Ke et al. 2012).

Table 17.2 Features of magnetic nanoparticle-based biosensors (adopted from Herpoldt et al. 2017)

Mode	Sensor type	Key MNP parameters	Limit of detection	References
Volumetric detection	NMR	High transverse relaxivity (T2)	<1 pM protein; 102 CFU/ml bacteria	Min et al. (2012) and Chen et al. (2015)
	Magnetic susceptometer	Brownian relaxation; colloidal stability	4 pM DNA	Strömberg et al. (2014) and Grossman et al. (2004)
Surface sensor	GMR	High magnetic moment, size	<10 fM protein	Gaster et al. (2011) and Hall et al. (2013)
	Hall-effect	High magnetic moment	Single cell	Issadore et al. (2012)
	Nanozyme	High catalytic activity	<pg/ml protein	Kim et al. (2015), Xia et al. (2015) and Duan et al. (2015)

Carbon nanotubes are hollow tubes made of either a single carbon sheet with (SWCNT) or concentric carbon sheets of different diameters forming multiwalled carbon nanotubes (MWCNT) with sp2 bonding (Rivas et al. 2007). Carbon nanotubes enhance the electrochemical reactivity of important biomolecules and promote the electrontransfer reactions of proteins. These have excellent electronic properties depending upon electron valency of carbon atoms. Carbon nanotube-modified electrodes have been shown to be useful in high-density surface loading of biomolecules (e.g., nucleic acids) due to their large surface area and hollow structure and hence, alleviating surface fouling effects (Wang et al. 2003). The combination of their wire-like structure and high electroactivity to the surface adsorbates allows radial diffusion affecting surface-to-noise ratio and leads to the development of highly sensitive electrochemical biosensors ranging from amperometric enzyme electrodes to DNA hybridization biosensors.

Nanowires are 1D structures with a length-to-width ratio greater than 1000. The movement of electrons in these wires is free to travel along the length, but their motion in the other two directions is quantum confined, thus differentiating them from 3D wires. The analyte binding on the surface of recognition element transduces a change in the charge density including electric field on the surface of the nanowire resulting in real time and quantitative detection. The antibodies-coated conductive TiO_2 nanowire (TiO_2-NW) bundles have been fabricated for selective detection of *Listeria monocytogenes.*

17.4 Types of Nanobiosensors

Nanobiosensors can be classified based on the signals generated due to the interaction between biological recognition element and the analyte as below:

17.4.1 Mechanical Nanobiosensors

There are two types of mechanical nanobiosensors, viz. cantilevers and acoustic mechanical. Cantilevers can be either surface-stress sensors or dynamic-mode sensors. The interaction between the recognition element and target molecules may be electrostatic, van der Waals, hydrogen bonding, etc., which causes surface stress leading to the deflection of the mechanical element. The degree of deflection is proportional to the analyte concentration. The dynamic-mode sensors oscillate with a resonance frequency which changes when the analyte is loaded on cantilever, and is the basis for detection. The acoustic devices work on bulk acoustic waves (BAW) or surface acoustic waves (SAW). The first one is sensitive to the changes in the resonant frequency of a piezoelectric crystal in response to changes in surface-adsorbed mass, while second responds to shear surface waves.

17.4.2 Optical Nanobiosensors

Optical biosensors exploit light absorption, fluorescence, luminescence, reflectance, Raman scattering and refractive index of a sample under irradiation by ultraviolet, visible or infrared light (Bănică 2012a). Two common approaches used for analyte detection through optical nanobiosensor are fluorescence and surface plasmon resonance-enabled spectroscopies. Fluorescence spectroscopy measures the emission of an excited fluorophore after it returns to its ground state. Fluorescent nanobiosensor uses QDs (quantum dots) or polymer nanoparticle or dye-doped silicon probes due to their more photostability than traditional fluorescent dyes (Vikesland and Wigginton 2010; Bae et al. 2012). The fluorescence signals may be "turn-on" or "turn-off" depending upon conformational change in the sensor due to interaction with the analyte of interest.

Surface plasmon resonance spectroscopy is based on the localized surface plasmon resonance (LSPR) of metal nanomaterials (Bănică 2012b; Wei et al. 2015) when it is irradiated with polarized light. The shift in LSPR is proportional to the mass of analyte on the surface.

17.4.3 Electrochemical Biosensors

Electrochemical nanobiosensors relay on the measurement of changes in current or potential that result from the interaction between an analyte and an electrode. Different techniques have been used to record these changes and include cyclic voltammetry, chronoamperometry, chronopotentiometry, impedance spectroscopy and various field-effect transistor-based methods (Grieshaber et al. 2008). Nanobiosensor designs are based on the modification of solid electrode (e.g., platinum, gold, silver and graphite) with nanocarbons (e.g., carbon nanotubes and graphene), nanoparticles or functionalization with biological recognition elements (e.g., antibodies and aptamers) (Sadik et al. 2009). Direct spatial contact between the nanomaterial-enabled electrode and the recognition element amplifies signal several times and improves signal-to-noise ratios compared to traditional electrochemical techniques (Grieshaber et al. 2008; Sanvicens et al. 2009). The detection limits for smaller particles improved due to their higher diffusivity and lower steric hindrance (Garcia-Aljaro et al. 2010).

17.4.4 Magnetic Biosensors

Magnetic nanoparticles, viz. superparamagnetic (Fe_3O_4), greigite (Fe_3S_4), maghemite (g-Fe_2O_3) and various types of ferrites ($MeOFe_2O_3$, where Me=Ni, Co, Mg, Zn, Mn, etc.), have been extensively used in nanobiosensors. Magnetic nanobiosensors are most suitable for detection in biological samples because of the low background magnetic signal (Koets et al. 2009). Magnetic-relaxation switches in superparamagnetic iron oxide nanoparticles are the underlying principle for detection mechanism, which results due to clustering of individual nanomagnetic probes into larger assemblies following binding with an analyte. Analyte binding induces enhanced dephasing of the spins of protons in the surrounding medium. The subsequent alteration in the spin–spin (T2) relaxation is detected by magnetic resonance relaxometry (Reddy et al. 2012; Kaittanis et al. 2006). Magnetic nanobiosensors have been used to detect proteins (Zhang et al. 2013), nucleic acids (DNA and mRNA) and viruses (Perez et al. 2003).

17.5 Environmental Application of Nanobiosensors

For environmental applications, nanobiosensors are more field and user-friendly over conventional analytical techniques due to their portability, miniaturization and work. Nanobiosensors can be used as monitoring tools in the assessment of chemical/biological/ecological quality of various environmental ecosystems via in situ detection of inorganic/organic pollutants and pathogens (Andrea et al. 2011). The nanomaterials such as metal NPs (gold, silver, cobalt, etc.), CNT, magnetic NPs and QDs have been intensively used in biosensors for detection of various contaminants (Prasad et al. 2017), which are summarized in the following text.

17.5.1 Pesticides

Pesticides often target specific enzyme, and hence, many forms of pesticide detection are based on analyzing the activity of respective enzyme either directly or indirectly. Many of the nanobiosensor methods developed for the detection of pesticides are based on the action of the enzyme acetylcholinesterase (AChE) (Kuswandi et al. 2008; Kuswandi and Swandari 2007). Pesticides such as organophosphates and carbamates prevent hydrolysis of acetylcholine by inhibiting enzyme AChE through binding to a serine moiety within the active site of the enzyme (Kuswandi et al. 2008; Kuswandi and Swandari 2007). Other compounds such as heavy metals and detergents can be interfering agent in this process as they can also selectively inhibit the enzyme (Nagatani et al. 2007; Kuswandi and Mascini 2005). In AChE-based nanobiosensor, for the detection of monocrotophos (LOD 10 nm), gold nanoparticles and multiwalled carbon nanotubes (MWCNTs) were used to modify glassy carbon electrodes through flow-based system. The carbon nanotubes contain chitosan to increase the immobilization level and improve the stability of AChE. The gold nanoparticles and MWCNTs promote electron transfer and catalyze the electro-oxidation of thiocholine in their amperometric biosensor (Norouzi et al. 2010). The detection of two pesticides methyl parathion and chlorpyrifos was established by Viswanathan et al. (2009) through the fabrication of a electrochemical biosensor by integrating the special features of ssDNA-SWCNT (single-walled carbon nanotubes), PANI (polyaniline) and AChE. Aniline was polymerized electrochemically and coated on vertically aliened thiol-terminated ssDNA-SWCNTs self-assembled on gold electrode. The AChE was immobilized on PANI matrix. Thin PANI film on SWCNT acts as good sensor for enzyme by-product acetic acid. The redox activity of PANI thin film on SWCNTs increases as a result of changes in pH near the electrode surface by enzymatic reaction (Fig. 17.2). This nanobiosensor can detect both the pesticides up to $1 \times 10 - 12$ M.

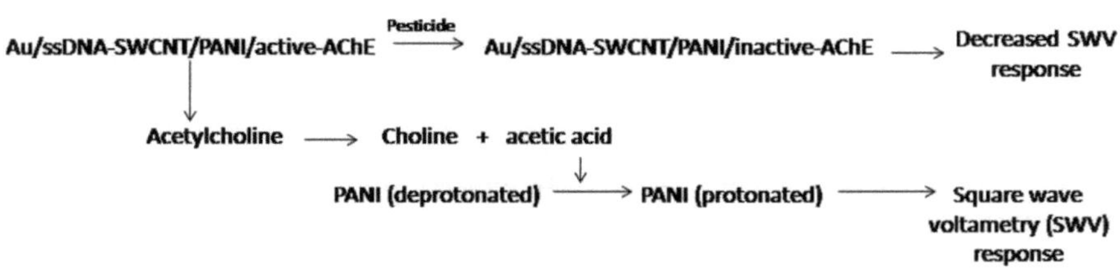

Fig. 17.2 Schematic representation of electrochemical biosenser (Au/ssDNA-SWCNT/PANI/active-AChE) working principle (Viswanathan et al. 2009)

A fluorescence detection scheme was chosen as the transduction scheme for monitoring of the organophosphorus pesticides dichlorvos and paraoxon at very low levels with liposome-based nanobiosensors. The enzyme acetylcholinesterase was effectively stabilized within the internal nanoenvironment of the liposomes. The pH-sensitive fluorescent indicator pyranine was also immobilized within the liposomes for the optical transduction of the enzymatic activity. Increasing amounts of pesticides leads to the decreased hydrolysis of the acetylcholine due to increased inhibition of enzymatic activity and thus to a decrease in the fluorescent signal of the pH indicator. The decrease of the liposome biosensors signal is proportional to the concentration of dichlorvos and paraoxon down to 10^{-10} M levels. Zou et al. (2019) constructed a novel bienzyme electrode [Cl/iron porphyrin (FePP)-modified MWCNTs/AChE/glassy carbon electrode], which included AChE and mimetic oxidase FePP for the detection of monocrotophos. The AChE electrodes generally have low stability and high over-potential; in this electrode, FePP is covalently bound to the AChE carrier via ionic liquid for increased electrode sensitivity and stability. Nanobiosensor was fabricated by using silica nanoparticle co-encapsulated organophosphate hydrolase with a pH-responsive fluorophore for paraoxon detection limit of 34 lM (Ramanathan et al. 2009).

17.5.2 Heavy Metals

Heavy metals may occur in the aquatic environment in the ionic form, or as organic/inorganic soluble complexes adsorbed on the sediment or incorporating into the biota. The ions are non-biodegradables pollutants and hence persist in the environment for long leading to a greater risk to human health and ecosystem. The major sources of heavy metal pollution are industrial effluents, mining and agricultural practices. The problems of heavy metals contamination in water bodies establish an urgent need for fast, sensitive and reliable methods for detection and treatment of these compounds.

Chitosan is a well-known chelating agent for heavy metals, and the presence of Zn(II) and Cu(II) can cause colloidal instability and loose aggregation (agglomeration) of gold nanoparticles. This phenomenon causes a rapid color change that is directly related to the heavy metal concentration. Pb(II) ions with a limit of 100 nm to 200 μM have been detected following an aggregation-dissociation protocol (Liu and Lu 2003) using the DNAzyme-directed assembly of gold nanoparticles cleaves in the presence of Pb(II) and results in a blue to red color change.

Cantilever nanobiosensors have emerged as a modern technology for the analysis and monitoring of heavy metals. Rigo et al. (2019a) developed cantilever nanobiosensors based on urease enzyme activity. Urease enzyme activity reduction proportional to the concentrations of different heavy metals can be used for quantitative analysis of these metals in solution. The cantilever was functionalized with chemical modification on the surface (70 nm of gold, 16-MHD) using self-assembled monolayers (SAM); the thiol group (-SH) located at one end of the 16-MHD adsorbs to the gold surface, while the carboxyl group present at the other end is available. To covalently immobilize the urease enzyme on this cantilever surface, the carboxyl groups of 16-MHD were coupled to the primary amine groups present in the structure of the enzyme with the EDC and NHS crosslinking agents. Then, NHS is added to increase the efficiency bound with the primary amine present in the enzyme by the carboxylic acid. The heavy metals present in the water solution bind to the active site groups of the urease enzyme by reacting with the sulfhydryl groups, causing a stress tension on the cantilever surface, resulting in a deflection that was measured by the voltage change of the cantilever nanobiosensor. Cantilever nanobiosensors with phosphatase alkaline were also developed by Rigo et al. (2019b) and applied to detect heavy metals (Pb, Ni, Cd, Zn, Co and Al) in water. The nanobiosensor surface was functionalized with the self-assembled monolayers (SAM) technique using N-hydroxysuccinimide (NHS), N-(3-dimethylaminopropyl)-N0-ethylcarbodiimide (EDC), 16-mercaptohexadecanoic acid and phosphatase alkaline enzyme. The sensing layer, with 80 nm of thickness, is deposited on the cantilever surface with a uniform morphology. The nanobiosensor has a fifteen days stability with a detection limit in the ppb range. Quantum dots have already been reported as potential nanobiosensors for the detection of heavy metal ions in the soil, water and food. Khachatryan and Khachatryan (2019) developed a nanosensor for the detection of Pb^{2+} and Cu^{2+} ions using novel starch-based nanocomposites embedded spherical ZnS quantum dots (10–20 nm) capped with L-cysteine. They were characterized using IR and UV spectra, photoluminescence and TEM/SEM photographs. The decreased emission intensity of the photoluminescent spectral bands was proportional to the concentrations of Pb^{2+} and Cu^{2+} ions in the solution.

17.5.3 Microbes

Several biosensors using different nanomaterials have been developed for the detection of various microbes in environmental samples. A nanobiosensor was developed for selective detection of Salmonella by using anti-*Salmonella* polyclonal antibodies immobilized by streptavidin–biotin binding or covalent binding to the quantum dot surface (QDs) (Kim et al. 2013). Anti-Salmonella antibody-coated beads were used to separate and concentrate cells from sample. The separated cells were labeled with antibody-conjugated QDs to form complexes. The fluorescent signals from complexes were measured with fluorometer.

A label-free immunoassay was developed by Bu et al. (2019) to detect pathogenic bacteria in water by introducing diversely positive charges functionalized AuNPs ((+) AuNPs). The (+) AuNPs can be loaded on negatively charged bacteria via electrostatic interaction, leading to a color labeling to target bacteria. Afterward, the (+) AuNPs-bacteria complex can be selectively recognized by monoclonal antibody (McAb) immobilized on test (T)-line. Under optimum conditions, the proposed LFS could be used for detecting *Salmonella enteritidis* (*S. enteritidis*) and *Escherichia coli* O157 (*E. coli* O157) with a superior sensitivity 10^3 CFU/mL and 10^4 CFU/mL,

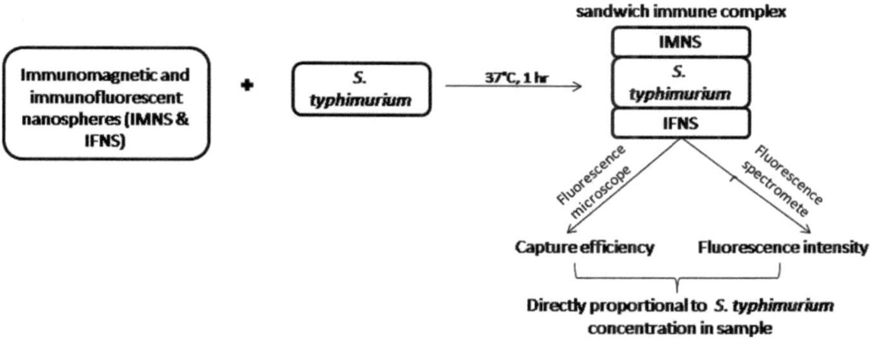

Fig. 17.3 Schematic representation of one-step detection of *S. typhimurium* using antibody-functionalized magnetic nanospheres (IMNS) and fluorescent nanospheres (IFNS) (Wen et al. 2013)

respectively. A nanoparticle-based electrochemical biosensor was developed for rapid and sensitive detection of plant pathogen DNA (Lau et al. 2017). The assay involved rapid isothermal amplification of target pathogen DNA sequences by recombinase polymerase amplification (RPA) followed by gold nanoparticle-based electrochemical assessment with differential pulse voltammetry (DPV).

Amine-functionalized MNPs have enabled rapid and cost-effective capture of bacterial pathogens in several matrices (Huang et al. 2010). Wen et al. (2013) demonstrated a one-step method for the sensitive detection of *S. typhimurium* using antibody-functionalized magnetic nanospheres (IMNS) and fluorescent nanospheres (IFNS) as shown in Fig. 17.3. The IMNS/IFNS were mixed and incubated at 37 °C for 60 min. with *S. typhimurium*-spiked solution to achieve a detection limit of 10 CFU/ mL^{-1} in a 1-h period with high specificity.

Wang et al. (2016) developed novel bioprobes based on biotin-labeled single-stranded DNA (ssDNA) aptamers conjugated with streptavidin-bound RuBPY-doped silica NPs for detection of *S. typhimurium.* Shams et al. (2019) developed ssDNA-GNRs biosensing system for the identification of both *Campylobacter jejuni* and *Campylobacter coli* species using specific *cadF* gene. The mechanism of detection was based on prodigious capability of gold nanorods (GNRs) to sense changes in the local environment and interparticle distance.

17.6 Challenges and Future Perspectives

Progress in the development of nanobiosensors is gaining momentum, which has revolutionized the environmental monitoring. The overall mechanisms have become time-efficient, cost and user-friendly. The inventions of different nanostructures like those of quantum dots, nanoparticles for enzyme immobilization and hybrid nanostructures with multiple functionalities have been significantly improving the transduction mechanism. Future argues very well for these dynamic, versatile and quick recognition systems considering their multidimensional potential. The use of nanomaterials has enabled the development of novel chemical-biological conjugate biosensors for multidimensional performance upgradation. Though nanobiosensors are the most dynamic, versatile and quick recognition systems for efficient environmental monitoring, their durability for long-term and safe applications is not always practicable. The harsh and hazardous conditions of environmental samples would easily destruct the composition and/or structure of the nanosized catalysts. The stability matters a lot in environmental applications as metal leaching from nanosized metal-based catalysts and/or newly formed intermediates can cause secondary pollution. Nanopollution risks the demolition of ecosystems and human health. Therefore, the sensing technology should be coherently designed and judiciously applied taking into account all the activity, selectivity and stability. The induced different processes within the nanomaterial and environmental sample interface could be specific to size, shape, composition, surface feature or hybridizations of the materials. The chemistry of environmental processes, bio-chemical engineering and materials science should be brought together for the practical environmental applications of theoretical findings to optimize nanobiosensors performance. However, there is a wide scope for fundamental understanding, identification and manipulation at the nano-/biointerfaces to optimize their performance and stability in real-world operating conditions, hence leading to the commercialization of these technologies.

17.7 Conclusion

Great efforts in the development of nanomaterials and the corresponding nanobiosensors have brought up enormous opportunities to conventional sensing technologies. The significant advantage is rapid results outcome because the approach to increase signal rather than the target analytes has revolutionized the paradigm of detection. This leads to the speediness in environmental monitoring and management, hence the betterment of ecosystems and public health.

Conflicts of Interest/Competing Interests:
The authors have no potential conflicts of interest.

References

Algarra M, Campos BB, Alonso B, Miranda MS, Martínez ÁM, Casado CM, Esteves da Silva JCG (2012) Thiolated DAB dendrimers and CdSe quantum dots nanocomposites for Cd(II) or Pb(II) sensing. Talanta 88:403–407

Aluri GS, Motayed A, Davydov AV, Oleshko VP, Bertness KA, Sanford NA, Rao MV (2011) Highly selective GaN-nanowire/TiO_2-nanocluster hybrid sensors for detection of benzene and related environment pollutants. Nanotechnology 22(29):295503. https://doi.org/10.1088/0957-4484/22/29/295503

Andrea MS, Silva LM, Melo AF (2011) Biosensor for environmental applications. In: Somerset V (ed) Environmental biosensors. IntechOpen, Rijeka, pp 3–16
Aragay G, Pino F, Merkoci A (2012) Nanomaterials for sensing and destroying pesticides. Chem Rev 112:5317–5338
Bae SW, Tan W, Hong J-I (2012) Fluorescent dye-doped silica nanoparticles: new tools for bioapplications. Chem Commun 48:2270–2282
Bănică F-G (2012a) Nanomaterial applications in optical transduction. Chemical sensors and biosensors. Wiley, Chichester, pp 454–472
Bănică F-G (2012b) What are chemical sensors? Chemical sensors and biosensors. Wiley, Chichester, pp 1–20
Bala R, Swami A, Tabujew I, Peneva K, Wangoo N, Sharma RK (2018) Ultra-sensitive detection of malathion using quantum dots-polymer based fluorescence aptasensor. Biosens Bioelectron 104:45–49. https://doi.org/10.1016/j.bios.2017.12.034
Bu T, Jia P, Liu J, Liu Y, Sun X, Zhang M, Tian Y, Zhang D, Wang J, Wang L (2019) Diversely positive-charged gold nanoparticles based biosensor: a label-free and sensitive tool for foodborne pathogen detection. Food Chem X:100052. https://doi.org/10.1016/j.fochx.2019.100052
Busetti F, Badoer S, Cuomo M, Rubino B, Traverse P (2005) Occurrence and removal of potentially toxic metals and heavy metals in the wastewater treatment plant of fusina (Venice, Italy). Ind Eng Chem Res 44(24):9264–9272
Chao MR, Chang YZ, Chen JL (2013) Hydrophilic ionic liquid-passivated CdTe quantum dots for mercury ion detection. Biosens Bioelectron 42:397–402
Chen H, Zuo X, Su S, Tang Z, Wu A, Song S, Zhang D, Fan C (2008) An electrochemical sensor for pesticide assays based on carbon nanotube-enhanced acetycholinesterase activity. Analyst 133(9):1182. https://doi.org/10.1039/b805334k
Chen Y, Xianyu Y, Wang Y et al (2015) One-step detection of pathogens and viruses: combining magnetic relaxation switching and magnetic separation. ACS Nano 9(3):3184–3191
Duan D, Fan K, Zhang D et al (2015) Nanozyme-strip for rapid local diagnosis of Ebola. Biosens Bioelectron 74:134–141
Duan G, Zhang Z, Zhang J, Zhou Y, Yu L, Yuan Q (2007) Evaluation of crude toxin and metabolite produced by *Helminthosporium gramineum* Rabenh for the control of rice sheath blight in paddy fields. Crop Prot 26(7):1036–1041
Gan TT, Zhang YJ, Zhao NJ, Xiao X, Yin GF, Yu SH, Wang HB, Duan JB, Shi CY, Liu WQ (2012) Hydrothermal synthetic mercaptopropionic acid stabled CdTe quantum dots as fluorescent probes for detection of Ag(+). Spectrochim Acta Part A Mol Biomol Spectrosc 99:62–68
Garcia-Aljaro C, Cella LN, Shirale DJ, Park M, Munoz FJ, Yates MV, Mulchandani A (2010) Carbon nanotubes-based chemiresistive biosensors for detection of microorganisms. Biosens Bioelectron 26:1437–1441
Gaster RS, Xu L, Han SJ et al (2011) Quantification of protein interactions and solution transport using high-density GMR sensor arrays. Nat Nanotechnol 6(5):314–320
Grieshaber D, MacKenzie R, Vörös J, Reimhult E (2008) Electrochemical biosensors—sensor principles and architectures. Sensors 8:1400–1458
Grossman HL, Myers WR, Vreeland VJ et al (2004) Detection of bacteria in suspension by using a superconducting quantum interference device. Proc Natl Acad Sci USA 101(1):129–134
Gui R, An X, Su H, Shen W, Chen Z, Wang X (2012) A near-infrared-emitting CdTe/CdS core/shell quantum dots-based OFF–ON fluorescence sensor for highly selective and sensitive detection of Cd^{2+}. Talanta 94:257–262
Hall DA, Gaster RS, Makinwa KAA, Wang SX, Murmann B (2013) A 256 pixel magnetoresistive biosensor microarray in 0.18 mm CMOS. IEEE J Solid-State Circuits 48(5):1290–1301
Herpoldt KL, Loynachan CN, Stevens MM (2017) 3.29 nanomaterials for biological sensing. Compr Biomater II, 635–656.
Hu T, Xu J, Ye Y, Han Y, Li X, Wang Z, Sun D, Zhou Y, Ni Z (2019) Visual detection of mixed organophosphorous pesticide using QD-AChE aerogel based microfluidic arrays sensor. Biosens Bioelectron 136:112–117. https://doi.org/10.1016/j.bios.2019.04.036

Huang YF, Wang YF, Yan XP (2010) Amine-functionalized magnetic nanoparticles for 854 rapid capture and removal of bacterial pathogens. Environ Sci Technol 44:7908–7913

Issadore D, Chung J, Shao H et al (2012) Ultrasensitive clinical enumeration of rare cells ex vivo using a micro-hall detector. Sci Transl Med 4(141):141ra92

Jia X, Li L, Yu J, Gao X, Yang X, Lu Z, Zhang X, Liu H (2018) Facile synthesis of BCNO quantum dots with applications for ion detection, chemosensor and fingerprint identification. Spectrochim Acta Part A Mol Biomol Spectrosc 203:214–221. https://doi.org/10.1016/j.saa.2018.05.099

Kaittanis C, Naser SA, Perez JM (2006) One-step, nanoparticle-mediated bacterial detection with magnetic relaxation. Nano Lett 7:380–383

Ke J, Li X, Shi Y, Zhao Q, Jiang X (2012) A facile and highly sensitive probe for Hg(II) based on metal-induced aggregation of ZnSe/ZnS quantum dots. Nanoscale 4:4996–5001

Kim G, Park SB, Moon JH, Lee S (2013) Detection of pathogenic Salmonella with nanobiosensors. Anal Methods 5(20):5717. https://doi.org/10.1039/c3ay41351a

Kim M, Kim MS, Kweon SH et al (2015) Simple and sensitive point-of-care bioassay system based on hierarchically structured enzyme-mimetic nanoparticles. Adv Healthcare Mater 4(9):1311–1316

Khachatryan G, Khachatryan K (2019) Starch based nanocomposites as sensors for heavy metals—detection of Cu2+ and Pb2+ ions. Int Agrophys 33(1):121–126. https://doi.org/10.31545/intagr/104414

Koets M, van der Wijk T, van Eemeren JTWM, van Amerongen A, Prins MWJ (2009) Rapid DNA multi-analyte immunoassay on a magneto-resistance biosensor. Biosensing Bioelectron 24:1893–1898

Koneswaran M, Narayanaswamy R (2009) l-Cysteine-capped ZnS quantum dots based fluorescence sensor for Cu^{2+} ion. Sens Actuators, B Chem 139:104–109

Kuswandi B, Mascini M (2005) Enzyme inhibition based biosensors for environmental monitoring. Curr Enzym Inhib 1:11–21

Kuswandi B, Fikriyah CI, Gani AA (2008) An optical fiber biosensor for chlorpyrifos using a single sol–gel film containing acetylcholinesterase and bromothymol blue. Talanta 74:613–622

Kuswandi B, Swandari NW (2007) Simple and sensitive flow injection optical fibre biosensor based on immobilised enzyme for monitoring of pesticides. Sensors Transducers 76:978–986

Kuswandi B (2018) Nanobiosensor approaches for pollutant monitoring. Environ Chem Lett 17:975–990. https://doi.org/10.1007/s10311-018-00853-x

Kvenvolden KA, Cooper CK (2003) Natural seepage of crude oil into the marine environment. Geo-Mar Lett 23(3–4):140–146

Lau H, Wu H, Wee E et al (2017) Specific and sensitive isothermal electrochemical biosensor for plant pathogen DNA detection with colloidal gold nanoparticles as probes. Sci Rep 7:38896. https://doi.org/10.1038/srep38896

Li C, Wei C (2017) DNA-templated silver nanocluster as a label-free fluorescent probe for the highly sensitive and selective detection of mercury ions. Sens Actuators B Chem 242:563–568. https://doi.org/10.1016/j.snb.2016.11.091

Liu J, Lu Y (2003) A colorimetric lead biosensor using DNAzyme-directed assembly of gold nanoparticles. J Am Chem Soc 125:6642–6643

Malik P, Katyal V, Malik V, Asatkar A, Inwati G, Mukherjee TK (2013) Nanobiosensors: concepts and variations. ISRN Nanomater 9.

Maruthupandi M, Thiruppathi D, Vasimalai N (2020) One minute synthesis of green fluorescent copper nanocluster: the preparation of smartphone aided paper-based kit for on-site monitoring of nanomolar level mercury and sulfide ions in environmental samples. J Hazard Mater 392. Article 122294. https://doi.org/10.1016/j.jhazmat.2020.122294

Min C, Shao H, Liong M, Yoon TJ, Weissleder R, Lee H (2012) Mechanism of magnetic relaxation switching sensing. ACS Nano 6(8):6821–6828

Nagatani N, Takeuchi A, Hossain MA, Yuhi T, Endo T, Kerman K et al (2007) Rapid and sensitive visual detection of residual pesticides in food using acetyl-cholinesterase-based disposable membrane chips. Food Control 18:914–920

Norouzi P, Pirali-Hamedani M, Ganjali MR, Faridbod F (2010) A novel acetylcholinesterase biosensor based on chitosan-gold nanoparticles film for determination of monocrotophos using FFT continuous cyclic voltammetry. Int J Electrochem Sci 5:1434–1446

Oliveira TMBF, Fátima Barroso M, Morais S, de Lima-Neto P, Correia AN, Oliveira MBPP, Delerue-Matos C (2013) Biosensor based on multi-walled carbon nanotubes paste electrode modified with laccase for pirimicarb pesticide quantification. Talanta 106:137–143. https://doi.org/10.1016/j.talanta.2012.12.017

Pedrosa VA, Paliwal S, Balasubramanian S, Nepal D, Davis V, Wild J, Ramanculov E, Simonian A (2010) Enhanced stability of enzyme organophosphate hydrolase interfaced on the carbon nanotubes. Colloids Surf B: Biointerfaces 77(1):69–74. https://doi.org/10.1016/j.colsurfb.2010.01.009

Perez JM, Simeone FJ, Saeki Y, Josephson L, Weissleder R (2003) Viral-induced self-assembly of magnetic nanoparticles allows the detection of viral particles in biological media. J Am Chem Soc 125:10192–10193

Prasad R, Bhattacharyya A, Nguyen QD (2017) Nanotechnology in sustainable agriculture: recent developments, challenges, and perspectives. Front Microbiol 8:1014–1014

Ramanathan M, Luckarift HR, Sarsenova A, Wild JR, Ramanculov ER, Olsen EV, Simonian AL (2009) Lysozyme-mediated formation of protein-silica nano-composites for biosensing applications. Colloids Surf B Biointerfaces 73:58–64. https://doi.org/10.1016/j.colsurfb.2009.04.024

Reddy LH, Arias JL, Nicolas J, Couvreur P (2012) Magnetic nanoparticles: design and characterization, toxicity and biocompatibility, pharmaceutical and biomedical applications. Chem Rev 112:5818–5878

Rigo AA, de Cezaro AM, Muenchen DK, Martinazzo J, Brezolin AN, Hoehne L, Steffens J, Steffens C (2019a) Cantilever nanobiosensor based on the enzyme urease for detection of heavy metals. Braz J Chem Eng 36(4):1429–1437. https://doi.org/10.1590/0104-6632.20190364s20190035

Rigo AA, de Cezaro AM, Muenchen DK, Martinazzo J, Manzoli A, Steffens J, Steffens C (2019b) Heavy metals detection in river water with cantilever nanobiosensor. J Environ Sci Health B. https://doi.org/10.1080/03601234.2019.1685318

Rivas GA, Rubianes MD, Rodriguez MC, Ferreyra NF, Luque GL, Pedano ML, Miscoria SA, Parado C (2007) Carbon nanotubes for electrochemical biosensing. Talanta 74(3):291–307. https://doi.org/10.1016/j.talanta.2007.10.013

Sadik OA, Aluoch AO, Zhou A (2009) Status of biomolecular recognition using electrochemical techniques. Biosens Bioelectron 24:2749–65. https://doi.org/10.1016/j.bios.2008.10.003

Sanvicens N, Pastells C, Pascual N, Marco MP (2009) Nanoparticle-based biosensors for detection of pathogenic bacteria. Trends Anal Chem 28:1243–1252

Shahbazi R, Salouti M, Amini B, Jalilvand A, Naderlou E, Amini A, Shams A (2018) Highly selective and sensitive detection of *Staphylococcus aureus* with gold nanoparticle-based core-shell nano biosensor. Mol Cell Probes 41:8–13

Shams S, Bakhshi B, Tohidi Moghadam T et al (2019) A sensitive gold-nanorods-based nanobiosensor for specific detection of *Campylobacter jejuni* and *Campylobacter coli*. J Nanobiotechnol 17:43. https://doi.org/10.1186/s12951-019-0476-0

Strömberg M, Gómez Z, de la Torre T, Nilsson M, Svedlindh P, Strømme M (2014) A magnetic nanobead-based bioassay provides sensitive detection of single- and biplex bacterial DNA using a portable AC susceptometer. Biotechnol J 9(1):137–145

Sung TW, Lo YL (2012) Highly sensitive and selective sensor based on silica coated CdSe/ZnS nanoparticles for Cu^{2+} ion detection. Sens Actuators B Chem 165:119–125

Swierczewska M, Lee S, Chen X (2011) The design and application of fluorophore-gold nanoparticle activatable probes. Phys Chem Chem Phys 13:9929–9941. https://doi.org/10.1039/c0cp02967j

Vikesland PJ, Wigginton KR (2010) Nanomaterial enabled biosensors for pathogen monitoring—a review. Environ Sci Technol 44:3656–3669

Viswanathan S, Radecka H, Radecki J (2009) Electrochemical biosensor for pesticides based on acetylcholinesterase immobilized on polyaniline deposited on vertically assembled carbon nanotubes wrapped with ssDNA. Biosens Bioelectron 24(9):2772–2777. https://doi.org/10.1016/j.bios.2009.01.044

Wang J, Kawde A, Mustafa M (2003) Carbon-nanotube-modified glassy carbon electrodes for amplified label-free electrochemical detection of DNA hybridization. Analyst 128:912–916. https://doi.org/10.1039/B303282E

Wang P, Zhao L, Shou H, Wang J, Zheng P, Jia K, Liu X (2016) Dual-emitting fluorescent chemosensor based on resonance energy transfer from poly(arylene ether nitrile) to gold nanoclusters for mercury detection. Sens Actuators B Chem 230:337–344. https://doi.org/10.1016/j.snb.2016.02.041

Wei H, Abtahi SH, Vikesland PJ (2015) LSPR-based colorimetric and SERS sensors for environmental analysis. Environ Sci 2(2):120–135

Wen CY, Hu J, Zhang ZL, Tian ZQ, Ou GP, Liao YL, Li Y, Xie M, Sun ZY, Pang DW (2013) One-step sensitive detection of *Salmonella typhimurium* by coupling 860 magnetic capture and fluorescence identification with functional nanospheres. Anal Chem 85:1223–1230

Xia X, Zhang J, Lu N et al (2015) Pd-Ir core-shell nanocubes: a type of highly efficient and versatile peroxidase mimic. ACS Nano 9(10):9994–10004

Yu C, Irudayaraj J (2007) Multiplex biosensor using gold nanorods. Anal Chem 79(2):572–579. https://doi.org/10.1021/ac061730d

Yunus IS, Harwin Kurniawan A, Adityawarman D, Indarto A (2012) Nanotechnologies in water and air pollution treatment. Environ Technol Rev 1(1):136–148. https://doi.org/10.1080/21622515.2012.733966

Zhang Y, Yang Y, Ma W, Guo J, Lin Y, Wang C (2013) Uniform magnetic core/ shell microspheres functionalized with Ni^{2+}—iminodiacetic acid for one step purification and immobilization of his-tagged enzymes. ACS Appl Mater Interfaces 5:2626–2633

Zou B, Chu Y, Xia J (2019) Monocrotophos detection with a bienzyme biosensor based on ionic-liquid-modified carbon nanotubes. Anal Bioanal Chem 411:2905–2914. https://doi.org/10.1007/s00216-019-01743-z

Chapter 18
Prospects and Challenges of Bio-Nanomaterials for Wastewater Treatment

Meenakshi Sati, Vishwanath Sharma, Anup Jyoti Goswami, Krishna Giri, and Gaurav Mishra

Abstract Water is a precious and limited resource on earth which play pivotal role in life, agriculture, industries and ecological processes. Toxic chemical substances released from industrial processes, agricultural activities, gray water and sewage are the major water pollutants. Due to increasing human population, industrial and agricultural operations, the magnitude of water pollution has been increased significantly. Release of untreated or partially treated effluent in the environment is the major source of water pollution. Pollution of drinking water sources and natural water bodies have negative effects on human health, living organisms and ecological systems. The existing physical and chemical methods of wastewater treatment are neither eco-friendly nor cost-effective, under such circumstances pollutants removal from contaminated water warrant alternative and eco-friendly technologies. Bioremediation is an alternative to these technologies where plants, microorganisms and organic materials are used for treatment and removal of toxic chemicals and metal ions from polluted water. With the emergence of nanoscience and its concept of thinking big and working small, nanotechnology-based treatment methods have proved to be the most effective and eco-friendly approach to combat water pollution. Development of nano-based bioremediation technologies gives rise to a novel, much rapid and efficient remediation technology termed as nanobioremediation. Nanobiremediation utilizes microorganisms and plants to synthesize bio-nanomaterials which have potential to clean wastewater generated from large-scale industrial processes. It is advantageous over other treatment methods because of pollutants removal without posing any toxic effects to the microorganisms and also enhances microbial activity in the contaminated environment. Although much research has been carried out on application of bioremediation for wastewater treatment, however, a little is known about bio-based

M. Sati
Forest Research Institute (Deemed to be) University, Dehradun 248 195, Uttarakhand, India

V. Sharma · A. J. Goswami · K. Giri (✉) · G. Mishra
Rain Forest Research Institute, Jorhat 785 001, Assam, India
e-mail: krishna.goswami87@gmail.com

K. Giri · G. Mishra
Indian Council of Forestry Research and Education, Dehradun 248 006, Uttarakhand, India

J. P. N. Rai and S. Saraswat (eds.), *Nano-biotechnology for Waste Water Treatment*, Water Science and Technology Library 111,
https://doi.org/10.1007/978-3-031-00812-2_18

nanomaterials in wastewater treatment. This chapter aims to comprehend the recent advances and application of bio-nanomaterials for wastewater treatment and scope of environmental pollution control.

Keywords Bio-nanomaterials · Water pollution · Nanobioremediation

18.1 Introduction

Water is an essential resource for survival and development of living beings on earth. Availability of clean and safe water can boost the economy of any country, but due to the rapid industrialization, population growth, extensive agricultural practices and urbanization have led to the spread of enormous amounts of pollutants in water which ultimately deteriorate the quality of water. Pollution is one of the major causes of the declining quality of the environment in today's world. Water pollution is the serious issue of global concern for living being health, threat to biodiversity and aquatic ecosystem. Wastewater contains many hazardous pollutants which originate from sewage, agricultural and industrial processes. Treatment of wastewater using physical, chemical and biological methods is essential to protect living organisms and the environment. Wastewater treatment is a process of separation of pollutants from polluted water using different processes before releasing it in the environment. Various techniques such as electrochemical, advanced oxidation processes and valorization have been used for wastewater treatment (Gupta and Shukla 2020). These techniques have higher operational cost and generate several byproducts of environmental concern. Use of efficient, eco-friendly and economically viable technologies can help to achieve the goal of environmental sustainability (Chauhan et al. 2020). The development of nanotechnology and nanoscience has opened new avenues for the remediation of water pollutants. Bioremediation of water pollutants using bionanoparticles is one of the eco-friendly and economically feasible approach. Nanotechnology is among the fastest growing areas of scientific research and technology development across the world. Nanoparticles have found applications in bioremediation of heavy metal pollution, solid waste, hydrocarbons and uranium contamination, groundwater and wastewater remediation (Chauhan et al. 2020). Nanoparticles have unique capabilities to remove toxic substances from wastewater and also provide a healthy substrate for microbial activity which accelerate the process of environment clean-up (Sherry et al. 2017).

Nanoparticles used for bioremediation are biologically synthesized from plant extracts, fungi and bacteria. Traditionally, nanoparticles produced by chemical and physical methods are very costly. Therefore, the need of a cost-effective pathways for biosynthesis of nanoparticles arose. Synthesis of nanoparticles by using plant extracts and microorganisms is a cost-effective and alternative method than the conventional methods. Nanoparticles are highly reactive particles because of high surface to volume ratio. Nanoparticles can be either used as activated nanoparticles or as a carrier, due to smaller size and high surface to volume ratio. Due to quantum effect nanomaterials require less activation energy to make the chemical reactions feasible and faster. Nanoparticles also exhibit surface plasmon resonance which is helpful for the detection of toxic material (Sherry et al. 2017; Rizwan et al. 2014). Small size of nanoparticles makes them most appropriate purification agents. Biologically synthesized iron nanoparticles have been used in remediation of wastewater due to their redox potential, magnetic susceptibility and non-toxic nature (Bolade et al. 2020). The different types of bio-nanomaterials such as zinc, iron, silver, gold, copper, etc. synthesized from plants, animals and microbes have been used at the nanoscale for clean-up technology as well as wastewater treatment.

18.2 Nanobioremediation

Integration of bio-based nanomaterials for removal of pollutants from wastewater is known as nanobioremediation. The rapid urbanization, industrialization and modern agricultural practices (excess use of fertilizers and pesticides) cause environmental pollution and adverse effect the living organisms. The remediation of wastewater pollutants using nanoparticles is more efficient than other methods due to smaller size, high surface area to volume ratio and higher activity of nanoparticles (Baruah et al. 2019). Green synthesis of nanomaterials from microorganisms and plant extracts has proved a path toward the eco-friendly approach for remediation of pollutants. Nanobioremediation approach includes a wider range of potential applications with low cost and minimum negative impacts on the environment (Rizwan et al. 2014) for treatment of groundwater and wastewater (Yogalakshmi et al. 2020), heavy metals, hydrocarbons (Gisi et al. 2017) and polluted soil sediments (Bharagava et al. 2020).

Presently, nanomaterials have been used for removal of toxic pollutants from contaminated sites in the environment. The basic principle of nanobioremediation is defined as the degradation of organic wastes using nanocatalysts as a medium which allows them to penetrate deep inside the contaminants and facilitate microbes to act on such substances without affecting the surrounding (Yadav et al. 2017). Nanomaterials bring the pollutants (complex organic compounds) to a level where the pollutants can be degraded easily into environmentally innocuous by products (Pandey 2018). In this process, a proper interaction between nanoparticles and microorganism is essential for nanobioremediation of toxic pollutants. Tan et al. (2018) reported that interactions among nanomaterials, microbes and contaminants depend on size and shape of nanomaterials, surface coating, chemical nature of the nanomaterials, contaminant, media, pH, reaction temperature and type of organism used in synthesis. Stability of nanomaterials is major issue for synthesis of nanoparticles. Microorganisms have the potential to reduce the size of metal ions at nanoscale and produce nanoparticles/materials because of extracellular enzymes. Bacteria have special affinity for metals and this unique metal binding property makes them useful for nanobioremediation. Apart from bacteria, yeast and fungi are also being used for biosynthesis of nanoparticles. The purpose of using fungal cells is to harness the fungal proteins for large amount of nanoparticle synthesis. Biosynthesis of nanoparticles is generally done by the following bottom-up approach which is the result of redox reaction. The phyto-chemicals with antioxidants or microbial enzymes are usually responsible for reduction of metal compounds into their respective nanoparticles size.

18.3 Green Synthesis of Nanoparticles

Green synthesis of bio-nanomaterials involves plants (Philip 2010) and microorganisms (Ahmad et al. 2003; Roh et al. 2001; Mukherjee et al. 2001) which is an eco-friendly and non-toxic way of nanosynthesis with a wide range of shapes, size, compositions and physicochemical properties (Mohanpuria et al. 2008). Because of its cost-effectiveness, eco-friendly nature, controlled toxicity and rapid reaction speed, etc. green synthesis of nanomaterials is the most advantageous approach over other conventional methods. For large quantities, plants mediated nanoparticle synthesis is a more valuable and straightforward approach in comparison with the microbial mediated synthesis as the employment of microorganisms for nanoparticle synthesis requires complex procedures like isolation, purification and handling of cultures (Iravani 2011; Thakkar et al. 2010). However, the faster growth rate makes microorganisms more advantageous over other biological entities.

Synthesis of nanoparticles from plant resources required green reducing agents (Salehi et al. 2016; Nikalje 2015) which are obtained from phytochemical extracts (Boisselier and Astruc 2009). Phytoextracts from plants like *Sinapis arvensis*, *Abelmoschus esculentus*, *Aloe vera*, *Punica granatum*, *Cinnamomum camphora* and *Melia azedarach* can be utilized for the biosynthesis of Silver, Gold, Iron, Copper, Zinc and Palladium nanoparticles (Lam et al. 2018; Chaturvedi and Verma 2015; Kumar et al. 2011; Dashora and Sharma 2018; Laokul and Maensiri 2009). Plant extracts possess a combination of biomolecules including; amino acids, proteins, enzymes, alkaloids, polysaccharides, saponins, tannins, terpenoids, phenolics, vitamins and flavonoids (Marshall et al. 2007; Castro et al. 2011) that can be incorporated during nanoparticle synthesis (Table 18.1). These bioactive polyphenols have been extracted from leaves, stems, roots, flowers, fruits, fruit peels, seeds, gums and waste of vegetation (Yew et al. 2020). Extracellular enzymes secreted by microorganisms can be utilized for microbe mediated bionanomaterial synthesis (Tripathi et al. 2015; Calderon and Fullana, 2015). The bottom-up approach of nanobiomaterial synthesis from various bioresources is given in Fig. 18.1.

Silver nanoparticles can be biosynthesized from the phytoextracts of various plants like *Sinapis arvensis*, *Lantana camara*, *Trigonella foenumgraecum*, *Artemisia nilagirica*, *Nerium oleander* (Lam et al. 2018), etc. while gold nanoparticles are synthesized from the plant extracts of *Abelmoschus esculentus*, *Angelica*, *hypericum*, *Eucalyptus* and *Mentha*, (Chaturvedi et al. 2015). Copper nanoparticles are synthesized from leaf extracts of *Ocimum tenuiflorum*, *Nerium oleander*, *Ricinus communis*, etc. Zinc nanoparticles and Palladium nanoparticles have been biologically synthesized using plant extracts from *Camellia sinensis* and *Coffea arabica*, *Cinnamomum camphora*, *Melia azedarach*, *Delonix regia and Evolvulus alsinoides* (Pandey 2018).

Table 18.1 Green synthesis of nanoparticles using plant extract for wastewater treatment

Nanoparticles	Plant extract	Application of nanoparticles	References
Nano-zerovalent copper	Test a extract of *Anacardium occidentale*	Removal of uranium from wastewater	Chandra and Khan (2020)
Magnetite NPs	Crude latex of *Jatropha curcas* and leaf extract of *Cinnamomum tamala*	Removal of methylene blue dye, Cu(II) and Co(II) from aqueous solution	Das et al. (2020)
Zinc oxide NPs	Aloe vera and Cassava starch	Removal of copper ion	Primo et al. (2020)
Nano-zerovalent copper	Aqueous fruit extract of ripened *Duranta erecta*	Reduction of azo dyes (congo red and methyl orange)	Ismail et al. (2019)
Copper nanoparticles	*Eichhornia crassipes* (water hyacinth) extract	Detection of hazardous H_2O_2	Roy et al. (2019)
Copper nanoparticles	Leaf extract of *Camelia sinensis*	degradation of dyes	Ahmed et al. (2019)
Copper nanoparticles	Extract of *Cynomorium coccineum*	Adsorption of methylene blue dye	Sebeia et al. (2019)
Copper nanoparticles	Leaf extract of *Nerium oleander*	Interaction with organic dyes	Sebeia et al. (2019)
TiO_2 Nanoparticles	*Jatropha curcas* leaf extract	Photocatalytic treatment of tannery wastewater	Goutam et al. (2018)
Silver nanoparticles	Leaf extract of *Ficus Benjamina*	Removal of Cd(II) from contaminated solution	Al-Qahtani (2017)
Silver nanoparticles	*Piliostigma thonningii* aqueous leaf extract	Heavy metal removal activity in laboratory simulated wastewater	Shittu and Ihebunna (2017)
Fe Nanoparticles	Green tea (*Camellia sinensis*) and pomegranate (*Punica granatum*) leaf extract	Organic carbon removal from textile wastewater	Ozkan et al. (2017)
Au Nanoparticles	*Lagerstroemia speciosa* plant leaf extract	Photocatalytic reduction of organic pollutants	Choudhary et al. (2017)
Silver nanoparticles (AgNPs)	*Penciillium Citreonigum* Dierck and *Scopulaniopsos brumptii Salvanet-Duval*	AgNPs showed an excellent antibacterial property on gram-positive and gram-negative bacterial strains	Moustafa (2017)

(continued)

Similarly bacteria, yeast, fungi, algae and actinomycetes are used for the green synthesis of different nanoparticles. Microorganisms have the potential to reduce size of metal ions which lead to the synthesis of nanoscale materials. Pandey (2018) reported that microorganisms secrete extracellular enzymes which are being used for the synthesis of relatively pure nanoparticles while Koul et al. (2018) reported that

Table 18.1 (continued)

Nanoparticles	Plant extract	Application of nanoparticles	References
Silver nanoparticles	Testa extract of *Anacardium occidentale*	Reductive degradation of carcinogenic azo dyes like congo red and methyl orange	Edison et al. (2016)
Silver nanoparticles	*Ocimum sanctum* and *Artemisia annua* leaf extract	Reduction in pH, odor, color, dissolved oxygen (DO) and biochemical oxygen demand (BOD)	Sardar and Khatoon (2016)
Iron oxide NPs	Tangerine peel extract	Removal of cadmium ions from contaminated solution	Ehrampoush et al. (2015)

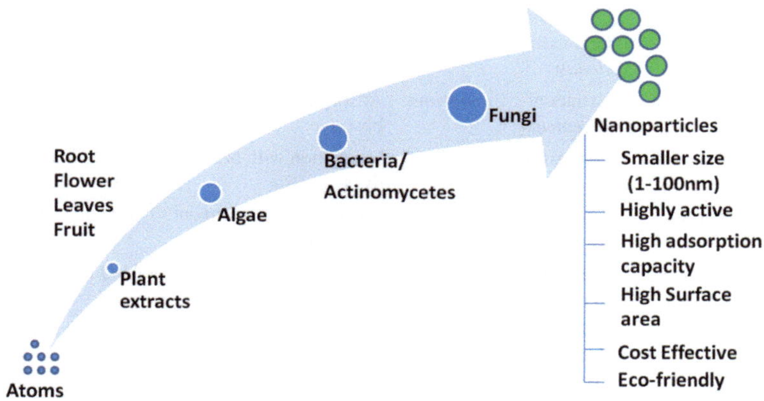

Fig. 18.1 Bottom-up approach for green synthesis of nanoparticles

bacteria have been most commonly used for the production of iron nanoparticles. In a study by Bharde et al. (2005) *Actinobacteria* sp. was used to produce spherical iron nanoparticles under aerobic environment. Greigite (Fe_3S_4) and maghemite ($\gamma{-}Fe_2O_3$) nanoparticles were also synthesized using the same bacterial spp. Thermophilic bacterium *Thermoanaerobacter* has also been used for the production of magnetite nanoparticles. Sahayaraj (2012) reported that microbes belonging to the *Streptomyces, Verticillium, Lactobacillus, Fusarium, Aspergillus, Pseudomonas, Corynebacterium, Aeromonas, Bacillus, Desulfovibrio, Plectonema* and *Rhodopseudomonas* genus have been used for the green synthesis of nanoparticles. Apart from bacteria, many fungi are also being used for the synthesis of nanoparticles and because of the secretion of a large amount of protein they proved to be better at producing a larger amount of nanoparticles as compared to bacteria (Raliya et al. 2013; Gurunathan et al. 2015; Yadav et al. 2018).

18.4 Nanoparticles for Water Disinfection and Purification

Nanotechnology has potential applications in wastewater treatment (Athirah et al. 2019). Noman et al. (2019) observed inactivation of antibiotic-resistant *Escherichia coli* (Gram-negative) and *Staphylococcus aureus* (Gram-positive) bacterial presence in graywater using Zn/Cu nanoparticles biosynthesized from a novel fungal strain. *Aspergillus iizukae* EAN605. Similarly, Ag nanoparticles biosynthesized using *Penciillium citreonigum Dierck* and *Scopulaiopsos brumptii Salvanet* have been found effective against *E. coli, Staphylococcus sps.* and *Pseudomonas sps.* removal from wastewater (Moustafa 2017) with complete decontamination at 676.0 mg/L NP concentration and 120 minute contact time. The nano-silver coated foam has been reported to remove both the gram-positive and gram-negative bacteria from contaminated water. Several researchers have synthesized Ag NPs from bacteria (Saifuddin et al. 2009), yeast (Pimprikar et al. 2009), fungi (Rai et al. 2016), plants and plant extracts (Anandalakshmi et al. 2016) for water disinfection (Table 18.2).

18.5 Nanoparticles as Sensors and Detectors for Water Pollutants

Organic and inorganic contaminants present in wastewater, like heavy metals, aromatic compounds, organophosphates, toxins and dyes, which are highly toxic and carcinogenic needs prior detection due to the toxicity posed by these pollutants even at trace levels. Detection of these pollutants with simple and sensitive metal ion sensors needs further advancement as these are non-biodegradable and can accumulate in the food chain, which poses a severe threat to the environment and human health. In general, a simple colorimetric sensor is preferable, and it can be made more advanced by the application of green chemistry which uses biogenic synthetic protocols based on non-toxic, multifunctional reactants derived from natural sources. For example, biomolecules, unicellular microorganism and higher plants, for a highly sensitive and selective detection of pollutants (Joshi and Kumar, 2018). In a study, fruit extracts of *Citrus limon* and *Citrus limetta* were used for biosynthesis of silver (Ag) and gold (Au) nanoparticles and used for Hg+ detection at pH range from 3.2 to 8.5. (Ravi et al. 2013). Ha et al. (2014) synthesized Au NPs from *Xanthoceras Sorbifolia* for sensing of Cr^{3+} ion with a detection limit of 3 μM. Plant extract of *Allium sativum* was used to synthesize Ag NPs for the detection of polyaromatic hydrocarbon (PAH) such as phenanthrene, pyrene and anthracene (Abbasi et al. 2014). For the detection of NH_3 (1 ppb) biosynthesized Au NPs from *Cyamopsistetrago Aloba* were used (Pandey et al. 2013). Biosynthesized Ag NPs from pods of *Phaseolus vulgaris* was used in photodegradation of mordant black and Congo red (Sunkar et al. 2013). All these tested bionanosensors provide a promising approach for the selective and sensitive detection of various pollutants in the contaminated water.

Table 18.2 Use of nanomaterials in wastewater treatment

Nanomaterials	Water pollutants	Organism and biological systems used	Effect	References
Microorganism-immobilized nanocellulose composites [Bacteria-decorated nanocellulose (BDN)]	Diuron herbicide	*Arthrobacter globiformis* D47	>90% pollutant removal in 48 hours	Liu et al. (2018)
Unzipped carbon nanotube (CNT), single-walled CNT and multi-walled CNT	Organophosphates and heavy metals	Enzyme organophosphate hydrolase	22% removal of heavy metals	Fosso-Kankeu et al. (2014)
Iron oxide nanopowder (nano-Fenton process) (<50 nm)	Azo dye C.I. Direct Yellow 86	–	81–99% decolorization of wastewater	Kos et al. (2014)
Nanoscale zerovalent iron (nZVI) and palladized nZVI	Polybrominated diphenyl ethers (PBDEs)	*Sphingomonas* sp.	67% of deca-BDE was transformed to the lower BDEs	Kim et al. (2012)
Glutathione (GSH) or cysteine-capped ZnS nanocrystals (3–5 nm)	p-nitrophenol (pNP) and Acid Orange 7	Enzymatic degradation by bacteria	80% of the AO7 was degraded within 15 minutes	Torres-Martínez et al. (2001)

18.6 Recent Advances in Bio-Nanomaterial and Wastewater Treatment

Nanotechnology offers some of the promising techniques for wastewater treatment which includes photocatalysis, nanofiltration and nanosorbents (Bora and Dutta 2014). Nanocatalyst, nanoadsorbents and nanomembranes are mostly being used for wastewater treatment. A variety of nanoparticles have been successfully reported for water and wastewater treatment such as zerovalent metal nanoparticles (Ag nanoparticles, Fe nanoparticles, Zn nanoparticles), metal oxide nanoparticles (TiO_2 nanoparticles, iron oxide nanoparticles, ZnO nanoparticles), carbon nanotubes and nanocomposites (Mueller and Nowack, 2009). Goutam et al. (2018) reported that photocatalytic treatment/degradation of tannery wastewater using green synthesized TiO_2 nanoparticles revealed promising results. Nano-based photocatalysis employed by Devatha et al. (2016), Wang et al. (2014), Ghaly et al. (2011) and Bordes et al. (2015) has also reported the potential of nanoscience in wastewater treatment. Nanosorbents have been widely used for the removal of microbes, organic dyes and heavy metal pollutants from water and wastewater. Studies have been conducted on the use of nanoparticles as nanoadsorbent in the wastewater treatment for iron oxide as nanoadsorbent for the removal of various pollutants from wastewater, Fe-La composite oxide for the removal of As (III) from wastewater (Nassar 2012; Zhang et al. 2014). For the treatment of textile dye, Banerjee et al. (2014) synthesized silver nanocomposites using leaf extract of *Ocimum tenuiflorum* (Black Tulsi). Cadmium removal from wastewater by iron oxide nanoparticles synthesized using peel extract of tangerine was reported by Ehrampoush et al. (2015). Rosales et al. (2017) reported the degradation of textile dye by zerovalent iron nanoparticles synthesized from two different extracts, green tea (*Camellia sinensis*) and Rooibos (*Aspalathus linearis*). Similarly, *Jatropha curcas* leaf extract was successfully utilized for the synthesis of titanium dioxide (TiO_2) nanoparticles for photocatalysis of treated wood waste (Goutam et al. 2018). Biosynthesized iron oxide nanoparticles using *Padina pavonica Thivy* and *Sargassum acinarium* were successfully entrapped in calcium alginates beads and used for Pb adsorption (El-Kassas et al. 2016). Many researchers across the globe are working on eco-friendly techniques such as green synthesis of nanoparticles for proving solution to the problem of wastewater.

High reactivity, larger surface area, better disposal capability, eco-friendly and non-toxic nature are some of the advantages of bio-nanomaterials for wastewater treatment. These properties can be utilized for adsorption and reduction of various harmful pollutants in the water. Examples of bio-based nanoparticles used for water remediation include Fe NPs, Ag NPs, TiO_2 NPs, Silver nanocomposite hydrogel (SNC), ZnO, CuO, Co_3O_4, Nickel oxide and Cr_2O_3, Nano-zerovalent iron (nZVI), etc. (Goutam et al. 2018; Ozkan et al. 2017; Shittu and Ihebunna 2017; Devi et al. 2016; Shanker et al. 2016; Hoag et al. 2009). In a study, iron nanoparticles (FeNPs) were synthesized via a green method using loquat *(Eriobotrya japonica)* leaves aqueous extract as a renewable reducing agent which showed more than 90% Cr (VI) adsorption at a wide range of initial Cr (VI) concentrations (50–500 mg/L) (Onal et al. 2019). Mehrotra et al. (2017) employed protein capped zerovalent iron NPs synthesized using yeast, for complete degradation of organophosphorus insecticide (Dichlorvos). At an optimum experimental condition with 2000 mg /L nanoparticles dosage, 1000 μl H_2O_2, 99.9% degradation was observed within 60 min of reaction. *E. coli*-mediated biogenic synthesis of gold NPs served as an efficient, eco-friendly and cost-effective bionanocomposite for the complete degradation of 4-nitrophenol.112 R. In a study, *Tilia* leaf extract was used for the synthesis of copper nanoadsorbent for remediation of aquatic systems contaminated with Ibuprofen, Naproxen and Diclofenac (Dalal et al. 2019).

18.7 Advantages and Limitations of Bio-Nanomaterials

Though nanoparticles have shown potential in treating contaminated sites with higher effectiveness. However, there are several problems associated with the use of nanomaterials for environmental clean-up. The major limitations are loss of reactivity with time, transportation and their effect on microorganisms (Mohammed et al. 2017; Lunge et al. 2014). Studies suggest that Iron nanoparticles exhibit, a loss in reactivity level after a certain period of time, a blocking effect in the soil by clogging the pores of soil and restricting the passage of fluids. The use of stabilizers, such as lactate, enhances the mobility of iron nanoparticle in turn facilitating their improved transport in soil (Warner et al. 2011). Various studies under controlled conditions have been executed on the consequence of nanoparticles on microorganisms and the results thus found were conflicting (Bai et al. 2011). Some of the studies have shown inhibitory effects on microorganisms like *Staphylococcus aureus* and *Escherichia coli* (Simeonidis et al. 2015; Qu et al. 2013 and Demming 2011). Few other studies have shown the stimulating effect of nanoparticles as electron donors on microorganisms such as bacteria and methanogens (Dong et al. 2015). A few nanoparticles with certain concentrations are hazardous, toxic and can thus damage human health and the environment (Cornelis et al. 2014). Carbon nanotubes in the form of asbestos can cause lung cancer when inhaled in insufficient quantities through treated water (Buzea et al. 2007).

Moreover, the synthesis of nanoparticles utilizes conventional methods which include physical and chemical methods. The physical methods of nanoparticle synthesis comprise high energy irradiation, lithography or laser ablation, while chemical synthesis utilizes photochemical reduction, electrochemistry and chemical reduction methods. These conventional methods require huge amount of energy for the successful operation of physical methods whereas synthesis of nanoparticles through chemical methods poses detrimental effects on the environment due to the use of toxic chemicals in the synthesis process and the formation of lethal byproducts. However, with the application of biological methods or green synthesis approach for the synthesis of nanoparticles, these limitations can be effectively overcome.

Biologically synthesized nanoparticles have been lucratively applied in the treatment of wastewater with minimal effluent discharge. Bionanoremediation is a sustainable approach to nanoremediation as the nanoparticles are non-toxic in nature, eco-friendly and can be synthesized by a simple green chemistry approach using plants and microbes. Thus, the eco-friendly approach of biologically driven nanotechnology restricts the risk of producing toxic intermediates and end-products. Undoubtedly biologically synthesized nanomaterials have proved more beneficial for human health and environment prospects; however, it is also very important to consider the challenges associated with them (Ali et al. 2017).

Long-term exposure of bio-nanomaterials should be evaluated before application. In spite of the basic amalgamation strategy, the use of solution extract volume, temperature, solvent type, pH, strength of forerunner and functional groups from plant metabolites ought to be optimized to avoid any alteration in the magnetic behavior and saturation magnetization value of the bio-nanomaterials. To maintain the stability of bio-nanomaterials for efficient treatment of wastewater, efforts are required to formulate the nanoparticle morphology and saturation magnetization value. For selective and wide range of pollutant removal, bio-nanomaterials should be synthesized with multiple functional groups by manipulating the plant metabolites and synthesis method. In addition, the properties of the synthesized nanoparticles must be studied systematically in comparison with their chemical counterparts. For commercial application cost–benefit analysis should be performed as there is no information on this aspect. Until date, the biological synthesis of metallic nanoparticles has been mostly carried out at laboratory scale. The industrial scale optimization is required for large-scale production.

18.8 Future scope of Nano-Biomaterials for Water Pollution Control

Distribution of water on the Earth's surface is enormously irregular with only 3% fresh water available in rivers, glaciers, groundwater and in the atmosphere while the remaining 97% is found in the ocean. Providing clean and affordable water for human consumption is a huge challenge in the present scenario when the population is increasing at a faster rate. Besides, water supply is worsened by developmental activities, global climate change and water quality deterioration due to environmental pollution. Currently, many conventional and non-conventional wastewater treatment technologies are available to remove pollutants from various types of wastewaters, but high operating cost, high energy requirement, reduced efficiency and non-eco-friendly nature limit the extensive use of these methods (Bora and Dutta 2014). On contrary, nanotechnology offers a great potential for the development of next-generation water and wastewater treatment technologies and could replace the conventional methods (Goutam et al. 2018). Nanoparticles (NPs) are increasingly applied for wastewater treatment and purification due to their high reactivity and degree of functionalization, large surface area, size-dependent properties, etc. Nanotechnology finds major application in the water pollution management like detection of water pollutants, wastewater treatment and purification.

18.9 Conclusion

Water is a distinctive feature that makes the earth a habitable and superior place than other planets. The continuous decrease in freshwater sources due to never-ending water demand has led the water insufficient to fulfill the human water requirements like drinking, sanitation, developmental activities, etc. Moreover, the contamination of water bodies with pollutants, have made the situation worse, under such circumstances novel ways of water treatment needs to be explored in order to recycle water and to eliminate high load of contaminants from water reservoirs. Conventional water treatment methods include use of chemicals which generate by products and require further disposal mechanisms. However, the biological methods utilizing microorganisms and plants for wastewater treatment are eco-friendly and non-toxic in nature but have a slow operating rate. Nanotechnology is advantageous over chemical methods of treatment because of its faster reaction rate. Nanomaterials possess unique properties like high reactivity and sensitivity, accumulation on substrates to form films, high surface to volume ratio, high adsorption rate, etc. which makes them much more suitable for wastewater treatment. Due to these properties, nanomaterials proved effective against various organic and inorganic pollutants, heavy metals and harmful microorganisms present in contaminated water. Moreover, the conventional methods for nanoparticles synthesis have certain problems like higher energy requirements for the successful operation of physical methods whereas synthesis of nanoparticles through chemical methods pose detrimental effects on the environment due to the use of toxic chemicals in the synthesis process and formation of lethal byproducts. Existing wastewater treatment technologies are capable of removing organic and inorganic pollutants from wastewater, but these methods are energy-intensive and uneconomical due to their inability to fully decontaminate wastewater as well as to be retained again for further remediation. Biologically synthesized nanoparticles have been lucratively applied in the treatment of wastewater with minimal effluent discharge. These nanoparticles act as an adsorbent for the remediation of water pollutants evidently due to their physicochemical properties, nanosize, controlled growth and surface alteration. The carbohydrates, proteins and enzymes present in these biological entities act as a surfactant and capping agents which reduces the use of harmful chemical surfactants. Nanobioremediation is a sustainable approach to nanoremediation as the nanoparticles are non-toxic in nature, eco-friendly and can be synthesized by a simple green chemistry approach using plants and microbes. The bio-based nanoparticles offer a promising alternative to conventional nanoparticles in wastewater treatment, as these are biocompatible, economical, eco-friendly and energy-intensive. Thus, Nanobioremediation or bio-based synthesized nanoparticles can become an essential component of wastewater treatment systems in the future. Further research is necessary on this direction to commercialize the use of bio-based nanoparticles for pollutant removal from wastewater.

References

Abbasi M, Saeed F, Rafique U (2014) Preparation of silver nanoparticles from synthetic and natural sources. Mater Sci Eng 60:1–9

Ahmad A, Senapati S, Khan MI, Kumar R, Sastry M (2003) Extracellular biosynthesis of monodisperse gold nanoparticles by a novel extremophilic actinomycete. *Thermomonospora Sp.* Langmuir 19(8):3550–3553

Ahmed A, Usman M, Liu QY, Shen YQ, Yu B, Cong HL (2019) Plant mediated synthesis of copper nanoparticles by using Camelia sinensis leaves extract and their applications in dye degradation. Ferroelectrics 549(1):61–69, 13

Ali I, Peng C, Naz I, Khan ZM, Sultan M, Islam T, Abbasi IA (2017) Phytogenic magnetic nanoparticles for waste water treatment: a review. RSC Adv 7:40158–40178

Al-Qahtani KM (2017) Cadmium removal from aqueous solution by green synthesis zero valent silver nanoparticles with Benjamina leaves extract. Egypt J Aquat Res 43:269–274. https://doi.org/10.1016/j.ejar.2017.10.003

Anandalakshmi K, Venugobal J, Ramasamy V (2016) Characterization of silver nanoparticles by green synthesis method using *Pedalium murex* leaf extract and their antibacterial activity. Appl Nanosci 6:399–408

Athirah A, Al-Gheethi AAS, Noman EA, Mohamed RMSR, Kassim AHM (2019) Centralised and decentralised transport systems for greywater and the application of nanotechnology for treatment processes. Management of greywater in developing countries. Springer, Cham, pp 227–244

Bai H, Liu Z, Sun DD (2011) Highly water soluble and recovered dextran coated Fe3O4 magnetic nanoparticles for brackish water desalination. Sep Purif Technol 81(3):392–399

Banerjee P, Sau S, Das P, Mukhopadhyay A (2014) Green synthesis of silver - nanocomposite for treatment of textile dye. Nanosci Technol 1(2):1–6

Baruah A, Chaudhary V, Malik R, Tomer VK (2019) Nanotechnology based solutions for wastewater treatment. Nanotechnol. Water Wastewater Treat 337–368. https://doi.org/10.1016/b978-0-12-813902-8.00017-4

Bharagava RN, Saxena G, Mulla SI (2020) Introduction to industrial wastes containing organic and inorganic pollutants and bioremediation approaches for environmental management. In: Bioremediation of Industrial Waste for Environmental Safety. Springer, Singapore.

Bharde A, Wani A, Shouche Y, Joy PA, Prasad BLV, Sastry M (2005) Bacterial aerobic synthesis of nanocrystalline magnetite. J Am Chem Soc 127:9326–9327

Boisselier E, Astruc D (2009) Gold nanoparticles in nanomedicine: preparations, imaging, diagnostics, therapies and toxicity. Chem Soc Rev 38:1759–1782

Bolade OP, Williams AB, Benson NU (2020) Green synthesis of iron-based nanomaterials for environmental remediation: A review. Environ. Nanotechnol Monit Manag 13:100279

Bora T, Dutta J (2014) Applications of nanotechnology in wastewater treatment-a review. J Nanosci Nanotechnol 14:613–626

Bordes MC, Vicent M, Moreno R, Garcia-Montano J, Serra A, Sanchez E (2015) Application of plasma-sprayed TiO_2 coatings for industrial (tannery) wastewater treatment. Ceram Int 41:14468–14474

Buzea C, Pacheco II, Robbie K (2007) Nanomaterials and nanoparticles: Sources and toxicity. Biointerphases 2(4):17–71

Calderon B, Fullana A (2015) Heavy metal release due to aging effect during zero valent iron nanoparticles remediation. Water Res 83:1–9

Castro L, Blazquez ML, Munoz JA, Gonzalez F, Garcıa-Balboa C, Ballester A (2011) Biosynthesis of gold nanowires using sugar beet pulp. Process Biochem 46:1076–1082
Chandra C, Khan F (2020) Nano-scale zerovalent copper: green synthe-sis, characterization and efficient removal of uranium. J Radio-Anal Nucl Chem. 324:589–597. https://doi.org/10.1007/s10967-020-07080-1
Chaturvedi V, Verma P (2015) Fabrication of silver nanoparticles from leaf extract of *Butea monosperma* (Flame of Forest) and their inhibitory effect on bloom-forming cyanobacteria. Bioresour Bioprocess 2:18
Chauhan R, Yadav Hari OS, Sehrawat N (2020) Nanobioremediation: A new and a versatile tool for sustainable environmental clean-up-Overview. J of Mat and Env Sci 11(4):564–573
Choudhary BC, Paul D, Gupta T, Tetgure SR, Garole VJ, Borse AU, Garole DJ (2017) Photocatalytic reduction of organic pollutant under visible light by green route synthesized gold nanoparticles. J Environ Sci 55:236–246
Cornelis G, Hund-Rinke K, Kuhlbusch T, Van den Brink N, Nickel C (2014) Fate and bioavailability of engineered nanoparticles in soils: A review. Crit Rev Environ Sci Technol 44(24):2720–2764
Dalal ZH, Reda H, Mostafa F, Al-H, (2019) Green-synthesized copper nano-adsorbent for the removal of pharmaceutical pollutants from real wastewater samples. Heliyon 5(8):e02339
Das C, Sen S, Singh T, Ghosh T, Paul SS, Kim TW, Jeon S, Maiti DK, Im J, Biswas G (2020) Green synthesis, characterization and application of natural product coated magnetite nanoparticles for wastewater treatment. Nanomaterials 10(8):1615. https://doi.org/10.3390/nano10081615
Dashora A, Sharma K (2018) Green synthesis of nanoparticles and their applications. Adv Sc Eng and Med 10(6):523–541
De Gisi S, Minetto D, Lofrano G, Libralato G, Conte B, Todaro F, Notarnicola M (2017) Nano-scale Zero Valent Iron (nZVI) treatment of marine sediments slightly polluted by heavy metals. Chem Eng Trans 60:139–144
Demming A (2011) Big challenges and nanosolutions. Nanotech 22(29):1–2
Devatha CP, Thalla AK, Katte SY (2016) Green synthesis of iron nanoparticles using different leaf extracts for treatment of domestic waste water. J Clean Prod 139:1425–1435
Devi GK, Kumar PS, Kumar KS (2016) Green synthesis of novel silver nanocomposite hydrogel based on sodium alginate as an efficient biosorbent for the dye wastewater treatment: prediction of isotherm and kinetic parameters. Desalin Water Treat 57(57):27686–27699
Dong H, Zhao L, Zhang L, Chen H, Gao C, Winston HWS (2015) High-flux reverse osmosis membranes incorporated with NaY zeolite nanoparticles for brackish water desalination. J Memb Sci 476:373–383
Edison TNJI, Atchudan R, Sethuraman MG, Lee YR (2016) Reductive-degradation of carcinogenic azo dyes using *Anacardium occidentale testa* derived silver nanoparticles. J Photochem Photobiol B 162:604–610
Ehrampoush MH, Miria M, Salmani MH, Mahvi AH (2015) Cadmium removal from aqueous solution by green synthesis iron oxide nanoparticles with tangerine peel extract. J Environ Health Sci Eng 13:84
El-Kassas HY, Aly-Eldeen Mohamed A, Gharib Samiha M (2016) Green synthesis of iron oxide (Fe_3O_4) nanoparticles using two selected brown seaweeds: characterization and application for lead bioremediation. Acta Oceanol Sin 35:89–98
El-Sheekh MM, El-Kassas HY (2016) Algal production of nano-silver and gold: their antimicrobial and cytotoxic activities: a review. J Genet Eng Biotechnol 14(2):299–310
Fosso-Kankeu E, Mulaba-Bafubiandi AF, Mishra AK (2014) Prospects for immobilization of microbial sorbents on carbon nanotubes for biosorption: bioremediation of heavy metals polluted water. In: Mishra AK (ed) Application of nanotechnology in water research. Wiley, Hoboken, NJ, USA, pp 213–235
Geetha N, Geetha TS, Manonmani P, Thiyagarajan M, (2014) Green synthesis of silver nanoparticles using *Cymbopogan Citratus* (Dc) Stapf. Extract and its antibacterial activity. Aus J Basic Appl Sci 8(3):324–331

Ghaly MY, Jamil TS, El-Seesy IE, Souaya ER, Nasr RA (2011) Treatment of highly polluted paper mill wastewater by solar photocatalytic oxidation with synthesized nano TiO_2. Chem Eng J 168:446–454

Goutam SP, Saxena G, Singh V, Yadav AK, Bharagava RN, Thapa KB (2018) Green synthesis of TiO_2 nanoparticles using leaf extract of Jatropha curcas L. for photocatalytic degradation of tannery wastewater. Chem Eng J 336:386–396. https://doi.org/10.1016/j.cej.2017.12.029

Gupta GK, Shukla P (2020) Insights into the resources generation from pulp and paper industry wastes: challenges, perspectives and innovations. Bioresour Technol 297:122496. https://doi.org/10.1016/j.biortech.2019.122496

Gurunathan S, Kim E, Han JW, Park JH, Kim JH (2015) Green chemistry approach for synthesis of effective anticancer palladium nanoparticles. Molecules 20(12):22476–22498

Ha W, Yu J, Wang R, Chen J, Shi YP (2014) Green colorimetric assay for the selective detection of trivalent chromium based on *Xanthoceras sorbifolia* tannin attached to gold nanoparticles. Anal Methods 6:5720–5726

Hoag GE, Collins JB, Holcomb JL, Hoag JR, Nadagoudab MN, Varma RS (2009) Degradation of bromothymol blue by 'greener' nano-scale zero-valent iron synthesized using tea polyphenols. J Mater Chem 19:8671–8677

Iravani S (2011) Green synthesis of metal nanoparticles using plants. Green Chem 13:2638–2650

Ismail M, Gul S, Khan MI, Khan MA, Asiri AM, Khan SB (2019) Green synthesis of zerovalent copper nanoparticles for efficient reduction of toxic azo dyes congo red and methyl orange. Green Process Synth 8(1):135–143

Joshi P, Kumar D (2018) Role of bionanomaterial-based devices in water detoxification. In: Chaudhery MH, Ajay KM (eds) Nanotechnology in environmental science. Wiley-VCH Verlag GmbH & Co. KGaA, Weinheim, Germany

Kim YM, Murugesan K, Chang YY, Kim EJ, Chang YS (2012) Degradation of polybrominated diphenyl ethers by a sequential treatment with nanoscale zero valent iron and aerobic biodegradation. J Chem Technol Biotechnol 240:525–532

Kos L, Sójka-Ledakowicz J, Michalska K, Perkowski J (2014) Decomposition of azo dye C.I. Direct Yellow 86 by the fenton process in the presence of nanoparticles of iron oxides. Fibres Text East Eur 5:114–120

Koul B, Taak P (2018) Biotechnological strategies for effective remediation of polluted soils. Springer, pp. 240. https://doi.org/10.1007/978-981-13-2420-8

Kumar KP, Paul W, Sharma CP (2011) Green synthesis of gold nanoparticles with Zingiber officinale extract: characterization and blood compatibility. Process Biochem 46(10):2007–2013

Lam SJ, Wong EHH, Boyer C, Qiao GG (2018) Antimicrobial polymeric nanoparticles. Prog Polym Sci 76:40–64

Laokul P, Maensiri S (2009) Optoelectronics and advanced materials-rapid communications. J of Optoelectronics and Adv Mat 11(6):857–862

Liu J, Morales-Narváez E, Vicent T, Merkoçi A, Zhong GH (2018) Microorganism-decorated nanocellulose for efficient diuron removal. Chem Eng J 354:1083–1091

Lunge SS, Singh S, Sinha A (2014) Magnetic nanoparticle: synthesis and environmental applications. Int Conf Chem Civil Environ Eng (CCEE2014), Nov 18–19. 2014, Singapore.

Marshall AT, Haverkamp RG, Davies CE, Parsons JG, Gardea-Torresdey JL, Van Agterveld D (2007) Accumulation of gold nanoparticles in *Brassic juncea*. Int J Phytoremed 9:197–206

Mehrotra N, Tripathi RM, Zafar F, Singh MP (2017) Catalytic degradation of dichlorvos using biosynthesized zero valent Iron nanoparticles. IEEE Trans Nanobiosci 16(4):280–286

Mohammed L, Gomaa HG, Ragab D, Zhu J (2017) Magnetic nanoparticles for environmental and biomedical applications: A review. Particuology 30:1–14

Mohammed TM (2017) Removal of pathogenic bacteria from wastewater using silver nanoparticles synthesized by two fungal species. Water Sci 31(2):164–176

Mohanpuria P, Rana NK, Yadav SK (2008) Biosynthesis of nanoparticles: Technological concepts and future applications. J Nanopart Res 10:507–517

Moustafa MT (2017) Removal of pathogenic bacteria from wastewater using silver nanoparticles synthesized by two fungal species. Water Sci 31(2):164–176

Mueller NC, Nowack B (2009) Nanotechnology developments for the environment sector. Report of the Observatory NANO

Mukherjee P, Ahmad A, Mandal D, Senapati S, Sainkar SR, Khan MI, Parishcha R, Aiayumar PV, Alam M, Kumar R, Sastry M (2001) Fungus-mediated synthesis of silver nanoparticles and their immobilization in the mycelia matrix: A novel biological approach to nanoparticles synthesis. Nano Lett 1:515–519

Nassar NN (2012) Iron oxide nanoadsorbents for removal of various pollutants from wastewater: an overview. In: Amit B (eds) Application of adsorbents for water pollution control. Bentham Science Publishers, pp.81–118

Nikalje AP (2015) Nanotechnology and its applications in medicine. Med Chem 5(2):5081–5089

Noman E, Al-Gheethi A, Mohamed RMSR, Talip BA (2019) Myco-remediation of xenobiotic organic compounds for a sustainable environment: a critical review. Top Curr Chem 377(3):17

Onal ES, Yatkın T, Aslanov T, Ergüt M, Ozer A (2019) Biosynthesis and characterization of Iron nanoparticles for effective adsorption of Cr(VI). Int J Chem Engg 1–13

Ozkan ZY, Cakirgoz M, Kaymak ES, Erdim E (2017) Rapid decolorization of textile wastewater by green synthesized iron nanoparticles. Water Sci Technol 77(2):511–517

Pandey G (2018) Prospects of nanobioremediation in environmental cleanup. Orient J Chem 34(6):2828–2840

Pandey S, Goswami GK, Nanda KK (2013) Green synthesis of polysaccharide gold nanoparticle nanocomposite: an efficient ammonia sensor. Carbohydr Polym 94:229–234

Philip D (2010) Green synthesis of gold and silver nanoparticles using *Hibiscus rosa sinensis*. Phys E 42:1417–1424

Pimprikar PS, Joshi SS, Kumar AR, Zinjarde SS, Kulkarni SK (2009) Influence of biomass and gold salt concentration on nanoparticle synthesis by the tropical marine yeast *Yarrowia lipolytica NCIM 3589*. Colloids Surf B 74:309–316

Primo JO, Bittencourt C, Acosta S, Sierra-Castillo A, Colomer JF, Jaerger S, Teixeira VC, Anaissi FJ (2020) Synthesis of zinc oxide nanoparticles by eco-friendly routes: adsorbent for copper removal from wastewater. Front Chem 8:571790. https://doi.org/10.3389/fchem.2020.571790

Qu X, Alvarez PJJ, Li Q (2013) Applications of nanotechnology in water and wastewater treatment. Water Res 47(12):3931–3946

Rai M, Ingle AP, Gaikwad S, Padovani FH, Alves M (2016) The role of nanotechnology in control of human diseases: perspectives in ocular surface diseases. Crit Rev Biotechnol 36:777–787

Raliya R, Rathore I, Tarafdar JC (2013) Development of microbial nanofactory for zinc, magnesium and titanium nanoparticles production using soil fungi. J Bionanosci 7:59

Ravi SS, Christena LR, SaiSubramanian N, Anthony SP (2013) Green synthesized nanoparticles for colorimetric sensing of Hg^{2+}in aqueous solution in wide pH range. Analyst 138:4370–4377

Rizwan MD, Singh M, Mitra CK, Morve RK (2014) Ecofriendly application of nanomaterials: Nanobioremediation. J Nanopart 1:1–7

Roh Y, Lauf RJ, McMillan AD, Zhang C, Rawn CJ, Bai J, Phelps TJ (2001) Microbial synthesis and the characterization of metal-substituted magnetites. Solid State Commun 118:529–534

Rosales E, Meijide J, Pazos M, Sanromán MA (2017) Challenges and recent advances in biochar as low-cost biosorbent: from batch assays to continuous-flow systems. Bioresour Technol 246:176–192

Roy K, Ghosh CK, Sarkar CK (2019) Rapid detection of hazardous H_2O_2 by biogenic copper nanoparticles synthesized using *Eichhornia crassipes* extract. Microsyst Technol 25(5):1699–1703

Sahayaraj K (2012) Bionanomaterials: synthesis and applications. In: Proceedings of the first national seminar on new materials research and nanotechnology (NSNMRN'2012); 12–14 September, Government Arts College, Ooty, Tamil Nadu, 2012.

Saifuddin N, Wong CW, Yasumira AA (2009) Rapid biosynthesis of silver nanoparticles using culture supernatant of bacteria with microwave irradiation. J Chem 6:61–70

Salehi S, Shandiz SAS, Ghanbar F (2016) Phytosynthesis of silver nanoparticles using Artemisia marschalliana Sprengel aerial part extract and assessment of their antioxidant, anticancer, and antibacterial properties. Int J Nanomedicine 11:1835–1846

Sardar M, Khatoon N (2016) Green and sustainable silver nanoparticles for waste water treatment. J Microb Biochem Technol 8:2(Suppl)

Sebeia N, Jabli M, Ghith A (2019) Biological synthesis of copper nanoparticles, using *Nerium oleander* leaves extract: characterization and study of their interaction with organic dyes. Inorg Chem Commun 105:36–46

Shanker U, Jassal V, Manviri R (2016) Catalytic removal of organic colorants from water using some transition 2 metal oxide nanoparticles synthesized under sunlight. RSC Adv 6:94989–94999

Sherry DA, Prakash P, Thamaraiselvi K (2017) Nanobioremediation Technologies for Sustainable Environment. In: Prashanthi M, Sundaram R, Jeyaseelan A, Kaliannan T (eds) Bioremediation and sustainable technologies for cleaner environment. Environmental Science and Engineering. Springer, Cham. https://doi.org/10.1007/978-3-319-48439-6_2

Shittu KO, Ihebunna O (2017) Purification of simulated waste water using green synthesized silver nanoparticles of *Piliostigma thonningii* aqueous leave extract. Adv Nat Sci Nanosci Nanotechnol 8:1–9

Simeonidis K, Kaprara E, Samaras T, Angelakeris M, Pliatsikas N, Vourlias G (2015) Optimizing magnetic nanoparticles for drinking water technology: the case of Cr (VI). Sci Total Environ 535:61–68

Sunkar S, Nachiyar V, Karunya A (2013) Phytogenic silver nanoparticle synthesis with potential antibacterial activity and dye degrading ability. Res J Pharm Biol Chem Sci 4:1085–1097

Tan W, Peralta-Videa JR, Gardea-Torresdey JL (2018) Interaction of titanium dioxide nanoparticles with soil components and plants: Current knowledge and future research needs-a critical review. Environ Sci Nano 5:257–278

Thakkar KN, Mhatre SS, Parikh RY (2010) Biological synthesis of metallic nanoparticles. Nanotechnol Biol Med Nanomed 6:257–262

Tiwari DK, Behari J, Sen P (2008) Application of nanoparticles in waste water treatment. World App. Sci J 3(3):417–433

Torres-Martínez C, Kho R, Mian OI, Mehra RK (2001) Efficient photocatalytic degradation of environmental pollutants with mass-produced ZnS nanocrystals. J. Colloid Interface Sci.

Tripathi V, Fraceto LF, Abhilas PC (2015) Sustainable clean-up technologies for soils contaminated with multiple pollutants: Plant-microbe-pollutant and climate nexus. Ecol Eng 82:330–335

Wang T, Jin X, Chen Z, Megharaj M, Naidu R (2014) Green synthesis of Fe nanoparticles using eucalyptus leaf extracts for treatment of eutrophic wastewater. Sci Total Environ 466–467:210–213

Warner MG, Warner CL, Addleman RS, Yantasee W (2011) Magnetic nanomaterials for environmental applications. Nanotechnologies for the Life Sciences. Wiley-VCH Verlag GmbH & Co. KGaA, Weinheim, Germany, pp 5–15

Yadav D, Kumar P, Kapur M, Mondal MK (2018) Phosphate removal from aqueous solutions by nano-alumina for the effective remediation of eutrophication. Environ prog and sustain energy 00(00): 1–9

Yadav KK, Singh JK, Gupta N, Kumar VA (2017) Review of Nanobioremediation Technologies for Environmental Cleanup: A Novel Biological Approach. J of Mat and Env Sci 8(2):740–757

Yew YP, Shameli K, Miyake M, Khairudin NBBA, Mohamad SEB, Naiki T, Lee KX (2020) Green biosynthesis of superparamagnetic magnetite Fe3O4 nanoparticles and biomedical applications in targeted anticancer drug delivery system: A review. Arab J Chem 13(1):2287–2308

Yogalakshmi KN, Das A, Rani G, Jaswal V, Randhawa JS (2020) Nano-bioremediation: A new age technology for the treatment of dyes in textile effluents. In: Bioremediation of industrial waste for environmental safety. Springer, Singapore.

Zhang W, Fu J, Zhang G, Zhang X (2014) Enhanced arsenate removal by novel Fe–La composite (hydr) oxides synthesized via coprecipitation. Chem Eng J 251:69–79

MIX
Papier aus verantwortungsvollen Quellen
Paper from responsible sources
FSC® C105338

If you have any concerns about our products,
you can contact us on
ProductSafety@springernature.com

In case Publisher is established outside the EU,
the EU authorized representative is:
Springer Nature Customer Service Center GmbH
Europaplatz 3, 69115 Heidelberg, Germany

Printed by Libri Plureos GmbH
in Hamburg, Germany